Internet of Things

Technology, Communications and Computing

Series Editors

Giancarlo Fortino, Rende (CS), Italy

Antonio Liotta, School of Computing, Edinburgh Napier University, Edinburgh, The Netherlands

The series Internet of Things - Technologies, Communications and Computing publishes new developments and advances in the various areas of the different facets of the Internet of Things. The intent is to cover technology (smart devices, wireless sensors, systems), communications (networks and protocols) and computing (theory, middleware and applications) of the Internet of Things, as embedded in the fields of engineering, computer science, life sciences, as well as the methodologies behind them. The series contains monographs, lecture notes and edited volumes in the Internet of Things research and development area, spanning the areas of wireless sensor networks, autonomic networking, network protocol, agent-based computing, artificial intelligence, self organizing systems, multi-sensor data fusion, smart objects, and hybrid intelligent systems.

Indexing: *Internet of Things* is covered by Scopus and Ei-Compendex **

Niranjan Suri • Konrad Wrona • Zbigniew Zieliński
Editors

Military Applications of Internet of Things

Architectures, Security, Reliability, and Interoperability for MIoT

Springer

Editors
Niranjan Suri
IHMC
Pensacola, FL, USA

US Army DEVCOM Army Research Laboratory
Adelphi, MD, USA

Konrad Wrona
NATO Communications and Information Agency (NCIA)
The Hague, The Netherlands

Military University of Technology
Warsaw, Poland

Zbigniew Zieliński
Military University of Technology
Warsaw, Poland

ISSN 2199-1073 ISSN 2199-1081 (electronic)
Internet of Things
ISBN 978-3-032-03377-2 ISBN 978-3-032-03378-9 (eBook)
https://doi.org/10.1007/978-3-032-03378-9

This work was supported by North Atlantic Treaty Organization.

This Springer imprint is published by the registered company Springer Nature Switzerland AG
The registered company address is: Gewerbestrasse 11, 6330 Cham, Switzerland

Preface

The North Atlantic Treaty Organization (NATO) Science and Technology Organization (STO) exists to foster collaborative research among NATO member and other partner nations on topics of mutual interest for national defense. In this context, the Information Systems Technology (IST) panel within the STO established the IST-147 Research Task Group (RTG) on Military Applications of Internet of Things (IoT) to focus on the rapidly developing and evolving domain of IoT and its impact on national defense.

Given the rapid advancement in IoT technology and the pervasive adoption and insertion of this technology into all aspects of modern society, the objective of the research group was to examine the multitude of ways by which IoT would impact different processes in the military, from logistics to communications to medical care to Intelligence, Surveillance, and Reconnaissance (ISR) activities.

We, as members of the RTG, believe that the research conducted by this group identifies the potential for IoT to create transformational impact on all aspects of the military, and hence set about documenting the research findings of the RTG in this book.

We believe that this book helps readers improve their understanding of the impact of IoT on military systems, networks, security, and operational concepts, including exploring the benefits of IoT to multiple military scenarios. The book also identifies some of the key challenges with integrating IoT capabilities into the military domain and systems, and provides ideas for solutions.

We would like to thank the following external experts who contributed to the activities of our RTG and the creation of this book:

- Dr. Peter Lenk, NCIA, The Netherlands—Reviewer
- Prof. Marek Amanowicz, NASK National Research Institute, Poland—Mentor
- Dr. Jan Chudzikiewicz, Military University of Technology, Poland—Member
- Cpt Stefan-Ciprian Arseni, Romanian Army, Romania—Member
- Steen Dam-Jensen, Ebicon, Denmark—Member
- Dr. Davide Fabbri, IDS Corporation, Italy—Member
- Dr. Krzystof Gierlowski, Gdańsk University of Technology, Poland—Member

- Dr. Tomasz Gierszewski, Gdańsk University of Technology, Poland—Member
- Dr. Mateusz Krzysztoń, NASK National Research Institute, Poland—Member
- Prof. Ewa Niewiadomska-Szynkiewicz, Warsaw University of Technology, Poland—Member
- Goran Olsen, Ebicon, Denmark—Member
- Ian Owens, M.Sc., University of Cranfield, UK—Member
- Bert-Jan te Paske, M.Sc., TNO, The Netherlands—Member
- Dr. Andrzej Sikora, NASK National Research Institute, Poland—Member
- Krzysztof Skurzak, M.Sc., NATO Allied Command Transformation, USA—Collaborator
- Students of the Military University of Technology, Poland—Contributors
- Robin Hoffman, Florida Institute for Human & Machine Cognition (IHMC)—Copy editor

Pensacola, FL, USA Niranjan Suri
The Hague, The Netherlands Konrad Wrona
Warsaw, Poland Zbigniew Zieliński
February 2025

Contents

Acronyms

3D	3-Dimensional
3DES	Triple Data Encryption Standard
5G	5th Generation Cellular Networks
ABAC	Attribute-Based Access Control
ABE	Attribute-Based Encryption
AC	Authorization Center
ACT	Allied Command Transformation
ADR	Adaptive Data Rate
ADS-B	Automatic Dependent Surveillance-Broadcast
AES	Advanced Encryption Standard
AI	Artificial Intelligence
AIS	Automatic Identification System
AIV	Adaptive, Information-centric, and Value-based model
AJP	Allied Joint Publication
ALV	Autonomous Land Vehicle
AMQP	Advanced Message Queuing Protocol
AMR	Automatic Meter Reading
API	Application Programming Interface
APP	Allied Procedural Publication
ARL	Army Research Laboratory
ARM	Architectural Reference Model
ASW	Anti-Submarine Warfare
ATAK	Android Assault Toolkit
AUV	Autonomous Underwater Vehicle
BAN	Body Area Network
BIST	Built-In Self-Test
BLE	Bluetooth Low Energy
BLS	Boneh–Lynn–Shacham signature scheme
BW	Bandwidth
BYOD	Bring Your Own Device
C2	Command and Control

C4ISR	Command, Control, Communications, Computers for Intelligence, Surveillance, and Reconnaissance
CA	Certification Authority
CAC	Cryptographic Access Control
CAN	Controller Area Network
CBC	Cipher Block Chaining
CBRN	Chemical, Biological, Radiological, and Nuclear
CH	Cluster Head
C-I-A	Confidentiality, Integrity, and Availability
CIMIC	Civil–Military Cooperation
CIS	Communications and Information System
CoAP	Constrained Application Protocol
COP	Common Operational Picture
CONOPS	Concept of Operations
COTS	Commercial Off The Shelf
CP	Command Post
CP-ABE	Ciphertext Policy Attribute Based Encryption
CPR	Content-based Protection and Release
CR	Coding Rate
CRIO	Consumption Ready Information Object
CSA	Cyber Situational Awareness
CSN	Collecting Sensor Node
CSS	Chirp Spread Spectrum
CTP	Common Tactical Picture (Page 209, Line 932)
DA	Discovery Agent
DDA	Distributed Diagnostic Agent
DAC	Discretionary Access Control
DARPA	Defense Advanced Research Projects Agency
DCGS	Distributed Common Ground System
DCS	Data-Centric Security
DDS	Data Distribution Service
DDoS	Distributed Denial of Service
DevSecOps	Development, Security, and Operations
DIA	Distributed Intelligent Agent
DISA	Defense Information System Agency
DISSUB	Distressed Submarine
DL	Distributed Ledger
DL	Description Logic
DLA	Defense Logistics Agency
DM	Directional Modulation
DML	Data Management Layer
DNS-SD	DNS Service Discovery
DoD	Department of Defense
DOGA	Deny or Grant Access
DoS	Denial of Service

DOTMLPFI	Doctrine, Organization, Training, Materiel, Leadership and education, Personnel, Facilities, and Interoperability
DR	Data Rate
DSL	Digital Subscriber Line
DVB	Digital Video Broadcasting
EC	European Commission
ECB	Electronic Codebook
ECC	Elliptic Curve Cryptography
ECG	Electrocardiogram
EDA	European Defence Agency
EEPROM	Electrically Erasable Programmable Read-Only Memory
EK	Endorsement Key
EKG	Elektrokardiogramm (German for ECG)
ELIA	Light Information Equipment via the Automatic Dependent Surveillance-Broadcast (ADS-B)
eMBB	Enhanced Mobile Broadband
EMG	Electromyography
EMP	Electromagnetic Pulse
EMS	Emergency Medical Services
ENISA	European Union Agency for Cybersecurity
EO	Electro-Optical
ERP	Enterprise Resource Planning
ETSI	European Telecommunications Standards Institute
EU	European Union
FANET	Flying Ad-hoc Network
FCAPS	Fault, Configuration, Accounting, Performance, and Security
FFI	Norwegian Defence Research Establishment
FKIE	Fraunhofer Institute for Communication, Information Processing and Ergonomics
FMN	Federated Mission Networking
FTA	Finnish Transport Authority
GCM	Galois/Counter Mode
GEO	Geosynchronous Earth Orbit
GIS	Geographic Information System
GSM	Global System for Mobile Communications
GPS	Global Positioning System
GPX	GPS Exchange Format
GX	Global Xpress
HADR	Humanitarian Assistance and Disaster Relief
HD	High Definition
HF	High Frequency
HMAC	Hash-based Message Authentication Code
HMI	Human-Machine Interface
HVAC	Heating, Ventilation, and Air Conditioning
IaC	Infrastructure-as-Code

ICRC	International Committee of the Red Cross
ICT	Information and Communications Technology
IDS	Intrusion Detection System
IEC	International Electrotechnical Commission
IEEE	Institute of Electrical and Electronics Engineers
IES	Information Exchange Specification
IGC	Integrated Data Environment/Global Transportation Network Convergence
IIoT	Industrial Internet of Things
IMSBridge	Information Management Systems Bridge
IMT	Iridium Messaging Transport
IO	Information Object
IoBT	Internet of Battlefield Things
IoE	Internet of Everything
IOP	Interoperability Profiles
IoT	Internet of Things
IoT-A	Internet of Things Architecture
ISL	Intelligent Service Layer
IP	Internet Protocol
IR	Infrared
ISL	Intelligent Service Layer
ISM	Industrial, Scientific, and Medical
ISO	International Organization for Standardization
ISR	Intelligence, Surveillance, and Reconnaissance
IST	Information Systems Technology
IT	Information Technology
ITU	International Telecommunication Union
JAUS	Joint Architecture for Unmanned Systems
JP8	Jet Propellant 8
JSON	JavaScript Object Notation
KGR	Key Generating and Renewing
KP-ABE	Key Policy Attribute Based Encryption
KS	Key Server
LBAC	Lattice-Based Access Control
LEO	Low Earth Orbit
LIDAR	Light Detection and Ranging
LoRa	Long-Range Radio
LoRaWAN	Long-Range Wide-Area Network
LPWAN	Low-Power Wide-Area Network
LTE	Long-Term Evolution
LTE-M	Long-Term Evolution Machine Type Communication
mCOP	Mobile Common Operational Picture
M2M	Machine-to-Machine
MA	Medium Assurance
MAC	Mandatory Access Control

MANET	Mobile Ad-hoc Network
MARGOT	Multi-domain AsynchRonous Gateway Of Things
MASCAL	Mass Casualty
MCM	Mine Counter Measures
mDNS	multicast Domain Name System
MedEvac	Medical Evacuation
MEO	Medium Earth Orbit
MILES	Multiple Integrated Laser Engagement System
MIM	MIP Information Model
MIoT	Military Internet of Things
MIP	Multilateral Interoperability Program
mMTC	Massive Machine-Type Communications
MQTT	Message Queuing Telemetry Transport
MTOC	Mobile Tactical Operations Center
MTTF	Mean Time to Failure
MTTR	Mean Time to Repair
NAEMT	National Association of Emergency Medical Technicians
NATO	North Atlantic Treaty Organization
NB-IoT	Narrowband IoT
NCIA	NATO Communications and Information Agency
NCCoE	National Cybersecurity Center of Excellence
NCMS	NATO Core Metadata Specification
NFC	Near-Field Communication
NGI	Next Generation Internet
NGVA	NATO Generic Vehicle Architecture
NIST	National Institute of Standards and Technology
NMR	N Modular Redundancy
NMS	Network Management Station
NNEC	NATO Network Enabled Capability
NRDM	Navigation Reference Data Model
NSK	Node Security Key
NTK	Node Transfer Data Key
NVRAM	Non-Volatile Random-Access Memory
OAS	OpenAPI Specifications
OData	Open Data Protocol
OGC	Open Geospatial Consortium
OLP	Object-Level Protection
OODA	Observe, Orient, Decide, and Act
OSI	Open Systems Interconnection
OT	Operational Technology
OWASP	Open Web Application Security Project
OWL	Web Ontology Language
PAL	Perception and Actuation Layer
PCR	Platform Configuration Registers
PIR	Passive Infrared

PKI	Public Key Infrastructure
PoC	Proof of Concept
PP	Public Parameters
PPG	Photoplethysmography
PUF	Physically Unclonable Function
QoE	Quality of Experience
QoS	Quality of Service
QRNG	Quantum Random Number Generator
RADAR	Radio Detection and Ranging
RAID	Redundant Array of Independent Disks
RAM	Random-Access Memory
RAML	ReSTful API Modeling Language
RBAC	Role-Based Access Control
RC4	Rivest Cipher 4
REA	Rapid Environmental Assessment
recon	Reconnaissance
REST	Representational State Transfer
RFC	Request for Comments
RFID	Radio Frequency Identification
ROC	Recovery-Oriented Computing
ROS	Robot Operating System
RPi	Raspberry Pi
RSA	Rivest-Shamir-Adelman algorithm
RTG	Research Task Group
SA	Situational Awareness
SALSA	Smart Adaptive Long- and Short-range Acoustic network
SATCOM	Satellite Communications
SB	Switch Broadband
SDK	Software Development Kit
SF	Spreading Factor
SHA-256	Secure Hash Algorithm 2–256 bit
SK	Secret Key
SMC	Service Management and Control
SMS	Short Message Service
SNMP	Simple Network Management Protocol
SN	Sensor Node
SNR	Signal-to-Noise Ratio
SOA	Service-Oriented Architecture
SOAP	Simple Object Access Protocol
SOSA	Sensor, Observation, Sample, and Actuator
SPF	Sieve, Process, and Forward
SPI	Serial Peripheral Interface
SRAM	Static Random Access Memory
SRWC	Short Range Wireless Communication
SSN	Semantic Sensor Network

STANAG	NATO Standardization Agreement
STO	Science and Technology Organization
SVM	Support Vector Machine
SWaP	Size, Weight, and Power
SWE	Sensor Web Enablement
SWRL	Semantic Web Rule Language
TCCC	Tactical Combat Casualty Care
TCP/IP	Transmission Control Protocol/Internet Protocol
TFA	Traffic Flow Analysis
TIDE	Technology, Information, Decision, and Execution
TLS	Transport Layer Security
TN	Tactical Network
TOC	Tactical Operations Center
TPM	Trusted Platform Module
UAV	Unmanned Aerial Vehicle
UDDI	Universal Description, Discovery, and Integration
UGV	Unmanned Ground Vehicle
UGS	Unattended Ground Sensor
UHF	Ultra High Frequency
UI	User Interface
UK	United Kingdom
UMV	Unmanned Maritime Vehicle
UN	United Nations
URLLC	Ultra-Reliable Low-Latency Communications
US	United States
USTRANSCOM	US Transportation Command
UTC	Universal Time Coordinated
UxV	Unmanned Vehicle
VE	Virtual Entity
VFT	Virtualization and Fault Tolerance
VHF	Very High Frequency
VM	Virtual Machine
VoI	Value of Information
WADL	Web Application Description Language
WBAN	Wireless Body Area Network
WiMAX	Worldwide Interoperability for Microwave Access
WSDL	Web Service Description Language
WSN	Wireless Sensor Network
WS-Notification	Web Services Notification
WLAN	Wireless LAN
XACML	eXtensible Access Control Markup Language
XLG	XML Labelling Guard
XML	Extensible Markup Language

List of Figures

List of Tables

Part I
Motivation and Use Cases

Chapter 1
Motivation for Military Applications of Internet of Things

Niranjan Suri and Konrad Wrona

1.1 Introduction

The Internet of Things (IoT) may be defined as the empowerment of everyday objects with ubiquitous computing and communications. The *things* can range from very small passive devices such as Radio Frequency Identification (RFID) tags to fully programmable computer platforms. Furthermore, the ability to gather data from a myriad of information sources, and to analyze these data, can provide powerful insights into everything from the performance of systems to the utilization of resources to the behavior patterns of individuals, crowds, and societies.

As IoT continues to permeate all aspects of modern society, it also directly or indirectly impacts all aspects of military operations. The use of sensors and actuators in military applications is not new, but the dual use of IoT technologies and their pace of evolution can be disruptive for the military. To face adversaries effectively and cost-efficiently, military forces also need to embrace IoT technology and integrate it with their CIS and military doctrine, such as [8, 14]. IoT has an impact both at the command level, for example, by improving the performance and agility of command and control, optimizing the decision process, and controlling information, and at the tactical level, for example, by increasing tactical agility, delivering optimized effects, and avoiding unnecessary effort by

N. Suri (✉)
US Army DEVCOM Army Research Laboratory (ARL), Adelphi, MD, USA

Florida Institute for Human and Machine Cognition (IHMC), Pensacola, FL, USA
e-mail: nsuri@ihmc.org

K. Wrona
NATO Communications and Information Agency (NCIA), The Hague, The Netherlands

Military University of Technology, Warsaw, Poland
e-mail: konrad.wrona@ncia.nato.int

N. Suri et al. (eds.), *Military Applications of Internet of Things*, Internet of Things,
https://doi.org/10.1007/978-3-032-03378-9_1

personnel [7, 15]. IoT can significantly enhance surveillance and reconnaissance by augmenting traditional military capabilities with IoT-based sensing. IoT impacts communication, networking, and information management by vastly expanding the number of devices and *things* that are networked and communicate with each other. IoT increases the breadth, depth, accuracy, and speed with which military Command and Control (C2) can develop and maintain Situational Awareness (SA). One of the earliest applications of IoT has been asset tracking and logistics management, both critical functions within the military that can be significantly improved by leveraging advances in IoT.

Humans are key elements of military IoT deployment, not only as passive users but also by extending IoT as a producer of data and a consumer of control information. IoT has also strongly influenced healthcare—especially from the perspective of physiological monitoring. This is an area where the military could benefit, as health monitoring and conditioning are extremely important during military operations.

The ongoing progress in implementation and deployment of IoT leads to a migration toward Internet of Everything (IoE), which also encompasses people and processes. This migration is particularly important for the military, as humans are an important part of the decision-making processes in military systems. In this context, this means further developing military IoT to integrate existing C2 military systems, soldiers, and military IoT systems and also non-military IoT device networks, e.g., smart city CCTV. However, integrating heterogeneous sensors and systems poses several challenges from a military perspective due to the diversity of technology solutions, environmental constraints, component fidelity levels, and security considerations.

There are also many secondary implications and benefits of IoT to the military. Size, Weight, and Power (SWaP) continues to be a challenge for military operators, with much of the computing and communications equipment tending to be bulky and power-intensive. There are multiple reasons for this current state, including security concerns and ruggedness. However, one benefit of IoT has been the immense focus within the commercial sector on building small, low-power, convenient-to-use devices with long battery lives that stand up to routine use (and misuse). This convenience has been essential for the adoption of IoT technologies by end users and could also benefit the military. The exponential growth of IoT in the commercial sector has led to smaller and more energy-efficient embedded processors in addition to new communications technologies such as 5th Generation Cellular Networks (5G), Narrowband IoT (NB-IoT), and Long-Range Radio (LoRa), all of which can potentially benefit military communications.

Another consequence of the pervasiveness of IoT is the ability to gather and analyze large volumes of data. These data originate from many devices with different sensing capabilities and, more importantly, different modalities. Some devices might be simple and generate simple data, e.g., a light being on or off, the presence of something tagged with an RFID tag, motion detected in a room, or the power consumption of some appliance. Other devices may be more complex and continuously generate massive volumes of data, e.g., video feeds or radar images. Networking these devices and sharing these data from multiple perspectives pervasively provide significant military benefits. The ability to archive these data over time, integrate these data, and analyze them by looking at temporal trends, as well as correlations across multiple modalities, provides insights into the physical world and the behavior of entities within it in a manner that has simply not been possible until the advent of IoT. This type of data analytics, either on data gathered within the military domain (e.g., from smart, IoT-equipped military vehicles) or from data accumulated from the surrounding environment (e.g., Smart Cities—see below), can provide new efficiencies and unique insights to military operations.

Finally, an equally important development in commercial IoT has been the application of IoT to create smart cities and smart environments. The concept behind smart cities is to deploy pervasive IoT sensing and actuation capabilities within urban environments to provide enhanced services to citizens. Application areas range from smart transportation to environmental monitoring to smart grids. Smart cities empower residents to not only improve their quality of life but also provide an opportunity for the military—especially in the context of Civil–Military Cooperation (CIMIC) and Humanitarian Assistance and Disaster Relief (HADR) operations. IoT capabilities in smart cities can significantly increase the coverage and reduce the latency in the development of SA. This is particularly true for HADR operations, in which smart cities could immediately provide awareness of damage and destruction, in addition to the condition and needs of the population.

However, military application or use of IoT is not a panacea and comes with its own set of challenges. The first challenge is the impact on the military's Concept of Operations (CONOPS). Traditionally, the military is used to owning all of the sensing, computing, and communications assets that they deploy, which engenders a certain degree of confidence in the information that is generated or provided by these assets. However, relying on civilian IoT solutions can be problematic, as the military decision-making process must take into account that the underlying information is being generated by sensors not owned by the military, processed by services not supervised and verified by the military, and transported by networks that the military has not secured. However, if the CONOPS can be modified, civilian IoT can be a new source of information from which the military could benefit. Furthermore, the more trusted IoT (and non-IoT) capabilities owned and deployed by the military can always be used to supplement civilian IoT. Doing so will enable the military to harness substantial numbers of low-cost sensor data and support the attainment of information superiority. However, incorporating such external data sources requires appropriate data analytics measures to detect anomalies in the obtained data and accurately assess its authenticity and trustworthiness.

In the above, we have mainly discussed the opportunities and challenges related to using IoT technology *for good* [4]. However, low-cost IoT technology is equally available to both regular military forces and adversaries. Access to some of these sensing and actuating capabilities can diminish, or even cancel, the longstanding information superiority of NATO and Western World forces. Asymmetric adversaries can particularly benefit from IoT to level the technological ground and increase its opportunistic, agile, and decentralized mode of operation. IoT also introduces many new adversarial and illegitimate use cases related to criminal and terrorist activities [19]. Possible adversarial scenarios range from blackmail based on information obtained from insecure, breached, or deliberately planted malicious IoT devices to life-threatening availability attacks on medical life-support infrastructure. For organized crime, IoT devices can be used both as entry vectors and as specific targets for cybercriminal activities, such as ransomware, and as a source of reconnaissance for more traditional criminal activities, such as burglaries and robberies. From a terrorist's perspective, IoT devices can be used, for example, for reconnaissance, to trigger traditional explosives, or as a means of delivering deadly attacks [5, 10].

1.2 Interplay Between Commercial and Military IoT

Given the historical development of military and commercial/public (civilian, in general) systems and the unique security challenges that military systems must address, we do not envision that public IoT solutions will be directly integrated with military systems. But military tactical and enterprise systems should be able to access public IoT information. Therefore, one objective is to repurpose civilian IoT information into useful military information, using civilian IoT as an additional source of open-source intelligence. In this way, public IoT information concerning, for example, traffic conditions, energy consumption, and environmental conditions could inform military decision-making. The other reason for considering civilian IoT technologies for military applications is that the civilian world increasingly drives Information and Communications Technology (ICT) innovation [16].

Therefore, to maintain a technological edge over our adversaries while keeping up with the speed of technological change and adapting to a reduced defense budget, it is advisable to exploit civilian Commercial Off The Shelf (COTS) solutions as much as possible in support of military applications. In fact, most of the modern military ICT systems use the same basic digital components as the civilian systems.

One importance of the ability to use COTS IoT components where possible is highlighted by companies and innovators usually seeing little benefit in catering to stringent military requirements, as defense is a small and demanding customer relative to commercial markets. Another chief constraint on military adoption of new technologies is the current budget limitations. Even if new technologies generate significant long-term savings, they often have significant up-front costs associated with developing and deploying new devices and applications. The use of COTS components can lower the cost and implementation risks significantly. Moreover, to create new military IoT solutions, the military must partner with innovators, who often do not accept the culture of secrecy and bureaucracy involved in complying with defense acquisition processes. Relying on COTS could not only reduce the need for these partnerships but also enable the rapid development of low-cost proof-of-concept solutions, embracing the culture of experimentation and failure, thus reducing the risk of large-scale deployment of new and unfamiliar technology. The reduced risk of security vulnerabilities, integration failures, and functional imperfections is essential in the military acquisition environment, as defense personnel are reluctant to change from solutions they know, tested in operations, and have already incorporated into their doctrines, organizational processes, and tactics to new and less certain technology. They need to be convinced not only that the change will bring benefits but also that the risk of something going wrong during its deployment is manageable. The ability of the military to develop specialized solutions is also often undermined by restrictions on intellectual property rights for inventions developed in part with defense funding. Export controls may discourage private innovators from working with the military to create customized solutions. In some countries, when the military buys a new technology, it is classified as military or dual-use technology and becomes subject to export controls. Private tech companies may not want to cut off their access to the larger and potentially more lucrative consumer IoT market by partnering with the military for product development.

The progressing digitization of forces will likely drive many applications of IoT to operational systems, either custom-built or shared with civilian partners, in order to:

- Accelerate and improve logistics flows
- Reduce cost of managing stocks and maintaining the materials
- Strengthen infrastructure security
- Enhance and accelerate medical treatment
- Shorten the Observe, Orient, Decide, and Act (OODA) loop

The main topic of this book is how to integrate and leverage civilian IoT infrastructure and capabilities within the military environment. In addition to technical interoperability challenges (e.g., at the network level), significant security challenges must be addressed. For example, mission-critical (and life-dependent) decisions made by military commanders at all levels require adequate understanding of the quality and reliability of information obtained from sensors that lead to those decisions. Commanders are used to dealing with non-perfect information—the phenomenon often referred to as the *fog of war*. However, it is essential to enable commanders to understand the relative reliability and veracity of the information so that they can form an opinion of the accuracy and trustworthiness of sensor data. To some extent, this problem is similar to civilian applications; however, mission-critical decisions of military applications require a higher level of understanding the nature and characteristics of the data underlying the decision-making. When relying on IoT data from civilian infrastructures, sensors owned and deployed by unknown and untrusted entities capture, transmit, and process data on systems and networks that may be easier to compromise by adversaries. Therefore, there is a need to identify and discount data that could be bogus, either intentionally or unintentionally. Security challenges and potential solutions are other key topics addressed in this book. The reader may notice that we have decided to largely omit in our work the important topic of the IoT data analytics. We believe that this topic deserves a larger study on its own. However, performing a proper analysis and reasoning on IoT data are not possible without first implementing the fundamental enablers for Military Internet of Things (MIoT) discussed in the following chapters.

The possible transition toward increasing the use of civilian IoT in the military environment brings multiple challenges. As mentioned earlier, civilian IoT solutions are commonly developed for permissive environments, whereas military operations typically take place in adversarial environments in which jamming and cyber-threats are usually present. Finally, there are also constraints regarding the speed and availability of training of military personnel, as well as a mismatch between life cycles of military equipment and that of IoT technology. A traditional military procurement program spans 10–15 years, which is significantly slower than the pace of evolution of civilian ICT. The agility in adoption of new technologies is where the adversaries currently have the asymmetric advantage over established military organizations. However, overcoming all the above constraints and achieving widespread use of military IoT could reduce this gap and support future collaborative warfare scenarios [7] that integrate command and control, personnel, and equipment into a continuously connected and secure data ecosystem.

It is essential to develop a suitable strategy and means to control the spread of IoT devices within the military Communications and Information System (CIS) and to ensure their effective administration. This is a formidable Service Management and Control (SMC) challenge that consists of quickly identifying, locating, and patching or taking off line devices affected by relevant vulnerabilities. Although conventional security approaches are largely applicable to IoT, they need to be enhanced to deal effectively and cost-efficiently with the sizeable number of IoT devices that must be monitored and controlled and also to implement adequate detection techniques for specific IoT attacks.

The large amount of new situational data obtained from IoT and required to be processed by soldiers can be an added value as much as an incapacitating burden. In this context, an important concept is the Value of Information (VoI) [17], essential for prioritizing useful data and filtering out superfluous data.

1.3 Related Work

The use of IoT for military applications enhances the concept of *network-centric warfare* [21], which postulates the integration of sensors and actuators, network connectivity, data dissemination, and the decision process [1].

Modern military CIS interconnects a large number of custom-developed sensors such as radar, sonars, cameras, and electromagnetic wave detectors, installed in aircraft, satellites, Unmanned Aerial Vehicles (UAVs), ships, ground stations, and even those carried by soldiers. In the USA, the Distributed Common Ground System (DCGS) [9] is the primary Command, Control, Communications, Computers for Intelligence, Surveillance, and Reconnaissance (C4ISR) data-integration platform, which aggregates sensor data and provides common situational awareness about friendly and enemy forces.

Several existing combat systems rely on tightly integrated sensors and actuators. For example, the United States (US) Navy Aegis Combat System is an integrated fire control system for surface ships that incorporates fully automated firing capabilities based on radar sensor inputs [2]. Similarly, the Predator UAVs can combine information obtained from their sensors with that from DCGS to identify targets and launch missiles that are also sensor-equipped for increased targeting precision. Tomahawk Land Attack Missiles can send video footage to the command post via a satellite link and can be redirected in flight to a new target that is preprogrammed or based on new Global Positioning System (GPS) coordinates. Many training and simulation systems use a combination of sensors and actuators, such as the Multiple Integrated Laser Engagement System (MILES), which offers a game-like laser environment for the warfighters.

On the logistics side, US Transportation Command (USTRANSCOM) Integrated Data Environment/Global Transportation Network Convergence (IGC) collects logistics data and allows users to enter and track orders and monitor inventory. Pallets in transit between major hubs are tagged with active and passive RFID. The Defense Logistics Agency (DLA) uses sensors to monitor fuel levels in tanks in distribution hubs and aggregates input from multiple sensors to adjust readings of fuel volume given environmental conditions, such as temperature.

The use of COTS mobile devices for defense applications is the subject of several pilot and deployment programs, both within the North Atlantic Treaty Organization (NATO) and in individual nations. The US Army Net Warrior program and the Defense Information System Agency (DISA) Mobility Program have developed hardened Android devices for various classes of users, ranging from infantry units to Department of Defense (DoD) executives. The US Air Force has been using commercial iPads for various applications, such as electronic flight bags and aircraft loading plans. The US Army Research Laboratory (ARL) started a fundamental research program on IoT and its intersection with military operations, particularly Army operations. Founded in 2017, the Internet of Battlefield Things (IoBT) Collaborative Research Alliance is a 10-year research program that began by addressing four key challenges:

- Dynamic composition of multiple IoT elements to achieve specific mission-derived goals, which include the pivotal challenge of discovering available IoT devices, information sources, and algorithmic capabilities and constructing pipelines of these primitive elements to create mission-relevant and mission-enhancing capabilities
- Autonomic capabilities that exhibit autonomous properties to enable intelligent adaptation in the face of ever-changing battlefield conditions, especially when confronting adversarial activity
- Distributed analytics, which addresses the unique challenge within military environments to exploit hybrid cloud/edge computing models rather than relying on centralized and enterprise cloud architectures, especially in the face of limited and disrupted communications

- Cyber-physical security, a cross-cutting issue for many IoT systems that examines the risks associated with the permeation of IoT devices within our ecosystem and potential mitigations to address this challenge

Other overarching themes of the program include understanding and addressing the challenges of machine learning in distributed (and not enterprise cloud) settings in the face of adversarial activity and discovering and exploiting not just blue but also gray IoT assets, that is, the millions of commercial IoT devices that are deployed and owned by third-party entities such as municipal governments, private corporations, and even individuals. More information about the IoBT program is available at https://iobt.illinois.edu.

Several innovative IoT solutions have been developed as part of a digital nursery at the smart French Air Force Base Evreux [7]. Light Information Equipment via the Automatic Dependent Surveillance-Broadcast (ADS-B) (ELIA) is a cooperative air traffic control monitoring system. The ADS-B can be seen as a civilian equivalent of the military's location radar and allows multilateration, location by emissions, and responses to tag interrogations. It requires a precise measurement of the arrival time of the signals and a suitable geometric arrangement of the system formed by the aircraft and the beacons. The ELIA hardware is light and low cost and offers an additional means of obtaining relevant situational awareness. It allows data transmission to personnel and tool installation nationally, within hostile territory, or on an aircraft. The project team has used only COTS components and open-source software, costing less than 100 euros. Despite being vulnerable to jamming attacks, ELIA demonstrates that anyone, including our adversaries, can collect the same useful data, including the real-time location of civil and military aircraft, on a modest budget. GeoCourtix is another project that, at a similarly low cost, provides a device that allows geolocation of personnel and transmits the location information via a secure channel. It was implemented as a proof of concept for budget enhancement solutions to existing military equipment. Using both a Raspberry Pi (RPi) microcontroller programmed to receive GPS signals and display the positions received on a map and another Moteino microcontroller capable of sending these GPS data, military personnel can be located in real time. The GeoCourtix platform also allows adding other sensors, such as a heart sensor or a camera, and transferring the acquired data via a secure channel.

The NATO Parliamentary Assembly has assessed the opportunities and risks related to the deployment of IoT [18]. The study extensively discusses the significant risks associated with integrating IoT with critical infrastructure and reiterates many observations and recommendations of earlier studies [22]. These risks are increasing because IoT devices increasingly manage and service essential infrastructure. Therefore, the probability that unsecured IoT devices and services will expose critical infrastructure to vulnerabilities is increasing.

The Ukraine–Russia conflict provides a recent example of nation-state cyberattacks exploiting IoT devices [3]. Since the beginning of the war, both sides were reported to have exploited vulnerabilities in critical infrastructure and unsecured IoT devices. Compromised IoT provided attackers with more efficient methods of performing battlefield Intelligence, Surveillance, and Reconnaissance (ISR), leading to physical damage. By gaining access to the sensors, such as cameras, an adversary can obtain access to sensitive data, e.g., live video feeds, which they can subsequently use for intelligence or attack. Before and during the conflict, both sides deployed botnets in the opponent's CIS. IoT devices are easy targets as they are often forgotten when performing regular cyber hygiene and are relatively easy to discover and analyze by skilled adversaries. Although there is little unclassified data available on military use of IoT botnets, based on the price listings available on the Dark Web, it can be assumed that there is a large number of pre-deployed IoT botnets waiting for use and featuring tens of thousands of compromised devices.

1.4 Scope of the Book

The military domain is large, and there are many applications of IoT within this domain. Even a cursory examination of traditional military functions shows that IoT can significantly affect military operations and personnel in the context of communications, surveillance, reconnaissance, logistics, health, and Civil–Military Cooperation (CIMIC). It is impossible to cover all the applications of the IoT in all military domains within the scope of a single book, and we do not intend to do so. Instead, this book focuses mainly on the subset of domains and IoT capabilities that were investigated within the 4-year effort of the NATO Information Systems Technology (IST) 147 Research Task Group (RTG) on Military Applications of Internet of Things. The focus of our investigation has been on examining commercial IoT devices, architectures, and their capabilities in the land domain. This does not imply that IoT could not contribute interesting capabilities to maritime, air, space, or cyber domains. We believe that our discussion on IoT use in the land domain provides representative conclusions for the perspectives and challenges of IoT implementation in other military domains.

Similarly, we do not intend to address all possible applications of continuously expanding civilian IoT solutions to the military land domain. The IEEE P2413 Working Group [11] aims to provide a foundational approach to six areas of civilian IoT: Smart Manufacturing, Smart Grid, Smart Buildings, Intelligent Transport, Smart Cities, and Healthcare. For practical reasons, the IST 147 group had to choose a subset of these areas for its analytical and experimentation activities, with a particular focus on technologies related to smart cities and healthcare. Furthermore, new types of devices and use cases are continuously introduced in the civilian world (and the number of things connected to the Internet now exceeds the number of people connected to the Internet). For example, complex modern equipment, such as a car, can have tens of devices (sensors, actuators, etc.) [6, 12], and the ability to collect near real-time sensor information from this equipment enables new business models such as condition-based maintenance [20] or the provision of equipment, e.g., aircraft engines, as a service, rather than transferring ownership [13]. Many of the new ideas and use cases that originate in the civilian world can be adapted to a military environment and, as shown by recent conflicts, can contribute to the execution and success of defensive and offensive military operations [22]. Therefore, one of the objectives of this book is to stimulate and inspire the readers to explore the vast opportunities offered by IoT for military purposes and CIMIC rather than to provide a ready-to-use cookbook to apply these technologies. In particular, although we review several proposed architectures for IoT applications and make some recommendations regarding their use, it should be noted that there is no single best-fit architecture and, as shown in the subsequent chapters, some specific adjustments could be needed to ensure efficiency and manageability of a military solution. Furthermore, in our discussions, we focus primarily on technical architectures, leaving aside some pertinent operational questions related to the management and maintenance of the IoT network and devices. In particular, although we discuss some of the issues related to key management, many other questions, e.g., related to how devices are patched with firmware updates, such as what type of network monitoring is needed, and what controls are required over the operation of the network, are left largely open and need to be considered when planning an operational deployment of IoT in a military environment.

The structure of the book is as follows. Chapter 2, which is part of this first section of the book on Motivation and Use Cases, discusses several application areas within the military domain where IoT could play a role.

The second part of this book, Challenges and Solutions, is the core of the book and discusses several key topics. It begins with Chap. 3 providing a review of various IoT architectures and then provides a discussion of specific architectural challenges that are likely to arise in the military domain. This is followed by Chap. 4, which discusses various foundational technologies for IoT. Chapter 5 then describes a few middleware prototypes that were built as part of the activities of this group. The second part is rounded out by a discussion on security considerations in Chap. 6 and then by a discussion on reliability and fault tolerance in Chap. 7.

The third part of the book, Experimental Results and Recommendations, rounds out the book by providing a description of the various experiments and demonstrations that have been conducted by the group in Chap. 8, followed by a brief discussion of conclusions and future directions in Chap. 9.

References

1. Alberts, D.S., Garstka, J.J., Stein, F.P.: Network Centric Warfare: Developing and Leveraging Information Superiority, 2nd edn. DoD C4ISR Cooperative Research Program (1999)
2. Bath, W.G.: Overview of platforms and combat systems. J. Hopkins APL Tech. Dig. **35**(2), 90–94 (2020)
3. Broomhead, B.: Defending against attacks on vulnerable IoT devices. Vulnerabilities & Threats Dark Reading (2023). https://www.darkreading.com/vulnerabilities-threats/defending-against-attacks-on-vulnerable-iot-devices
4. Bughin, J., Hazan, E., Allas, T., Hjartar, K., Manyika, J., Sjatil, P.E., Shigina, I.: Tech for Good: Smoothing disruption, improving well-being. Tech. rep., McKinsey Global Institute (2019)

5. Bunker, R.J.: Terrorist and insurgent Unmanned Aerial Vehicles: use, potentials, and military implications. Tech. rep., Strategic Studies Institute, US Army War College (2015)
6. Cai, Z., Wang, A., Zhang, W.K.: 0-days & mitigations: roadways to exploit and secure connected BMW cars. In: Proceedings of Black Hat USA Conference (2019)
7. Chaire de Cyberdefense et Cybersecurite: Les objets connectés et le monde militaire. In: Seminaire interarmees. Cesson Sévigné (2017)
8. État-major de l'armée de Terre: Action Terrestre Future: Demain se gagne aujourd'hui. Tech. rep., Ministère des Armees - Armee de Terre, Paris, France (2016)
9. Haase, E.: Distributed Common Ground System-future moving into the 22nd century today. Joint Forces Q. **77**(2), 109–113 (2015)
10. Hoenig, M.: Hezbollah and the Use of Drones as a Weapon of Terrorism. Public Interest Report, Federation of American Scientists (2014). https://fas.org/publication/hezbollah-use-drones-weapon-terrorism/
11. IEEE P2413: IEEE Draft Standard for an Architectural Framework for the Internet of Things (IoT). IEEE P2413/D0.4.5, December 2018, pp. 1–264 (2018)
12. Jeong, J.: IPv6 Wireless Access in Vehicular Environments (IPWAVE): Problem Statement and Use Cases. Request for Comments (RFC) 9365, Internet Engineering Task Force (IETF) (2023)
13. Jørgensen, S., Pedersen, L.J.T.: Service-Logic Rather than Product-Logic, pp. 89–102. Springer, Berlin (2018). https://doi.org/10.1007/978-3-319-91971-3_7
14. Lenk, P.: NATO as an information platform. In: 2019 International Conference on Military Communications and Information Systems (ICMCIS) (2019)
15. Lenk, P.: Some thoughts regarding data-centric consultation, command and control. In: 2023 International Conference on Military Communications and Information Systems (ICMCIS) (2023)
16. Reuven, N.: The shift in technological innovation from the defense sector to the civilian sector. BESA Center Perspectives Paper 2205, The Begin-Sadat Center for Strategic Studies, Bar-Ilan University (2023)
17. Suri, N., Benincasa, G., Lenzi, R., Tortonesi, M., Stefanelli, C., Sadler, L.: Exploring value-of-information-based approaches to support effective communications in tactical networks. IEEE Commun. Mag. **53**(10), 39–45 (2015)
18. Tonin, M.: The Internet of Things: Promises and Perils of a Disruptive Technology. Tech. rep., NATO Parliamentary Assembly Science and Technology Committee (2017)
19. Tzezana, R.: Scenarios for crime and terrorist attacks using the Internet of Things. Eur. J. Fut. Res. **4**(18), 1–7 (2016)
20. Williams, J., Davies, A., Drake, P.: Condition-based Maintenance and Machine Diagnostics. Springer, Berlin (1994)
21. Zheng, D., Carter, W.: Leveraging the Internet of Things for a more efficient and effective military. Tech. rep., Center for Strategic & International Studies, Washington, DC (2015)
22. Zheng, D.E., Carter, W.A.: Leveraging the Internet of Things for a More Efficient and Effective Military. Rowman & Littlefield, Lanham (2015)

Chapter 2
Selected Application Areas

Konrad Wrona, Michael Street, Michał Dyk, Manas Pradhan, and Mauro Tortonesi

2.1 Introduction

This chapter discusses some application scenarios for the use of IoT in a military environment that were considered by the STO IST-147 RTG at the start of the activity. This initial exploration activity was meant to be as broad as possible and resulted in the enumeration of a diverse number of application domains for IoT in the military. Given the duration and resources available to the group, exploring all of these diverse application areas was impractical. Hence, the group primarily focused on Humanitarian Assistance and Disaster Relief and the benefits of IoT in that context, as will be seen in the later chapters. However, this chapter does describe the broad set of scenarios that were considered as a way of highlighting the wide impact that IoT can have (and is having) on military operations across the board.

The application scenarios for the use of IoT in a military environment revolve around a variety of different types of devices that could be introduced into the military domain:

K. Wrona (✉)
NATO Communications and Information Agency (NCIA), The Hague, The Netherlands

Military University of Technology, Warsaw, Poland
e-mail: konrad.wrona@ncia.nato.int

M. Street
NATO Communications and Information Agency (NCIA), The Hague, The Netherlands

M. Dyk
Military University of Technology, Warsaw, Poland

M. Pradhan
DKSR, Berlin, Germany

M. Tortonesi
University of Ferrara, Ferrara, Italy

N. Suri et al. (eds.), *Military Applications of Internet of Things*, Internet of Things,
https://doi.org/10.1007/978-3-032-03378-9_2

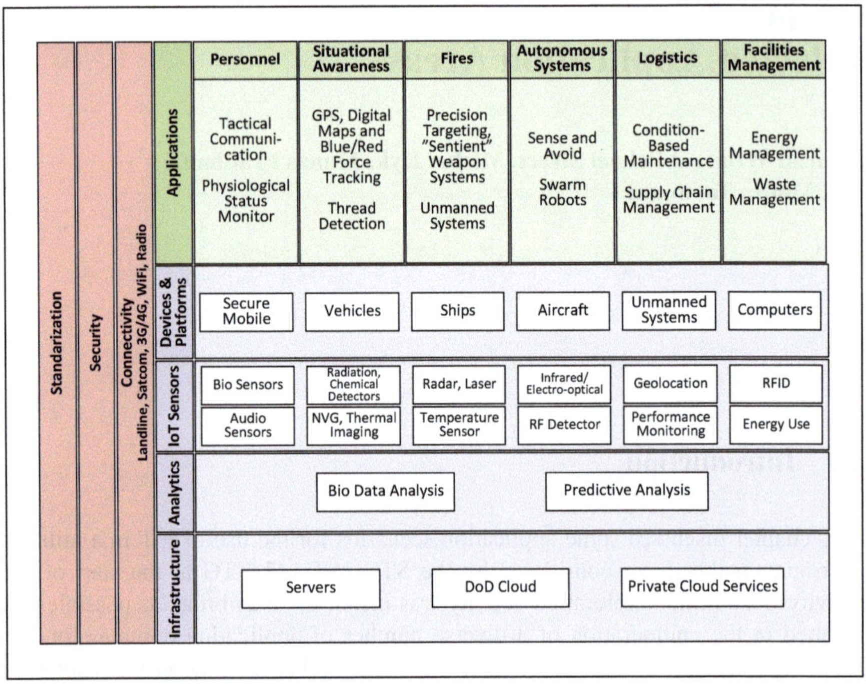

Fig. 2.1 An overview of military applications of IoT [14]

- IoT devices custom-developed for military purposes
- Commercial Off The Shelf (COTS) IoT devices managed by forces and used in a military setting for routine operations, such as logistics and healthcare
- Bring Your Own Device (BYOD) COTS IoT devices brought and used individually and independently by military personnel, for example, a smartwatch used by the military in operation
- Civilian IoT devices deployed in the theater, belonging to and operated by third parties but interconnected with military systems to enhance situational awareness and produce effects, for example, a smart city infrastructure

An overview of various military applications of the IoT is presented in Fig. 2.1.

2.2 Civil–Military Cooperation for Resilient Smart Cities

Urban and hybrid warfare have transformed elements of military operations. The ability to sense hostile activity—and precursors of hostile activity—in an urban environment is critical in protecting cities against both terrorist and state attacks. However, detecting such activity in urban environments is a highly challenging task.

Smart cities are already providing their digital citizens with intelligent resource utilization solutions for energy, water, mobility, and parking spaces, as well as a new generation of real-time and time-critical, location-, social-, and context-aware services for health care, entertainment, and social good [5]. Most (if not all) of these applications use the functions of IoT devices operating as a capillary network of sensors that provide a constant stream of information [1].

Traditional analytic solutions based on transferring all the data to the cloud, processing them using big data methodologies and tools, and returning the results to interested users are often too slow for applications with strict latency constraints and too burdensome for the network infrastructure. Instead, the smart city scenario is particularly well suited for the adoption of distributed processing approaches such as fog computing (see also Chap. 3) in which information processing services are executed on edge devices in proximity to raw data sources, information consumers, or both [8].

The widespread adoption of IoT technologies in smart city scenarios raises significant security issues. In fact, the multitude and pervasiveness of Information Technology (IT) services provided by smart cities present a massive attack surface whose protection requires the development of new, more scalable, and more resilient cybersecurity solutions.

Security has generally been recognized as a fundamental pillar for the implementation of smart city platforms. Industry and academia have focused on designing cybersecurity solutions for smart city scenarios that take advantage of existing experience in protecting critical infrastructure, cyber-physical systems, and large-scale IT systems.

Recently, the military has also begun to investigate the opportunities and challenges brought about by smart environments, including the vital role that the IoT can play in battlefield scenarios [9]. In particular, the civilian IoT infrastructure can become a significant source of situational awareness information during military operations in urban environments [11]. Smart city infrastructure, such as traffic monitoring systems, smart utility networks, public transportation systems, and video surveillance networks, originally designed to improve the quality of life of citizens, might become a dual-use technology, improving situational awareness in military operations by significantly augmenting information obtained from purposely built and deployed military sensors.

The use of civilian assets has the potential to be particularly important in Humanitarian Assistance and Disaster Relief (HADR), counterterrorism, and mass protection scenarios. In HADR operations, interconnecting purpose-built ad hoc communication systems deployed by rescuers with the surviving, even if degraded IoT infrastructure, could contribute to significant improvement in coverage, accuracy of a common operational picture, and reduction of deployment time. These improvements contribute to the cost-effectiveness of disaster response operations, but more importantly, they can significantly increase the ability of rescuers to save lives. Similarly, counterterrorism operations would benefit immensely from access to the monitoring assets of smart cities, such as traffic cameras, road conditions, pollution sensors, and crowd movement patterns. Note that in this book, we do not address the exploitation of the IoT capabilities located in areas controlled by an adversary. We focus on IoT capabilities that are in friendly or neutral areas, but we understand that they might still be compromised by an adversary, raising issues of trust.

2.2.1 Public Safety Applications

In the following scenario, we consider an example of a large public event in a smart city as a use case for analyzing how IoT assets could contribute to providing more comprehensive and intelligent Civil–Military Cooperation (CIMIC) in support of public safety and security services to citizens.

In the public event scenario, a city section is closed off to motorized traffic and accessible only to pedestrians and authorized vehicles. In anticipation of a large gathering, military and police forces, as well as Emergency Medical Services (EMS) personnel, are deployed to ensure the order and safety of the public. Information sources, such as traffic cameras, normally dedicated to daily monitoring of the city, will funnel captured data feeds into a secure and custom-configured fog computing platform for analysis.

To support security operations, multiple applications are deployed to the smart city fog computing platform concurrently, each leveraging a specific set of information obtained from the city's IoT infrastructure. Such applications can take advantage of the low-latency processing allowed by the fog and the computational capabilities of the cloud.

For example, a crowd monitoring application could use information from video camera feeds and from the number of personal devices connected to the network to provide a relatively accurate and up-to-date estimate of the number of people present in the festival area. This information can help identify the optimal distribution of safety personnel and resources, such as ambulances and water distribution points.

In this scenario, as depicted in Fig. 2.2, a first-order security monitoring service, operating with the highest possible priority and without associated resource consumption limits, could continuously scan video feeds collected from accessible cameras, including both city-owned traffic and surveillance cameras, as well as privately owned cameras contributed by volunteers, to identify any potential anomalies. The service could leverage coarse-grained anomaly detection algorithms with relatively low computational requirements, running software components on devices in the fog layer to process data obtained from sensors with the lowest possible latency. The low-latency, coarse results provided by this first-order service could help police and military forces identify obvious or evident security threats, such as a person wielding a weapon or a group of people starting a brawl as quickly as possible.

The first-order service could also provide functions in support of EMS personnel delivering medical services. For example, it might try to identify possible health emergencies, such as heat strokes and dehydration, through the fusion of data collected from IoT sensors (video feeds, temperature and humidity monitors, etc.), wearable devices (heartbeat, sweat, and other physiological sensors), and mobile devices (pace monitors, fall detector apps, etc.).

The first-order service could be complemented by a second-order service that performs information processing at the cloud layer. The considerable resources available in the cloud would allow the service to run more powerful and fine-grained classification and detection algorithms, possibly against a database of known threat actors, in order to help the police identify less evident and potentially more dangerous security risks for counterterrorism purposes. This service would take advantage of a different trade-off between latency and computational resource availability.

Finally, both services could disseminate general public information and integrate with commercial applications. More specifically, they could provide helpful information to citizens by distributing traffic information or suggesting what public transport connection to take when leaving the festival area. Other applications can offer services that integrate with IoT sensors, for example, identifying impromptu performances from street artists or potentially interesting shopping offers, and directing users to their locations.

Each service used to support the public event scenario could be associated with a specific policy on access control, priority, and resource consumption limits. At the same time, the collection and storage of this type of information will bring its own challenges in the context of privacy and protection of personal data.

At the architectural and workflow levels, taking inspiration from the microservices paradigm, the services could be implemented on top of smaller, loosely coupled software components that could coordinate to achieve the desired results. This would provide significant benefits, both in terms of ease of migration of service components to better suited computing platforms—and in turn to make the best use of the available computation and communication resources, of reuse of the output of an upstream software component as input for several other components—as in the case of the Anomaly recognition software component in Fig. 2.2, which provides information to both the instances of the logistic support and security alert software components running at the fog layer.

All the above applications operate on a wide range of data types, requiring different processing. Developing a heterogeneous application ecosystem could be significantly facilitated using an appropriate information model and a corresponding information-centric and value-based service framework. This could help address the challenges related to processing the deluge of raw data continuously generated in a smart city environment. A related requirement is implementing a comprehensive management platform for a plethora of different services running in federated cloud and fog environments. This management platform needs to support the allocation of services to suitable edge devices or to the cloud, in line with other constraints such as current network conditions.

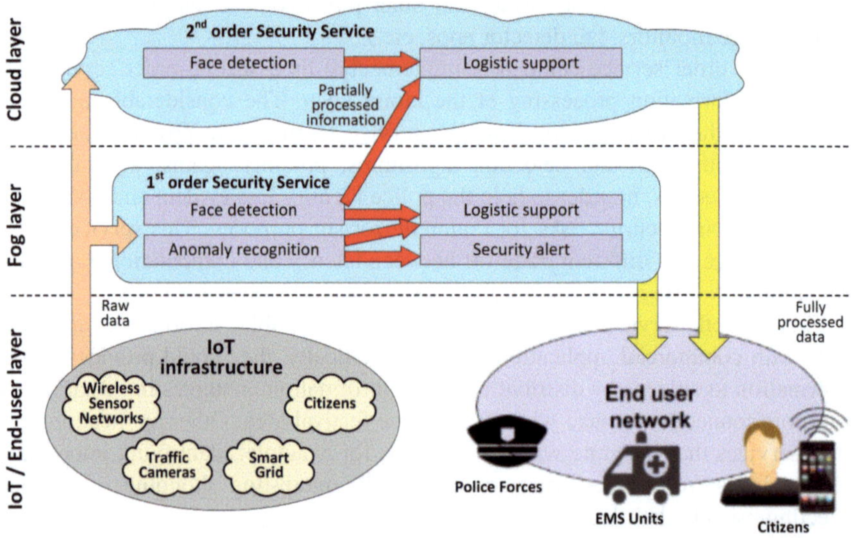

Fig. 2.2 Multilevel processing of information in a fog computing scenario

2.2.2 Humanitarian Assistance and Disaster Relief

Disaster recovery operations are extremely challenging and place significant demands on multiple resources, including local and international emergency response personnel, nongovernmental organizations, and the military. In the immediate aftermath of a disaster, one of the most pressing requirements is for comprehensive Situational Awareness (SA) so that civilian and military resources, including personnel and supplies, may be prioritized to have the most impact and help those in the most need. As the recovery operations continue, SA needs to be continuously updated based on changing conditions in the affected areas. Many sources of information provide SA, including reporting by the victims of a disaster, as well as observations made by responding civilian and military personnel. In this context, SA can be significantly enhanced via information obtained from IoT devices, especially in a smart city environment. There is a lot of potential to exploit smart city IoT capabilities to help disaster recovery operations.

When responding to a large-scale disaster, multiple civilian agencies often have to collaborate and share resources and information with military forces to quickly understand the situation on the ground and develop and execute a recovery operation [10]. The term Humanitarian Assistance and Disaster Relief (HADR) describes such operations, and examples of CIMIC in support of HADR include the Haiti Earthquake of 2010 and the Fukushima nuclear disaster following the Tsunami in Japan in 2011. The advent of smart cities with widely deployed IoT technology within the city's infrastructure will influence the potential future of CIMIC in HADR operations, as well as improve the daily operations of the city and help citizens and residents experience enhanced services and capabilities offered by such technologies.

Setting up a smart city infrastructure offering resilience required for CIMIC and HADR includes installing a broad range of resilient sensors, actuators, effectors, and communications equipment. This equipment might be COTS IoT devices or existing legacy devices that fit into the various use cases and scenarios. For example, for a smart city traffic monitoring system, traffic light posts might incorporate cameras to monitor the traffic flow at junctions. At the same locations, there might be sensors to monitor the level of pollution and current weather conditions. These and other information sources help the city plan the flow of traffic to balance pollution and reduce traffic congestion but can also provide indispensable SA for HADR.

In the event of a natural disaster phenomenon such as an earthquake, the IoT devices within the city (e.g., the traffic cameras, weather, and pollution sensors) can be re-tasked to identify emergency conditions such as toxic gases in the air, people trapped in vehicles or rubble, or the status of the power grid. Knowledge of the situation (such as population concentrations) immediately prior to a disaster can also help focus rescue operations and otherwise provide situational awareness.

In addition, the agencies responding to the emergency can bring their own IoT sensors and communication equipment to start monitoring the ground conditions. These could be statically deployed at strategic locations or even mounted on various types of unmanned vehicles. Responding agencies can then share this information with local authorities to improve their situation management. Integrating these varying information sources is critical to providing and maintaining joint SA for civilian and military forces and enabling effective HADR operations.

From a logistics perspective, civilian IoT solutions can integrate with military logistics and resource planning systems in support of both the supply chain administration and also operational and tactical planning. As one example, consider a HADR scenario where the military could leverage hospitals' logistics and supply chain management. As the military is performing triage, it could consider specific conditions at local hospitals, such as the available capacity and the availability of specific medications, medical instruments, tools, or medical expertise. Likewise, as the military handles the transporting and disseminating of critical supplies, detailed and up-to-date awareness about local stockpiles and needs can help optimize the overall logistics operation.

2.3 Smart Military Base

A military base is similar to a small town that houses military personnel, equipment, hospitals, schools, office buildings, police stations, airports, visitors, and much more. A large community of military and nonmilitary personnel performs multiple tasks much like in a small city ecosystem.

Making military bases and camps smart could reduce many of the responsibilities and workload that requires human resources. This *smartness* could also increase connectivity and information sharing and slash operating costs while improving safety planning and reducing the response time to undesired events at the bases. Jobs like perimeter security could be automated, freeing security forces for more essential tasks [2, 12]. Smart cameras running Artificial Intelligence (AI)-based computer vision algorithms can actively monitor, track, and detect objects without involving deployed soldiers, thus alleviating the need to permanently station a soldier on guard. Radio Frequency Identification (RFID) sensor tags could identify personnel and civilians separately, allowing for precise authorization and authentication. Personnel arriving at gate perimeters and other entry points could put these identifying tags on their cars. The data collected from the array of sensors and actuators deployed in the smart bases could be pushed into the existing military Command and Control (C2) and base monitoring systems to increase the SA of the base personnel. In addition to traditional C2 systems, mobile devices such as smartphones and watches, for example, can use IoT to generate alerts, reports, and commands, which can be relayed directly to base personnel without interference from manual reporting systems such as radio or paperwork. Apart from tactical and security-related aspects, equipping bases with smart meters, switches, light sensors, and grid converters would also reduce electrical consumption by 35% or more [14].

Smart cameras and sensors can distinguish between authorized personnel, military personnel, and civilians, which can further help reduce traffic congestion around the bases, especially during on-base public events. Sensors at waste collection bins, garbage cans, and on-base utility stores can enable better waste management by recording waste levels and notifying authorities to replenish depleting inventory.

In cases of actual attacks or accidents at the base, base security systems can notify emergency response services to take appropriate action. These systems can also automatically notify required personnel and services to lock down facilities, for example, schools and hospitals. In addition, smartphone and other tactical systems can be used to send notifications out to base personnel ordering them to avoid or follow certain protocols in the affected areas.

To reduce costs, smart technologies such as 3-Dimensional (3D) printing can provide quicker and more cost-effective construction. For example, the US Navy's *Print the Fleet* is a 3D printing project for smart manufacturing that enables faster and cheaper construction of parts manufactured on bases or onboard ships [6].

In addition to operational use, deployable smart bases have the added benefit of being a much more suitable environment for training through simulation and for collecting lessons learned.

There are several areas where smart military bases can use civilian IoT, including base deployment, environment integration (early warning systems, reconnaissance efforts, and environmental data extraction), logistics and resupply, and sustainability. Each of these areas can have its own IoT architecture and services; however, all systems need to be integrated with a single data analytical layer to provide a comprehensive SA. The communication technologies required for such integration could vary from standard Ethernet communications to different types of wireless networking protocols. Also, data processing would most likely occur at the edge, in the individual systems, and the central data analytics system.

The following subsections describe deployment, monitoring, and sustainability aspects of the smart base in more detail.

2.3.1 *Deployment*

Advanced IoT systems consisting of robots, actuators, and sensors can significantly shorten the time and risks involved in setting up a new expeditionary/forward military base or expanding an existing one. IoT technology can assist personnel or even perform self-assembly of facilities in a specified location and configuration. Such automated building technology is already successfully used in civilian scenarios [4]. The use of automation during base deployment would not only reduce cost and time but also reduce threats to the health and safety of personnel, as forward operating bases are often deployed in or close to hostile environments.

Considering the possibility of a mobile smart base that can tap into a network of environmental sensors to simulate real conditions in closed environments, multiple interconnecting containers can constitute the smart base, each container with sensors for location, proximity, velocity, etc., that can provide enough data for a central controller to automatically align each container into assigned positions. This interconnection can reduce the number of personnel required during the process of deployment, ensuring an additional amount of time for technicians to prepare their mission upon arrival.

2.3.2 *Monitoring*

After a successful deployment, the smart base can either tap into an existing network of environmental sensors that can provide data for different services (e.g., early warning systems, reconnaissance capabilities, environmental information extraction for real-life simulations in controlled areas, etc.) or implement its own network of sensors by:

- Establishing a certain perimeter that drones can automatically monitor
- Implementing an early warning system using sensors that detect motion, heat, audio, or other characteristics and sending the data back to base for analysis and response
- Integrating with the environment different types of sensors that can extract data such as wind velocity and direction, outside temperature, sunlight intensity, environmental sounds, etc. These data can then accurately simulate real-world conditions but in closed environments, thereby allowing soldiers to be trained in safe and controlled areas.

Ubiquitous deployment of COTS and BYOD IoT devices can also enable better monitoring and control of resources, including detecting abnormal situations. For example, an airman jogging on the perimeter road recognizes the pungent smell of Jet Propellant 8 (JP8) and sees the fuel spill has spread across much of that part of the ramp. He activates an emergency call button on a perimeter smart light pole equipped for emergency communications and relays what he sees to the base Command Post (CP). The Security Forces Duty Officer in the connected base CP can see the spill from a smart light pole equipped with Electro-Optical (EO) and Infrared (IR) High Definition (HD) cameras and atmospheric sensors. He recognizes this is a major incident and initiates the response protocol. The fire department can also see the spill via the smart light system sensors.

2.3.3 *Sustainability*

Deployment of a smart base brings the additional economic and tactical benefit of cutting down energy consumption by using smart Information and Communications Technology (ICT) technologies in traditional electrical installations. In addition, alternative and renewable forms of energy can be more easily harvested and integrated with the base infrastructure. Many of these concepts are carried over from existing smart city concepts [7], although in the civilian environment, the focus is often more on limiting environmental impact than on ensuring self-sustainment and reducing dependence on external energy sources and delivery of fuels.

The US Army has erected a smart solar energy installation at Fort Stewart, Georgia, covering an area of over 1 square kilometer, allowing the base to operate independently of the traditional power grid. This renewable energy source allows the base to store energy and utilize it independently and in case of emergencies. Extending the concept, Fort Bragg, North Carolina, is pilot-testing driverless vehicle for transporting wounded personnel, to eventually use in battlefield situations. Additionally, bases are equipping themselves with smart technologies for water resource usage and construction [12], and the US Army has saved more than $150 million in costs enacting its vision for the smart energy program.

Automatic control of power-consuming resources could benefit from information about human presence in rooms and environmental data to execute preconfigured or administrator-activated rules that increase energy preservation and optimize resource use.

Integrating sensors into power-consuming activities can make these activities more sustainable by reducing energy or fuel consumption. Some examples include:

- Automatic detection of presence in a shower or bathroom for turning on water pumping and water heating
- Automatic heating/cooling of only occupied rooms or those previously reserved for scheduled activities
- Automatic light control (dimming and turning on/off) based on light placement, sources of natural light, and the type of ongoing activities in that area

- Detecting leaks or short circuits by placing sensors in the power grid, thereby reducing the risk of accidents and enhancing equipment lifetime

2.4 Logistics and Maintenance

Civilian logistics operations were significantly improved by IoT technologies such as RFID, GPS-based vehicle tracking, and even smart shelves in supermarkets that track product consumption and predict supply needs. Military logistics and resupply can exploit many of the innovations from the civilian use of IoT technologies, such as:

- RFID tagged packets for monitoring the supply chain
- RFID tags or other miniature sensors placed on food/water containers to monitor the amount consumed/remaining for automatically ordering resupplies when needed
- Fuel monitoring for patrolling or for trips made in emergencies[1]
- Special equipment monitoring, used in case of alerts (fire, attack, earthquake, flood, etc.). Essentially, every critical or important object/component of the logistics chain can be monitored and controlled for better resource usage, growth in performance, or abnormal conduct detection.

The use of sensors in various military equipment opens up logistics opportunities. To use sensor data for logistics purposes and to drive logistic opportunities, the Enterprise Resource Planning (ERP) system must be adapted to meet real-time demand.

Concerning maintenance, the use of sensor indicators will steer the need for maintenance toward maintenance on demand. Military equipment will often require a combination of maintenance on demand (sensor-driven) and maintenance based on plans (planned maintenance), as not all maintenance will be indicated by sensors. For example, consider a military vehicle with required (planned) vehicle testing according to a plan based on the vehicle's registration number. Maintenance on demand might indicate the need for a brake checkup before the vehicle is ready for vehicle testing. Supply will also benefit from maintenance on demand, in that it will meet the need for spares when and where they are needed. In addition, supply will benefit from consumption indications of, for example, ammunition and fuel. Vehicle storage conditions should include consideration of the onboard sensors and inform how a vehicle is stored and how to reactivate it, ensuring the sensors' continued

[1] As an example, during the North Atlantic Treaty Organization (NATO) mission in Afghanistan, at the Kandahar military base, a fuel truck drove around the base every day to top up each of the hundreds of diesel generators, whether they needed it or not. An automated system that allowed refueling only when needed would have saved considerable resources.

operability (e.g., test cycles). In addition, storage conditions must also ensure that the vehicle is protected from illegitimate tracking of sensor signals during storage.

As the dependence on IoT grows in such applications, it becomes more important to know via a failsafe or self-diagnosis mechanism when a sensor stops working, so personnel can exchange it for a properly functioning sensor. Otherwise, false positives (or false negatives) will decrease trust in IoT-based logistics and maintenance, resulting in these capabilities and services not being used to their full extent.

2.5 Smart Healthcare and Well-Being of the Personnel

Military personnel can leverage IoT and smart infrastructure to enable *smart healthcare*, both while stationed at a base and during active deployment. This is very relevant since the military personnel need to stay in shape and have their health monitored for active deployment anytime. This would require the personnel to have regular health checkups and assessments done on the fly to inform them about serious deployment-related health issues they might face, injured personnel being taken care of, etc. Military hospitals and medical facilities need to continuously care for the health and well-being of military personnel. Equipping these facilities with smart devices and technologies would make their work easier and more efficient. Smartphones, smartwatches, and other devices that fit the concept of BYOD can be used to message and inform personnel, collect relevant data remotely, and proactively assess the fitness of all personnel, pre-, during, and post-deployment. Analysis based on the collected data can be pushed to the hospitals and the command level based on the need-to-know requirements, for example, in the context of an upcoming mission deployment.

For example, the US Air Force has deployed Cisco's Smart+Connected healthcare solutions on a pilot basis. Base hospitals can remind personnel about their scheduled medical appointments using messaging services connected to their medical profiles. This has enabled hospitals to reduce costly *no-shows* by almost 32% [14].

Smart healthcare can also extend to military operational scenarios. In particular, during their missions, soldiers can carry sensors monitoring their vital signs, for example, pulse, breathing, body temperature, and body posture. Sensors transmit the gathered medical data to the soldier's smartphone and then relay it to mission control, where the soldier's health status is included in the situational awareness picture.

A typical military medical scenario related to deployment outside of a military base could be as follows: A tactical patrol operating in an urban environment encounters hostile activity and sustains some causalities. Local infantry support is operating within the vicinity (within 1 km). This scenario covers three stages:

1. Battlefield medical attention: Casualties must be treated with available local medical supplies, available either from the patrol or from other units operating

nearby which move to the incident to assist. Medical assessment may be provided either locally or remotely to ensure that battlefield casualties receive appropriate medical attention within reasonable timelines.
2. Casualties are transported to the operating base. Sensors provide continuous medical assessment during this journey and may provide data to the medical facility at the operating base.
3. Strategic medevac. Casualties are evacuated to a suitable medical facility outside the area of operations. Technology that minimizes the journey time, ensures the availability of patient information when necessary, and optimizes the availability of medical supplies will enhance this function.

IoT can support combat triage at a battlefield as described in [3]. Such critical operational application implies that all soldiers are equipped with various sensors that monitor their vital signs and that these sensors connect via a reliable network to medical or C2 personnel that can assess the situation during the conducted mission. A combat triage is typically carried out when something bad has happened to soldiers, for example, they are shot, or otherwise seriously injured, or there is a Mass Casualty (MASCAL) situation, where the need for medical treatment is greater than the capability. The triage scenario also assumes that in that first stage, a soldier's sensors are up and running and still connected with the Network of Things. This might not be always possible in reality, or at least not very likely, as the sensors and communication devices would likely have also suffered damage and might have even worsened the soldier's wounds. In most cases, you would strip the soldier of his gear to perform combat first aid, thus making triage from a distance seem unlikely. An interconnected IoT soldier health monitoring solution offers a reasonable and feasible way of handling the information given by the sensors, and together with AI, personnel could make some assessment of the soldiers' state of health. Given a MASCAL situation, the support outlined above might work, but a MASCAL situation is very rarely the case in operations today. The number of injured soldiers is normally very low, and the strategic medevac to a military surgical facility is normally sufficient though highly dependent on the delivery of on-site life-saving first aid.

Capturing the physiological state of the soldier and taking into account the complete health chain, for example, before injury, impact, triage, monitoring, care, and rehabilitation, are important and valuable. However, one must treat the obtained data cautiously, as it is difficult to define good health. A complete assessment should take into account real-time and historical data on human behavior, including a great variety of external signals, such as the voice characteristics, and internal ones, such as heart rate and vascular ripple. Taking into account such data usually requires rigor and strict laboratory conditions. For example, collected heart rate data may be inadequate if covering only a short period. Similarly, the voice has an identifiable basic frequency, but it can change with urgency and stress, etc. Therefore, it is important to carefully decide what data to measure, how to measure it, and how to integrate and meaningfully present the resulting measurements to soldiers and medical personnel. Smart healthcare stresses the importance of having high-quality measurement data that provides richly related metadata.

2.6 Reconnaissance

A typical tactical reconnaissance patrol consists of several phases, for example, assembly, decampment, mission, and return to the base.

During each phase of Reconnaissance (recon), various IoT assets may be utilized, via different communication means, especially for the backhaul link to the infrastructure/cloud at the base. The collected and exchanged IoT data may also vary during different phases, for example, frequency, precision, and information type.

During their route, the patrol passes several deployed outposts equipped with sensors monitoring the environment. For energy reasons, the outposts have no long-haul network connection to the base but offer a short-range communication technology (e.g., Wireless LAN (WLAN), Bluetooth, Near-Field Communication (NFC), or 5G) to read out and store the collected data as the patrol passes an outpost and detects short-range communication connectivity. When the patrol returns to the base, the data is forwarded to the smart base's networking infrastructure.

The development of smart systems has led to the increasing replacement of relatively dangerous and expensive human tactical patrols with Unmanned Aerial Vehicle (UAV) and Unmanned Ground Vehicle (UGV) reconnaissance. Several UAVs (e.g., quadcopters) equipped with cameras capable of image recognition are sent out for reconnaissance in advance of the patrol. When the UAV's image recognition algorithm detects a suspicious object, it generates an alarm message that is sent to the vehicles of the patrol and to mission control.

A relevant example is an operator positioned in a patrol vehicle remotely controlling a UGV deployed to explore an abandoned building. The UGV delivers a live video stream to the operator on the vehicle, allowing for the building's interior exploration.

Both UAV and UGV recon need adequate availability guarantees for communication channels to the CP, both to ensure the effective operation of the vehicle and to show gathered information to the operator and other members of the chain of command and analysts. Communication channels would preferably be real time, but in the case of autonomous vehicles could also include *store and forward* solutions that could provide commanders with still valuable situational awareness, even if not real time.

Each vehicle, whether human-operated or autonomous, is equipped with onboard sensors monitoring several vehicle status parameters such as oil level, fuel level, tire pressure, condition of brakes and shock absorbers, etc. The use of such sensors is already state of the art for civilian vehicles. Some of the vehicle's status data is relayed directly to mission control at the smart base for live monitoring to reduce unanticipated failures. Other data are transmitted to the cloud when the patrol returns to the base to facilitate condition-based maintenance and allow on-demand ordering of spare parts [13]. Personnel could also use the same sensor readings to collect statistics on vehicle use to optimize its utilization or ensure proper exploitation and adherence to procedures. Therefore, it is important to ensure an adequate level of accuracy and reliability of a sensor indication. Each vehicle reports its overall and subsystem status to mission control. Vehicle status-known as a health monitoring system on aircraft-is being built into more and more vehicles but is usually focused on informing the driver of the vehicle. Soldiers typically perform checks of their vehicles before going on patrol, during patrols (if time allows), and after patrols to prevent failures while patrolling. However, relaying this kind of information to mission control could give valuable information about the situation on site.

A remote tactical patrol is equipped to detect and respond to Chemical, Biological, Radiological, and Nuclear (CBRN) threats when operating at a distance of up to 50 km from an operating base that includes CBRN capabilities. Biosensors may be included in a soldier's equipment, monitoring the soldier's health continuously to indicate an infection or other reactions to biohazards to which the soldier himself might not react. The patrol includes remotely deployable, nonmobile CBRN sensors. UAVs hosting sensors may be available at the operating base. Both static and mobile sensor platforms carry sensors to detect leaks of various kinds. In this scenario, sensors on the remote patrol detect the presence of a biological threat and provide an early warning. Additional sensors are then deployed in the area to improve the resolution of the detected threat. The patrol returns to the operating base, and a rapidly deployable specialist CBRN outbreak team is sent out to conduct analysis and containment.

2.7 Risks Related to IoT Adoption in Military Domains

The previous sections have outlined a number of application areas within the military that could benefit from IoT. However, adoption of IoT also introduces some additional risks that need to be considered.

The heterogeneous character of the above scenarios and the involved IoT devices introduces several tough challenges on the way to a fully integrated future collaborative warfare data ecosystem capable of generating, sharing, analyzing, and creating value from the collected data and available actuation effects. Of particular concern is the impact of the *IoT transformation* on the cybersecurity posture of military systems. With increasing use of the same technology base within civilian and military systems, for example, TCP/IP, Bluetooth, and Wi-Fi, military systems lose an aspect of *security through obscurity*, as they are no longer *unique*. The convergence and dual use of technologies enable the military to benefit more quickly from technological progress and to leverage civilian R&D funding and training of technical personnel. But it also means that the systems used by the military are subject to widespread scrutiny and potential weaknesses and that vulnerabilities can be exploited more quickly. Moreover, in the civilian world, the trade-off between usability and security usually leans toward usability, leading, for example, to increasingly simplified and transparent connectivity and possibly reduced control.

Some specific usability risks related to an increased use of IoT assets in military operations include:

- Soldiers being overwhelmed with data and losing situational awareness about the physical environment as their cognitive focus is on trying to make sense of data delivered by various sensors and displays
- Increasing dependence on sensors and actuators that may not always be available during the operations, especially if the commercial nature of the devices makes them more susceptible to both adversarial activities such as jamming and random failures
- Loss of autonomy and authority at the local tactical level when the system is reduced to a simple set of sensors and weapons/actuators that are controlled and activated by remote and in-person command and control centers

Moreover, by multiplying the interactions between the digital and physical worlds and the connections among components, IoT can substantially increase the volume of information that needs to be transported and processed by military CIS. This has particularly important implications for tactical wireless networks with limited bandwidth and energy sources. Additionally, the wireless connectivity commonly used by IoT devices increases the risk of frequency congestion and exposes devices and data to confidentiality, integrity, and availability threats. For example, jamming and intercepting commercial IoT signals is easier, as frequencies and protocols are known, and multiple tools are openly available to perform such attacks. In addition, modified messages can be injected into wireless channels, compromising the integrity of data. To protect the confidentiality and integrity of IoT information flows, appropriate cryptographic mechanisms for encryption and authentication must be implemented. Although the focus of military information security is often on protection of data confidentiality, in the context of IoT and cyber-physical systems, ensuring availability and integrity of data might be of even greater importance, as not receiving critical information or receiving information that has been tampered with might have devastating consequences. An example of such a situation is a temperature sensor in a data center. If it "lies" to the monitoring system and reports that the temperature is 18°C, while in fact it is 40°C, the results can be catastrophic. Therefore, the integrity and availability of information from the IoT devices is a major concern that military users usually did not have to deal with in the past while relying on isolated, high-assurance networks and ruggedized, tamper-proof or tamper-resistant sensors and devices.

Each military application of IoT introduces some specific requirements for transmitted information (e.g., type, load, and receiving entities), suitable communication means, and security. These requirements are summarized in Sect. 2.8.

2.8 Conclusions

This chapter has outlined a diverse set of military application domains that could benefit from IoT.

Although IoT technology has several attractive military applications, current IoT systems have several limitations. Most of the systems are relatively closed, with limited interoperability for connectivity, use cases, and data distribution. IoT devices must have reduced weight and size and increased autonomy to be suitable for a wide spectrum of military use cases. They need simple and reliable control interfaces to support, not distract, military personnel in their functions. IoT devices need to support a common and open data model, multiple communication interfaces, and flexibility with respect to use cases. Military IoT applications need to be context-aware, capable of adapting the presentation of information to the specific environment and mission objectives, and help manage the volume and complexity of the data. The operation of IoT systems and the data distribution process should be automated as much as possible. Data processing and intelligence should be performed as close to the edge as possible to mitigate the impact of limited connectivity and enable disconnected operations. Another set of challenges is related to the management of IoT devices. Service Management and Control (SMC) aspects of military IoT systems, potentially consisting of a large number of heterogeneous devices that need to be managed over limited bandwidth interfaces and with little standardization of the management protocols, are going to be challenging. This highlights the important challenge in the deployment of IoT systems that is not only how to extract trustworthy data from the system but also how to manage and patch IoT devices to ensure their secure operations.

The next section and chapters within that section describe these challenges in more detail and also highlight potential building blocks and solutions to these challenges.

In the short term, even simple IoT applications could add value to the forces' daily routines. For example, an IoT device that counts the ammunition fired by a weapon, performs predictive maintenance, or can create maintenance parts thanks to 3D printers could simplify and optimize the logistics process. Terrestrial or aerial drones could be used not only for increased situational awareness and effect delivery but also to act as relays to overcome the limited radio range of ground troops.

As mentioned in the Introduction of this chapter, the STO IST-147 RTG was active for a limited duration and had limited resources and hence could not focus on all of the application domains described above. Instead, the focus was on examining existing IoT architectures and identifying misalignments when they are applied to the military domain. Furthermore, the group selected two specific problem domains—CIMIC for HADR operations and medical monitoring of military personnel to explore in more detail. These topic areas, and the results from demonstrations and experiments conducted to support these topic areas, are detailed later in this book.

References

1. Al-Fuqaha, A., Guizani, M., Mohammadi, M., Aledhari, M., Ayyash, M.: Internet of Things: a survey on enabling technologies, protocols, and applications. IEEE Commun. Surv. Tutor. **17**(4), 2347–2376 (2015)
2. Cisco: Smart and Connected Bases: An Opportunity for the USAF to Secure Tomorrow's Efficiency Through Today's Investment. White Paper, Cisco (2015)
3. Dyk, M., Chmielewski, M., Najgebauer, A.: Combat triage support using the Internet of Military Things. In: Proceedings of the Federated Conference on Computer Science and Information Systems (FedCIS), vol. 11, pp. 835–842 (2017)
4. Keating, S.J., Leland, J.C., Cai, L., Oxman, N.: Toward site-specific and self-sufficient robotic fabrication on architectural scales. Sci. Robot. **2**, eaam8986 (2017)
5. Khatoun, B., Rida, Y., Zeadally, S.: Smart cities: concepts, architectures, research opportunities. Commun. ACM **59**(8), 46–57 (2016)
6. Lessel, E., Beyer, B., Johnson, T.: Byting the Bullet: Now is the time for smart military bases. White Paper, Deloitte Center for Government Insights (2017)
7. Mathiesen, B.V., Lund, H., Connolly, D., Wenzel, H., Østergaard, P.A., Möller, B., Nielsen, S., Ridjan, I., Karnøe, P., Sperling, K., et al.: Smart energy systems for coherent 100% renewable energy and transport solutions. Appl. Energy **145**, 139–154 (2015)
8. Mukherjee, M., Shu, L., Wang, D.: Survey of fog computing: fundamental, network applications, and research challenges. IEEE Commun. Surv. Tutor. **20**(3), 1826–1857 (2018)
9. Suri, N., Tortonesi, M., Michaelis, J., Budulas, P., Benincasa, G., Russell, S., Stefanelli, C., Winkler, R.: Analyzing the applicability of Internet of Things to the battlefield environment. In: 2016 International Conference on Military Communications and Information Systems (ICMCIS), pp. 1–8. IEEE, Piscataway (2016)
10. Suri, N., Zielinski, Z., Tortonesi, M., Fuchs, C., Pradhan, M., Wrona, K., Furtak, J., Vasilache, D.B., Street, M., Pellegrini, V., Benincasa, G., Morelli, A., Stefanelli, C., Casini, E., Dyk, M.: Exploring smart city IoT for disaster recovery operations. In: 2018 IEEE 4th World Forum on Internet of Things (WF-IoT), pp. 463–468. IEEE, Piscataway (2018)
11. Tortonesi, M., Morelli, A., Govoni, M., Michaelis, J., Suri, N., Stefanelli, C., Russell, S.: Leveraging Internet of Things within the military network environment - challenges and solutions. In: 2016 IEEE 3rd World Forum on Internet of Things (WF-IoT), pp. 111–116 (2016)
12. URENIO: Smart cities provide the model for smart military bases - URENIO watch. https://www.urenio.org/2017/02/22/smart-cities-provide-model-smart-military-bases/. Accessed 06 Aug 2019
13. Zheng, D., Carter, W.: Leveraging the Internet of Things for a more efficient and effective military. Tech. rep., Center for Strategic & International Studies, Washington, DC (2015)
14. Zheng, D.E., Carter, W.A.: Leveraging the Internet of Things for a More Efficient and Effective Military. Rowman & Littlefield, Lanham (2015)

Part II
Challenges and Solutions

Chapter 3
Architectural Considerations

Christoph Fuchs, Manas Pradhan, Niranjan Suri, Mauro Tortonesi, Frank T. Johnsen, and Zbigniew Zieliński

3.1 Survey of Internet of Things (IoT) Architectures and Standardization Efforts

The term Internet of Things (IoT) has been included in the scientific and industrial vocabulary for more than two decades [1], and the general concept of a network of smart devices has existed for even longer. Despite the relatively long history of the digital age, there is still no commonly accepted definition of the concept of IoT. Furthermore, while several standards exist for individual elements, there is no common, internationally accepted standard that defines the components and interfaces used by IoT products. This absence has led to the situation where IoT solutions from different vendors use their own set of standardized or non-standardized technologies for communication, data representation, and service description, for example, making them non-interoperable. Interoperability, though,

C. Fuchs
Fraunhofer FKIE, Wachtberg, Germany

M. Pradhan
DKSR, Berlin, Germany

N. Suri (✉)
US Army DEVCOM Army Research Laboratory (ARL), Adelphi, MD, USA

Florida Institute for Human and Machine Cognition (IHMC), Pensacola, FL, USA
e-mail: nsuri@ihmc.org

M. Tortonesi
University of Ferrara, Ferrara, Italy

F. T. Johnsen
Norwegian Defence Research Establishment (FFI), Kjeller, Norway

Z. Zieliński
Military University of Technology, Warsaw, Poland

N. Suri et al. (eds.), *Military Applications of Internet of Things*, Internet of Things,
https://doi.org/10.1007/978-3-032-03378-9_3

is a fundamental and necessary concept for a network of smart devices that need to communicate dynamically with each other, share information with higher level entities, and, by using this combined information, provide additional value to users through IoT or other applications.

Due to the absence of common standards, the following questions arise when designing an architecture for an Internet of Military Things:

- What are the basic building blocks of IoT?
- How are they connected, and how do they communicate?
- How does the nature of connectivity and communications differ in the military domain?
- What standards exist and are most commonly used in civilian IoT?
- How is IoT evolving, and in what directions (w.r.t. technology, policy, etc.)?

Before addressing the above questions, the following subsections provide an overview of the most prominent architectural approaches to IoT in the civilian sector and the standardization efforts from major organizations and joint efforts of large industry players. This foundation is essential to examining whether these existing architectural principles would apply to the military domain and determining where the logical interconnection points would be between the civilian and military domains.

Following this review, specific requirements for Military Internet of Things will be discussed, along with some potential architectural designs.

3.1.1 European Telecommunications Standards Institute (ETSI) Machine-to-Machine (M2M) Architecture

The European Telecommunications Standards Institute (ETSI) [12] is the recognized European body for standardizing telecommunications, broadcasting, and other electronic communications networks and services. ETSI is also active in the standardization of IoT technologies. The focus of ETSI in the area of IoT lies in the standardization of Machine-to-Machine (M2M) communications. In this context, ETSI has published three specification documents for a standardized M2M communications platform, defining the requirements [14], the functional architecture [13], and the interface descriptions [15]. M2M communication is a focus of this standardization effort (as well as the next one presented—the oneM2M effort), as M2M communication is one of the differentiating factors between IoT and more traditional, legacy, networked systems.

The functional architecture described in the ETSI specification makes a high-level distinction between the *device and gateway domain* and the *network domain* as depicted in Fig. 3.1. The **device and gateway domain** consists of M2M devices, M2M area networks, and M2M gateways. An M2M device is defined here as a

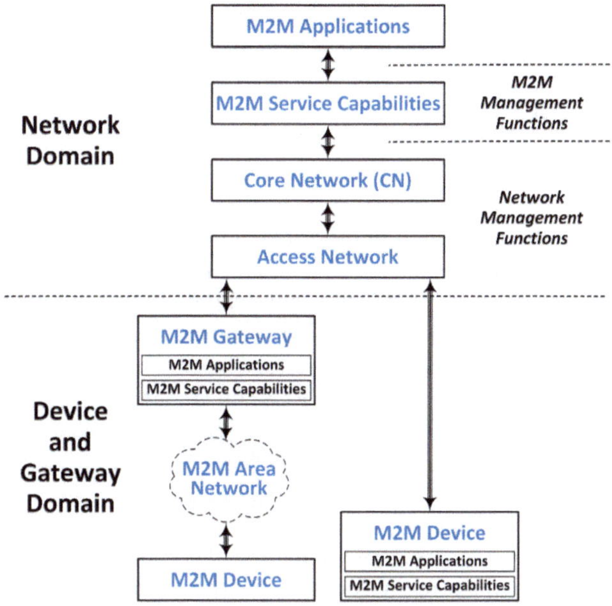

Fig. 3.1 ETSI M2M functional architecture. Based on [13]

device running an M2M application that offers M2M services and can communicate with other M2M devices. The M2M device can connect to the network domain in two ways: The device can either be "directly connected" via an access network or use an M2M gateway as an intermediary. The M2M area network provides the gateway connection using a wired or wireless communication technology. Typical examples of communication technologies used in the M2M area network as per the ETSI specification are ZigBee® [45], Bluetooth® [4], or KNX [29] (an open standard for commercial and domestic building automation). The M2M gateway is part of the M2M network and also connects to the network domain via an access network. It thus serves as a proxy to connect M2M devices to the network domain. In addition, the gateway itself may run M2M applications. It can offer services exclusively for its connected devices that differ from the service capabilities available in the network domain.

The main elements of the **network domain** include the *access network*, *core network*, *M2M service capabilities*, and *M2M applications*. The network and M2M management functions, which enable the management of the main elements, provide additional support to the network domain. The access network provides the communication link between the M2M device and gateway domain and the core network. Typical technologies used for the access network are Digital Subscriber Line (DSL), Wireless LAN (WLAN), cable modem, powerline communications, cellular radio access networks (2G to 5th Generation Cellular Networks (5G)), and satellite communication. The core network essentially provides Internet Protocol

(IP) connectivity between the gateways and interconnection with other networks. It also connects the network elements that implement the M2M service capabilities. These elements provide M2M functions through a set of open interfaces that different M2M applications can use.

The scope of the ETSI M2M standardization effort is limited to the M2M service capabilities in the M2M devices, at the gateways, and in the network domain. For the communication technologies used in the core and access networks and for device-to-device communication in the M2M area network, the architecture refers to existing standards and technologies as mentioned in the earlier examples. To standardize the access to the M2M service capabilities at the different elements of the functional architecture, the ETSI M2M specifications define several reference points describing the available procedures and operations, respectively [15]. For accessing data, the ETSI M2M adopts a RESTful architecture [17] in which information is represented as resources. Additional capabilities include identification and authentication, store and forward communication, secure channel transport, and device management.

ETSI started developing its M2M architecture in 2009. The standard was published in two releases, the core release 1 in 2011 and the extended release 2 at the end of 2013. ETSI has fused its standardization work within the M2M context into the *oneM2M* architecture and standard, described in the following section.

3.1.2 oneM2M Architecture and Standard

oneM2M [33] is a global partnership founded in 2012 to develop standards for M2M communications to enable the large-scale implementation of IoT. It consists of 14 partners. Besides ETSI, seven other international standards development organizations and representatives and consortia from the industrial sector form the consortium. The goal of oneM2M is to develop technical specifications for a common M2M service layer that allows the interconnection of various hardware and software to enable comprehensive cross-vendor or cross-platform M2M applications. Besides the service layer, oneM2M also considers other aspects of M2M communications, including use cases, security and privacy, information models, and data management. oneM2M uses five working groups to perform the standardization work covering requirements, architecture, protocols, security and management, abstraction, and semantics.

oneM2M has released a comprehensive set of technical specifications giving a quite detailed definition of a standard for M2M communication in IoT, which includes a description of the overall architecture, interfaces, security aspects, and communication protocols [32].

As a general concept, oneM2M proposes a layered model for supporting end-to-end M2M services, including an *application layer*, a *common services layer*, and an underlying *network services layer*. The application layer comprises the M2M applications and business logic functions. The common services layer consists

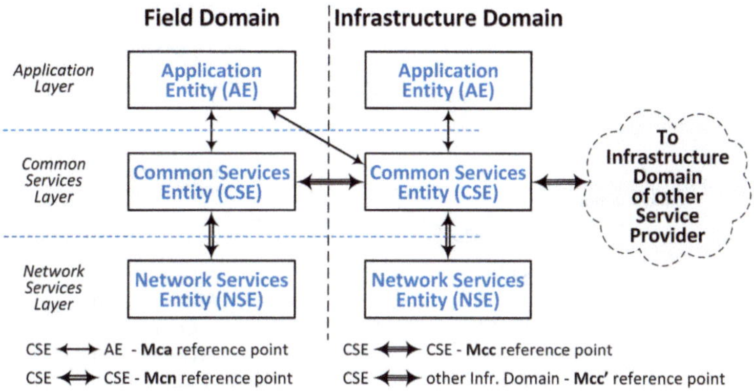

Fig. 3.2 Functional architecture of the oneM2M specification. Based on [34]

of service functions that enable M2M applications such as device management, discovery, and policy enforcement. Finally, the network services layer provides transport services, connectivity between entities, and other service functions (e.g., network-based location service).

Figure 3.2 illustrates the oneM2M functional architecture. For each of the three layers, application layer (A), common services layer (CS), and network services layer (NS), the architecture depicts two entities, one for the so-called **field domain** on the left half of the figure and one for the **infrastructure domain**. These domains semantically correspond to the device and gateway domain on the left and the network domain on the right when mapping to the ETSI M2M architecture described in the previous section. Generally, entities from all three layers can reside either in the field or in the infrastructure domain. The oneM2M architecture also defines reference points between the entities, as shown in the figure. Information exchange between entities is realized by defining an interface or protocol for the corresponding reference point.

3.1.3 Internet of Things Architecture (IoT-A)

The European Commission (EC) sponsored the European IoT-A project within its 7th Framework Program from 2010 to 2013 [11]. The project's goal was to develop an Architectural Reference Model (ARM) for IoT and define the key building blocks of such an architecture. The purpose of outlining principles for the technical design of protocols, interfaces, and algorithms was to achieve interoperability of heterogeneous IoT systems. Several European research institutions, universities, and industry partners participated in the project.

Along with developing an ARM described in [2, 3], the project also focused on security, addressing, management, and protocol-level interaction of the various

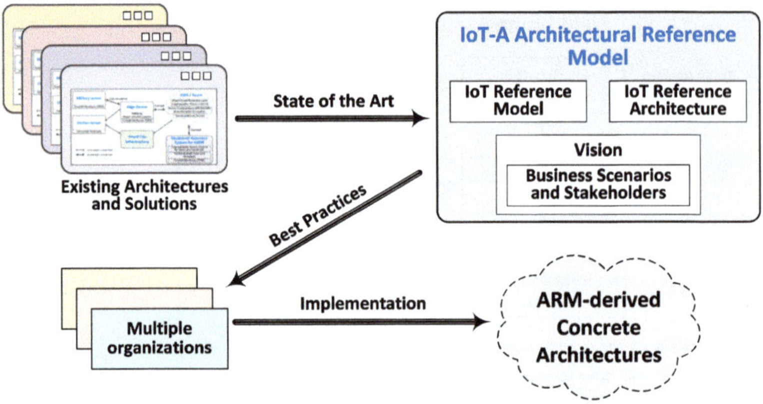

Fig. 3.3 IoT-A Architectural Reference Model building blocks. Based on [3]

components of the architecture. The ARM combines top-down reasoning about architectural principles and design guidelines. It furthers this reasoning with an experimental paradigm supported by simulation and prototyping to explore the technical consequences of architectural design choices. These essential building blocks and design choices allow end users to deal with conflicting functionality, performance, deployment, and security requirements. The IoT-A architecture is somewhat general, but it describes a way to define compliant IoT architectures that can be tailored to a specific need. The ARM can be visualized as a matrix derived from a robust and exhaustive analysis of the State-of-the-Art IoT architectures. It represents a super-set of all possible functionalities, mechanisms, and protocols derived from the analyzed architectures. Using the ARM as the base, it becomes possible for IoT system architects to select the protocols, functional components, and architectural options desired to build their concrete IoT systems. The process used for defining the ARM and an overview of its basic building blocks is depicted in Fig. 3.3.

The IoT ARM itself consists of four parts:

- *Vision* provides the rationale and discusses the underlying assumptions considered when designing the ARM. The ARM's usage, the business scenarios, and the stakeholders are also addressed in this section.
- *Business scenarios defined as requirements* includes various requirements set by the stakeholders that act as drivers for designing the architecture. The concrete architecture devised can be validated against selected business scenarios.
- *IoT Reference Model* consists of the IoT *Domain Model*, the IoT *Information Model*, and the IoT *Communication Model*. With this set, the reference model provides an abstract level for understanding the IoT domain. The Domain Model provides the top-level description, the Information Model explains how the IoT information has to be modeled, and the Communication Model describes the communication between the various IoT devices and the Internet.

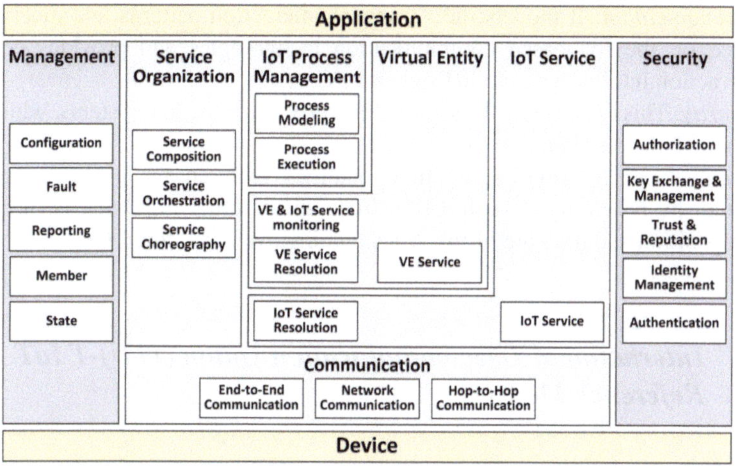

Fig. 3.4 Functional decomposition viewpoint of the IoT-A Reference Architecture. Based on [3]

- *IoT Reference Architecture* describes the views and perspectives of the various analyzed architectural aspects that concern the IoT stakeholders. The reference architecture is very abstract in nature and is intended to be used as a reference to build compliant IoT architectures.

Figure 3.4 shows the Functional View diagram of the IoT reference architecture and the nine Functionality Groups (FGs). The Application FG and the Device FG are out of scope of the IoT-A Reference Architecture, and the Management FG and the Security FG support all other FGs.

The various FGs presented in Fig. 3.4 are:

- *IoT Process Management*: This FG represents the integration of the IoT ARM with traditional process management systems. The FG aims to provide the functional concepts and interfaces necessary to augment existing (analyzed business) processes with the requirement specifications of the IoT domain.
- *Service Organization*: This is the communication hub for several other FGs and provides the concept of abstraction through "Services." These services provide communication within the IoT ARM and are organized to provide composing and orchestrating services.
- *Virtual Entity (VE)*: This FG enables discovering and looking up services that can provide information or allows interaction with VEs through sets of functions. It also describes functionalities required for dynamically finding and managing associations and monitoring their validity.
- *IoT Service*: It enlists the range of IoT services along with their discovery, lookup, and name resolution mechanisms.

- *Communication*: It models the variety of interaction patterns and mechanisms based on the IoT-based communication technologies and provides common interaction interfaces to the IoT Service FG.
- *Security*: This FG ensures the security and privacy of the IoT systems, which tend to be IoT-A compliant.
- *Management*: It provides the functionalities required for governing an IoT system by providing cross-functionalities across other FGs; that is, it enables composition and tracking of functions that involve other core FGs.

3.1.4 International Telecommunication Union (ITU)-T IoT Reference Model

The International Telecommunication Union (ITU) is a specialized agency constituted by the United Nations (UN) for the domain of telecommunications, information, and communication technologies. The ITU-T [25], a permanent part of ITU, is responsible for studying and standardizing global telecommunications standards.

The ITU-T proposed the recommendation ITU-T Y.2060 [26] to provide an overview of IoT and an IoT reference model to showcase the various functionalities and services encompassing the IoT domain. Figure 3.5 shows the layers defined by the reference model:

- *Application Layer*: This layer represents the set of IoT applications that leverage the capabilities of IoT devices.
- *Service Support and Application Support Layer*: This layer consists of:

    1. *Generic Support Capabilities*: Common capabilities generally used by different IoT applications (e.g., data processing). These might be invoked by specific support capabilities, for example, to build more sophisticated and specialized capabilities.

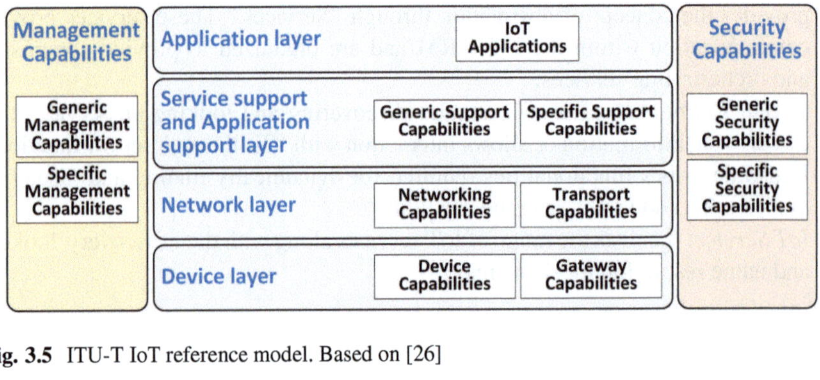

Fig. 3.5 ITU-T IoT reference model. Based on [26]

2. *Specific Support Capabilities*: Higher level capabilities that support the requirements of specific applications

- *Network Layer*: This layer consists of:

1. *Networking Capabilities*: Control functions required for network connectivity, for example, mobility management function
2. *Transport Capabilities*: Capabilities required for transport of IoT service- and application-specific data along with the IoT-related control and management information

- *Device Layer*: This layer consists of:

1. *Device Capabilities*: Capabilities of IoT devices that have either direct or indirect links with communication networks. Devices using direct communication links do not need gateway capabilities, whereas devices using indirect communication links need a gateway to reach the host/core network. It also details the capabilities of devices to perform ad hoc networking and implement automatic sleep and wake-up mechanisms.
2. *Gateway Capabilities*: Capabilities of gateway devices at the device layer supporting multiple interfaces through wired or wireless networks such as Bluetooth (wireless) or Controller Area Network (CAN) (wired). At the network layer, gateways may use communication technologies, such as 3G or Long-Term Evolution (LTE) networks.

- *Management Capabilities*: These capabilities entail the management functions for Fault, Configuration, Accounting, Performance, and Security (FCAPS) classes. Generic management capabilities include the management of devices, local network topology, and traffic and congestion. Specific management capabilities enable applications with specific requirements, such as smart homes monitoring electric energy transmissions in and out of the house.
- *Security Capabilities*: These consist of generic security capabilities that are independent of the applications, such as authorization and authentication mechanisms, and specific security capabilities that are used for application-specific requirements, such as security mechanisms for mobile payment.

3.1.5 Eclipse IoT

The Eclipse IoT working group [8] of the Eclipse Foundation is a collaboration between organizations and companies with the vision to build interoperable IoT solutions based on existing standards. To promote the development of such IoT solutions, the group provides open-source implementations of various existing standards. These standards cover different aspects of an IoT architecture, including protocols for device-to-device communication and device management and inter-

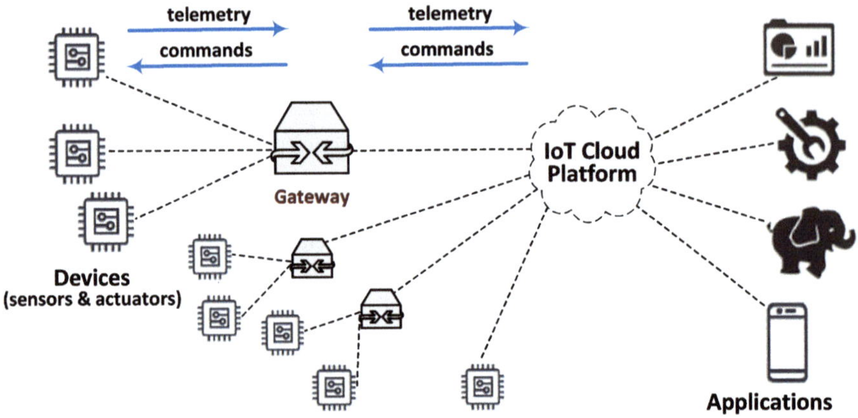

Fig. 3.6 Eclipse IoT: Typical IoT solution (Image source [9])

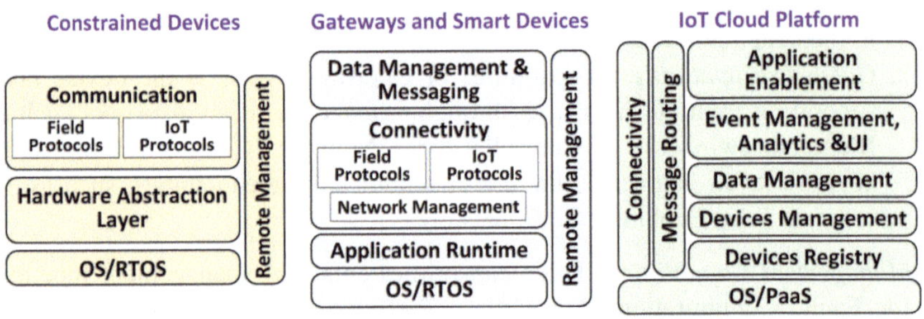

Fig. 3.7 Eclipse IoT: the three software stacks required for IoT architectures. Based on [9]

faces for gateways and servers. In addition, the software suite also offers solutions for data aggregation, security, event management, and data analysis [10].

In a white paper [9], the working group describes a general IoT architecture by identifying and defining three types of software stacks that are required by any IoT solution. The typical architecture for an IoT solution is divided into three major domains, as depicted in Fig. 3.6. These domains cover the single *devices* like sensors and actuators that may not only communicate with each other but also typically use a *gateway* to connect to a back-end server hosting an *IoT cloud platform*.

Figure 3.7 shows the software stacks derived for each of these domains. For the **device domain**, which incorporates sensors and actuators that are typically resource-constrained, a rather minimal stack is described. The main tasks here are to gather data or interact with the physical world and communicate with other devices or the gateway, often using low-power wireless communication technology. The communication protocols are divided into "field protocols," which denote the protocols used on the physical, data link, and network layers such as Bluetooth, Z-Wave [46] and CAN bus [5], and "IoT protocols," which incorporate higher layer communication protocols responsible for session establishment and reliability assurance, such as Message Queuing Telemetry Transport (MQTT) [31] and Constrained Application Protocol (CoAP) [37].

The software stack for a **gateway** includes more functions, such as network management, since the gateway often serves as a hub for attached devices and coordinates their connectivity with each other and with an external network. Other functions that can be implemented at the gateway are data aggregation and processing, local storage and persistence to provide "data at the edge," and protocol translation between incompatible devices.

The **IoT cloud platform** hosts the software and services that process the data from all devices and provide access to it, thus enabling specific IoT applications. At the IoT cloud platform, the data from different device classes using different protocols and data formats can be combined, achieving horizontal integration of available data. It serves as central data storage with high scalability and provides a centralized device registry and management of devices that are part of IoT solutions.

Finally, **Applications** represent what end users ingest through the cloud platform, through the various enabling services, and access Application Programming Interfaces (APIs). The applications map the user requirements and configurations to the service parameters, which are used to form API calls. These API calls create and invoke the data from the cloud platforms, fulfilling user requests.

In addition to the three software stacks for the devices, gateways, and IoT cloud platform domains, the Eclipse IoT working group describes a number of cross-stack functions. These functions provide security and ontology features, tools, and Software Development Kits (SDKs) that should be considered and addressed across the different stacks.

The group lists several key characteristics for IoT stacks as design guidelines for implementing IoT solutions, including:

- Loose coupling of software stacks to support the combination of stacks from different suppliers in each domain
- Modularity within each stack to facilitate feature integration of external suppliers
- Platform independency
- Reliance on open standards to ensure interoperability
- Clear definition of APIs for each stack to enable integration with other applications

3.1.6 IEEE P2413

The IEEE P2413 Working Group [22] of the IEEE Standards Association is a coalition of more than 25 industry members active in the IoT sector. IEEE-SA's Industry-Strategic IoT team initiated the group in 2014, focusing on bridging the market needs within the developing IoT technology landscape. The group's main goal is to accelerate IoT market growth by enabling interaction among the different application domains of IoT (e.g., consumer, commercial, industrial, and public sector) and the unification of platforms. They plan to achieve this by increasing the system compatibility, interoperability, and functional interchangeability of IoT deployments. To unify the creation of standards and architectures in the world of IoT, the group developed an evolving architectural framework that captures the commonalities across different domains and serves as a basis for the instantiation of concrete IoT architectures [20].

In December 2018, the group released a "Draft Standard for an Architectural Framework for the Internet of Things (IoT)" [21]. The draft standard specifies the approach to the IoT application domain through descriptions of:

1. **IoT Domains**: This section describes the application domains of IoT, including *Smart Manufacturing, Smart Grid, Smart Buildings, Intelligent Transport, Smart Cities, and Healthcare*. Each domain details how IoT technologies are deployed and specifies the key deployment drivers, stakeholders and their roles in the domains, and the services involved for application use cases. Then, it interconnects these individual components to show the basic layouts of the Information and Communications Technology (ICT) infrastructure.
2. **Domain Commonalities and Shared Concerns**: This part takes the common components of each domain and shows how they are shared between the application use cases identified for the IoT domains. For example, common stakeholder concerns include how actuation using the IoT technologies would affect the physical or legacy ICT capabilities that exist for the domains. Stakeholders may also be concerned about the interoperability with existing Operational Technology (OT) systems and whether the data volume and semantics can be interpreted and handled throughout the life cycle of the new ICT deployment.
3. **Architecture Framework**: In this part, the various elements needed to construct an IoT architecture framework are presented. The section presents the view of architects from a software engineering perspective and details how appropriate viewpoints are used to develop solutions to address specific concerns based on the target system. Such viewpoints include the stakeholders involved and their concerns, the meta-models required to describe the operational structure of the target system, the capabilities and conventions needed to describe the components and their effects on the system, the corresponding models and correspondence rules, and the life cycle of the target system.
4. **IoT Architecture Examples**: This section describes example architectures for satisfying the target system and use cases based on the architecture framework viewpoints discussed above. Figure 3.8 presents such a representative architecture, which includes modern ICT computing needs for Edge Computing, Cloud Computing, and Application Enabler platforms. The IoT domains presented at the top of the figure can all take advantage of this reference architecture to meet their needs. In fact, such an architecture can be referred to for any ICT deployment, including the scope of military applications for IoT. The physical layer comprised of *sensors* and *actuators* deployed for any operational use case, such as an Intelligence, Surveillance, and Reconnaissance (ISR) operation, could use the IoT-based *Communication Network* and the *Edge Computing*

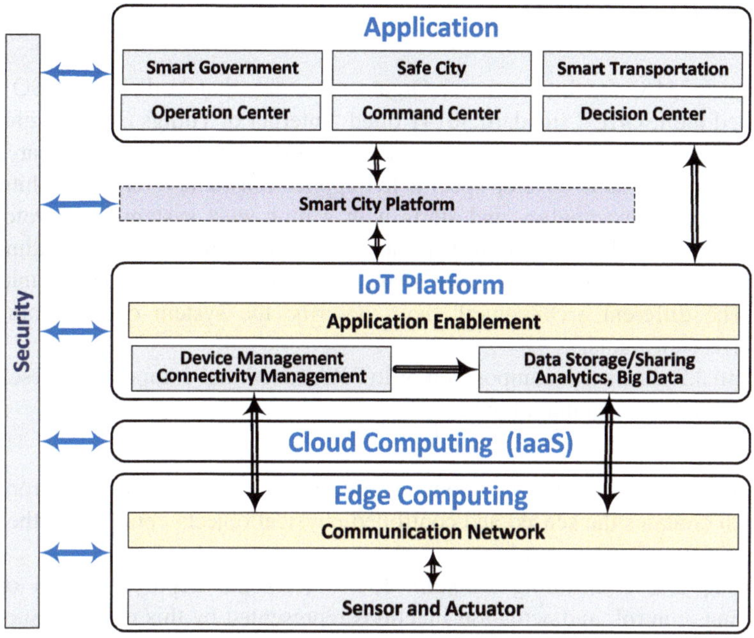

Fig. 3.8 Components of an IoT system with optional edge computing, cloud computing, and application platform. Based on [21]

platforms to provide filtered and analyzed data to higher *Cloud Computing* layers. Doing this would offload the computation closer to the source, relieving the network of bandwidth constraints and providing better QoS adherence. Additionally, a central cloud deployment for tactical operations is not always the solution. There are always accessibility and performance advantages to keeping data at the end-platform level. The direct connection of the *Edge Computing platforms* to the *IoT platform* would serve this purpose. The *IoT platform* can take care of management, analytics, and storage functionalities to enable IoT-specific applications. Furthermore, these aggregated data can be shared with *Command and Control (C2) platforms*, such as *Decision Making* and *Actuation*, for example. For any military deployment, all these components must work securely to ensure the safety and security of the target operation. The *Security* requirement is an intrinsic part of each IoT component involved, which means that there must be inherent security mechanisms for data encryption, access methods and policies, privacy, integrity, trust, and reliability.

3.1.7 *ISO/IEC 30141*

In August 2018, the International Organization for Standardization (ISO) [23] published the ISO/IEC standard 30141 titled "Internet of Things (IoT)—Reference Architecture" [24]. The document describes a standardized IoT reference architecture that can be used to develop specific IoT applications. The reference architecture uses a common vocabulary and starts with a high-level system-based reference model that is broken down further into five architecture views, including the functional view, system view, user view, information view, and communication view. The different architectural views describe the system details from their respective viewpoints.

Figure 3.9 shows a decomposition of IoT RA functional components, describing the intra- and cross-domain functions.

The intradomain functions include:

- *Physical Entity Domain*: This domain represents the actual physical environment, which contains the sensed and controlled physical objects upon which the other higher layer domains act.
- *Sensing and Controlling Domain*: IoT objects are deployed based on the sensing, control, and actuation functions represented by this domain. Such IoT objects span hardware, firmware, device drivers, software elements, involved logic and processes, and system behavior and states. Additionally, this domain also includes network access mechanisms that enable access to the underlying sensors, actuators, controllers, gateways, and edge systems.
- *Application Domain*: This domain represents the range of functions needed to implement application logic from the end user's perspective. This includes requirements mapped to specific business functionalities for service providers.

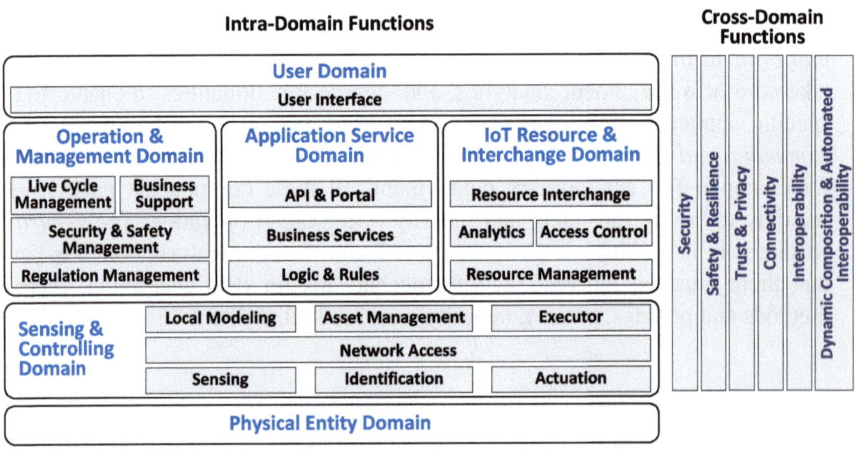

Fig. 3.9 IoT RA functional view. Based on [24]

The functionalities presented by service providers have components such as logic and rules, service sets, and APIs.

- *Operation and Management Domain*: The functions responsible for life cycle management supporting business operations, security and safety management, and regulation management are represented here. As management functions from core business structures, directives are issued to the control systems or the corresponding devices. Life cycle management of the various deployed IoT components and their operations include functions for provisioning, monitoring, maintenance, diagnostics, optimization, billing, etc.
- *Resource and Interchange Domain*: The functions needed for the interchange of IoT resources for the deployed systems are represented by this domain. Such functions include resource management, analytics, resource interchange, access control, etc. These resource-sharing functions enable the sharing of intelligence, knowledge, information, data, control, etc., throughout the deployment, which may also include non-IoT systems.
- *User Domain*: This domain represents functions and interfaces, such as Human-Machine Interfaces (HMIs), with which the end user interacts and allows users to access IoT services, including information on how to use them. The users subscribe and receive the application domain services through the interfaces.

The cross-domain functions include:

- *Security*: The security functions ensure that information exchange occurs with Confidentiality, Integrity, and Availability (C-I-A) and reliability. Security policies are an integral function of system design, as they further affect functions such as system configuration, policy, software and firmware updates, and other life cycle management operations.
- *Safety and Reliance*: This function is closely related to the security and fault tolerance of the system. IoT presents possibilities for autonomous operations such as self-healing, self-configuring, self-organizing, and self-protection by the deployed systems themselves. Such functions allow IoT systems to self-induce their safety and reliance.
- *Trust and Privacy*: Various stakeholders have various requirements related to the trust and privacy functionalities represented in an IoT system. Privacy is achieved by protecting sensed and transmitted data using APIs and portals needed for monitoring, information resource interchange, HMI, etc. By protecting the confidentiality of the data and validating them, trust is established, which can be further enhanced with reputation-based approaches.
- *Connectivity*: IoT systems engage heterogeneous networks based on use cases and, thus, varied network-enabled capabilities. Such a heterogeneous integration of IoT components belonging to different networks or using different technologies is needed to achieve seamless connectivity across the deployed system.
- *Interoperability*: Interoperability functions enable the exchange of IoT data with a common interpretation of the information. The syntactic interoperability of the IoT systems allows information exchange in a common data format overlaid on a common protocol to structure the data. Semantic interoperability, on the

other hand, enables the correct common interpretation of the meaning of the symbols/information in the messages.
- *Dynamic Composition and Automated Interoperability*: This domain provides a flexible service composition method, allowing for dynamic integration of IoT components at run-time. Such dynamic components, supported by the semantic interoperability of the information exchange, enable the building of adaptable services.

Such a functional architecture represents an abstract view of how any ICT deployment involving IoT systems can be developed and deployed. It shows the entire life cycle of the deployment, where individual functions can be abstracted to form interoperable and loosely coupled service descriptions. These service descriptions can be used in their domain sphere and modeled independently while interfacing with the other interacting layers, both horizontally and vertically. These domains depend on each other, providing the necessary functions for the entire deployment to work as a functional unit.

3.2 Deployment and Integration of IoT in Military

3.2.1 General Requirements

Military IoT applications need to meet some specific requirements [43]:

- Processing of Existing Data: Sensors and embedded devices collect massive amounts of information, but data is often never processed or analyzed. As a result, most of the information is never used. Significant investments in infrastructure and analytical software are needed to transfer, store, and analyze the enormous volume of data generated by IoT devices. Without digital processing, the sheer volume of data is overwhelming, causing "data overload," which hurts efficiency and operations rather than strengthening them.
- Increased Automation: Fully IoT-enabled applications leverage sensors and digital analytics to generate automated responses. Current IoT-related technologies deployed by the military lack automation. Automating processes can significantly improve the efficiency of operations by reducing the number of humans needed to complete a task, limiting opportunities for human error, and improving reaction times. Current command and control systems often require a human being to analyze the data and convert them into a usable form manually. The dependency on manual processing can cause significant delays in getting important information to those who can act on it. It can also mean that important information is missed because a human analyst may not have the capacity to find, understand, and analyze all available data. This can lead to suboptimal or even wrong courses of action being proposed or incorrect decisions being made, resulting in stalled or failed missions or forcing commanders to make decisions

without all the relevant data. Automation of equipment and vehicles extends the reach of cyber-threats into the physical domain, allowing them to go beyond disrupting systems and causing confusion to inducing physical damage. Many military leaders are wary of putting the lives of soldiers in the hands of computers and prefer to have analog and manual backup options in case technology fails. This cautiousness raises issues for the adoption of IoT, which often generates efficiencies by reducing the role of humans in processes to leverage the speed and accuracy of digital systems.

- Device and Network Security: One of the best ways to improve the security of complex networks is to limit the number of nodes that an attacker can access from any given entry point. However, this approach to secure network architecture is fundamentally at odds with the IoT, which generates much of its value from the integration of systems and data. Traditional approaches to security, such as compartmentalized architecture, advanced encryption, and multiple authentication, can slow down or prevent data exchange between devices and systems on the network, require more computing power on devices to decrypt data or authenticate them regularly, or require human input for authentication.
- Electronic Warfare: Adversaries can use relatively simple methods such as RF jamming to block these signals and render the devices unable to communicate with the backbone infrastructure. Wireless connections also increase the risk of exposing the positions of troops through RF emissions.
- Interoperability: Many current military systems are designed to prioritize protection of need-to-know over responsibility to share. Leveraging the full value of IoT is about maximizing the number of nodes and connections in the data ecosystem, requiring a common set of standards, protocols, and security policies to allow systems to talk to each other.
- Capability Development Approach: In order to enable use of the IoT technology in military operations, it is important to focus on the full spectrum of Doctrine, Organization, Training, Materiel, Leadership and education, Personnel, Facilities, and Interoperability (DOTMLPFI), in particular including training, as well as organizational and doctrinal implications and impediments. For example, the commanders can struggle to understand how to apply new technologies to old challenges. Career service members who have established routines for decades to accomplish their mission may resist change. This effect is exacerbated by the mixed record of military IT systems, which promise impressive new capabilities and dramatic improvements in efficiency in theory but are difficult to operate or have limited functionality in practice.

The broader development and deployment of IoT applications in the military will take time; however, there are areas where the military can generate significant savings from IoT using existing technologies and business practices available in the commercial sector.

- Condition-Based Maintenance: Allied military forces should retrofit their fleet of vehicles with onboard sensors to monitor engine performance and parts status,

facilitating condition-based maintenance and on-demand ordering of parts, and reducing unanticipated failures.
- Real-Time Fleet Management: Adopting IoT devices for real-time fleet management, including geolocation, status, fuel efficiency, and weight and cargo sensors, could reduce fuel costs and increase fleet utilization.
- Inventory Management: The deployment of RFID tags and standardized barcodes to track individual supplies down to the tactical level can provide real-time supply chain visibility and allow the military to order parts and supplies on demand.
- Base Management and Energy Efficiency: Smart thermostats reduce consumer heating and cooling costs and could also generate significant savings for defense organizations.

The key enablers for the adoption of IoT technology in military applications include:

- Extend Connectivity: The military should invest in resilient, flexible capabilities to extend Internet connectivity in denied areas, including high-altitude communications relay platforms, nano-satellites, and, where appropriate, piggybacking on commercial communications systems, such as Low Earth Orbit (LEO) constellations, e.g., StarLink [35], and 5G.
- Common Security Overlays for COTS Devices and Applications: The military should invest in developing and applying new security techniques to commercial, off-the-shelf (COTS) devices and applications, including those hosted in the cloud, focusing on investing in scalable security measures instead of securing individual systems.
- Develop Common Standards and Protocols: The key to delivering IoT capabilities is a suite of common standards and protocols that will enable the integration of new IoT devices into military CIS and leverage existing systems in innovative ways.
- Pursue Innovative Ways to Access Innovation: The military should adopt commercial best practices for technology development and procurement to enable enhanced collaboration with the private sector to field, maintain, and update IoT systems quickly with the newest technologies. The military needs to engage with the tech community to identify how COTS IoT solutions can enhance military missions rather than solely relying on custom solutions built to predetermined requirements.
- Technology Testbed: NATO with its member and partner nations should consider creating a collaborative technology testbed and a live-training environment to identify and experiment with technologies that could transform the way the military accomplishes its mission and to serve as a link between military personnel and technology developers.
- Adopt Agile Development, Security, and Operations (DevSecOps): Commercial companies commonly use agile software development, which emphasizes constant bilateral communication between users and developers. DevSecOps describes how to continuously deploy code through the use of modern tools and automated processes, with specific focus on security assurance. Continuous

Integration and Continuous Delivery encourage and support frequent code check-in, version control, test automation, continuous low-risk releases, and feedback, often through a number of electronic tools. These methodologies allow for frequent adjustments and updates to requirements and capabilities, in contrast to traditional military contracts that emphasize formal reporting and bureaucratic oversight.

- Platform as a Service: Where possible, the military should contract with private providers to deliver web-based services without building and maintaining them themselves, thereby creating a more flexible framework for the provider to adjust systems to accommodate military requirements and update systems with the newest capabilities.

3.2.2 Technical Challenges

Military environments pose unique challenges concerning resource limitations, especially concerning communications, and also in terms of computation and power. Most military networks, especially those that support deployed/dismounted soldiers operating in a battlefield environment (often called tactical networks), do not rely on existing communications systems (e.g., LTE, 5G, etc.) that require any centralized or preexisting capabilities. Instead, for security and resiliency purposes, they operate in a decentralized fashion, typically using the limited communications resources they have brought with them. Many of these systems use Mobile Ad-hoc Networks (MANETs) and operate in a potentially adversarial context subject to interference and jamming. The result is that these networks tend to have limited bandwidth and unreliable connectivity and exhibit high or variable latency. Furthermore, for security purposes, these networks are secured and isolated without any connectivity to the commercial Internet. Typically, their connectivity extends only through their command hierarchy (e.g., to a military operations center).

Furthermore, dismounted soldiers need to minimize their Size, Weight, and Power (SWaP) footprint and, therefore, are also limited in terms of their power and processing capabilities. Their computational devices tend to be either a phone or a tablet platform with limited connectivity over their tactical networks to their next hop, typically a military vehicle or an airborne platform. These devices, in turn, connect via a (sometimes slightly) more robust communication link to the operations center, as shown in Fig. 3.10.

For successful deployment of IoT within the military environment, the challenges discussed in the following subsections should be considered and addressed.

3.2.2.1 Decentralized Infrastructures for Data Analytics

Cloud computing represents a key enabling technology for future IoT applications. Most solutions advocated so far follow the traditional centralized cloud approach,

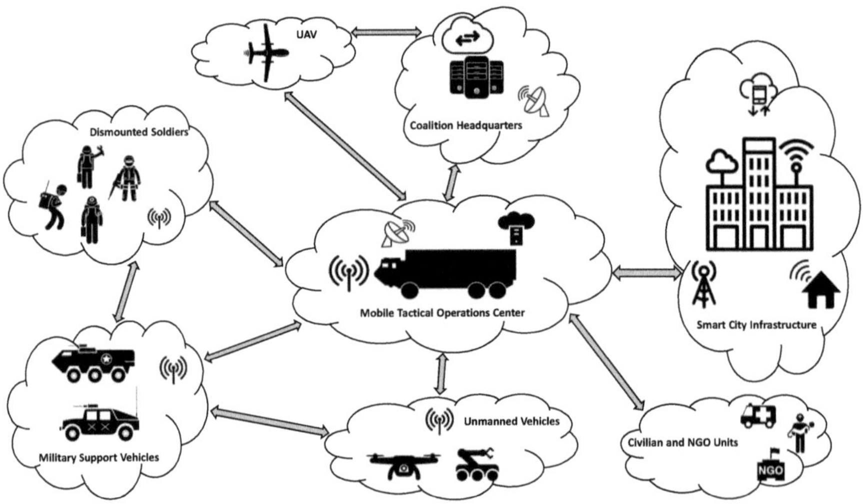

Fig. 3.10 Strawman deployment architecture for IoT in the military domain

in which every single instance of raw data generated by the IoT is transmitted to the cloud for analysis. Operating on a large amount of IoT input data, which is typically augmented (e.g., using open data), is an approach that is potentially (but not necessarily) capable of highlighting nontrivial correlations, patterns, or anomalies in the data.

However, the centralized cloud approach also has several drawbacks regarding time and resource requirements. The transmission of raw data from the IoT to the cloud considerably strains the (typically wireless) network infrastructure. Additionally, the processing phase for such a large amount of data represents a complex and computationally expensive procedure, which typically utilizes sophisticated big data methodologies and tools. Finally, a non-negligible delay exists between the time of IoT data generation and when the results of the analysis are available. These aspects can be of critical importance in Military Internet of Things (MIoT) scenarios.

For MIoT technology to be viable in dynamic battlefield environments, large-scale data ingestion and analysis must be carried out in near real time. Here, military-specific challenges come from time constraints on data analysis and potential connectivity challenges.

One of the luxuries of the commercial IoT environment is ubiquitous connectivity and the ability to rely on large, centralized cloud data centers. Even in the presence of cellular or wireless networks, once the device reaches a base station or an access point, the connectivity is ubiquitous, making it straightforward for any IoT device to reach a cloud-based resource. As a result, most IoT devices rely on uploading the bulk of their data to a cloud, typically owned or managed by the device provider. This architecture allows the provider to essentially take ownership of the data, process them as required, and disseminate the results to the consumer (while, at the same time, potentially exploiting other commercial opportunities for the data).

However, as mentioned earlier, most battlefield environments rely on tactical wireless networks, and any approach that requires a centralized cloud-based infrastructure is not likely to work effectively. Tactical networks are often limited in bandwidth, and the available bandwidth is carefully allocated to service various mission-critical data exchanges. Furthermore, connectivity is not as ubiquitous as in the commercial environment. Even if a device has a high bandwidth link to a local resource, it is not likely that all devices will be able to have good connectivity to the same cloud-based platform. Therefore, a challenge still to be addressed is developing a decentralized infrastructure to support IoT on the battlefield.

Decentralizing computational resources by simply creating multiple (and local) small clouds (or cloudlets) is insufficient if the overall approach is still to send raw data from IoT devices to a local cloud for processing, since bandwidth is perhaps one of the most precious resources in a tactical network environment.

3.2.2.2 Limited Network Access

In addition to the challenge of limited bandwidth, military networks are severely limited by frequent disconnections, partitioning, and fluctuations in radio channel conditions. This can result in variations in sensor availability and constraints on sensor usage that are not as apparent in commercial settings. For example, very simple sensor devices may not function as designed if they have limited local storage (and processing) and rely on offloading their data to or via a gateway continuously. They may not be able to periodically reach their gateway, or the gateway itself may be unable to reach other parts of the network infrastructure. Hence, architectures must be carefully designed to avoid data loss, for example, by supporting patterns such as disruption-tolerant networking [30], which introduces additional complexities, especially for simple IoT devices [39].

Finally, military networks are also purposefully isolated from the commercial Internet, potentially causing additional complications in the deployment of software updates and security fixes. Most IoT devices often provide a mechanism to update their firmware by downloading it from a gateway, which in turn is connected to the Internet. This is not feasible in a partitioned and isolated network, and military networks must find other mechanisms, or the result will be IoT devices that are vulnerable to identified security exploits in the commercial world.

3.2.2.3 Interoperability

Given the variety of functions served by military hardware, adequate interoperability among devices is often not achieved. This problem potentially increases when integrating devices across coalition partners (e.g., between the United States (US), United Kingdom (UK), and European Union (EU)) or incorporating potentially useful devices in an area of operations (e.g., like those in a smart city deployment).

One of the most popular approaches to improve interoperability is to utilize Service-Oriented Architectures (SOAs). SOAs typically exploit well-defined interfaces and common messaging protocols to expose their capabilities and dynamically share information between multiple services. SoAs are particularly attractive for rapidly and dynamically enabling interoperability through aspects such as service reuse, compatibility with dynamic workflows, and rapid configuration and reconfiguration. From the perspective of military systems, SOAs in the tactical domain aim to address interoperability challenges specific to Command, Control, Communications, Computers for Intelligence, Surveillance, and Reconnaissance (C4ISR) and hence are well suited to the interoperability needs of MIoT. In addition, SOAs could also help leverage commercial IoT capabilities, which tend to be primarily SOA and cloud-based.

In comparing IoT systems—and more broadly, SOA—from military and commercial perspectives, many unique challenges exist in the military infrastructure to threaten system interoperability. First, both military computers and sensor networks will have a longer service life than their commercial counterparts [7], resulting in a greater need to maintain support for legacy devices and protocols. Additionally, the use of different hardware designs and data standards can affect cohesion in MIoT infrastructures, leading to "stovepiped" systems [42].

3.2.2.4 Trust and Security

Trust and security are additional considerations for developing military applications of IoT. From a command perspective, equipment failures can compromise both intelligence gathering and planned operations. Additionally, military equipment can be subject to sabotage or compromise by adversary activities, resulting in service interruptions or propagation of misinformation [38, 41].

IoT components create systems that should meet all the security requirements for "traditional" systems. These requirements include the C-I-A triad comprising the following elements:

- *Confidentiality:* Ensuring that sensitive data is only accessible by authorized objects (e.g., by authorized users) and kept out of the reach of unauthorized objects. Security mechanisms such as user names, passwords, access control lists, and encryption can be used to implement confidentiality. Additionally, it is common to categorize the data depending on the damage that could occur if it falls into the wrong hands.
- *Integrity:* Ensuring that data remains in a format that is true and consistent with its original purpose. The recipient must receive the data that he was supposed to receive in an unchanged form, as generated by the author. Only authorized persons can edit the data, and it must remain in its original state. Security mechanisms such as data encryption and determining the value of the hash function can be utilized to implement integrity. Finally, note that changes in data can also occur due to Electromagnetic Pulse (EMP) or server failure.

- *Availability:* Ensuring that data and other resources are available to those who need them. Methods such as hardware maintenance, patching software, and network optimization can be used to implement availability. Additionally, mechanisms such as redundancy, failover, Redundant Array of Independent Disks (RAID) constructions, and high availability clusters are utilized to mitigate serious consequences in the event of hardware problems. Finally, specialized hardware solutions can protect against downtime and inability to access data due to malicious actions, such as Denial of Service (DoS) attacks.

Trust is a much more context-dependent concept than confidentiality, integrity, and availability. The need to trust is caused by incomplete information and, in this respect, is related to risk management. Therefore, trust can be defined as the willingness of one party to take a risk based on a subjective belief that another party will exhibit expected behavior under uncertainty due to conflicting or incomplete evidence [6]. This willingness usually depends on a given situation and is often based on an assessment of previous experience with a specific counterpart. In the context of military applications of IoT, the typical situation that requires trust occurs when data obtained from sensors is used to inform and influences the decision process [19]. This is due to the fact that it is generally impossible to provide guarantees about the accuracy and reliability of the sensor data obtained. This particularly applies to federated environments, such as Civil–Military Cooperation (CIMIC). Trust can be increased if the originator of the sensor data or the actuator commands can be properly identified and authenticated so that trust in data can be bound to the trustworthiness of its origin. This requires the implementation of appropriate authentication and identity management mechanisms within the MIoT.

3.2.2.5 Sensor and Device Utilization

Constraints on power are one of the perennial problems in a battlefield context. Unlike in a commercial environment, connecting sensors and other IoT devices to stable power sources is not always possible. It is also not easy to recharge them periodically, as commercial device users often do. In the military domain, IoT sensors and devices are likely to be powered by batteries or perhaps solar power. In either case, the expectation is that the devices will last for extended periods of time (or at least for the duration of the mission). Therefore, sensors and devices need to use power efficiently, and the system/user needs to use the devices judiciously. For example, a solar-powered device may have a relatively short duty cycle before recharging. Similarly, it is often impractical to swap batteries in deployed devices. Even in the case of body-worn devices, it is impractical to expect soldiers to carry extra batteries in addition to their current equipment, which implies that the system must be aware of and manage the demands placed on those devices.

Furthermore, military equipment will usually need to be designed to operate in more extreme environmental conditions than commercial equivalents, as outlined in MIL-STD-810G [40]. It is reasonable to expect that the needed ruggedization of military devices may impose constraints on their operational capabilities (e.g., power cell size or transmission capability).

3.2.2.6 Integration of Civilian IoT with the Existing Military Systems

An interoperable architecture must be developed to integrate civilian/commercial IoT systems and military environments successfully. In particular, the architecture must address the challenges outlined in the previous section to enable soldiers operating in the military domain to deploy commercial IoT systems. In another use case, the architecture must allow soldiers to tap into existing information sources in the civilian IoT domain, such as a smart city (cf. Sect. 2.2), and integrate those sources (and potentially devices) into the military communications architecture and information space.

To integrate commercial IoT-based devices or civilian IoT information sources into the military communications and information domain, the Information Systems Technology (IST)-147 group proposed a high-level architecture, as described in [13]. The architecture accounts for the common key aspects of existing architectures identified in Sect. 3.1. Most importantly, it accounts for separating the device and network infrastructure domains and the interconnection of both using gateway components. A complete translation and transformation of the military ICT infrastructure to civilian or open IoT architectures is technically and operationally unfeasible. As a result, any such IoT integration needs the *gateway* concept. Furthermore, these gateways would allow the reuse of existing heterogeneous military standards to interface with the new military IoT domain.

The proposed architecture is shown in Fig. 3.11 and consists of instances of IoT device groups, each using its own IoT architecture or a set of IoT communication protocols. Specialized IoT gateways perform the task of connecting the different device groups to the military infrastructure. In the proposed architecture, these gateways differ from the gateways used within each group of devices, which often simply implement IP network connectivity to the currently used wireless M2M communication technology. The proposed IoT gateways perform data mapping and protocol translation between different standards or architectures in a common format used in the operational domain. In addition, the functions for data preprocessing, aggregation, and caching enabled by edge computing can be implemented here, thus addressing the specific challenges of the military domain described in the previous section.

IoT data integration into the military domain, carried out by the IoT gateways, can occur at different levels of the communication and information processing hierarchy. For example, the IoT gateway can directly access the raw sensor data using the appropriate M2M communication standard. Alternatively, the IoT gateway can communicate to the hub gateway of an IoT device group using a provided API. At this stage, many architectures give access to preprocessed sensor data. The coupling to civilian IoT information sources available on the Internet can be realized even further up the hierarchy in the infrastructure domain using high-level

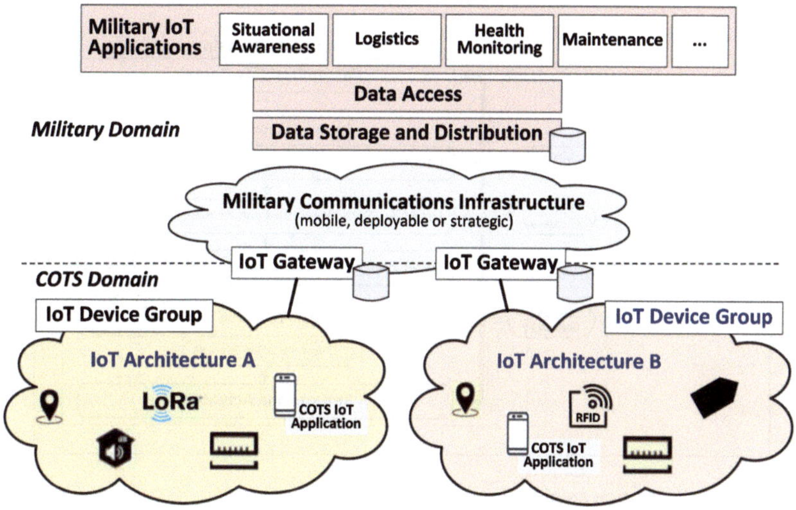

Fig. 3.11 High-level MIoT architecture

protocols and services. Figure 3.12 illustrates how sensors from a smart city can be used in various ways. The right side of the figure shows the basic levels included in architecture within the smart city domain, and from bottom to top, these are [36]:

- Level 1: Sensors collect raw data at this level. These data may be in a standardized format, but depending on the sensor, they may also be in a proprietary format, complicating the process of utilizing the information.
- Level 2: The data collected at Level 1 is processed here using various algorithms.
- Level 3: Various data sources are integrated and reasoned over at this level, using technologies such as the Semantic Web.
- Level 4: Level 3 data is used to generate alerts, control actuators, or issue commands to integrated devices. Such alerts can also be sent to recipient devices of interested partners in smart city infrastructure. Examples of such partners are city citizens, third-party service providers, and city administrative units.

The left side of Fig. 3.12 illustrates how systems in the military domain can utilize the information from the smart city domain. The figure shows that information can be obtained through gateways, which can be located on Level 1, 3, or 4, depending on the type of information needed. In general, obtaining information at lower levels means more processing must be done by systems within the military domain, whereas the higher levels provide processed information.

As a first step toward a Proof-of-Concept (PoC) implementation of this architecture, we have started at the bottom, investigating how to obtain data on Level 1.

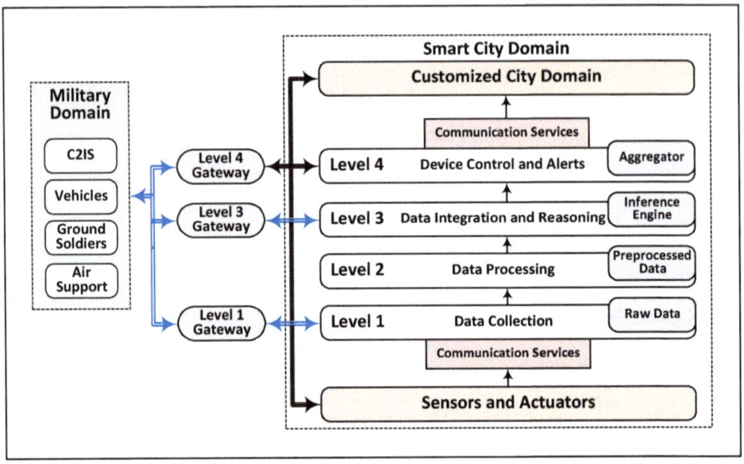

Fig. 3.12 Data integration architecture. According to [18]

From a military viewpoint, the main question is: How can the original data from the smart city sensors be obtained and used? We address this issue by breaking it down into four key questions:

1. Which architectural approach should the sensor system take? The approach must enable a network of sensors that not only allows for raw sensor data sharing (Level 1 Gateway) but also supports higher level functionality, such as data processing. The latter will enable information sharing at Levels 3 and 4 later.
2. How can sensors be discovered? Not being a normal part of the smart city infrastructure, the military must be able to discover available sensors in order to retrieve the data.
3. How can sensors be described? Once the available sensors are discovered, each sensor needs an overall description to assess its capabilities and characteristics.
4. How can sensor data be retrieved? Fetching data from a sensor requires an application programming interface (API) for the sensor.

We address these four key questions in Chap. 5, but first, we discuss alternative approaches for sensor systems integration and identify the most suitable approach in our scenario. Following this discussion, Sect. 4.5 provides details on questions 2 and 3. We investigate how to *describe* sensors in a standardized and machine-readable way and how to share information about sensor services to *discover* them automatically and assess their capabilities. This is accomplished by identifying existing standards that support this functionality and then describing how to use these standards to ensure interoperable discovery between sensors and information consumers.

After discovering the sensors, we also need to support information retrieval, where information consumers access the data offered by the sensors directly or after data processing. To support as many sensors and consumers as possible, it is essential to leverage best practices for performing information exchange in large-scale distributed systems. A common approach in modern systems is to use Representational State Transfer (REST) [16] APIs, which are widely supported by both commercial sensor systems and potential consumer devices, such as smartphones. Section 4.6 further discusses how this approach can support information retrieval from the sensor systems. As discussed in [28], fog computing should be considered an intermediate layer between sensors and end systems.

3.3 Generic Architecture of the MIoT

The MIoT system can be viewed as a collection of uniquely identifiable distributed objects that are interconnected and collaborative within the existing military communication infrastructure. Each object can be equipped with a microcontroller, several sensors or actuators (S), and transmitter-receiver devices, which we refer to as Sensor Node (SN). For efficiency and reliability/robustness reasons, we assume that it will be partitioned among disjoint groups/clusters with a limited number of objects based on communication capabilities. The general structure of the military IoT network is illustrated in Fig. 3.13.

We also assume that the MIoT area of interest comprises a large number of self-organized sensor nodes. Sensor nodes usually consist of various physical sensors that gather environmental data or executive elements, a microcontroller processing the data, and a wireless interface to communicate with other nodes.

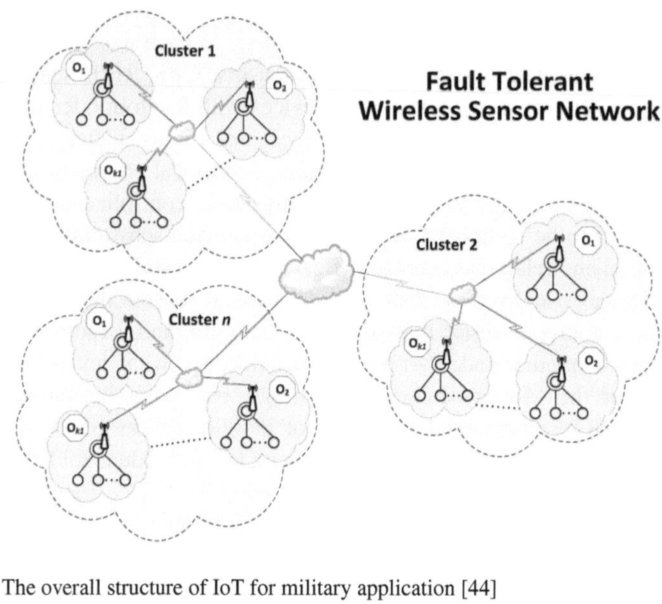

Fig. 3.13 The overall structure of IoT for military application [44]

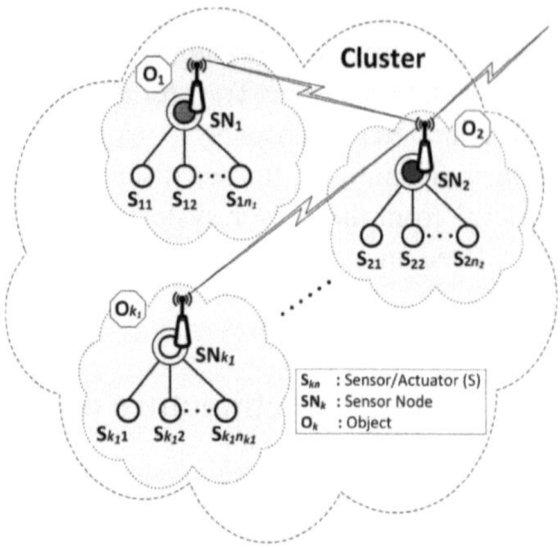

Fig. 3.14 Structure of a sensor nodes cluster [44]

We will focus on a single cluster, which includes several objects that can perform similar usable functions (Fig. 3.14). An example object could be a wired or wireless body area network installed on a soldier's body.

The concrete MIoT solution architecture depends on the specific military scenario and the application area. However, we generally adopt a high-level generic architecture for military applications based on characteristics and requirements from various military scenarios. The generic layered MIoT architecture reference model is presented in Fig. 3.15. According to Fig. 3.15, we will distinguish four main layers.

Perception and Actuation layer: The lowest layer presented in Fig. 3.15 is related to the physical IoT sensor/actuator nodes. The primary goal of this layer is to obtain real-time "sensed" data from the environment of each sensor node, including RFID tags, 5G war-fighters phones, military sensors, GPS, RFID sensors, devices embedded with sensors, etc. It consists of devices with low to moderate computing and communication capability. These devices are usually battery-operated and typically receive data from the environment, perform local processing on the data, and then transmit the result to gateways or physically controlled objects. We can expect this layer to be composed of many self-organized sensor devices that work in collaboration and relay messages to the gateways/base stations.

Sensor devices usually consist of several physical sensors that gather environmental data, a microcontroller that processes the data, and a radio interface that communicates with other nodes. At present, it is not clear what devices or platforms will be part of the Military Internet of Things, but almost certainly they will include UAV, UGV, unattended stationary or mobile sensors (sensors networks), Unattended Ground Sensor (UGS), infantry vehicles, individual soldier equipment sensors that monitor the vital functions of the warfighter (e.g., Electrocardiogram (ECG), heart rate, respiration rate, and skin temperature), helmets with built-in sensors to help diagnose brain injuries, and wearable computer technologies.

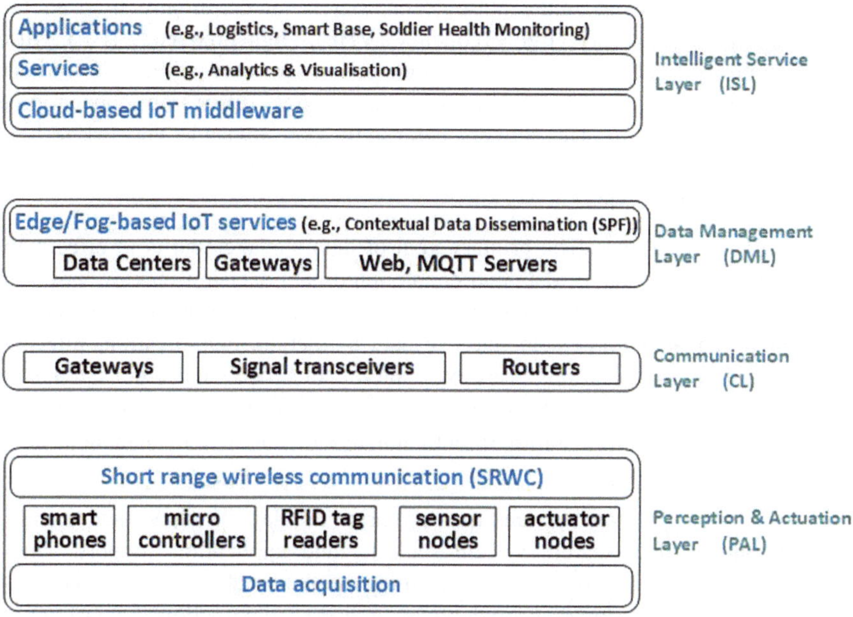

Fig. 3.15 Layered MIoT architecture reference model

The perception layer may be divided into data acquisition and short-range wireless transmission parts. The data acquisition portion collects information using RFID, sensors, and other technologies. Short Range Wireless Communication (SRWC) concentrates the intercommunication of information on a small scale, for example, in the case of individual dismounted soldier sensors. An example solution using SRWC would be the WBANs used in civilian health care systems. In military applications WBANs could be intelligent wireless monitoring systems consisting of wearable and implantable computing devices on or in the body of the warfighter.

Communication Layer The purpose of this layer is to transmit the information received from the perception layer to any particular information processing system using existing military communication networks, including access networks (5G, Wi-Fi, ad hoc networks, etc.) or core networks (Internet/intranet). The communication layer comprises devices (routers, signal transceivers, etc.) responsible for reliable data transmission. The communication layer combines the information received from various devices into a heterogeneous network to classify and transport the collected data and information. This layer can help heterogeneous networks access the Internet/Next Generation Internet (NGI) ubiquitously through a wide range of wireless links, including LoRa, Wi-Fi, ZigBee, Worldwide Interoperability for Microwave Access (WiMAX), Global System for Mobile Communications (GSM), and satellite. In the military domain, there is considerable heterogeneity among the communication technologies that transmit data between communicating entities. Additionally, communication technologies not commonly used in the civil domain, such as Very High Frequency (VHF), High Frequency (HF), Ultra High Frequency (UHF), or satellite communication, are used within the military domain, resulting in network characteristics that strongly differ from conventional wireless communication. The network layer is the core transport network to load information transmission and provides a base transmission network for a wide range of applications and services.

Data Management Layer The primary goal of data management is to effectively process massive amounts of data and enable intelligent decision-making based on network behavior. The core of data management is automated processing technology, which can help implement massive data extraction, classification, filtering, identification, and data mining. However, intelligent processing is not enough to detect malicious information. Therefore, designing an effective trust and repudiation management mechanism is an important challenge for the data management layer in IoT.

Intelligent Service Layer Services that comprise this layer include data storage, data computing, data search, system security, and application support. These services would typically be provided by middleware, usually based on the SOA approach. In the MIoT, middleware deployment is essential to allow the abstraction of heterogeneous network technologies used. Specialized fault-tolerant services required by MIoT must be identified, defined, and integrated into SOA.

We believe that the Perception and Actuation and Communication Layers are where key reliability issues in the MIoT can occur. Furthermore, the Perception Layer is more vulnerable to an external attack, such as an external attack against a battery of Wireless Sensor Networks (WSNs), which can also influence MIoT reliability. Therefore, establishing lightweight and effective intrusion detection and recovery mechanisms to improve system robustness is another important security problem. Meanwhile, establishing a trust and reputation model to evaluate suspicious node behavior can significantly reduce the influence of malicious nodes. Conventional intrusion detection methods need to be modified and improved for application to the IoT due to certain limitations, including resource-constrained devices, the limited memory and battery capacity of nodes, and specific protocol stacks. For example, a lightweight Intrusion Detection System (IDS) for anomaly detection in IoT proposed in [27] aimed at detecting a common type of Distributed Denial of Service (DDoS) attack. This work proposed a lightweight attack detection strategy that uses a supervised machine learning-based Support Vector Machine (SVM) to detect an adversary that tries to inject unnecessary data into the IoT network. The proposed IDS focused on two prominent issues: the attribute of the receiving data to classify the signal and the machine learning-based classifier. An SVM-based classifier was utilized for classification purposes, with input given as two or three simple features.

As discussed earlier, the MIoT reference architecture should also encompass security mechanisms, ensuring confidentiality, integrity, availability, and trustworthiness of IoT data. Some of the security challenges and possible solutions are discussed in Chaps. 6 to 8.

3.4 Conclusions

The IoT architectural activities described above are mainly progressing independently, without any integration (with the exception of ETSI and oneM2M that work together successfully). Each effort defines IoT from a slightly different perspective and tries to address slightly different requirements. Not every architecture has the need for M2M communication (like P2413). On the other hand, some architectures focus only on M2M. There is still no leading standard architecture or organization. IoT-A attempts to interconnect different approaches; however, it is currently a European project and may or may not be accepted worldwide. Lastly (although perhaps not as important), there is still no agreement on a precise definition of IoT.

Several of the IoT architectures considered in Sect. 3.1 pursue a similar concept of separation: In a typical deployment, one or more IoT device domains with direct M2M communications exist. It is also typical to have an infrastructural domain consisting of a core network and additional services, enabling the construction of IoT applications using the information gathered in the device domains. Device domains are sometimes connected to the infrastructural domain using dedicated gateways. The high-level architecture from ETSI (cf. Sect. 3.1.1) defines the separation between the network domain and the device and gateway domain. Direct communication between IoT devices over an M2M area network is conceptually located in the *device and gateway domain*. An access network can establish a backhaul connection from the devices to an infrastructural core network (residing within the *network domain*). In the ETSI architecture, *M2M applications* based on M2M service capabilities can be implemented at different levels and locations, either directly within a device, at the gateway, or within the network domain.

A similar separation between communicating devices and the infrastructural network can be found in the IoT reference model described by the ITU-T (cf. Sect. 3.1.4). This model distinguishes between four layers: the device, network, service support, and application layer. Here, the devices and gateways conceptually reside at the *device layer*, whereas the infrastructural network, including transport capabilities, is located at the *network layer*.

Most of these commercial IoT architectures also assume the existence of an enterprise cloud at the back end, as well as a reliable high-bandwidth communications network that connects the gateways (or, sometimes the IoT devices themselves) to this enterprise cloud. Although communications within the device domain may be limited (e.g., using low-power long-range standards such as LoRa, Narrowband IoT (NB-IoT), Sigfox, Bluetooth, Z-Wave, etc.), the gateways themselves connect to the enterprise cloud via the commercial Internet or some other high-speed link (e.g., LTE, 5G, Satellite Communications (SATCOM)). Furthermore, the enterprise cloud is not resource-constrained and is assumed to provide the necessary scalability for data storage and processing to support the desired IoT user-facing applications.

References

1. Ashton, K.: That 'Internet of Things' thing. RFID Journal **22**, 97–114 (2009)
2. Bassi, A., Bauer, M., Fiedler, M., Kramp, T., van Kranenburg, R., Lange, S., Meissner, S. (eds.): Enabling Things to Talk: Designing IoT solutions with the IoT Architectural Reference Model. Springer, Berlin (2013)
3. Bauer, M., Boussard, M., Bui, N., Carrez, F., Jardak, C., Loof, J.D., Magerkurth, C., Meissner, S., Nettsträter, A., Olivereau, A., Thoma, M., Walewski, J., Stefa, J., Salinas, A.: D1.5 - Final architectural reference model for the IoT v3.0. Project deliverable, IoT-A (2013)
4. Bluetooth Special Interest Group: Bluetooth Special Interest Group Website. https://bluetooth.com/
5. CAN in Automation Group: CAN in Automation Group Website. https://www.can-cia.org/about-us/
6. Cho, J.H., Chan, K., Adali, S.: A survey on trust modeling. ACM Comput. Surv. **48**(2) (2015). https://doi.org/10.1145/2815595
7. Demko, E.: Commercial-off-the shelf (cots): a challenge to military equipment reliability. In: Proceedings of the Annual Reliability and Maintainability Symposium, pp. 7–12 (2016)
8. Eclipse IoT Working Group: Website. https://iot.eclipse.org/
9. Eclipse IoT Working Group: The three software stacks required for IoT architectures (2016). https://iot.eclipse.org/community/resources/white-papers/iot-architectures/
10. Eclipse IoT Working Group: Open Source Software for Industry 4.0. White Paper, Eclipse (2017). https://iot.eclipse.org/community/resources/white-papers/industry40/
11. European Commission: Internet of Things Architecture - Project Information. https://cordis.europa.eu/project/rcn/95713/factsheet/en
12. European Telecommunications Standard Institute (ETSI): ETSI Website. https://www.etsi.org/
13. European Telecommunications Standard Institute (ETSI): Machine-to-Machine communications (M2M); Functional architecture, Technical Specification, ETSI TS 102 690, V2.1.1 (2013)
14. European Telecommunications Standard Institute (ETSI): Machine-to-Machine communications (M2M); M2M service requirements, Technical Specification, ETSI TS 102 689, V2.1.1 (2013)
15. European Telecommunications Standard Institute (ETSI): Machine-to-Machine communications (M2M); MLA, DLA and MLD interfaces, Technical Specification, ETSI TS 102 921, V2.1.1 (2013)
16. Fielding, R.: Architectural styles and the design of network-based software architectures. Doctoral dissertation, University of California, Irvine (2000)
17. Filho, O.F.F., Ferreira, M.A.G.V.: Semantic web services: a RESTful approach. In: IADIS International Conference WWW Internet, pp. 169–180. IADIS (2009)
18. Gaur, A., Scotney, B., Parr, G., McClean, S.: Smart city architecture and its applications based on IoT. Proc. Comput. Sci. **52**, 1089–1094 (2015). The 6th International Conference on Ambient Systems, Networks and Technologies (ANT-2015), the 5th International Conference on Sustainable Energy Information Technology (SEIT-2015)
19. Głowacka, J., Krygier, J., Amanowicz, M.: A trust-based situation awareness system for military applications of the Internet of Things. In: 2015 IEEE 2nd World Forum on Internet of Things (WF-IoT), pp. 490–495 (2015). https://doi.org/10.1109/WF-IoT.2015.7389103
20. IEEE P2413: Report from April 2016 WG meeting (2016)

21. IEEE P2413: IEEE Draft Standard for an Architectural Framework for the Internet of Things (IoT). IEEE P2413/D0.4.5, December 2018, pp. 1–264 (2018)
22. IEEE Working Group P2413: Standard for an Architectural Framework for the Internet of Things. http://grouper.ieee.org/groups/2413/
23. International Organization for Standardization (ISO): Website. https://www.iso.org
24. ISO/IEC JTC 1/SC41 Technical Committee: Internet of Things (IoT) – Reference Architecture (2018)
25. ITU-T: ITU-T Website. https://www.itu.int/
26. ITU-T: Recommendation ITU-T Y.2060 (06/2012): Overview of the Internet of Things. Tech. rep., ITU (2012)
27. Jan, S.U., Ahmed, S., Shakhov, V., Koo, I.: Toward a lightweight intrusion detection system for the Internet of Things. IEEE Access **7**, 42,450–42,471 (2019). https://doi.org/10.1109/ACCESS.2019.2907965
28. Johnsen, F.T., Bloebaum, T.H., Brannsten, M.R., Lund, K., Aubert, D.E.: Using open standards for utilizing IoT sensors in a smart city scenario. In: International Command and Control Research and Technology Symposium (ICCRTS) (2018)
29. knx.org: Website. https://www.knx.org/knx-en/for-professionals/
30. McMahon, A., Farrell, S.: Delay- and disruption-tolerant networking. IEEE Int. Comput. **13**(6), 82–87 (2009). https://doi.org/10.1109/MIC.2009.127
31. MQTT.org: MQTT.org Website. https://mqtt.org/
32. oneM2M partners: oneM2M Published Specifications. http://www.onem2m.org/technical/published-drafts
33. oneM2M partners: oneM2M Website. https://www.onem2m.org/
34. oneM2M partners Type 1: oneM2M Technical Specification TS-0001; Functional Architecture, V 3.12.0 (2018). http://www.onem2m.org/technical/published-drafts
35. Osoro, O.B., Oughton, E.J.: A techno-economic framework for satellite networks applied to Low Earth Orbit constellations: Assessing Starlink, OneWeb and Kuiper. IEEE Access **9**, 141,611–141,625 (2021). https://doi.org/10.1109/ACCESS.2021.3119634
36. Pradhan, M., Fuchs, C., Johnsen, F.T.: A survey of applicability of military data model architectures for smart city data consumption/aggregation. In: IEEE 4th World Forum on Internet of Things (2018)
37. Shelby, Z., Hartke, K., Bormann, C.: The Constrained Application Protocol (CoAP). Request for Comments (RFC) 7252, IETF (2014)
38. Snyder, D., Powers, J., Bodine-Baron, E., Fox, B., Kendrick, L., Powell, M.: Improving the Cybersecurity of U.S. Air Force Military Systems Throughout Their Life Cycles. Tech. Rep. RR-1007-AF, RAND Corporation (2015)
39. Suri, N., Benincasa, G., Tortonesi, M., Stefanelli, C., Kovach, J., Winkler, R., Kohler, U.R., Hanna, J., Pochet, L., Watson, S.: Peer-to-peer communications for tactical environments: observations, requirements, and experiences. IEEE Commun. Magazine **48**(10), 60–69 (2010). https://doi.org/10.1109/MCOM.2010.5594678
40. United States Department of Defense: MIL-STD-810G, Department of Defense Test Method Standard for Environmental Engineering Considerations and Laboratory Tests. https://www.atec.army.mil/publications/mil-std-810g/mil-std-810g.pdf (2008). Accessed 19 Jan 2021
41. Whitmore, A., Agarwal, A., Da Xu, L.: The Internet of Things - a survey of topics and trends. Inf. Syst. Front. **17**(2), 261–274 (2014)
42. Zheng, D., Carter, W.: Leveraging the Internet of Things for a more efficient and effective military. Tech. rep., Center for Strategic & International Studies, Washington, DC (2015)
43. Zheng, D.E., Carter, W.A.: Leveraging the Internet of Things for a More Efficient and Effective Military. Rowman & Littlefield, Lanham (2015)

44. Zieliński, Z., Chudzikiewicz, J., Furtak, J., Głębocki, P.: Integrating some security and fault tolerant techniques for military applications of Internet of Things. In: 2016 IEEE 3rd World Forum on Internet of Things (WF-IoT), pp. 117–122 (2016). https://doi.org/10.1109/WF-IoT.2016.7845507
45. Zigbee Alliance: ZigBee Alliance Website. URL https://zigbeealliance.org/
46. Z-Wave Alliance: Z-Wave Alliance Website. https://z-wavealliance.org/. Accessed 10 Mar 2021

Chapter 4
Building Blocks: Sensing, Actuating, and Communicating

James R. Michaelis, Niranjan Suri, Aaron Cohen, Vincenzo Pellegrini, Mika Helsingius, Manas Pradhan, Michal Marks, Konrad Wrona, and Zbigniew Zieliński

4.1 Introduction

In prior multi-domain operations research [44], the IoT has emerged from a collection of sensing, actuation, and networking technologies. In particular, the current IoT landscape has strongly leveraged work on cyber-physical systems,

J. R. Michaelis
US Army DEVCOM Army Research Laboratory (ARL), Adelphi, MD, USA

N. Suri (✉)
US Army DEVCOM Army Research Laboratory (ARL), Adelphi, MD, USA

Florida Institute for Human and Machine Cognition (IHMC), Pensacola, FL, USA
e-mail: nsuri@ihmc.org

A. Cohen
US Naval Research Laboratory (NRL), Washington, DC, USA

V. Pellegrini
IDS Corporation, Pisa, Italy

M. Helsingius
Finnish Defence Forces, Riihimäki, Finland

M. Pradhan
DKSR, Berlin, Germany

M. Marks
NASK National Research Institute, Warsaw, Poland

K. Wrona (✉)
NATO Communications and Information Agency (NCIA), The Hague, The Netherlands

Military University of Technology, Warsaw, Poland
e-mail: konrad.wrona@ncia.nato.int

Z. Zieliński
Military University of Technology, Warsaw, Poland

N. Suri et al. (eds.), *Military Applications of Internet of Things*, Internet of Things,
https://doi.org/10.1007/978-3-032-03378-9_4

wireless networking technologies, and distributed computing. To determine the viability of IoT architectures to meet the needs of military operations such as CIMIC and HADR, an ongoing assessment of constituent technologies must be conducted to:

- Establish their fitness for use in specific operational settings
- Assess their capability to meet the needs of particular federated architectures
- Identify potential vulnerabilities to environmental/adversary actions

This section provides an overview of IoT sensing, actuation, and communications technologies investigated in support of NATO IST-147 RTG activities.

4.2 Sensing Modalities Overview

From a C2 perspective, information dominance is achieved by real-time collection and processing of information to enable quicker decisions and actions than opposing forces. Current-generation IoT assets offer many sensing capabilities to meet varying operational needs. Table 4.1 provides an overview of common sensing types, modes, and processing requirements for an embedded designer to consider prior to selecting an embedded system to meet a particular operational use case. Descriptions for each listed sensing type are as follows:

- **Radio Frequency (RF) Sensing** consists of direction finding of transmitters, intercepting and decoding transmissions, and radar processing for finding objects. As IoT edge processing increases, RF sensing, with its high processing

Table 4.1 Sensing overview for the IoT embedded devices

Sensor domain	Sensing modes/types	Processing requirements
Radio frequency	Spectrum utilization, object detection (radar), direction finding, flight tracking (ADS-B, GPS)	High sample rates
Optical	Object detection (image recognition), distance detection (computer stereo vision)	High sample rates
Acoustic	Speech/language, object detection (sonar), direction finding	Low to high sample rates
Environmental	Temperature, Humidity, Pressure	Low sample rates
Medical	Fitbit (heart rate), connected hearing aids, other connected medical sensing	Low to medium sample rates

requirements, becomes feasible. Common RF signals of interest for decoding include ADS-B for flight tracking and GPS for edge node location and decoding RFID for military logistics and support operations.

- **Optical Sensing** consists of object/person detection and identification. In a military sense, this can be used in (1) a counter UAV system for friendly or adversarial UAV recognition or (2) face recognition for security or entry verification.
- **Acoustic Sensing** consists of direction finding, speaker recognition, language decoding, and sonar/ultrasound for object detection. Many acoustic sensors are low cost and require significantly less processing capability than RF sensors to handle the common audio sample rates. In synthetic aperture applications, such as imaging of seabed mines from unmanned underwater vehicles, the data that need to be processed per processing cycle can be even higher, since multiple pings need to be combined to produce a single output.
- **Environmental Sensing** consists of measuring temperature, humidity, and/or pressure. There are several low-cost sensors available capable of taking these measurements. From a military perspective, these readings may indicate localized regions of interest to investigate in the event of natural or man-made events.
- **Medical Sensing** consists of human health devices. Examples of these devices include hearing aids, fitbits, and other medical monitoring devices worn or implanted. Typically, these devices require lower sampling rate devices when compared to radio frequency devices. Many civilian devices are connected to cellular phones and through the phones to the Internet. From a military perspective, these readings when combined from a large group could indicate localized regions of interest to further investigate in the event of natural or man-made events.

4.3 Actuation in C2 Systems

In parallel with advances in sensing technology, IoT-based actuators have seen continuous capability increases paired with cost reduction. Commercially available IoT actuators can vary significantly in capability and form factor, with many options available to address academic and industrial use cases [28]. A key advantage of IoT-based actuators over traditional variants lies in their potential to integrate with "cyber-physical systems" through appropriate network connectivity, enabling remote actuation coupled with features such as monitoring the state of actuators, conducting maintenance, and managing remote software updates. In the manufacturing and industry sector, the Industry 4.0 paradigm has advanced the concept of smart factories augmented with Industrial IoT (IIoT) infrastructure. Under Industry 4.0, cyber-physical systems become a core component of the large-scale manufacturing process, automating and optimizing manufacturing and decision-making [21].

Table 4.2 lists common actuator types, their operation methods, and related examples.

Table 4.2 Actuator types, operation methods, and examples

Type of actuator	Operation method	Example
Pneumatic actuators	Compressed air to generate motion	Smart pneumatic valves used in bottling industries
Hydraulic actuators	Liquid to generate motion	Hydraulic presses used in building aircraft
Electric actuators	External power source, such as a battery, to generate motion	Robots using linear actuator arms
Thermal actuators	Heat source to generate motion	Overheating protection for opening flaps of HVAC systems

Actuators can be deployed in several configurations depending on identified use cases, including:

- **Single-actuator systems** such as simple switches to turn on and off the lights at a smart base
- **Multi-actuator systems** such as robot arms with multiple degrees of freedom to sort and load inventory at a logistics base
- **Composite systems** with sensors and actuators, such as a land robot with a Light Detection and Ranging (LIDAR) sensor surveying the environment and using electromechanical motors to drive around

In addressing continued civilian IoT market growth, commercial manufacturers of IoT actuators have led the way in design improvement and cost reduction, as well as innovation of new capabilities and form factors. From a military perspective, Commercial Off The Shelf (COTS) IoT actuators can address a variety of operational needs at a significantly lower cost than military-developed actuators [40]. By assessing particular mission needs (e.g., base perimeter defense and tactical operations support), militaries can selectively piggyback on COTS-based solutions and integrate them into their infrastructures. Assessing the usage of COTS IoT actuators in broader military infrastructures represents a key research objective in the emergent IoT-based Internet of Battlefield Things (IoBT) paradigm [44]. Below are a set of associated military use cases for IoT-based actuators:

Smart Bases

As discussed in Chap. 2, Sect. 2.3, military smart bases are rapidly evolving into practice. IoT-based actuators can help address many goals for smart bases, including:

1. Automating various tasks which typically require human intervention
2. Reducing energy consumption
3. Perimeter security and surveillance
4. Rapid action and alerts in cases of attacks or mishaps on-site

To manage energy consumption, composite actuators equipped with weather and environment monitoring sensors and SOCs can help save energy by automatically turning on and off several devices, such as HVAC systems, using electromechanical actuators coupled with the sensing systems. In addition, smart electronic inverters can monitor energy usage at the base and switch between the electric grid and renewable sources deployed at the base (such as solar modules) to help keep electricity consumption from the grid down and even push electricity into the grid from solar farms. Similarly, for perimeter security and surveillance, COTS-based cameras running facial and object recognition algorithms can be integrated with automated gates to reduce the need for human intervention and attention at bases.

ISR Mission Support

In support of HADR operations, Intelligence, Surveillance, and Reconnaissance (ISR) missions can be aided by COTS drones and land robots based on their capabilities in desired operational environments (e.g., cities and urban areas). As COTS-based robots become more popular and commercially accepted in the public domain, their associated costs have steadily decreased, while their capabilities have increased. Currently available COTS robots come with a growing variety of sensors and actuators combined with increasingly powerful SOCs to handle computational requirements [9]. Furthermore, commercial manufacturers are aligning on standards and conventions for the implementation and integration of drone hardware and software, allowing military forces to take advantage of these COTS capabilities and exploit them as standalone systems, in cluster formations, or as complements to specialized military robots. Furthermore, COTS robots come with various connectivity and controller options that provide flexibility in both usage and deployment. With these design features, COTS robots can carry out missions in closed and compact environments to record video, RADAR, and LIDAR feeds and transmit them back to the operations centers to support decision-making and planning. They can also deliver medical payloads and rations to injured people where conventional delivery methods do not work. An example of a COTS IoT drone conducting an ISR mission is presented in Chap. 8.

Autonomous Vehicles

Various projects have evaluated the military use of autonomous vehicles in a variety of capacities:

1. Land-based systems, as featured in the Defense Advanced Research Projects Agency (DARPA) Autonomous Land Vehicle (ALV) project [22]
2. Air-based systems, including Unmanned Air Vehicles (UAVs) such as Predator Drones first introduced in the 1990s [4]
3. Underwater systems, covering Autonomous Underwater Vehicles (AUVs) first investigated by the University of New Hampshire's Marine Systems Engineering Laboratory in the 1970s [58]

Since their initial investigation in the 1970s, significant progress has been made in designing and operating autonomous vehicles in the military context. Technologies for automation have become cheaper and more accessible, enabling commercial IoT manufacturers to develop solutions for the autonomous vehicle domain based on IoT. COTS-based autonomous vehicles have previously demonstrated viability in supporting HADR operations, such as the UAVs and Unmanned Maritime Vehicles (UMVs) deployed after the Fukushima nuclear disaster [31]. These autonomous platforms enabled surveillance and reporting on damaged reactor buildings, which were not possible to view through alternate means, such as satellites, and without putting human life in danger. By extension, such platforms would likely prove helpful for Chemical, Biological, Radiological, and Nuclear (CBRN) defense operations.

UAV swarms have been increasingly used in both the civilian and military spheres. In particular, UAV swarms have drawn attention in the military domain. The use of UAV swarms with increased autonomy of its elements is of particular importance. It is expected that autonomous UAV swarms will become a key element of future military operations. Autonomous UAV swarms are crucial for future military operations and civilian applications. The obvious benefit of using UAV swarms is to increase the efficiency of operations, speed up their execution, and increase the probability of success. In the civilian sphere, UAV swarms most often perform tasks that include inspecting infrastructure, supporting site construction, all SAR (Search and Rescue) type tasks, transporting goods, monitoring a designated area, supporting crops and agricultural areas, and countering and fighting the effects of natural disasters. Emerging new applications of UAV swarms include their involvement in the Smart Cities framework. Drones that are part of a swarm used as part of a smart city can perform various roles depending on the current situation and its needs. For example, it could be monitoring and reporting incidents on the road, including replacing police forces in some road control activities and supporting the infrastructure needed for a smart city. Drone swarms can also be used to build flying ad hoc networks (FANETs), which do not require ground infrastructure where the constructed network is to operate. This is a variant of mobile ad hoc networks (MANETs), which takes advantage of the characteristics of drones, particularly their mobility.

A special type of UAV drone application supports search and rescue or HADR operations. HADR operations are extremely challenging, and situational awareness is one of the most pressing requirements in the immediate aftermath of a disaster. UAV swarms equipped with sensors such as daylight and infrared cameras, infrared detectors, rangefinders, sensors for atmospheric conditions, contamination, etc. provide an essential platform for situational awareness. As part of rescue operations involving the military, drone swarms can indeed help in quickly obtaining situational awareness. They can also be used to establish ad hoc networks over areas without dedicated infrastructure and in many other situations. Military applications include various types of reconnaissance and strike systems, search and rescue operations and MedEvac-type systems, systems for reconnaissance and interception of enemy drones in public spaces, and disruption of operating swarms with hostile intentions.

Generally, a UAV swarm's tasks are assigned to one of three categories [8]: organization in space, collective displacement, or collective action and decision-making. The tasks of the first group involve the swarm obtaining some spatial features. An example would be the desired distance between drones. Tasks in this category primarily include collection at a point (aggregation), dispersal, area coverage, and formation. The characteristic of these tasks is that the goal is to change the swarm itself. The collective displacement group includes all tasks involving avoiding obstacles or finding scattered objects. The most commonly considered tasks of this group in the literature are exploration, search (foraging) involving the collection of objects on the map, group navigation involving the formation and modification, if necessary, of an array heading in the direction chosen by the swarm, and group transportation involving the movement of objects, which requires the involvement of at least several UAVs for this purpose. The last group consists of tasks involving decision-making by swarm members regarding the entire swarm. Drones make decisions based on their perception of the environment and information obtained from other UAVs. Tasks in this category include consensus setting involving an equal decision by the entire swarm, task assignment involving assigning each drone in the swarm a certain sentence from a list of available sentences, determining the location of swarm elements, etc.

Due to the nature and complexity of drone swarm systems, the design process is carried out in multiple stages, often through hierarchical functional decomposition at different levels of detail [8]. To increase its efficiency, it must be supported with computer tools. Strategies for cooperative control of multiple UAVs are usually divided into top-down and bottom-up. The top-down approach is mainly based on the hierarchical solution by decomposing the original problem into several subproblems and solving the subproblems through negotiation and cooperation between individual drones. The optimization algorithms used to solve subproblems include exact optimization algorithms and heuristic algorithms. Exact algorithms require huge computational resources and cannot be used dynamically during swarm operations. Several heuristic algorithms have recently been proposed that solve these problems with less effort. These algorithms reduce the computational effort but are still computationally complex for online scheduling in dynamic environments with complex constraints. Because of the high computational workload and poor response time of top-down methods, bottom-up approaches [59] are most often used. In addition, the existing bottom-up approaches solve the problem of planning search-and-attack missions for a swarm UAV based on an intelligent self-organized algorithm [59] by allowing consideration of such factors as (1) limiting the maximum range of the UAV and (2) avoiding collisions between UAVs or between UAVs and obstacles. The proposed intelligent algorithms for the problem of planning search and attack missions using multiple UAVs take into account not only maneuverability limitations, collision avoidance limitations, and threat avoidance limitations but also range limitations. Each UAV can detect the battlefield environment online and make decisions independently. Therefore, when a UAV joins or leaves the battlefield, existing UAVs can adjust their actions to realize the cooperation of multiple UAVs.

4.4 Communications Technologies for Military Internet of Things (MIoT)

Commercial IoT has both leveraged a multitude of existing wireless communications technologies and ushered in some specific new technologies driven by the specific needs of IoT. Given that the primary goal of IoT is to convert and connect everyday devices to the network, wireless communications in various forms are widely employed (Figs. 4.1 and 4.2). These include Wi-Fi and Bluetooth, which are very common for many types of devices, not just IoT, and also some that are more specific to IoT and sensor networks, such as ZigBee and Z-Wave. Most of these wireless technologies were designed for short- to medium-range wireless communications (e.g., indoor or close proximity).

Deployment of IoT in more widespread areas (e.g., in city or rural environments) also led to the development of two other categories of technologies—those based on Cellular communications (e.g., NB-IoT and Long-Term Evolution Machine Type Communication (LTE-M)) and those based on specific modulation techniques that were designed for low-power, low-bandwidth, but long-range communications (SigFox and Long-Range Radio (LoRa)). A good survey of the various communications protocols and options is provided in Section IV.B of [7].

The IST-147 RTG had an opportunity to perform more extensive experimentation using LoRa as a communications technology for IoT devices. LoRa was selected as it was a protocol specifically designed to support IoT devices and as such offered interesting properties in terms of long-range transmission and low power utilization. The group decided to investigate LoRa further as it seemed a promising technology that did not require pre-deployed infrastructure (such as LTE-M, NB-IoT, and SigFox) and yet could be utilized for a variety of deployment scenarios such as large-scale sensor networks. A broader analysis of the applicability of 5th Generation Cellular Networks (5G) technology to military scenarios involving IoT has been performed within the NATO STO IST RTG 187 on 5G Technologies Application to NATO Operations [61].

4.4.1 IoT Communications via LoRa

LoRa is a physical layer protocol acquired by Semtech in 2012 in support of commercial IoT applications, with foundational research conducted in 2009 by Cycleo [10] to address use cases, including Automatic Meter Reading (AMR). LoRa uses a spread spectrum modulation scheme, achieving long-range transmission with very low power. It is based on Chirp Spread Spectrum (CSS) modulation, where the frequency increases (upchirp) or decreases (downchirp) with time, a technique also used in radar and sonar [56]. The advantages of chirp modulation include a large link budget, resilience to interference, and resistance to multipath effects. The

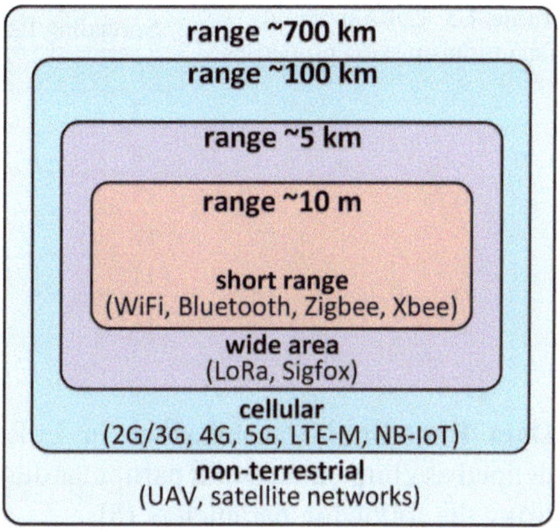

Fig. 4.1 Range of various IoT communication technologies (Based on [51])

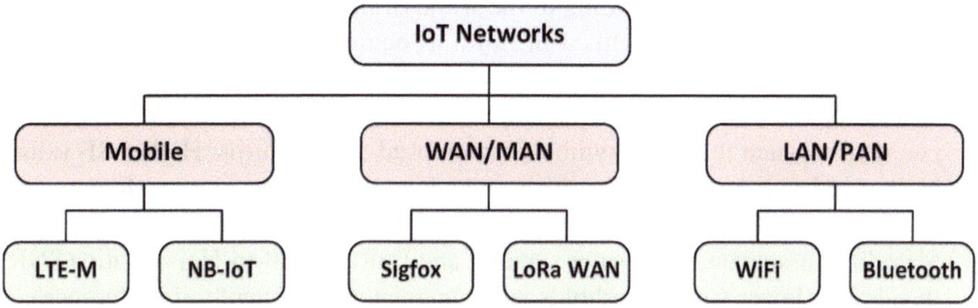

Fig. 4.2 Classification of IoT communication technologies (Based on [51])

LoRa modulation is patented and proprietary, though reverse engineering studies have been conducted [20, 43].

4.4.1.1 LoRa Basics

Transmission Range While LoRa transmission coverage in the range of 2–5 km has been demonstrated in urban environments [29], distances exceeding 10 km have been achieved in open areas with established transmitter-to-receiver line of sight. To date, the longest established transmission range for LoRa was achieved in 2017 at 704 km from a balloon at an altitude of 38.8 km using a transmission power of 25 mW (14 dBm) [49].

Transmission Bands LoRa radios are normally used in license-free Industrial, Scientific, Medical (ISM) bands, which vary in frequency range according to region. At present, LoRa leverages the 868 MHz and 433 MHz bands in Europe, the 915 MHz band in North America, and the 923 MHz band in Asia. As a result, there can be a large variation in permitted frequencies and maximum transmission power among different countries.

Table 4.3 LoRa spreading factor transmission properties

Spreading factor	Chirps/symbol	LoRa demodulator SNR
6	64	−5 dB
7	128	−7.5 dB
8	256	−10 dB
9	512	−12.5 dB
10	1024	−15 dB
11	2048	−17.5 dB
12	4096	−20 dB

Data Encoding Transmissions in LoRa are commonly expressed as *symbols*, defined as chirp signals of a particular duration. Symbols can be configured in LoRa using the following parameters [5]:

- *Bandwidth (BW)*, corresponding to the width of the band occupied by the carrier signal. The chirp rate is defined in LoRa as equivalent to bandwidth, such that one chirp is transmitted per second per hertz.
- *Spreading Factor (SF)*, defined as the base-2 logarithm of the number of chirps per symbol, such that each symbol is composed of 2^{SF} chirps. Higher SF values correspond to longer duration chirps, which are more robust to interference and can be transmitted at greater ranges. Table 4.3 provides a reference listing of SF values, associated chirps per symbol, and LoRa Signal-to-Noise Ratio (SNR) thresholds (lower value thresholds indicate greater robustness to interference).
- *Coding Rate (CR)*, which is applied toward forward error correction techniques used by LoRa to ensure transmission integrity

When combined, these factors define the nominal bitrate R_b for LoRa transmissions as $R_b = SF * BW/2^{SF} * CR$, where CR is defined as 4/N, where N can take values of 5, 6, 7, or 8, and N-4 gives the number of bits dedicated to error detection/correction. Here, increasing SF by one while keeping the same BW will double the duration of a symbol transmission, effectively reducing the achieved bitrate. Some examples of bitrates are shown in Table 4.4. Achievable bitrates are limited, ranging from tens of bits to a few kilobits per second. Further, duty cycle limitations imposed by LoRa make the protocol unable to support continuous transmission. Therefore, while LoRa is not usable for applications such as audio or video transmission, it is well suited for cases that require sending small packets of data periodically (e.g., numeric values corresponding to sensor readings).

4.4.1.2 LoRa Networking Protocols

LoRa defines the physical transmission layer, ready to be used for point-to-point communications or small star networks. More extensive networks can be built by combining LoRa with media access control (MAC) layers. The most popular one

Table 4.4 Sample LoRa bitrates and corresponding transmission parameters

Bandwidth (kHz)	Spreading factor	Coding rate	Nominal R_b (bps)
7.8	12	4/5	18
10.4	12	4/5	24
15.6	12	4/5	37
20.8	12	4/5	49
31.2	12	4/5	73
41.7	12	4/5	98
62.5	12	4/5	146
125	12	4/5	293
250	12	4/5	586
500	12	4/5	1172

is LoRaWAN, defined by LoRa Alliance [24], with other alternatives, including DASH7 from Haystack and Symphony Link from Link-Labs [6, 23].

While each LoRa networking protocol leverages LoRa at the Open Systems Interconnection (OSI) physical layer, they can differ significantly in higher layer implementations. These design variations can impact protocol properties such as latency, capacity, multi-hop network capability and support for over-the-air firmware updates. The general drawback of using LoRa networking protocols is that they might limit the size of messages, the number of messages per day, available frequencies, or other capabilities. With LoRa, these are important considerations as the total bitrates are very low. Hence, the user must compare different alternatives carefully and find the best solution for his or her specific needs. On the other hand, the benefit of using LoRa networking protocols is that they provide a standardized data format, and there are independent providers for products like sensor nodes and gateways.

Using LoRa networking protocols makes it possible to connect many access points through a backbone network. Higher OSI layers supported over the networking protocol take care of the unique identification of clients, handover between different access points, any necessary changes in frequency, and cryptographic functions.

Further, the client can change the access point depending on network traffic or when it is moving. If there is a regional or national third-party service provider, it is possible to use similar clients over a large geographic area. These services have a gateway from the LoRa network to the Internet, and standardized web interfaces can access the data. Existing telecommunications companies provide these services, and the community-based "The Things Network" is also popular in many areas [48].

Typically, the maximum transmission time on one channel is limited, as indicated by the duty cycle limitation. The duty cycle is measured over 1 hour, can be 0.1%, 1%, or 10% (3.6s, 36s, or 360s per hour), and varies between different areas and frequency bands. In some cases, radios with a "listen before talk" feature (to check channel availability before transmitting) can transmit for more extended periods of

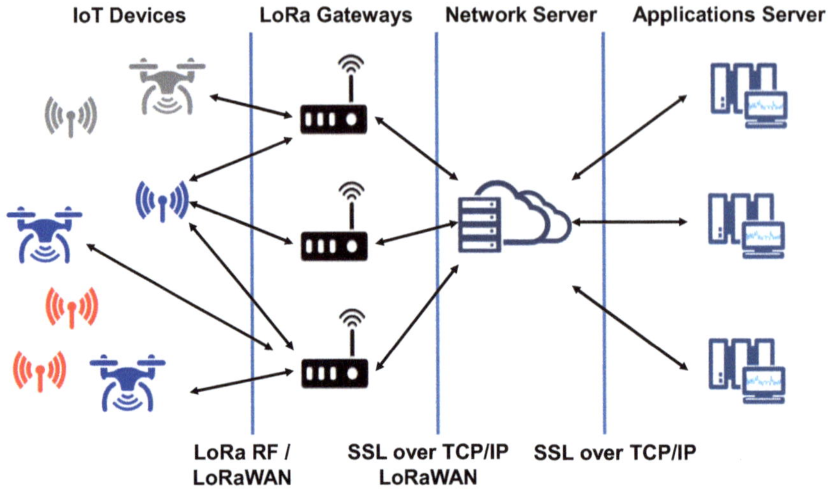

Fig. 4.3 LoRa/LoRaWAN system architecture

time on certain bands. However, the user must verify the rules for each country and frequency band.

LoRaWAN (Long-Range Wide-Area Network) refers to an LPWAN (Low-Power Wide-Area Network) protocol defined by the LoRa Alliance that spans the MAC, network, and application layers of the ISO/OSI model (layers 2, 3, and 7, respectively) [26]. The corresponding systems architecture defined by LoRa/LoRaWAN is depicted in Fig. 4.3.

At the network edge, IoT devices use LoRa to directly send messages to one or more local gateways, which in turn are connected to a centralized network server. The result is a star-on-star network topology, where the network server and the gateways are the centers of the stars. The network server is responsible for handling duplicated packets coming from multiple gateways and selecting the best gateway through which to send packet acknowledgments and payloads to a specific end device, whenever downlink communication is required. In addition, the network server performs security checks and can control some parameters of end devices, including data rate and transmission channel [25]. In turn, application servers connect to the network server to send and receive payloads from end devices.

LoRaWAN devices support up to three different classes of operation [5]. Class-A is designed mainly for uplink transmissions, where devices remain in sleep mode until they have a packet to transmit and have a small downlink transmission window immediately after, achieving the greatest battery lifetime among classes of operation. Class-B devices establish downlink reception windows by negotiating a ping interval for receivers to listen for incoming packets, providing a trade-off between battery efficiency and downlink packet delivery latency. Finally, Class-C devices operate with their receivers always on, enabling gateways to perform downlink transmissions any time except during uplink transmissions. The Class-C operation mode is designed for devices directly connected to power supplies, due to greater energy consumption by receiver operation.

4.4.1.3 Developer Considerations

From the perspective of IoT infrastructure development, the benefits of using LoRa are simple structure, high payload efficiency, low protocol overhead, and high customizability. Such features have resulted in the widespread adoption of LoRa in environments with varying degrees of local infrastructure development. For example, the WAZIUP EU-Africa Project has used LoRa to develop simple, low-cost IoT networks for deployment in African countries, addressing use cases that include environmental monitoring, agriculture management, and animal tracking [54].

Since 2012, Semtech has been marketing and licensing LoRa chipsets for use in commercial IoT devices. The core, which consists of 50k gates, can be easily integrated into existing radio devices. A standard Serial Peripheral Interface (SPI) [57] enables the transmission of data and control codes to and from these chips. Although Semtech first created the basic LoRa chipset implementation, they have licensed manufacturing to many other companies. Commercially available modules differ in details, such as the size of data buffers. The simple ones work on a single channel at a time, whereas the more expensive ones, intended for gateways, can operate on several channels simultaneously. The products made by HopeRF are popular for many inexpensive implementations and experimental projects [16]. It is easy to create simple single-channel gateways. However, real LoRaWAN gateways must always listen to a minimum of eight channels simultaneously, as the protocol demands that they must be able to change channels during operation.

To support software development with LoRa, several open-source C and C + + libraries have been created [27, 39], which can be extended or modified as needed by a developer.

4.4.2 Satellite Communications (SATCOM) Solutions

Traditional SATCOM solutions have been bulky and designed for maritime use. However, with advances in error correction coding, signal processing, and transceiver design, newer SATCOM solutions have been produced that fit into smaller form factors for airborne platforms, including UAVs. With the advances commercializing the LEO Satellite Constellations, modern smartphones will soon be connected through satellites when out of range of cellular towers.

There are several common commercial SATCOM systems used extensively today. The first group consists of satellite constellations in LEO. Iridium NEXT [32] is a low Earth orbit satellite system with a constellation of 66 satellites to provide full global coverage. Starlink is a low Earth orbit satellite Internet system with a

Table 4.5 Satellite Internet constellations overview

Constellation name	Orbit	Number of satellites	Transceiver notes	Throughput
AST SpaceMobile	LEO	1 experimental	Smartphone	>=10 Mbps
Chinasat	GEO	1 operational Chinasat-26	Compatible with Viasat transceivers	>=100 Gbps
Globalstar	LEO	25 operational	Transceiver weighs 7.6 g (0.27 oz)	Low speed data Phone
Iridium-NEXT	LEO	76 operational	Transceiver weighs 30 g	up to 8 Mbps
Lynk Global	LEO	8 operational 10 planned	Standard cellular phone	Low text messaging
O3b mPOWER	MEO	6 operational 11 planned	Large radome	Multiple Gbps
OneWeb	LEO	634 operational	11.8 kg	Nearly 200 Mbps
Orbcomm OG2	LEO	12 operational	Size of credit card	
Project Kuiper	LEO	3,236 planned	Portable antenna will be 7-inches square and weigh 1 lb	planned up to 400 Mbps
Sfera	MEO	640 planned	No details found	
Starlink	LEO	>4500 launched	2.9 kg (6.4 lb) without the included kickstand	50 to 200 Mbps
Viasat-2	GEO	1 operational	23.2 kg (51.3 lb)	260 Gbps

significant number of operational satellites. Their satellite Internet system has been used all over the world. The second group of commercial satellite constellations is in Geosynchronous Earth Orbit (GEO). INMARSAT is a geosynchronous orbit satellite system [12]. However, because of its distance from Earth, it requires larger transceivers, and the latencies tend to be higher, making this a less desirable solution for UAVs. Viasat-2 is another satellite system in GEO, but with a smaller footprint coverage because it relies on a limited number of satellites.

Table 4.5 provides an overview of the different satellite Internet constellations that are planned or operational in 2024. The throughput reported in Table 4.5 is for fixed transceivers and is likely less for mobile transceivers.

AST SpaceMobile Transceivers AST SpaceMobile has a direct to cellular phone satellite constellation in LEO consisting of the BlueWalker 3 test satellite with plans to expand to a full constellation of Bluebird satellites. Transceivers for their satellite constellation are fully compatible with modern cellular smartphones from major vendors. AST SpaceMobile plans at least 20 operational satellites for their Bluebird constellation to cover the most attractive market areas.

Chinasat Transceivers Chinasat entered the broadband internet service with its recent successful launch in 2023. Their service is expected to use 37.5–42.5 GHz for the space-to-Earth downlink and 47.2–51.4 GHz for the Earth-to-space uplink. In

2019, Viasat announced that Chinasat will pair with Viasat equipment in the aircraft. Further analysis of Viasat transceivers will be presented in the Viasat subsection.

Globalstar Transceivers Globalstar constellation consists of 25 LEO satellites for phone and low-speed data communications with coverage over land. A transmitter for low-data asset tracking can weigh as little as 6 g or 0.21 oz. Their low-data ST150m transmitter specifications list a low-data-rate device that weighs 7.6 g or 0.27 oz.

Iridium NEXT Transceivers The most promising transceiver on the market today is the FlytLINK, made by Thales. FlytLINK's specified dimensions are 32.11 cm × 19.76 cm × 6.07 cm with a payload weight of 3.2 kg (∼7.05479 pounds), making it small enough to fit in common UAV payload areas and below the weight limit for mid-size UAVs. The vendor claims that his transceiver supports Iridium Certus rates of 176 kbps (Data and Voice) and 708 kbps (Download) and 352 kbps (Upload) as of today. According to the vendor, the small transceivers from NAL Research are limited to 2.4 kbps on Iridium NEXT. Cobham provides transceivers that support Iridium Certus for maritime platforms but did not respond to a request for information on transceivers for UAVs. Cambridge consultants were reported to have developed the actual transceiver module that is used by Cobham and Thales to achieve the Iridium Certus data rates. This is not an end user product but could be a potential path to a lower Size, Weight, and Power (SWaP) solution for a UAV. Recently, Iridium has announced the introduction of Iridium Messaging Transport (IMT), a two-way service optimized for small- to moderate-sized messages sent over Iridium Certus and designed to make it easier to add satellite connections to IoT systems [55].

Lynk Global Transceivers Lynk Global's constellation is currently the only proven and commercially licensed satellite-direct-to-standard-phone service supported by what they are calling "Satellite-Cell-Towers." Their constellation consists of six satellites with a plan for 10 in LEO. A standard cellular phone is the transceiver. A commercial iPhone 12 weighs 164 g or 5.78 oz.

O3b mPOWER Transceivers O3b mPOWER is a satellite constellation in Medium Earth Orbit (MEO). It currently consists of six operational satellites launched between December 16, 2022 and November 12, 2023 with many more future launches planned. The MEO transceivers will be larger to achieve similar Signal-to-Noise Ratio (SNR) as a LEO system. This is due to a larger power amplifier and receiver dish.

OneWeb Transceivers OneWeb's satellite constellation consists of 634 satellites as of May 20, 2023 in LEO. The first user portable terminal was introduced around September 11, 2023. The FoldSat Leo transceiver is manufactured by Inster (Oesia Group) and weighs 11.8 kg.

Orbcomm Transceivers Orbcomm OG2 consists of 12 operational satellites out of a total of 18 launched satellites with 5 retired and 1 failure in LEO. Orbcomm OG2 is the world's first commercial satellite network dedicated to IoT. Orbcomm transceivers have a footprint the size of a credit card (https://www.orbcomm.com/PDF/datasheet/OG2-OG-ISAT-Satellite-Modems.pdf).

Project Kuiper Transceivers Project Kuiper is run by Kuiper Systems LLC, which is a subsidiary of Amazon established in 2019. The first two prototype satellites were launched on October 06, 2023. The planned constellation includes 3,236 satellites in low Earth orbit (LEO). As of March 2024, the Project Kuiper receivers feature a square antenna that is less than 11 inches in size and weighs less than 5 pounds. A portable antenna will be 7 inches square and weigh 1 pound with speeds maxing out at 100 Mbps.

Sfera Transceivers The Sfera constellation ("Sphere" in English) is a Russian Federation effort to create an alternative to OneWeb and Starlink. Approximately 640 satellites are planned to be launched in the span of 6 years from 2022 through 2028. Skif Demo (Skif-D) satellite, Sfera constellation's first Internet satellite, launched into Medium Earth Orbit (MEO) at around 8,070 km above Earth in October 2022. No details on size and weight were found on the Skif-D transceivers.

Starlink Transceivers The Starlink constellation consists of over 4500 operational satellites operating in low Earth orbit (LEO). The standard Starlink transceiver consists of an electronic phased array weighing 2.9 kg (6.4 lb) without the included kickstand or 3.2 kg (7 lb) with the included kickstand. Additional transceivers use similar electronic phased array antenna technology, with the newer models having a slightly larger size or footprint. The smallest "Mini" receiver was reported to be 11.4 inches by 9.8 inches, while the larger transceiver was reported to be 22.4 inches by 14.7 inches.

Viasat Transceivers The Viasat-1 constellation consists of one satellite in GEO. The Viasat-2 constellation consists of one satellite in GEO. Both constellations operate in the Ka-band. The transceivers for this band have considerable weight and size to achieve a high throughput with the GEO satellites. For example, the GAT-5510 terminal weighs 23.2 kg (51.3 lb).

INMARSAT Transceivers Newer INMARSAT transceivers, which support Global Xpress (GX) [17] data rates, must support the Ka-band to achieve this higher capacity throughput compared to legacy transceivers, which operate on the L-band [60]. The GX network consists of three satellite terminals in geosynchronous orbit. In addition, airborne terminals are available from Honeywell (JetWave). However, the JetWave terminal appears to exceed the common UAV payload weight limits and size dimensions. Concerning airborne tests, Boeing has proven GX data rates of 30 Mbps in a SATCOM-equipped helicopter. Although INMARSAT has the advantage of higher data rates, its significant size and weight are prohibitive for installation on most small- and medium-sized UAVs. There are some concerns with rain fade on the Ka-band. INMARSAT addresses this with backup via the L-band and additional processing to handle fading in their modems. INMARSAT offers Switch Broadband (SB) with lower data rate services (200 kbps). The Cobham Aviator UAV 200 terminal supports the 200 kbps data rate. This terminal weighs

1.45 kg with dimensions of 24 cm × 16 cm × 6 cm. This is a viable solution for common UAVs but achieves lower data rates than the Iridium system from Thales.

Comparison In general, GEO and MEO satellite transceivers are substantially larger than LEO due to the additional distance and its impact on SNR. The throughput tends to be reported to be less for mobile platforms than for fixed platforms. A common trade-off in transceiver design is to use the processing or coding gain to address lower SNR when reducing the physical size. Additional processing increases power requirements and may limit a UAV's flight time before recharging.

There is ongoing testing with Iridium NEXT for UAVs with a terminal within the size and weight restrictions of common UAVs. Supporting the INMARSAT GX terminals would require a larger UAV to support heavier and larger satellite terminal equipment. There is an INMARSAT SB terminal within the size and weight restrictions of the common UAVs that supports data rates up to 200 kbps. First, testing the Thales FlytLINK terminal using Iridium is recommended to see if it meets desired UAV requirements and to verify the stated performance. A second option is to test the Cobham Aviator UAV 200 terminal on INMARSAT to see if it meets the desired UAV requirements and the stated performance.

Orbcomm OG2 transceivers are the size of a credit card and market their service as an IoT service. A further comparison between their transceiver and generic smartphones is needed to determine the advantages or disadvantages of using either transceiver for UAV C2.

Significant investments, satellite launches, and testing have been ongoing to address the direct-to-smartphone market from companies such as Starlink, Lynk Global, and AST SpaceMobile. Additionally, Starlink gained much media exposure, as their Starlink Internet service was deployed in several active war areas where local infrastructure was repeatedly down.

4.4.3 JANUS

JANUS (STANAG 4748) is the first open standard for underwater digital acoustic communications, developed by NATO [15]. JANUS can be used for a wide range of applications such as Distressed Submarine (DISSUB), underwater Automatic Identification System (AIS), or simple chat/Short Message Service (SMS) communication [62]. Within the European Defence Agency (EDA) Smart Adaptive Long- and Short-range Acoustic network (SALSA) project, the Netherlands Navy, Norway, Germany, Sweden, Finland, and several industrial partners jointly developed a common protocol stack for underwater acoustic communication and networking based on JANUS. A network node with SALSA communication stack, installed on a ship, can communicate with different platforms like UAVs, buoys, and bottom nodes. Different military scenarios include Anti-Submarine Warfare (ASW), Intelligence, Surveillance, and Reconnaissance (ISR), Mine Counter Measures (MCM), or Rapid Environmental Assessment (REA).

4.4.4 5G and Beyond

Unlike earlier mobile networks where IoT solutions such as NB-IoT were designed as add-ons, 5G networks are designed to be a key IoT enabler [51]. The three main use cases of 5G, i.e., Enhanced Mobile Broadband (eMBB), Massive Machine-Type Communications (mMTC), and Ultra-Reliable Low-Latency Communications (URLLC), involve IoT scenarios. eMBB supports such IoT nodes as augmented and virtual reality devices and drones. mMTC supports IoT scenarios with a massive number of nodes sending small volumes of data to the private or public cloud. URLLC may be used to support mission critical and industrial IoT applications. Mission-critical IoT may involve a smaller number of nodes that exchange larger volumes of data, often requiring real-time communications, e.g., for autonomous vehicles, robots, or critical infrastructure control. Industrial IoT can be used to support smart manufacturing and industrial processes, often involving the use of digital twins and allowing optimization of various post-production processes, such as supply chain management and maintenance.

4.4.5 Physical Layer Security Through Directional Modulation

IoT devices could also take advantage of physical layer security in many relevant scenarios through Directional Modulation (DM). Directional modulation is a physical layer cryptographic technique that, if implemented according to architectures such as [37], can complement traditional encryption, typically achieved at higher OSI layers, by exploiting physical propagation on the radio channel to implement additional information security functions. Recent developments in the field have resulted in a prototype DM antenna system [37] that applies physical layer encryption on top of state-of-the-art radio communication interfaces by adding a dynamically phased array to existing radio chains without requiring additional modification or system-specific integration with the underlying radio components. Supported radio interfaces include, but are not limited to, terrestrial or satellite versions of Digital Video Broadcasting (DVB) standards, 802.11x systems, and the LTE access network interfaces. The behavior of the DM antenna system allows targeting only areas where communications should be enabled. The space surrounding the system is partitioned into different regions: the intended service zone, where connectivity is provided normally, and the interdicted zone, where all radio signals coming from the DM-enabled antenna are distorted to a degree of sufficient cryptographic security to be completely unusable and unable to provide significant information. Currently, the DM system has been applied to 802.11ac and

European Telecommunications Standards Institute (ETSI) DVB radios, although it could just also be applied to other waveforms such as LoRa. Space-based encryption can provide advantages within military and safety-critical IoT scenarios, as it can provide rejection of signals and data from sensors that either:

- Have been (accidentally or intentionally) moved from their intended location and could thus influence their host IoT network with non-pertinent or malicious data or
- Are deliberately transmitting from hostile locations in order to inject wrong/harmful information into the general awareness available to the command levels

It is worth noting that the physical layer already achieves rejection of harmful information, *without* requiring checks at higher network layers, re-keying procedures, explicit enabling/disabling of nodes, or human intervention, as would be the case with many other approaches. Furthermore, in specific IoT deployments, such as a smart city or a smart base, where many nodes compete for radio access and network capacity, the spatial filtering capability that DM inherently provides will help reject undesired data contributions immediately at the antenna stage, without letting useless content proceed further through the IoT protocol stack.

4.5 Describing and Discovering IoT Assets and Services

The pervasive network of sensors and IoT devices that permeates smart cities can prove to be critically advantageous by allowing rescuers to rapidly acquire situational awareness in highly unstable and dynamic environments, greatly simplifying the planning, organization, and decision-making phases. However, reaching that goal calls for IoT asset and service discovery solutions.

Despite the impressive results achieved by some outstanding initiatives, such as the one led by Forum Virium (https://forumvirium.fi/en/) in Helsinki, the lack of standardization in this area is widely recognized as a significant obstacle to the realization of IoT-based smart city applications. A uniform approach to describing and discovering sensor assets is needed to facilitate interoperability, including the discovery and ingestion of useful data sources.

4.5.1 Describing IoT Assets

The challenges identified in civilian use of IoT are *interoperability*, *complexity*, and *scalability* problems. Recent international standardization and research and development initiatives aim to address these challenges. In particular, these challenges are the same as those we have identified for military applications of IoT in IST-147.

Ontologies, that is, the formal definition of concepts in a domain of interest and the relationships between them, represent one of the most promising approaches to facilitate interoperability of IoT. Ontologies allow one to build a knowledge base that contains accurate descriptions of characteristics and semantics of IoT assets and is well suited for consumption by automated and high-level reasoning tasks.

We have surveyed the scientific and technical literature in this research area, describing several hundreds of ontologies, and identified the following seven as relevant for civilian IoT applications [19]:

- Semantic Sensor Network (SSN) ontology [14]
- oneM2M Base Ontology[1]
- Smart Appliances REFerence (SAREF) ontology[2]
- IoT-O [45]
- Spitfire [38]
- IoT-ontology[3]
- IoT-lite [3]

For the use and purpose of the NATO group during its working tenure, we carefully evaluated IoT-lite, IoT-ontology, and IoT-O for their ability to facilitate federation and interoperability between the NATO partner countries. More specifically, we used them as guidance to identify a topic structure for publish/subscribe communications based on the derived classes and ontology classification described in the ontology specifications. The resulting topic structure was designed to categorize IoT assets and data (sensors, actuators, and computing platforms) in the NATO domain. The experiment discussed in Sect. 8.5.2 provides more details on this topic.

In general, when evaluating IoT ontologies in the context of IST-147, we found that, despite some ontologies, especially SSN, which claim to cater to web application developers [14], most seem focused on the definition of a formal relationship model between sensors, sampling, unit types, etc. In addition, their designs appear to focus on allowing high-level reasoning development, with a relatively smaller amount of attention dedicated to interoperability in terms of, for example, service Application Programming Interface (API) descriptions.

We believe that an IoT asset-centric perspective, although at the cost of a less strict formal definition, represents a more promising approach to facilitate system integration. In this context, we had some level of success in describing sensors using SensorML [36], an earlier XML-based standard developed by the Open Geospatial Consortium (OGC) within its Sensor Web Enablement (SWE) initiative. The original motivation behind SensorML was to provide a means of processing observations without a priori knowledge of the sensor or processor characteristics,

[1] http://www.onem2m.org/ontology/Base_Ontology/

[2] https://sites.google.com/site/smartappliancesproject/

[3] http://ai-group.ds.unipi.gr/kotis/ontologies/IoT-ontology

thereby helping to avoid stovepipe systems when processing sensor data within different communities.

SensorML has several advantages: Its simplest form provides a standard way to describe sensors. Through a rich set of metadata, SensorML also provides a means for sensors to become discoverable. It can describe individual sensors, aggregated sensor capabilities at a gateway level, or both. Last but not least, the use of Extensible Markup Language (XML) makes it perhaps more approachable from the perspective of system integrators. These characteristics can be clearly seen by comparing the examples in Figs. 4.4, 4.5, and 4.6.

We hope that this analysis might be helpful in future studies on military IoT interoperability. However, it will eventually be up to NATO as a community to decide which ontology, if any, should be used and if so exactly how it should be applied. Note that new proposals for ontologies that address specific aspects, such

```
<?xml version="1.0" encoding="UTF-8"?>
<sml:PhysicalComponent gml:id="MY_SENSOR"
xmlns:sml="http://www.opengis.net/sensorml/2.0"
xmlns:swe="http://www.opengis.net/swe/2.0"
xmlns:gml="http://www.opengis.net/gml/3.2"
xmlns:gmd="http://www.isotc211.org/2005/gmd"
xmlns:gco="http://www.isotc211.org/2005/gco"
xmlns:xsi="http://www.w3.org/2001/XMLSchema-instance"
xmlns:xlink="http://www.w3.org/1999/xlink"
xsi:schemaLocation="http://www.opengis.net/sensorml/2.0
    http://schemas.opengis.net/sensorml/2.0/sensorML.xsd">
   <gml:description>
   Temperature sensor on historical building</gml:description>
   <gml:identifier codeSpace="uniqueID">
   sto.nato.int.1337</gml:identifier>
   <sml:outputs>
      <sml:OutputList>
         <sml:output name="temp">
            <swe:Quantity definition=
                "http://dbpedia.org/resource/Temperature">
               <swe:label>Environmental Temperature</swe:label>
               <swe:uom code="Cel"/>
            </swe:Quantity>
         </sml:output>
      </sml:OutputList>
   </sml:outputs>
   <sml:position>
      <gml:Point gml:id="stationLocation"
        srsName="http://www.opengis.net/def/crs/EPSG/0/4326">
        <gml:coordinates>44.8375 11.6195</gml:coordinates>
      </gml:Point>
   </sml:position>
</sml:PhysicalComponent>
```

Fig. 4.4 Example of sensor description using SensorML

```
@prefix rdf:  <http://www.w3.org/1999/02/22-rdf-syntax-ns#> .
@prefix rdfs: <http://www.w3.org/2000/01/rdf-schema#>.
@prefix xsd:  <http://www.w3.org/2001/XMLSchema#> .
@prefix sosa: <http://www.w3.org/ns/sosa/> .

<sensor/11-1337-2345> a sosa:Platform ;
  rdfs:label "Some temperature sensor"@en ;
  rdfs:comment "A temperature sensor outside historical
                                        building 1"@en ;
  sosa:hosts <sensor/11-1337-2345/TMPS01> .

<sensor/11-1337-2345/TMPS01> rdf:type sosa:Sensor ;
  rdfs:label "Somecompany TMPS01"@en ;
  sosa:observes <sensor/11-1337-2345/TMPS01/temperature> .
```

Fig. 4.5 Example of SOSA SSN sensor

```
@prefix sosa: <http://www.w3.org/ns/sosa/> .
@prefix xsd: <http://www.w3.org/2001/XMLSchema#>.
@prefix om:
<http://www.ontology-of-units-of-measure.org/resource/om-2/> .

<Observation/8716> a sosa:Observation ;
   rdfs:comment "Observation of the temperature outside
                                 historical building 1."@en ;
   sosa:hasFeatureOfInterest <historicalbuilding/1> ;
   sosa:hasResult [
      a om:Point ;
      om:hasScale om:CelsiusScale ;
      om:hasNumericalValue "25.7"^^xsd:double ] .
```

Fig. 4.6 Example of SOSA SSN observation

as interoperability [47] or data stream modeling [11], are still being investigated and proposed. Therefore, an extensive reevaluation of the state of methodologies and tools in ontologies for IoT applications will likely be needed in the near future.

4.5.2 Discovering Services

Being able to discover IoT assets (e.g., services that provide sensor information) in run-time is essential. Because sensors can break, run out of power, etc., it is important to be able to find the sensors that are actually available when one wants to consume data from them.

Many service discovery protocols exist, with benefits and drawbacks for each. The Technology, Information, Decision, and Execution (TIDE) community has promoted the multicast Domain Name System (mDNS) and DNS Service Discovery (DNS-SD) approach. NATO Allied Command Transformation (ACT) TIDE has developed several specifications on how to use mDNS for service discovery, along with approaches to tunnel information over wide-area networks and integrate it with other service discovery approaches such as Universal Description, Discovery, and Integration (UDDI). These specifications are available on Tidepedia [50] to registered users.

However, it is unrealistic to expect that NATO forces will be able to work in relatively controlled environments where all assets and services systematically adopt mDNS. Therefore, middleware solutions for IoT asset discovery that implement multiple discovery protocols should be adopted to maximize the chance of identifying all available assets—which is of the utmost importance in HADR and CIMIC operations. Viable solutions should also allow for contextual or localized asset discovery and support coalition-oriented information sharing. In addition, they should implement solutions to reduce the (often relatively high) latency intrinsically involved in discovery operations.

4.6 Retrieving Sensor Data

Two fundamentally different communication paradigms may be employed to share sensor data: a pull approach, typically implemented as the request/response pattern in which the client initiates a request to a server, and a push approach, typically implemented as a publish/subscribe pattern, in which the client sets up a subscription and data are later pushed according to the active subscriptions. These paradigms are described in Sects. 4.6.1 and 4.6.2, respectively.

4.6.1 Pull (API-Based) Approaches

It is essential to use open standards as a precursor to exploiting IoT information from a smart city, since the interoperability between data sources and consumers is necessary. The formats required depend on the type of sensor and whether it provides discrete or streaming data. Sensors offering streaming data need different formats from those delivering discrete data; for example, a sensor capable of video streaming (e.g., a surveillance camera) could use H.264 [18] coding, a standard for efficient video coding.

With IoT, discrete sensors are often exposed as Web APIs (Web services). Examples of open data formats are JavaScript Object Notation (JSON), typically used with Representational State Transfer (REST) Web services and XML, typically the foundation of Simple Object Access Protocol (SOAP) Web services. JSON is currently widely used to provide smart city data from sensors that present discrete data through a REST API. Furthermore, several specification formats can describe such APIs: The OpenAPI Specifications (OAS) [35], ReSTful API Modeling Language (RAML) [42], API Blueprint [1], Web Application Description Language

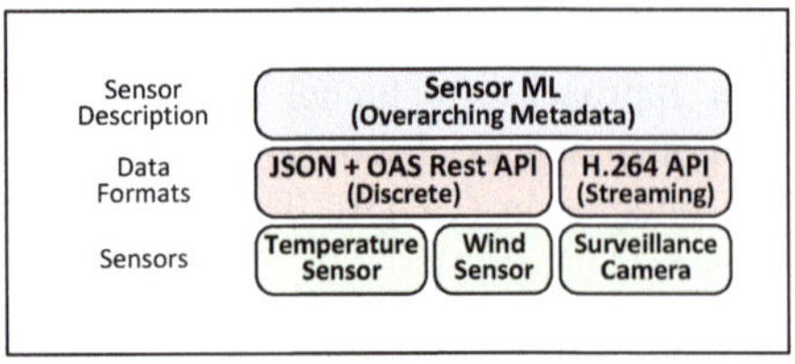

Fig. 4.7 Describing smart city sensor assets

(WADL) [52], Open Data Protocol (OData) [34], and Web Service Description Language (WSDL) [53]. The latter, WSDL, is mainly focused on SOAP Web services, as it was designed as an XML-based standard for describing a Request for Comments (RFC) service API implemented in SOAP. The others may all be used with a REST Web service approach, although OAS is gaining ground these days and is often used for describing REST APIs.

In our opinion, the single most important tool today is the Swagger framework, which implements OAS and provides a metadata API to describe the assets made available through the framework. This combination can be found in many smart cities. For example, the Finnish Transport Agency uses OAS to describe Digitraffic API [13]. This API provides access to sensors deployed throughout Finland. Sensors can be found based on their type (weather stations, traffic cameras, road surface condition sensors, visibility meters, precipitation meters, wind sensors, etc.), geographical location, current status, and other parameters, using the metadata provided.

Figure 4.7 provides an overview of how smart city assets can be described. In this example, both discrete and streaming sensors are available. These sensors must offer data according to a standard format and API. Furthermore, overarching metadata is necessary to allow uniform description and discovery of sensors.

4.6.2 Publish-Subscribe Approaches

While the request/response pattern requires the consumer to actively fetch data from the producer (i.e., the consumer pulls data), the publish/subscribe pattern implies that the consumer expresses interest in a given data type by setting up a *subscription*. The type of data of interest is typically specified using either topic or content filtering and is included in the subscription request. When new data of a given type (e.g., a new temperature reading) become available, the producer sends the new data to all consumers who have expressed interest in that type of data. This pattern is particularly well suited for IoT, as many sensors produce information more or less periodically, and thus a push pattern can considerably reduce network activity. In addition, it is common to use a *broker* between the producer and the consumer to unload administrative tasks from the producer, which is particularly important in IoT, as many sensors are relatively resource-constrained. The broker typically handles subscription management and dissemination so that

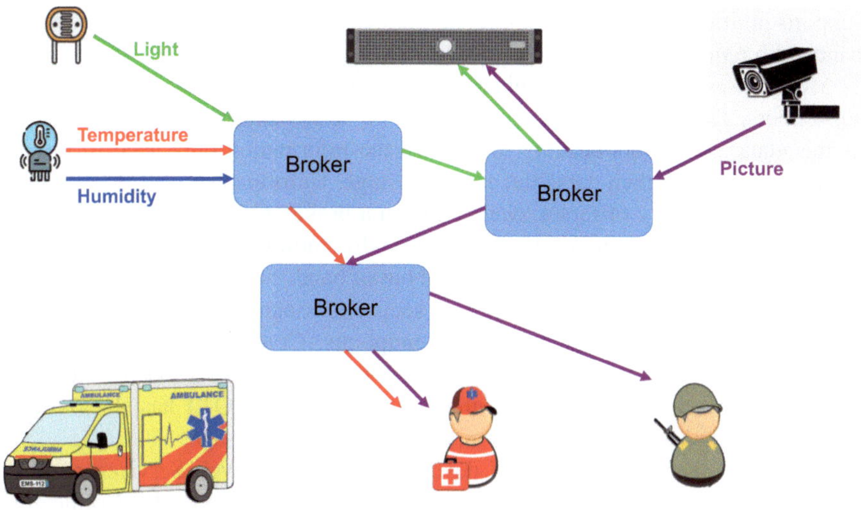

Fig. 4.8 Publish/subscribe information flow. Information is generated by producers, classified by topic, disseminated by brokers, and delivered to consumers

the producer only has to send new data to the broker, which then handles all further dissemination. Figure 4.8 shows a publish/subscribe information flow, where one or more publishers publish data to a broker. The broker uses a set of previously set-up subscriptions to determine which messages should be sent to a given subscriber.

There are several publish/subscribe standards. Three of the most well-known ones are Advanced Message Queuing Protocol (AMQP) [41], Message Queuing Telemetry Transport (MQTT) [30], and Web Services Notification (WS-Notification) [33].

AMQP is a communication protocol developed by JPMorgan Chase in the early 2000s to manage communication between software components in the banking institution's Information Technology (IT) systems. Initially designed for enterprise environments, with version 1.0, the protocol underwent a major "restyling" that made it suitable for adoption in IoT applications.

AMQP was designed as a flexible and (relatively) lightweight solution that implements broker-based message queue communications with different semantics: *direct exchange* and *headers exchange*, implementing a forwarding mechanism driven by message attributes (such as routing key and headers) and aimed at supporting round-robin load balancing tasks, *fanout exchange*, implementing one-to-many message forwarding designed for notifications, and *topic exchange*, implementing publish/subscribe communications—arguably the most interesting semantics for IoT applications.

MQTT is an extremely lightweight publish/subscribe messaging transport protocol designed for constrained environments. It is suitable for applications with a small code footprint, limited network bandwidth, low power usage, minimized data packets, and efficient data distribution to one or many receivers. It also supports multiple Quality of Service (QoS) policies to transport messages in various bandwidth requirement scenarios.

MQTT uses *topics* for disseminating information between data publishers and subscribers. Topics are typically string-based keywords attached to the information as metadata, so brokers needing to forward the information only need to inspect the metadata rather than parse the entire message. Unfortunately, MQTT does not support the discovery of topics available from a broker. Consequently, for a client to be able to make a meaningful subscription, information about the available topics must be either known in advance or shared out of band.

An MQTT messaging broker enables secure message transfers and the establishment of secure network connection endpoints. Confidentiality and integrity protection can be implemented using Transport Layer Security (TLS), thus supporting the security requirements of the military domain. MQTT has already been used in many industrial applications and has a well-proven usage.

WS-Notification is the standard developed within the Web Services protocol family designed to support publish/subscribe communications. WS-Notification heavily leverages XML at both the protocol message format and data encoding levels and is, therefore, more resource intensive (at both the bandwidth and the computation levels) than AMQP and MQTT. Consequently, WS-Notification is less suitable for adoption in IoT applications. Therefore, it would be best to pursue interoperability with systems using WS-Notification by implementing a dedicated bridging gateway when required.

We have evaluated all three standards in the context of several experiments described in Sect. 8.5.2.1. The results show that MQTT seems to represent the most promising publish/subscribe communication solution for NATO purposes.

IoT nodes can be deployed at virtually any location. A major limitation for designers of IoT systems that use pub/sub solutions is the limited capabilities of sensor nodes, especially in terms of computing power, power supply, and transmission range. These limitations can be addressed using communication solutions such as TCP-based MQTT based on TCP [2] and MQTT for Sensor Networks (MQTT-SN, based on UDP) [46] to organize data exchange between IoT nodes. Each of these protocols offers its own specific security solution, but they can all use TLS in different modes.

Using the MQTT and MQTT-SN protocols, it is possible to build an environment in which constrained devices can exchange data with the Internet. Thus, elements of such a network will be able to use efficient telecommunication links in all services utilized to secure data exchange commonly used on the Internet. An exemplary structure of such a network is shown in Fig. 4.9.

Sensor nodes (constrained devices) operate in the MQTT-SN domain, which uses UDP. The MQTT-SN broker plays a role of gateway between the MQTT-SN and MQTT domains. The gateway acts as a publisher/subscriber in the MQTT domain. A constrained device that has permanent access to a power source can play the gateway role. The MQTT domain uses the TCP protocol, and it is possible to use commonly available security mechanisms to generate and renew cryptographic

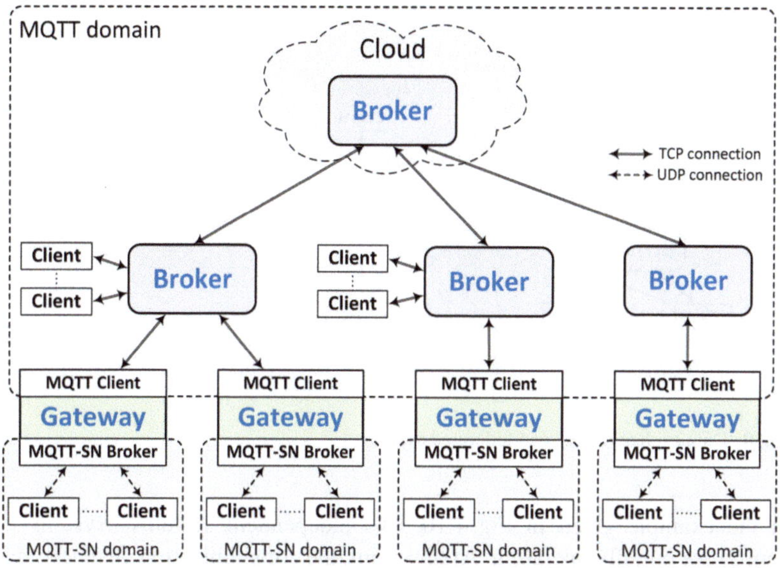

Fig. 4.9 A way of integrating the MQTT and MQTT-SN services

keys, authentication, security management, etc. In particular, it is possible to use the Certification Authority (CA).

The MQTT protocol offers several mechanisms to increase the security level of transmitted data. These include the QoS mechanism, user name and password-protected connections (this security, when used alone, is weak), the security of data exchange using TLS, and the use of certificates.

An interesting solution available in the MQTT protocol is the bridge mechanism. This mechanism allows data exchange between clients (pub/sub) that work directly with different brokers. Once this mechanism is initiated on one broker, communication with another (specific) broker is established using the pub/sub mechanism. That is, one broker becomes a publisher of the messages published by its client nodes and a subscriber of the messages published by clients of the other broker. In each broker's configuration files, one can define how the message topics the clients of the other broker publish are received. In addition, it is also possible to configure access restrictions to data transferred from the other broker. There is no restriction to configuring the bridge mechanism concurrently between different pairs of brokers. This mechanism makes it possible to build a federated data exchange between various systems that use the MQTT service in their resources. Figure 4.10 illustrates an example of the structure of such a federation system. Both the topic mapping and the message filtering mechanisms available in the MQTT service make it possible to build a system that can restrict access to individual messages and control data exchange. This solution also includes cryptographic data exchange protection mechanisms.

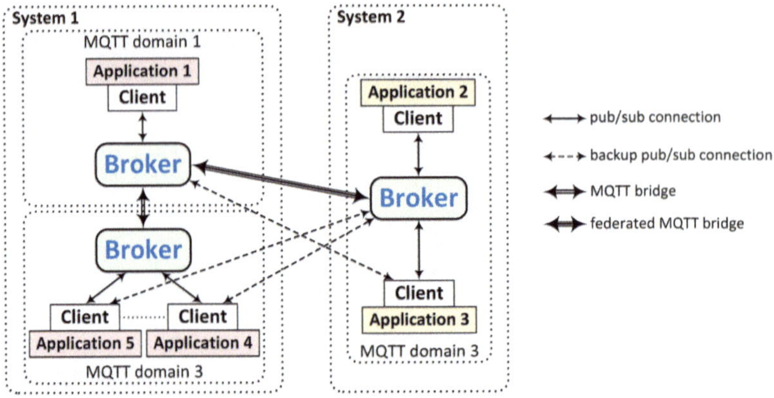

Fig. 4.10 Data exchange between federated systems

In the example given in Fig. 4.10, two independent systems (System 1 and System 2) are initially operating. At this point, the communication channels marked in the figure as "backup pub/sub connection" are unavailable for the clients of each system. In System 1, one MQTT bridge link is already used. After setting up the communication channel (marked in the figure as "federated MQTT bridge"), the brokers of both systems have the opportunity to inform their selected (i.e., not necessarily all) clients about the possibility of subscribing to topics published in the newly connected system and building backup links to the broker from outside their system. A positive effect of this approach will be to increase the reliability of the entire federated system. In the event of failure of one of the federalized brokers, data exchange between systems will still be possible.

4.7 Conclusions

This chapter provided background information on key building blocks that will likely be part of military applications and solutions for IoT. In particular, the focus was on communications technologies and on IoT-specific technologies such as LoRa, which could be beneficial to military applications. In particular, LoRa offers an interesting combination of low-cost, low-power, and long-range transmission capabilities that could be very useful for certain types of military IoT devices. While LoRa is low bandwidth, it is still well suited for many military sensing applications that just require either periodic reporting (e.g., fuel levels) or detections (e.g., radiation, movement, intrusion, etc.) These types of messages often require very little bandwidth and are well suited for LoRa (or technologies similar to LoRa). LoRa is also adequate for position reporting, for example, for blue forces or other assets. In fact, unless the sensor is using some specific modalities that are bandwidth intensive (e.g., audio, video, and RF), the capacity offered by LoRa is sufficient and, when combined with the other alternatives, makes it a promising technology.

This chapter also provided an overview of other communications modalities, including SATCOM. From the perspective of SATCOM, some of the more interesting technologies are Low Earth Orbiting systems, such as Starlink and OneWeb.

On the terrestrial side, 4G technologies including NB-IoT and LTE-M are promising, although the group did not have a chance to experiment with these specific capabilities. 5G, on the other hand, also provides interesting capabilities, especially from a high-bandwidth and low-latency perspective, and would be good for many IoT applications including autonomous vehicles and teleoperation. However, all of the cellular technologies require infrastructure, which could be provided by commercial operators or by custom military-owned network base stations.

From an underwater perspective, an interesting development has been the creation of the JANUS STANAG for underwater communications. While bandwidth and range remain a challenge, standardization is important and will encourage interoperability.

Unfortunately, standardization is one thing that does not exist when it comes to commercial deployment of IoT devices and services, for example, in Smart City environments. Commercial interests so far have led to different closed systems that do not implement any specific standard, which is a challenge both from the perspective of discovering these IoT devices and services and also to be able to access and utilize the data from them.

From a military perspective, COTS IoT sensors, actuators, and communications technologies can address a variety of operational needs at a significantly lower cost than military-developed equivalents [40]. By assessing particular mission needs, militaries can selectively piggyback on COTS-based solutions and integrate them into their infrastructures. Assessing the usage of COTS IoT technologies in broader military infrastructures represents a key research objective in the emergent IoT-based IoBT paradigm [44].

References

1. API Blueprint: API Blueprint Specification. https://apiblueprint.org/documentation/specification.html
2. Banks, A., Briggs, E., Borgendale, K., Gupta, R.: OASIS MQTT Version 5.0. Tech. rep., Organization for the Advancement of Structured Information Standards (2019)
3. Bermudez-Edo, M., Elsaleh, T., Barnaghi, P., Taylor, K.: IoT-Lite: A lightweight semantic model for the Internet of Things. In: 2016 Intl IEEE Conferences on Ubiquitous Intelligence Computing, Advanced and Trusted Computing, Scalable Computing and Communications, Cloud and Big Data Computing, Internet of People, and Smart World Congress (UIC/ATC/ScalCom/CBDCom/IoP/SmartWorld), pp. 90–97 (2016). https://doi.org/10.1109/UIC-ATC-ScalCom-CBDCom-IoP-SmartWorld.2016.0035
4. Blom, J.D.: Unmanned Aerial Systems: A Historical Perspective, vol. 45. Combat Studies Institute Press, Washington (2010)

5. Bor, M., Roedig, U.: LoRa transmission parameter selection. In: 2017 13th International Conference on Distributed Computing in Sensor Systems (DCOSS), pp. 27–34. IEEE, Piscataway (2017)
6. Burns, P.: LoRaWAN vs. Haystack (2016). http://haystacktechnologies.com/lorawan-vs-haystack/
7. Chettri, L., Bera, R.: A comprehensive survey on Internet of Things (IoT) toward 5G wireless systems. IEEE Internet Things J. **7**(1), 16–32 (2020). https://doi.org/10.1109/JIOT.2019.2948888
8. Cybulski, P., Zieliński, Z.: UAV swarms behavior modeling using tracking bigraphical reactive systems. Sensors **21**(2) (2021). https://doi.org/10.3390/s21020622. https://www.mdpi.com/1424-8220/21/2/622
9. Dauphin, L., Adjih, C., Petersen, H., Baccelli, E.: Low-cost robots in the Internet of Things: hardware, software & communication aspects. In: Proceedings of the 2017 International Conference on Embedded Wireless Systems and Networks, EWSN '17, pp. 284–289. Junction Publishing, New York (2017)
10. Design & Reuse: Cycleo unveils its first innovative semiconductor IP bringing unprecedented range to wireless data transmission (2009). https://www.design-reuse.com/news/21552/wireless-semiconductor-ip.html
11. Elsaleh, T., Enshaeifar, S., Rezvani, R., Acton, S.T., Janeiko, V., Bermudez-Edo, M.: IoT-stream: a lightweight ontology for Internet of Things data streams and its use with data analytics and event detection services. Sensors **20**(4), 953 (2020). https://doi.org/10.3390/s20040953
12. Evans, J.: Satellite systems for personal communications. Proc. IEEE **86**(7), 1325–1341 (1998). https://doi.org/10.1109/5.681367
13. Finnish Transport Agency: Digitraffic Road metadata API. https://tie.digitraffic.fi/api/v1/metadata/documentation
14. Haller, A., Janowicz, K., Cox, S.J., Lefrançois, M., Taylor, K., Le Phuoc, D., Lieberman, J., García-Castro, R., Atkinson, R., Stadler, C.: The modular SSN ontology: a joint W3C and OGC standard specifying the semantics of sensors, observations, sampling, and actuation. Semant. Web **10**(1), 9–32 (2018)
15. Hamilton, A., Barnett, J., Hobbs, A.M., Pelekanakis, K., Petroccia, R., Nissen, I., Galsdorf, D.: Towards secure and interoperable underwater acoustic communications: current activities in NATO IST-174 research task group. Prog. Comput. Sci. **205**, 167–178 (2022). https://doi.org/10.1016/j.procs.2022.09.018. https://www.sciencedirect.com/science/article/pii/S1877050922008833. 2022 International Conference on Military Communication and Information Systems (ICMCIS)
16. HopeRF: RFM95/96/97/98(W) - Low Power Long Range Transceiver Module V1.0 (2013). http://www.hoperf.com/upload/rf/RFM95_96_97_98W.pdf
17. Howell, A.: The future of next generation satellite broadband communications services: an operator's view. In: IET Seminar on Beyond Ka-Band: Meeting the Communication Bandwidth Requirements of the Future, pp. 1–10 (2011). https://doi.org/10.1049/ic.2011.0223
18. ITU-T: Advanced video coding for generic audiovisual services. ITU-T Recommendation H.264 (2003)
19. Johnsen, F.T., Bloebaum, T.H., Brannsten, M.R., Lund, K., Aubert, D.E.: Using open standards for utilizing IoT sensors in a smart city scenario. In: International Command and Control Research and Technology Symposium (ICCRTS) (2018)
20. Knight, M.: Reversing LoRa, Exploring Next-Generation Wireless. https://github.com/matt-knight/research/blob/master/2016_05_20_jailbreak/Reversing-Lora-Knight.pdf (2016). Accessed 22 Nov 2018
21. Lee, J., Bagheri, B., Kao, H.A.: A cyber-physical systems architecture for Industry 4.0-based manufacturing systems. Manuf. Lett. **3**, 18–23 (2015)
22. Leighty, R.D.: DARPA Autonomous Land Vehicle (ALV) summary. Tech. rep., Army Engineer Topographic Labs, Fort Belvoir, VA (1986)

23. LinkLabs: Symphony link™, a revolutionary wireless system for wide-area IoT networks. https://www.link-labs.com/symphony
24. LoRa Alliance: LoRa Alliance. https://lora-alliance.org/. Accessed 22 Nov 2018
25. LoRa Alliance Technical Committee: LoRaWAN 1.1 specification. Tech. rep., LoRa Alliance, 3855 SW 153rd Drive, Beaverton, OR (2017). Accessed 06 Mar 2023
26. LoRa Alliance Technical Marketing Workgroup 1.0: LoRaWAN - what is it? a technical overview of LoRa and LoRaWAN. Tech. rep., LoRa Alliance, 2400 Camino Ramon, San Ramon, CA (2015). Accessed 06 Mar 2023
27. McCauley, M.: RadioHead Packet Radio library for embedded microprocessors. http://www.airspayce.com/mikem/arduino/RadioHead/. Accessed 22 Nov 2018
28. Meyer, S., Ruppen, A., Magerkurth, C.: Internet of things-aware process modeling: integrating IoT devices as business process resources. In: International Conference on Advanced Information Systems Engineering, pp. 84–98. Springer, Berlin (2013)
29. Michaelis, J., Morelli, A., Raglin, A., James, D., Suri, N.: Leveraging LoRaWAN to support IoBT in urban environments. In: 2019 IEEE 5th World Forum on Internet of Things (WF-IoT), pp. 207–212. IEEE, Piscataway (2019)
30. MQTT.org: MQTT.org Website. https://mqtt.org/
31. Murphy, R.R.: International cooperation in deploying robots for disasters: lessons for the future from the Great East Japan Earthquake. J. Robot. Soc. Jpn. **32**(2), 104–109 (2014)
32. Noschese, P., Porfili, S., Di Girolamo, S.: ADS-B via Iridium NEXT satellites. In: 2011 Tyrrhenian International Workshop on Digital Communications - Enhanced Surveillance of Aircraft and Vehicles, pp. 213–218 (2011)
33. OASIS: WSN specifications. https://www.oasis-open.org/committees/wsn/
34. OData: OData – the best way to REST. http://www.odata.org/
35. OpenAPI Initiative: Homepage. https://www.openapis.org/
36. Open Geospatial Consortium (OGC): Sensor model language (SensorML). http://www.opengeospatial.org/standards/sensorml
37. Pellegrini, V., Principe, F., Guidi, R., Demauro, G., V.Martorelli, Cioni, R.: Cryptographically secure radios based on directional modulation. In: 2018 IEEE - ICASSP 2014, pp. 1–8 (2014)
38. Pfisterer, D., Romer, K., Bimschas, D., Kleine, O., Mietz, R., Truong, C., Hasemann, H., Kröller, A., Pagel, M., Hauswirth, M., Karnstedt, M., Leggieri, M., Passant, A., Richardson, R.: SPITFIRE: toward a semantic web of things. IEEE Commun. Mag. **49**(11), 40–48 (2011). https://doi.org/10.1109/mcom.2011.6069708
39. Pham, C.: A DIY low-cost LoRa gateway. http://cpham.perso.univ-pau.fr/LORA/RPIgateway.html. Accessed 22 Nov 2018
40. Pradhan, M., Gökgöz, F., Bau, N., Ota, D.: Approach towards application of commercial off-the-shelf Internet of Things devices in the military domain. In: Internet of Things (WF-IoT), 2016 IEEE 3rd World Forum on, pp. 245–250. IEEE, Piscataway (2016)
41. RabbitMQ: AMQP 0.9.1 Protocol Specification. https://www.rabbitmq.com/
42. RAML Workgroup: RAML – The simplest way to design APIs. https://raml.org/
43. RevSpace: DecodingLora. https://revspace.nl/DecodingLora
44. Russell, S., Abdelzaher, T., Suri, N.: Multi-domain effects and the Internet of Battlefield Things. In: MILCOM 2019-2019 IEEE Military Communications Conference (MILCOM), pp. 724–730. IEEE, Piscataway (2019)
45. Seydoux, N., Drira, K., Hernandez, N., Monteil, T.: IoT-O, a core-domain IoT ontology to represent connected devices networks. In: Blomqvist, E., Ciancarini, P., Poggi, F., Vitali, F. (eds.) Knowledge Engineering and Knowledge Management, pp. 561–576. Springer International Publishing, Cham (2016)
46. Stanford-Clark, A., Truong, H.L.: MQTT For Sensor Networks (MQTT-SN). Protocol Specification v.1.2. Tech. rep., International Business Machines Corporation (2013). https://www.oasis-open.org/
47. Tayur, V.M., Suchithra, R.: A Comprehensive ontology for Internet of Things (CoIoT). In: 2019 Second International Conference on Advanced Computational and Communication Paradigms (ICACCP), pp. 1–6 (2019). https://doi.org/10.1109/ICACCP.2019.8882936

48. The Things Network: Building a global Internet of Things network together. https://www.thethingsnetwork.org/. Accessed 23 Nov 2018
49. The Things Network: Ground breaking world record! LoRaWAN packet received at 702 km (436 miles) distance (2017). https://www.thethingsnetwork.org/. Accessed 23 Nov 2018
50. TIDE Community: TIDE specifications on mDNS. https://tide.act.nato.int/tidepedia/index.php?search=mdns
51. Vaezi, M., Azari, A., Khosravirad, S.R., Shirvanimoghaddam, M., Azari, M.M., Chasaki, D., Popovski, P.: Cellular, wide-area, and non-terrestrial IoT: a survey on 5G advances and the road toward 6G. IEEE Commun. Surv. Tutor. **24**(2), 1117–1174 (2022). https://doi.org/10.1109/COMST.2022.3151028
52. W3C: Web Application Description Language. https://www.w3.org/Submission/wadl/
53. W3C: Web Services Description Language (WSDL) Version 2.0 Part 1: Core Language. https://www.w3.org/TR/wsdl/
54. Waziup: Open innovation platform for IoT and big-data in sub-Saharan Africa. https://www.waziup.eu/. Accessed 23 Nov 2018
55. Weissberger, A.: Iridium introduces its NexGen satellite IoT data service (2022). https://techblog.comsoc.org/2022/12/22/iridium-introduces-its-nexgen-satellite-iot-data-service/
56. Wikipedia: Chirp. https://en.wikipedia.org/wiki/Chirp
57. Wikipedia: Serial peripheral interface. https://en.wikipedia.org/wiki/Serial_Peripheral_Interface
58. Wynn, R.B., Huvenne, V.A., Le Bas, T.P., Murton, B.J., Connelly, D.P., Bett, B.J., Ruhl, H.A., Morris, K.J., Peakall, J., Parsons, D.R., Sumner, E.J., Darby, S.E., Dorrell, R.M., Hunt, J.E.: Autonomous underwater vehicles (AUVs): their past, present and future contributions to the advancement of marine geoscience. Mar. Geol. **352**, 451–468 (2014)
59. Ziyang Zhen Dongjing Xing, C.G.: Cooperative search-attack mission planning for multi-UAV based on intelligent self-organized algorithm. Aerosp. Sci. Technol. **76**, 402–411 (2018). https://doi.org/10.1016/j.ast.2018.01.035. https://www.sciencedirect.com/science/article/pii/S1270963817301736
60. Zhuang, Q., Zheng, C.: Research on INMARSAT based on Ka band and applications. In: 2017 4th International Conference on Information, Cybernetics and Computational Social Systems (ICCSS), pp. 127–129 (2017). https://doi.org/10.1109/ICCSS.2017.8091398
61. Zmysłowski, D., Skokowski, P., Malon, Maślanka, Kelner, J.M.: Naval use cases of 5G technology. TransNav Int. J. Marine Navigation Safety Sea Transport. **17**(3), 595–603 (2023)
62. Zwanzig, C.: Interoperable and standardized underwater communication. In: Proceedings of the Undersea Defence Technology (UDT) (2020)

Chapter 5
Middleware Prototypes

Christoph Fuchs, Manas Pradhan, Niranjan Suri, Mauro Tortonesi, Frank T. Johnsen, and Janusz Furtak

5.1 Introduction

Chapter 3 presented an overview of some common architecture, such as oneM2M, ETSI, and IEEE P2413. Chapter 3 also discussed some specific challenges that must be addressed when deploying IoT in the military domain. This chapter discusses some prototype middleware implementations that target the military domain.

Realizing an ecosystem of smart city applications, such as those described in Chap. 2, presents many specific challenges. The first challenge is to implement *asset and service discovery*. In addition, there has to be an *interoperability* layer between the civilian and military systems. Information processing and dissemination platforms also need to be *federated* to allow secure information sharing between the IT systems used by police, civil protection, and military forces. Last but not least, the challenge of *taming the formidable deluge of data* generated by smart city services

C. Fuchs
Fraunhofer FKIE, Wachtberg, Germany

M. Pradhan
DKSR, Berlin, Germany

N. Suri (✉)
US Army DEVCOM Army Research Laboratory (ARL), Adelphi, MD, USA

Florida Institute for Human and Machine Cognition (IHMC), Pensacola, FL, USA
e-mail: nsuri@ihmc.org

M. Tortonesi
University of Ferrara, Ferrara, Italy

F. T. Johnsen
Norwegian Defence Research Establishment (FFI), Kjeller, Norway

J. Furtak
Military University of Technology, Warsaw, Poland

N. Suri et al. (eds.), *Military Applications of Internet of Things*, Internet of Things,
https://doi.org/10.1007/978-3-032-03378-9_5

and IoT assets is to provide actionable knowledge through accurate and low-latency analytics.

Although some of these research topics have received considerable attention in the scientific literature and industrial/commercial efforts, it should be noted that military applications introduce a significant number of additional challenges [25]. This is especially evident if we consider Humanitarian Assistance and Disaster Relief (HADR) operations, which require coordinated actions of different actors to alleviate and prevent human casualties and damage to assets in areas afflicted by large-scale disasters. The disaster recovery and mitigation process generally begins with the participation of state and nongovernmental local authorities such as police, fire and Emergency Medical Services (EMS) organizations, the Red Cross, amateur radio operators, and other volunteers. However, when disasters overwhelm local authorities, specific branches of federal agencies such as the coast guard and national reserves can participate in the relief operation and assist civilian forces by bringing assets and expertise generally available only to armed forces. Military units are well prepared to work in challenging environments and can provide adequate Command and Control (C2) and logistics assets to support relief activities in affected territories.

In Civil–Military Cooperation (CIMIC) operations, conditions that affect the city network during and after a disaster can limit access to a smart city sensing network. For example, large-scale disasters can cause link failure, network congestion, and variable latency, resulting in fragmentation of the network into a set of quasi- or completely disjointed network segments with limited capabilities [8]. The composition of these segments cannot be predicted before a disaster, which requires the ability to discover and exploit the available assets *on site and on demand.* Moreover, the sensing network environment after a disaster is extremely dynamic. Organizations can deploy new resources during an operation, and military forces can define militarized subnetworks within the public network, adding additional resources.

It should be noted that rescuers, who are usually not tied to the infrastructure of the city network, operate with limited knowledge of the IoT assets they can exploit to acquire situational awareness. This raises the issue of interoperability in data representation formats and communication protocols and (more fundamentally) with respect to asset discovery and access to available data sources.

Effective information sharing between units and other nodes in a Tactical Network (TN) is essential for the success of missions. However, the complexity of information sharing is higher in a multi-domain context, where different organizations must share data securely with selected parties over common links, relay nodes, and other network resources, without risking information leaks to unauthorized entities. Furthermore, unsupervised information sharing between locations and agencies can result in inefficient use of scarce network resources, which can congest communication links and negatively affect the situational awareness acquisition process. Additionally, balancing network resources to prioritize emergency traffic over regular information exchange requires network monitoring and information management.

The coexistence and co-deployment of military systems, civilian infrastructure, and commercial IoT solutions further exacerbate the issues of federation, interoperability, and trustworthiness [28]. The required integration, which is critical in HADR scenarios, introduces significant technical and organizational security and compatibility challenges.

Taming the IoT data deluge to build Situational Awareness (SA) is another challenge that requires a rethinking of information processing methodologies and tools. Traditional analytic solutions based on transferring all data to the cloud, processing them using big data methodologies and tools, and returning the results to interested users are too slow for applications with strict latency constraints and too burdensome for the network infrastructure.

Due to network fragmentation and the multi-partnered nature of the operations, centralized and cloud-based IoT platforms may not effectively respond to rescuers [23]. The distributed processing approach adopted by fog computing is particularly well suited for taming the deluge of information provided by IoT and crowdsourcing [4]. However, implementing big data applications in fog computing environments presents several problems. The most important one is arguably resource scarcity within the environment. Developers of fog services cannot assume to have enough resources to analyze all the incoming data using the full-fledged/sophisticated/fine-grained analytic techniques developed for cloud environments. Therefore, there has to be a trade-off: Either services discard some data, or they remodulate big data analytics so that when they run in the fog, they decrease their computational requirements, perhaps switching to coarser-grained but less computationally demanding algorithms. In turn, this requires novel computational and resource management models.

At the same time, some HADR scenarios might have a relatively low amount of computational resources at the fog level. In these cases, it is necessary to adapt applications to the available resource pool. Fortunately, toward that goal, two key factors can significantly speed up the deployment of specific IT infrastructure and services to support HADR operations [7]. First, the ubiquitous presence of Kubernetes as de facto orchestrator—both at the cloud and at the fog level—means that responders could leverage a standard orchestration API to quickly deploy dedicated IT assets on a wide range of platforms. Second, the increasing adoption of Infrastructure-as-Code (IaC) practices within NATO and military forces operated by the NATO nations allows the development and exploitation of customized and predefined deployment playbooks to quickly deploy dedicated IT assets on federated and coalition-based platforms. However, the efficient deployment of applications within a wide range of situations from the resource availability perspective—and in conditions that are fully reactive and do not allow any planning in advance—requires the adoption of adaptive architectures and programming models at the application level.

Finally, processing and dissemination have to consider information assurance aspects. Leveraging information from commercially deployed IoT systems in smart cities and other unknown and/or uncontrolled IT infrastructures raises potential information distortion issues in the gathered data. While that information might be very valuable from an SA perspective, it should not be treated as entirely reliable, and decision-makers should be explicitly informed about its possibly untrustworthy nature. Moreover, including new interfaces for data acquisition could increase the attack surface of the IT systems responsible for situational awareness. Therefore, appropriate security measures, such as data sanitization and content inspection, need to be implemented to prevent malicious use of these new interfaces by an adversary.

To address these challenges, purposely developed smart city middleware is needed to simplify the integration of external military authorities by rapidly providing access to city resources [30]. More specifically, middleware should provide properly designed proactive discovery procedures that can autonomously locate and register surviving segments of the IoT infrastructure and provide that information (and means of access) to rescuers. Finally, middleware should provide new service and resource management models suited for rapid SA acquisition in dynamic and resource-scarce environments such as HADR.

In the remaining sections of this chapter, we briefly introduce some of the middleware prototypes developed for the military domain and also offering interesting potential for civilian applications, e.g., in smart city environments.

5.2 Multi-domain AsynchRonous Gateway Of Things (MARGOT)

Multi-domain AsynchRonous Gateway Of Things (MARGOT) is a platform designed to enable the discovery of IoT resources in highly dynamic and disrupted networking environments, such as those commonly found in HADR operations [5] [6]. Leveraging a distributed and federated architecture and concurrently running multiple discovery mechanisms, MARGOT permits the identification of available IoT assets across separate domains and network segments. It also supports context-aware applications by providing them with a query interface that accepts parameters to refine the search criteria.

Figure 5.1 illustrates the MARGOT architecture, showing the main components: Federation Services, Discovery Agents (DAs), Information Processor, and the REST API and the interactions between them.

A MARGOT instance is responsible for the discovery of assets within a domain. The knowledge of all discovered IoT resources is then distributed across all MARGOT instances in the system, which exchange information via Federation Services. Federation is a key aspect of MARGOT and is covered in detail in the following subsection.

DAs are pluggable components that implement a single resource discovery solution and store information about the discovered assets in the local MARGOT database. DA implementations rely on specific protocols commonly used for IoT applications, such as CoAP and MQTT, and the corresponding discovery procedures to detect and identify IoT assets within the local network-or *domain* in MARGOT

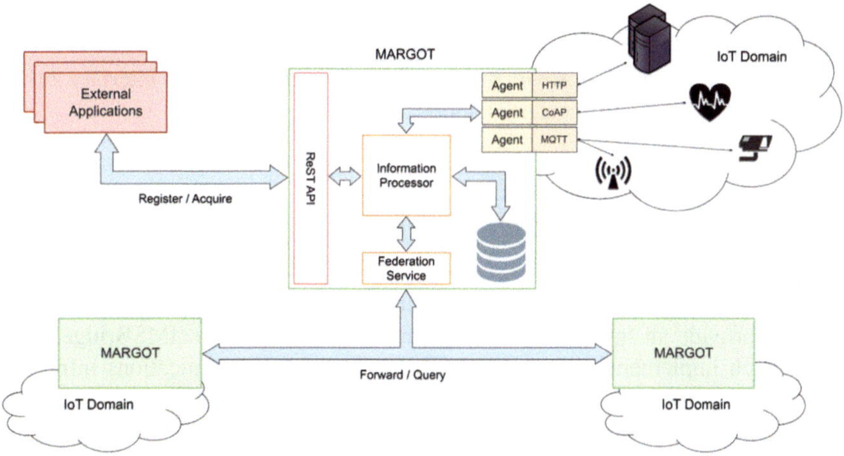

Fig. 5.1 The MARGOT architecture. Based on [5]

terminology. For example, a CoAP DA periodically broadcasts a request to the CoAP predefined URI `/.well-known/core` to obtain information from devices equipped with an active CoAP server. Alternatively, DAs can integrate with official (and centralized) smart city IoT resource lookup services, such as the 511 Traveler Information System (511NY) (https://511ny.org) and Airly (https://airly.eu), to browse the list of assets advertised by these services.

DAs can perform the discovery process in either of two ways: proactively, when the process is executed periodically or the implemented protocol supports some form of proactive discovery, or reactively, when specific events trigger discovery, such as a request issued by a client or coming from a federated MARGOT instance. Generally speaking, proactive behavior trades query latency for freshness of the information. Depending on the rate of user requests and the cost of the discovery process, proactive discovery could lead to a decrease or an increase in bandwidth consumption. To address this problem and provide more efficient proactive discovery strategies, DAs should allow MARGOT to tune parameters that control the discovery process, for example, the frequency of the process. Protocols that naturally support proactive discovery, such as CoAP and MQTT, can increase the efficiency of detecting new assets. However, they still require some form of periodic probing to ensure that previously discovered *Things* are still alive and reachable.

IP takes care of processing the data received from DAs, normalizing them, and storing them in the local database. During this process, IP also collects statistics that characterize domain and user requests, including the variability of the discovered IoT assets and frequently requested resources, and merges them with the statistics received from federated MARGOT instances. Lastly, IP manages the data exchange via Federation Services: It decides which IoT resources to publish in the MARGOT federation, forwards user queries if necessary, responds to queries received from remote MARGOT instances, and shares updated domain statistics.

Finally, the ReST API provides a coherent and comprehensive interface for the discovery of IoT assets for external applications.

5.2.1 Federating MARGOT Instances

MARGOT instances rely on Federation Services [13] to exchange data. Federation Services provide an Information Management Systems Bridge (IMSBridge) interface, which implements distributed publish/subscribe communications infrastructure that supports and simplifies information exchange in multi-domain scenarios. *Topics* control the routing of data over the Federation Services network. Topics are named abstractions (i.e., identified by unique strings) that can be thought of as independent communication channels over the network. Some topics are typically predefined via configuration files, but federates can also create new ones dynamically at run-time. New messages published within a topic are delivered to all federates subscribed to that topic, who will receive the message directly from the publisher or another IMSBridge. Clients can join or leave these channels at any time.

Each IMSBridge instance can specify policies that determine what information it can share with all other instances. This enables one domain's administrators to define local rules that implement that domain's security and data-sharing policies without coordinating with other domains' administrators. Sharing policies can be enforced using a security system based on Attribute-Based Encryption (ABE), which allows the combination of different security layers per message [19] and ensures that only clients possessing the correct keys can access the messages. Federates can use Attribute-Based Encryption (ABE) to encrypt the payload and metadata of messages before publication, but information related to the topic of publication will not be encrypted to guarantee correct routing.

Federation Services support other useful features for distributed multi-domain environments. *Distributed queries* allow federates to send and run queries on all or a subset of IMSBridge instances to retrieve published messages that match the client's criteria. Distributed queries take advantage of the metadata information specified during publication to select relevant messages. *Smart synchronization* enables two IMSBridge instances to exchange updated messages after a disconnection period, for example, caused by network problems or other issues. Depending on the nature of the messages, all or only the most recent updates will be exchanged.

MARGOT leverages the capabilities of Federation Services for all distributed actions, including the discovery of new instances, proactive querying and IoT data replication on federated instances, remote query execution, and the exchange of control information and usage and domain statistics. This enables users and clients to discover resources available across multiple IoT domains and allows the platform to adapt its behavior, for example, concerning caching and query forwarding, under

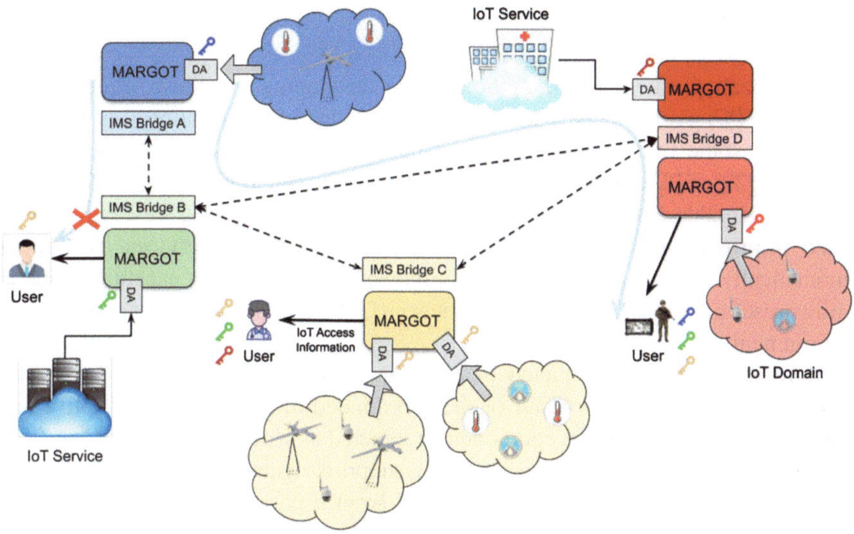

Fig. 5.2 Federated MARGOT instances. Based on [6]

certain circumstances. Finally, MARGOT leverages Federation Services' policy-based data sharing and ABE security to control access to IoT assets information.

Figure 5.2 shows a distributed MARGOT deployment connected via Federation Services. Each MARGOT in the figure is connected to a single IMSBridge that enables it to exchange data with other MARGOT instances through the Federation network; dashed arrows represent connections between IMSBridge instances that compose the Federation network. Users can connect to any MARGOT instance to obtain access information about IoT assets discovered by any federated MARGOT, according to the data sharing and security policies implemented by the involved IMSBridge instances. For example, on the right side of Fig. 5.2, a user has permission to access IoT asset information for three different domains (represented by the blue, green, and yellow keys next to him). The user can connect to his or her local MARGOT instance (light red box in the figure) to retrieve information about sensors in the blue domain. MARGOT translates the request received from the user into a Federation query, which disseminates from *IMSBridge D* across the Federation network and is solved locally by each federate. Finally, the answers are relayed back to the node from which the query originated. MARGOT then caches the information locally for future requests and generates a response for the user with the requested information. As a second example, another user has permission to access information about Things located in the yellow domain but does not have permission to access IoT information from the blue domain. Consequently, if this user issues a request for assets in the blue domain, he will not be able to decrypt the data received from MARGOT.

5.3 SPF Platform

The approach pursued by the authors while developing the Sieve, Process, and Forward (SPF) platform takes advantage of the Value of Information (VoI) concept to address some of the problems faced by IoT applications in smart city environments [29]. SPF adopts an innovative approach which advocates the adoption of an "acceptable lossiness" perspective to realize fog services. Leveraging VoI-based prioritization, SPF enables the development of fog services capable of automatically scaling their resource requirements to their current execution context while preserving high Quality of Experience (QoE) levels, even in resource-scarce environments.

More specifically, as depicted in Fig. 5.3, the processing function of an SPF fog service emerges as the result of the coordinated orchestration of adaptive and composition-friendly service components (depicted as violet rectangles in Fig. 2.2). This loose definition of fog services easily supports dynamic architectures, shown in Fig. 5.4, in which the single instances of service components can be migrated to different devices along the Cloud-IoT continuum according to the current execution context (service requirements, resource availability, user preferences, etc.). Developers will define and implement SPF fog services as a topology of service components, connected in a dynamic service fabric according to a service description that defines the service semantics and characteristics and the interactions between service components.

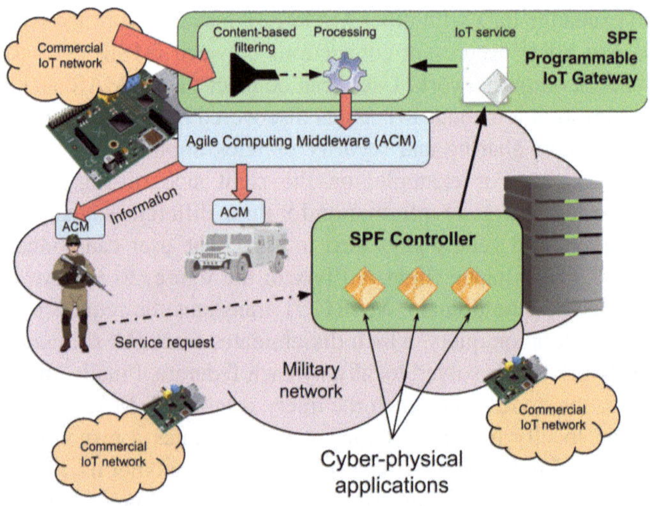

Fig. 5.3 Adoption of SPF in a mixed civilian/military scenario

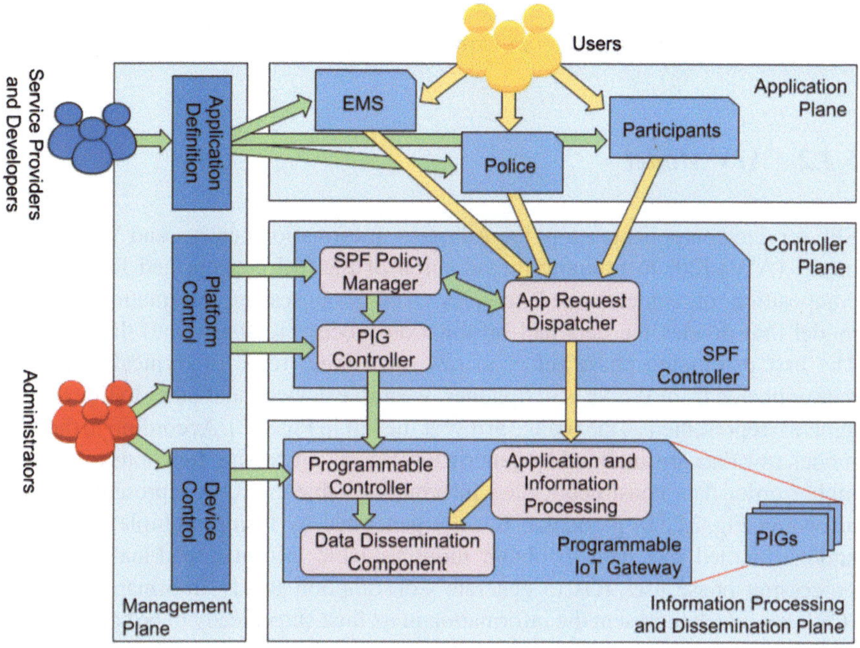

Fig. 5.4 The SPF architecture

5.3.1 Value of Information

The VoI concept measures the utility of information from a subjective and consumer-centric perspective. This concept can be traced back to Howard's seminal work that attempted to extend Shannon's information theory to consider both *the probabilistic nature of the uncertainties surrounding us and the economic impact that these uncertainties will have on us* [11]. The investigation of the utility that each discrete element of information provides to its consumer(s) has been an active research topic in economic and decision-making theories for the past 50 years. It continues to receive considerable attention [24] and holds interesting promises for several IoT application scenarios discussed in Chap. 2.

Classifying information according to the value it provides to its recipients represents an effective criterion for "pruning" the share of data whose processing is not allowed by the limited availability of fog resources. For example, sensing data that do not add significant value to the knowledge already built from the analysis of previous sensor readings would be characterized by a low and rapidly decreasing VoI. Thus, those data would have a very low probability of being selected for processing in (the very likely) case of resource shortages. Therefore, the classification of services and service components, according to the total VoI (T_{VoI})

they provide to end users, represents a natural and effective approach to implement self-adaptive services for fog computing applications.

5.3.2 AIV Model

SPF service components adopt the Adaptive, Information-centric, and Value-based model (AIV) [20] to promote the adoption of the VoI concept and facilitate the composition of content-based services. AIV proposes an information maturity model that divides the data and information processing stages into three phases. The first processing phase refers to raw data: input feeds of (typically sensing) data collected from WSNs, smartphones, wearable devices, and other IoT devices in general (represented as the orange arrow at the left in Fig. 2.2). According to the AIV model, raw data are analyzed by generic lower level processing functions to produce higher order data constructs called Information Objects (IOs), represented as red arrows in Fig. 2.2. In particular, IOs are generated by fusing multiple aggregated and/or distilled raw data to obtain more valuable information. Finally, the last processing phase uses IOs to generate Consumption Ready Information Objects (CRIOs), which represent the information in its final stage, ready to be consumed by users who requested them. This final phase is depicted as yellow arrows in Fig. 2.2. (Note that the generation of an IO could require many raw data objects, and a CRIO can result from the aggregation of multiple IOs provided by many different services.)

On top of the AIV model, each component of the service will calculate the total amount of VoI. Specifically, the VoI associated with each IO is calculated as a function of five factors: priority of service, (estimated) number of recipients of IO, decline in timeliness relevance (a factor taking into account the decrease in value for time-sensitive IOs by applying an estimated delivery time penalty), proximity relevance decay (a factor taking into account the decrease in value for location-aware IOs by applying an estimated delivery location penalty), and a service-specific VoI calculation function, which every service implementation is supposed to provide.

Of course, an IO has many potential recipients, and therefore, from the theoretical perspective, each user-specific copy of an IO should be assigned different VoI metadata. However, modeling each user and his set of interests in the system would be extremely difficult. To constrain complexity, many VoI solutions assume that, in practice, users can be grouped into homogeneous groups, e.g., according to their interest or role. As a result, VoI models can be defined for a given *(information type, user role)* tuple [7]. For example, IOs that contain video frames from traffic cameras will have different VoI models according to the role of consumers. At the same time, VoI is context-specific, and it is conceivable that IOs could have a highly fluctuating VoI during their lifetime, especially for longer-lived IOs. So far, researchers have focused mainly on quasi-static or short-term scenarios and proposed VoI estimation solutions leveraging machine learning-based classifiers [24] or Analytic Hierarchy

Process [9]. It would be desirable to further develop these solutions to specifically consider long-term scenarios.

Using T_{VoI} as a resource assignment criterion enables the SPF resource management function to naturally and seamlessly prioritize resource allocation to services that provide the highest value to their end users, either because they serve a considerable number of users or because they provide highly valuable IOs.

5.4 Use of COTS Computing Platforms

Figure 5.5 provides a high-level overview of some of the technical solutions proposed by IST-147. In particular, the group has evaluated several low-cost sensor platforms (e.g., Arduino [3], Pycom [21], and Waspmote [14]), edge devices (e.g., Raspberry Pi [22]), and public cloud services (e.g., AWS [2] and Azure [16]). They have also examined an extensive set of wireless technologies used within civilian IoT systems, ranging from ZigBee [31] to LoRa [15], in the context of their suitability for HADR and other military use cases. At the edge processing layer (i.e., compute resources offered by or co-located with sensors), IST-147 evaluated several COTS computing platforms. Some initially promising innovative hardware architectures did not become commercially viable solutions, including Adapteva's many-core Parallella board [18] and IBM's True North neuromorphic chip [1]. However, recent developments, such as Intel's Movidius Neural Compute

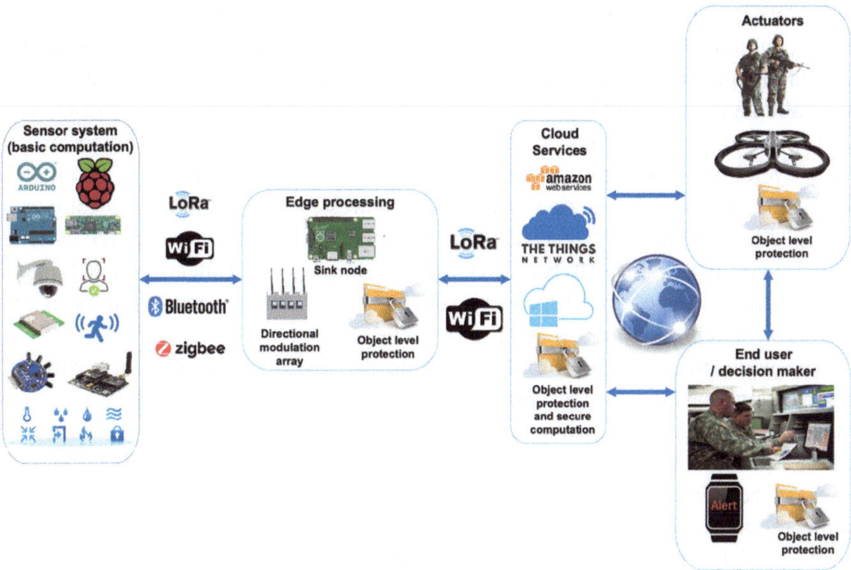

Fig. 5.5 A high-level system architecture investigated by the IST-147 group

Stick [12] and Google's Coral [10], seem promising in terms of providing low-cost edge computing options. More specifically, they are powerful solutions capable of running a (somewhat limited) selection of TensorFlow-based [27] machine learning models for inference on imagery on the order of a few milliseconds. In addition, the NVIDIA Jetson [17] line offers another similarly priced entry-level platform, the Jetson Nano, which is capable of running any machine learning application and presents designers with the option of vertical scalability to the higher end Jetson TX2, Jetson Xavier NX, Jetson Xavier AGX, and the more recent Jetson Orin AGX platforms. Evaluation of these solutions within IST-147 showed that they represent a solid foundation to build a limited but very useful spectrum of low-latency analytics for IoT applications.

5.5 Conclusions

The solutions discussed above often use models that represent a significant paradigm change with respect to the approaches currently practiced. As a result, there is still much work to be done to validate them as building blocks for the next generation of smart security applications. Performing some of this work is one of the objectives of NATO IST-176 Research Task Group (RTG), which is based on the results obtained by the IST-147 group. One of the main tasks of the group is to examine existing civilian and military standards and identify potential gaps that must be addressed to ensure interoperability in CBRN scenarios in smart environments. The results of this study were also presented to the NATO FMN community to be considered for incorporation into the FMN architecture and the future FMN spirals. Another important objective of the group is to propose a set of methods that allow secure and trusted fusion of data from various sources of information. The aim is to demonstrate and validate these methods through proof-of-concept trials in the context of HADR.

The developed solutions are currently being deployed on a distributed testbed with several physical locations for experimentation purposes. More specifically, we are in the process of realizing IoT infrastructures to host CIMIC applications in several cities, including Warsaw (Poland), The Hague (The Netherlands), Pensacola (FL, USA), and Ferrara (Italy). This will allow us to experiment with IoT applications that leverage a heterogeneous set of sensing and computing platforms and interact with a broad range of smart city services.

The widespread adoption of IoT and the increasing availability of smart city environments present compelling opportunities to increase public safety and security. However, their realization poses several challenges at various levels, including, but not limited to, IT service design, architecture, and integration. Groups in industry, academia, and the military are currently investigating these opportunities and challenges [26].

We envision that adopting IoT in the military will lead to a new era of dual use IoT solutions and the emergence of innovative and sophisticated civil-military cyber-physical systems. In particular, several ongoing initiatives target CIMIC in the context of HADR, counterterrorism, and public safety and security scenarios. These initiatives are increasingly looking toward IoT as an extremely valuable, although not entirely reliable, source of information for situational awareness purposes.

References

1. Akopyan, F., Sawada, J., Cassidy, A., Alvarez-Icaza, R., Arthur, J., Merolla, P., Imam, N., Nakamura, Y., Datta, P., Nam, G.J., Taba, B., Beakes, M., Brezzo, B., Kuang, J.B., Manohar, R., Risk, W.P., Jackson, B., Modha, D.S.: TrueNorth: design and tool flow of a 65 mW 1 million neuron programmable neurosynaptic chip. IEEE Trans. Comput. Aided Des. Integr. Circuits Syst. **34**(10), 1537–1557 (2015). https://doi.org/10.1109/TCAD.2015.2474396
2. Amazon Web Services (AWS): Cloud computing services. https://aws.amazon.com/. Accessed 03 Mar 2023
3. Arduino Team: Arduino. https://www.arduino.cc/. Accessed 03 Mar 2023
4. Buyya, R., Srirama, S.N. (eds.): Fog and Edge Computing: Principles and Paradigms. John Wiley & Sons, Hoboken (2019)
5. Campioni, L., Lenzi, R., Poltronieri, F., Pradhan, M., Tortonesi, M., Stefanelli, C., Suri, N.: Margot: dynamic IoT resource discovery for HADR environments. In: MILCOM 2019 - 2019 IEEE Military Communications Conference (MILCOM), pp. 809–814 (2019)
6. Campioni, L., Fontana, N., Morelli, A., Suri, N., Tortonesi, M.: A federated platform to support IoT discovery in smart cities and HADR scenarios. In: 2020 15th Conference on Computer Science and Information Systems (FedCSIS), pp. 511–519 (2020). https://doi.org/10.15439/2020F48
7. Campioni, L., Poltronieri, F., Stefanelli, C., Suri, N., Tortonesi, M., Wrona, K.: Enabling civil-military collaboration for disaster relief operations in smart city environments. Futur. Gener. Comput. Syst. **139**, 181–195 (2023). https://doi.org/10.1016/j.future.2022.09.020. https://www.sciencedirect.com/science/article/pii/S0167739X22003041
8. Ferdousi, S., Tornatore, M., Dikbiyik, F., Martel, C.U., Xu, S., Hirota, Y., Awaji, Y., Mukherjee, B.: Joint progressive network and datacenter recovery after large-scale disasters. IEEE Trans. Netw. Service Manag. **17**(3), 1501–1514 (2020)
9. Giordani, M., Higuchi, T., Zanella, A., Altintas, O., Zorzi, M.: A framework to assess value of information in future vehicular networks. In: Proceedings of the 1st ACM MobiHoc Workshop on Technologies, mOdels, and Protocols for Cooperative Connected Cars, MobiHoc '19. ACM, New York (2019). https://doi.org/10.1145/3331054.3331551
10. Google: Coral. https://coral.ai/. Accessed 03 Mar 2023
11. Howard, R.A.: Information value theory. IEEE Trans. Syst. Sci. Cyber. **2**(1), 22–26 (1966)
12. Intel: Intel Movidius Neural Compute SDK documentation. https://movidius.github.io/ncsdk/ncs.html. Accessed 03 Mar 2023
13. Lenzi, R., Benincasa, G., Casini, E., Suri, N., Morelli, A., Watson, S., Nevitt, J.: Interconnecting Tactical Service-Oriented Infrastructures with Federation Services. In: MILCOM 2013 - 2013 IEEE Military Communications Conference, pp. 692–697 (2013). https://doi.org/10.1109/MILCOM.2013.123
14. Libelium: Waspmote: The smart sensor platform to develop IoT projects. https://www.libelium.com/iot-products/waspmote/. Accessed 03 Mar 2023
15. LoRa Alliance: LoRa Alliance. https://lora-alliance.org/. Accessed 22 Nov 2018
16. Microsoft Azure: Cloud computing services. https://azure.microsoft.com/en-gb/. Accessed 03 Mar 2023
17. NVIDIA: Embedded systems developer kits & modules from NVIDIA Jetson. https://www.nvidia.com/en-us/autonomous-machines/embedded-systems/. Accessed 03 Mar 2023
18. Parallella: The Parallella board. https://parallella.org/. Accessed 03 Mar 2023

19. Poltronieri, F., Campioni, L., Lenzi, R., Morelli, A., Suri, N., Tortonesi, M.: Secure multi-domain information sharing in tactical networks. In: MILCOM 2018 - 2018 IEEE Military Communications Conference (MILCOM), pp. 1–6 (2018). https://doi.org/10.1109/MILCOM.2018.8599693
20. Poltronieri, F., Stefanelli, C., Suri, N., Tortonesi, M.: Phileas: a simulation-based approach for the evaluation of value-based fog services. In: IEEE International Workshop on Computer Aided Modeling and Design Communication Links and Networks, CAMAD (2018)
21. Pycom: Pycom - next generation Internet of Things platform. https://pycom.io/. Accessed 03 Mar 2023
22. Raspberry Pi Foundation: Teach, learn, and make with the Raspberry Pi foundation. https://www.raspberrypi.org/. Accessed 03 Mar 2023
23. Riberto, G., Govoni, M., Stefanelli, C., Suri, N., Tortonesi, M.: Leveraging civilian IoT infrastructures to support warfighting activities in urban environments. In: 2018 IEEE 4th World Forum on Internet of Things (WF-IoT), pp. 118–123 (2018)
24. Suri, N., Benincasa, G., Lenzi, R., Tortonesi, M., Stefanelli, C., Sadler, L.: Exploring value-of-information-based approaches to support effective communications in tactical networks. IEEE Commun. Mag. **53**(10), 39–45 (2015)
25. Suri, N., Tortonesi, M., Michaelis, J., Budulas, P., Benincasa, G., Russell, S., Stefanelli, C., Winkler, R.: Analyzing the applicability of Internet of Things to the battlefield environment. In: 2016 International Conference on Military Communications and Information Systems (ICMCIS), pp. 1–8. IEEE, Piscataway (2016)
26. Suri, N., Zielinski, Z., Tortonesi, M., Fuchs, C., Pradhan, M., Wrona, K., Furtak, J., Vasilache, D.B., Street, M., Pellegrini, V., Benincasa, G., Morelli, A., Stefanelli, C., Casini, E., Dyk, M.: Exploiting smart city IoT for disaster recovery operations. In: Proceedings of the IEEE World Forum Internet Things, WF-IoT 2018, pp. 458–463 (2018)
27. TensorFlow Community: TensorFlow. https://www.tensorflow.org/. Accessed 03 Mar 2023
28. Tortonesi, M., Morelli, A., Govoni, M., Michaelis, J., Suri, N., Stefanelli, C., Russell, S.: Leveraging Internet of Things within the military network environment - challenges and solutions. In: 2016 IEEE 3rd World Forum on Internet of Things (WF-IoT), pp. 111–116 (2016)
29. Tortonesi, M., Govoni, M., Morelli, A., Riberto, G., Stefanelli, C., Suri, N.: Taming the IoT data deluge: an innovative information-centric service model for fog computing applications. Future Gener. Comput. Syst. **93**, 888–902 (2019)
30. Wrona, K., Tortonesi, M., Marks, M., Suri, N.: Leveraging and fusing civil and military sensors to support disaster relief operations in smart environments. In: MILCOM 2019 - 2019 IEEE Military Communications Conference (MILCOM), pp. 790–797 (2019)
31. ZigBee Alliance: ZigBee Alliance Website. https://zigbeealliance.org/

Chapter 6
Security Considerations

Konrad Wrona and Janusz Furtak

6.1 Introduction

Security is one of the biggest challenges related to the widespread deployment of IoT. As a result, IoT security has been the subject of intensive research and development activities in recent years. In particular, the National Institute of Standards and Technology (NIST) has formed the *Cybersecurity for IoT* program that has been producing a set of guidance documents considering various aspects of the security of commercial IoT solutions [15]. This program has been accompanied by several joint activities with the NIST industrial partners performed by the National Cybersecurity Center of Excellence (NCCoE) [53]. One of these joint projects has recently released an initial guidance document on Trusted IoT Device Network Layer Onboarding and Lifecycle Management [19]. Similarly, the European Union Agency for Cybersecurity (ENISA) [18] has been developing IoT security practices and has proposed recommendations to operators, manufacturers, and decision-makers, focusing on different facets of the IoT ecosystem, including devices, network, and software development life cycle [17]. Several specialized research and development projects related to IoT security have also been implemented under various funding programs of the European Commission; a list of these projects is maintained by the European Research Cluster on the Internet of Things [34].

Furthermore, considerable technological progress has occurred in recent decades in IoT technology, including the introduction of software-defined technologies,

K. Wrona (✉)
NATO Communications and Information Agency (NCIA), The Hague, The Netherlands

Military University of Technology, Warsaw, Poland
e-mail: konrad.wrona@ncia.nato.int

J. Furtak
Military University of Technology, Warsaw, Poland

N. Suri et al. (eds.), *Military Applications of Internet of Things*, Internet of Things,
https://doi.org/10.1007/978-3-032-03378-9_6

cloud platforms, and miniaturized single-board computers. This has resulted in many Commercial Off The Shelf (COTS) components offering the functionality required in the military environment, thus reducing from a functional perspective the need for custom solutions. Using commercial IoT technology in the military environment offers an attractive and cost-effective alternative to developing custom sensing and monitoring solutions. Typically, such custom solutions are more expensive, take more time to deploy, and introduce increased implementation risk. However, they also offer fit-to-measure functionality and a higher level of security.

In this chapter, we provide a military perspective on the security of IoT applications and show how the proposed OLP [77] and Data-Centric Security (DCS) [73] paradigms can support IoT applications for future military operations.

6.2 Risks in Military Applications of IoT

There are several high-impact use cases for the deployment of IoT in a military environment. We briefly discuss some of the most important of these in the following. A more detailed discussion of the applicability of IoT to military operations can be found in [83].

The threat environment will depend on the exact nature of the operation, the function the IoT technology will perform, and its location within an operational area. A primary concern includes protecting sensor data against both interception and modification, particularly because IoT sensor information often triggers an action. The specific focus is to ensure the confidentiality and integrity of the data in the time frames relevant to the tactical environment, that is, in the range of hours or days rather than years or decades. Another important threat, particularly common in contested environments and military operations, is the jamming of wireless communication channels. Particularly common attacks consist of jamming or spoofing of GPS or other satellite navigation signals.

6.2.1 Situational Awareness

One of the most critical aspects of any military operation is proper situational awareness. Most armed forces already use a wide range of sensors and unmanned vehicles to gather intelligence. Incorporating civilian IoT solutions into military IT systems could improve a commander's available operational picture and provide substantially augmented overall situational awareness. However, achieving the positive effect of this increase depends on ensuring adequate availability and integrity of the information delivered by IoT systems. Therefore, COTS IoT systems must be carefully evaluated before being integrated with Cyber Situational Awareness (CSA) capability as reliable and trusted sensors.

6.2.2 Logistics

Using IoT, including sensors and RFID, is fundamental for improving the efficiency and effectiveness of logistics operations. This includes interoperability with third-party logistics systems, as many of the supplies required during military operations consist not only of military equipment but also of subsistence and medical materials for the forces. Using IoT systems could also contribute to the increased safety of logistics operations. Examples include preventing joint transportation of goods, such as chemical components, which could result in dangerous chemical reactions, or joint transportation of cryptographic equipment parts to prevent interception by an adversary [72]. However, it has been shown that the improper integration of RFID tracking solutions with the back-end system could lead to new attack paths in enterprise information systems [66]. Moreover, a lack of adequate confidentiality protection could enable an adversary to perform better targeted attacks on delivery convoys or use the information as a side channel for reasoning about planned military operations. Similarly, inadequate integrity and availability protection could be exploited by an adversary to severely impact the logistics operation, for example, by maliciously rerouting the goods. All of these risks should be considered when designing IoT-enabled logistics applications for the military.

6.2.3 Medical Care

Assistance in the treatment of medical conditions and injuries during combat operations is one of the most discussed applications of wearable and stationary IoT systems in the military environment. However, in addition to the high potential to improve the speed and accuracy of delivering, often lifesaving, medical treatment to soldiers, smart medical care systems can also introduce some new risks. Modern medical and health monitoring systems are typically equipped with wireless functionality, enabling devices to communicate with each other and medical back-end systems. However, these wearable or implantable medical systems can have serious security vulnerabilities, as demonstrated by Halperin et al. [29, 30], and Maisel and Kohno [45]. Moreover, it has recently emerged that some prominent public personalities were aware of the issues and were sufficiently concerned to disable remote connectivity in their implanted medical devices [44].

6.2.4 Smart Equipment

IoT can be applied to various types of military equipment, including vehicles, supplies, and weapons systems. However, many significant security flaws and vulnerabilities have been demonstrated in such network-enabled objects. In particular, several serious vulnerabilities have been identified in cars [10, 49, 52, 68], leading to massive recalls of vehicles. There are also known examples of the enemy exploiting weaknesses in the security of military cyber-physical systems [48]. Similarly, researchers have recently identified important security vulnerabilities in commercially available smart rifles [28].

6.3 IoT as a New Attack Surface

Like every technology, IoT potentially introduces a new attack surface in the military IT system. This attack surface consists of:

- IoT devices (i.e., sensors and actuators)
- Communication channels between the devices, as well as between the devices and the back-end system
- IoT-specific back-end applications
- Back-end data storage

The emergence of IoT and the proliferation of devices able to connect to military networks greatly expand the potential attack surface. Furthermore, IoT devices that are not properly managed and maintained after their initial deployment introduce a particular risk. Avoiding such risks requires an effective combination of technical and procedural capabilities. This is especially important because the lifespan of IoT hosting platforms, such as military ships, land vehicles, and airplanes, is counted in decades, whereas IoT components, in terms of both hardware and software, become quickly obsolete due to the rapid pace of innovation. Enabling secure continuous development and evolution of military CIS causes increasing difficulties, for example, in the context of traditional defense procurement and security accreditation processes. However, this is crucial to maintaining information superiority and technological advantage over adversaries. In the long term, due to its scale and heterogeneity, the proliferation of IoT components could lead to the presence of vulnerable or outdated devices, underscoring the increasing importance of defense-in-depth and novel security concepts, such as Zero Trust Architecture [67].

From an enterprise perspective, implementing the BYOD concept introduces new security challenges related to personal IoT devices. Although most of the current work on BYOD security policies and mechanisms focuses on relatively powerful traditional mobile devices, such as laptops and smartphones, employees also bring a large number of smart devices to the workplace, with all their related security risks and opportunities. How this situation could be exploited to increase enterprise security by taking advantage of additional sensing capabilities, rather than only opening new attack vectors for the enterprise system, is still an open question. Therefore, an ongoing challenge is to manage potentially complex security policies applicable to IoT BYOD. In this respect, formal approaches may offer an interesting solution.

6.3.1 *Attacker Model*

IoT systems used in military environments can be subjected to determined attackers with varying technical capabilities. Although some of the adversaries faced by NATO and NATO nations in their past operations are not technically sophisticated, this situation has recently changed rapidly. The ability of insurgents in Iraq to access unencrypted surveillance streams from American drones is a good example of a risk introduced by underestimating the capabilities of an adversary. In addition, many historically unsophisticated enemies have substantial financial resources and control large parts of the population. These assets allow them to acquire the required expertise externally through purchase or coercion.

Depending on the IoT use case, the attacker may be more likely (e.g., in the case of ground-seismic sensors) or less likely (e.g., in the case of unmanned vehicles) to have physical access to a device. However, it is reasonable to assume that the physical integrity of the device cannot be guaranteed. Therefore, devices should be equipped with an embedded secure element and a trusted execution environment [27] to protect cryptographic material and provide the ability to be wiped remotely.

One must also assume that the attacker has access to a wireless communication channel between IoT devices, as well as between IoT devices and the gateway to the back-end system. Therefore, these communication channels must be authenticated and protected against eavesdropping and, to the extent possible, against jamming.

The attacker's access to the IoT back-end system, typically located in a cloud, may vary depending on the exact location of the back-end system. For example, existing network defenses may reduce accessibility in the case of systems located within the private NATO cloud. However, systems hosted by third-party providers may use a public cloud or expose their interfaces to the public Internet.

Therefore, it is essential that the proper vulnerability assessment and penetration testing, including scenarios involving compromised IoT end devices, are performed regularly on the back-end system.

6.3.2 *Vulnerabilities and Threats*

A recent study [7] of 50 smart home devices found that none enforced strong passwords, used mutual authentication, or protected accounts against brute-force attacks. Finally, almost two out of ten mobile apps used to control the tested IoT devices did not use TLS to encrypt communications to the cloud. Furthermore, the tested IoT technology also contained many common vulnerabilities. In particular, some of the web portals used to control IoT devices had severe vulnerabilities that allowed unauthorized access to back-end systems.

The Open Web Application Security Project (OWASP) List of Top Ten IoT Vulnerabilities [51, 59] defines the attack surface of IoT applications:

1. Weak, Guessable, or Hardcoded Passwords
2. Insecure Network Services
3. Insecure Ecosystem Interfaces
4. Lack of Secure Update Mechanism
5. Use of Insecure or Outdated Components
6. Insufficient Privacy Protection
7. Insecure Data Transfer and Storage
8. Lack of Device Management
9. Insecure Default Settings
10. Lack of Physical Hardening

Specific threats related to military use of IoT include:

- *Location confidentiality:* detection of equipment or personnel location that may lead to compromise or destruction by an adversary—IoT devices, by definition, are detectable objects whose signals an adversary present in the radio perimeter could pick up
- *Availability:* the risk of interference by an adversary with wireless channels used by IoT devices or with the devices themselves
- *Data confidentiality:* the risk of information interception by an adversary
- *Data integrity:* the risk of information alteration by an adversary—the information emitted by the connected objects may be changed or falsified

Although these threats also exist in the civilian world, they are rarely systematically taken into account the pace of product development, simplicity of installation and use, low computational and communications overhead, and low cost often take precedence over the security-by-design requirements.

In addition, military operations are exposed to risks related to the increasing adoption of smart critical infrastructure, resulting in an increased attack surface on civilian populations and vital military support systems. IoT systems are vulnerable to a range of specific threats, including:

- Attacks on secondary functions such as power supply and clocks
- Side-channel attacks based on electromagnetic radiation, leading to localization, a confidentiality breach, or a compromise in the integrity of the results
- Cryptanalysis of authentication and encryption protocols weakened through optimizations required for implementation and operation on constrained devices
- Unauthorized modification of hardware components on devices operated in remote areas and in the absence of physical protection

In order to guard against software vulnerabilities, it is essential to implement:

- Ability to authenticate the device and to attest integrity of its specific configuration
- Ability to timely update IoT devices, also remotely
- Logging of events and incidents suitable for forensic investigations
- Software certification and validation with integrity and digital signature checks
- Minimum functional configuration of IoT devices defined by the specific use case

6.3.3 Privacy

The privacy implications of IoT have been a topic of research and public interest in Europe [74] and the USA [47]. Although privacy is not a primary concern during military operations, NATO and NATO nations must comply with the national privacy regulations applicable to deployed personnel. For example, national concerns and laws prevented biometrics from being used to authenticate national and NATO personnel during NATO operations in Afghanistan, where in the end the biometric systems were only used to authenticate the Afghan workers allowed on base. Therefore, it is crucial that the implemented IoT solutions do not unnecessarily infringe the privacy of personnel and that all personal information collected is adequately protected during transit and storage.

6.3.4 Security of Commercial IoT Solutions

Despite significant technological progress, security considerations with respect to COTS solutions are still relevant. Some security challenges have even been aggravated by the same technologies that have enabled the implementation of richer and more flexible functionality. For example, using IoT platforms offered by public cloud providers (e.g., Amazon Web Services IoT and Microsoft Azure IoT) provides an opportunity for a quick and low-investment implementation of data collection and fusion applications. Indeed, the availability protection provided by such large-scale public clouds is often superior and, in most cases, much more cost-effective than what is obtainable in custom-implemented closed military systems. However, the confidentiality and integrity protections offered by such public platforms are typically much lower than the protection offered by military systems. In particular, using public cloud services significantly increases the risk of disclosing potentially sensitive information to unauthorized parties and the risk that an adversary could maliciously modify information.

There are several reasons why COTS products and services can provide security measures that are not sufficient to meet military requirements. An essential consideration in a commercial environment, which is of less importance in a military environment, is the usability of security measures by a nonexpert user and the cost of a secure operation. Many currently offered commercial IoT solutions are designed with ease of use and straightforward configuration in mind rather than stringent security controls. This focus makes sense in the commercial world, where usability must be balanced with security to gain wide acceptance among users. In contrast, military personnel can be explicitly trained in the correct operation of secure solutions and typically have a considerable incentive to follow the defined procedures adequately. Related to usability is the cost issue of a secure deployment and operation. In particular, many commercial solutions tend to avoid or oversimplify key management procedures, often compromising the whole system's security, for example, by reusing a single key or small subset of keys or using an insufficient source of randomness for key generation. However, the military uses key management systems extensively. Although they always introduce substantial additional operational costs and therefore need to be optimized, they constitute a well-understood and integrated element of military operations. Commercial IoT devices typically provide a very constrained computational environment that might not be suitable for implementing more complex and sophisticated security measures or even standard measures used in the Internet. IoT nodes are very often mobile and use wireless communication links with low bandwidth and a short range. Such nodes usually have limited resources (particularly a small Random-Access Memory (RAM) with a size of several or several dozen KB) and low computing power. In addition, a permanent power supply source (i.e., from the power grid) is rare. Most often, it is batteries or accumulators that supply such nodes. The available energy resources are often inversely proportional to the bandwidth and range of the wireless link. In addition, the size of the energy resource can have a decisive impact on the size of installed RAM, the choice of processor, or the number and type of sensors [23]. Implementing a security mechanism in such constrained devices often requires some specific optimizations and trade-offs that, if not analyzed carefully, can introduce new vulnerabilities and weaken overall system security.

Another important security drawback inherent to public services is that they share some of the underlying infrastructure with other users and lack complete control over this infrastructure or the management of services. Although the level of sharing and lack of control may differ depending on the provider and acquired services, it cannot be fully eliminated when using public services. At the same time, due to cost considerations, the underlying technology and the processes implemented rarely offer high assurance of separation between different customers.

6.4 IoT as a Security Enabler

Despite the security and privacy risks introduced by IoT systems, the IoT can also provide security-relevant information and thus enable context-aware security mechanisms and support defense-in-depth principles. For example, context information received from IoT systems could be used as input to authentication mechanisms (e.g., using behavioral biometrics or dynamically adapting the level of deployed security measures based on the perceived threat level and operational picture). Our recent results on behavioral biometric authentication are promising [9]. Although our implementation was limited to the standard sensors included in a typical smartphone, the same approach could be applied to distributed scenarios, in which the user's behavior pattern is collected from multiple IoT devices, if sufficient trustworthiness of this information can be guaranteed.

6.5 Security Requirements for Military Applications of IoT

The basic security requirements concerning Confidentiality, Integrity, and Availability (C-I-A) for military applications of IoT systems do not differ substantially from the security requirements for any other IT systems deployed in the military environment. However, there are some additional requirements related to the distributed character of IoT, including authentication, non-repudiation, accountability, anti-tampering, and the need to track data provenance. In addition, the typically small memory resources of the sensor nodes, their limited energy resources, and their computational limitations also influence the security of IoT. We should also note that IoT elements are often mobile. Therefore, the requirements for Confidentiality, Integrity, and Availability (C-I-A) should extend to communication links. In particular, transmitted data portions should be encrypted and provided with control sequences to verify the integrity of the data.

The C-I-A security measures constitute the safety basis for IoT elements. However, they must be supplemented with additional measures, including:

- *Authentication:* Before providing it with any system resources, every element of the system must be subjected to an identity confirmation procedure. Identity confirmation is the basis for the authorization and implementation of a data access system, which is a crucial element in ensuring confidentiality.
- *Non-repudiation:* It refers to ensuring that the party to the contract or communication cannot deny the authenticity of its signature on the document or the message it originates from.
- *Accountability:* responsibility for the operations performed
- *Traffic Flow Analysis (TFA):* Traffic flow analysis aims to analyze the statistics of the observed traffic flows in order to deduce some sensitive information. TFA can be considered an important threat for military applications of IoT. In particular, if an adversary can establish which IoT nodes are most active before some activities, the adversary can prepare for certain events. For example, if wireless links associated with UAV control become very active, the adversary may deduce the need to take immediate cover. Therefore, protection against analysis of our IoT traffic and behavioral patterns should be taken into account in MIoT.

Another issue is that IoT devices can be exposed to direct access by unauthorized persons. Therefore, these devices should be equipped with built-in security functions that create a trusted environment to protect cryptographic material to ensure the integrity of the device and the ability to reset remotely. An essential element of any device (IoT element) should be mechanisms that constantly monitor the state of this device in terms of unauthorized modification of the hardware and software configuration. If anomalies are detected in this respect, the IoT element should provide information about such an event to other network elements. It can also take actions to block the device, erase sensitive data stored in the device's resources, or physically destroy the device.

6.5.1 Confidentiality

Confidentiality is an essential aspect of any military operation. In the case of IoT systems, confidentiality protection is required for all communication channels and data stored and processed both at the end nodes and in the back-end system. For example, in the case of wireless sensor networks or unmanned vehicles, compromising confidentiality may endanger both the forces and an operation's goals by giving clues about the military's plans and unintended support to the enemy by providing data streams and increased situational awareness. A real-life example of such a threat was observed in December 2009 when militants in Iraq obtained access to the downlink transmission used by the United States (US) drones [48].

6.5.2 Integrity

Ensuring data integrity is critical, as erroneous data can reduce or negate information superiority and cause an operation to fail. Adequate security assurance requires a fine-grained risk analysis specific to the use case. This may lead to trade-offs between the amount of potentially useful data that can be collected and ingested into military command and control CIS and the achievable level of security, reliability, and trustworthiness. It is also vital to put the data produced, collected, stored, and processed in context by preserving rich metadata on their provenance, trustworthiness, and meaning.

Integrity is crucial to many aspects of MIoT systems. It is vital that the information delivered to the C2 center by smart things has not been modified and that the trustworthiness of its source can be accurately evaluated so that the data can be incorporated into the operational picture with an appropriate confidence level and used for command decisions or ignored. However, it is equally important that the command and situational information provided to smart objects is of appropriate integrity. For example, it is stipulated that in December 2011, Iran successfully brought down a US reconnaissance RQ-170 Sentinel UAV by spoofing the GPS signals it received to navigate back to its launch point.

Integrity is not only a feature of information but also a system, where it refers to ensuring the preservation of a secure system state and configuration. Therefore, violating the integrity of the system can have a catastrophic impact on the C-I-A of the data processed by this system. An example of such an attack was observed in September 2011, when a virus-infected military system was used for keylogging C2 of a USUAV fleet at Creech Air Force Base in Nevada [31]. Consequently, these events led to the development of the DARPA research program in high-assurance cyber military systems [21, 63].

6.5.3 *Availability*

The full potential of IoT in military environments will not be achieved if the required availability of information delivered by sensors and devices cannot be ensured. Availability can be ensured in several ways. Ensuring high availability of individual devices and data links is one measure. Another measure is redundancy, which involves both duplication–deploying additional, similar devices–and diversity–using devices with different, heterogeneous technologies to provide equivalent information. Similarly, C2 information must be available to actuators and smart devices when required. The *sleep-deprivation attack* is a particular aspect of availability related to IoT. This type of attack targets specifically battery-powered devices, which are common among smart things, by preventing them from entering an energy-saving mode. This leads to depletion of battery power, and replenishing batteries may be difficult, if not impossible, in combat situations. Thus, zero-power technology and energy scavenging may be essential for the survival of IoT systems. These are the subjects of recent large DARPA research programs [14].

6.5.4 *Trust*

Trust in IoT devices is crucial to providing value to military users. Trust is broader than security. Although sensor security is a prerequisite for establishing trust in the data it provides, it is not the sufficient condition. A sensor can be fully secure, but the information it provides may still not be trustworthy—it can be an unreliable sensor or simply of bad quality. Similarly, the algorithms used to process the data, such as AI, may not provide trustworthy results even though they are secure. It is also unlikely that the entire IoT infrastructure of nodes, communication links, and gateways will have the required level of security that allows complete trust. This is particularly true for IoT devices that are not owned or operated by the military but by civilian entities, such as municipalities and individual users. Consequently, if military forces use IoT devices to enhance situational awareness, there is a risk that hostile forces will disable, disrupt, or deceive the network. The latter is the most serious case, as a hostile force could manipulate the information reported by the IoT network to maliciously guide commanders' actions.

6.5.5 *Safety and Security*

Although cybersecurity often supports operational safety, sometimes these requirements can conflict, for example, when an adequate cybersecurity response to an event would mean the immediate disconnection of a compromised or vulnerable but safety-critical system component. The balance between operational safety and cybersecurity requirements varies depending on the specific application and use case. Cybersecurity may be the most important factor when considering the interconnection of existing classified military CIS with new sensing and actuation capabilities offered by the IoT. Thus, a more balanced approach might be required, for example, in military base or vehicle automation scenarios. However, operational safety may be the most important when integrating complex systems or critical infrastructure hierarchies.

6.5.6 Data Provenance

An essential attribute determining the usefulness of the data is its provenance. If the data come from a reliable source, they are more useful than data from an unknown source. The evaluation of the credibility of the source should be associated with the authentication procedures of the data source. If the data source is a single sensor node, the identity of this node must be confirmed. However, sensor nodes can also be organized into clusters. In this case, it is enough to authenticate the cluster or a cluster representative, which provides data from the nodes making up the cluster to confirm the origin of the data.

It is possible to build a cluster in which each member is pre-authenticated before it can start its regular operation, using the local trust structure created on each sensor node equipped with a Trusted Platform Module (TPM). In this situation, the data from the sensor nodes within the cluster will be reliable. In addition, a node in the cluster can be selected to play a gateway role to transfer data from the cluster to other clusters. With local trust structures, it is possible to build a trust relationship between nodes acting as a gateway in different clusters, which means building a trust relationship between different clusters.

6.5.7 Cryptographic Mechanisms

As discussed above, although commercial solutions often provide a sufficient level of availability protection, in most cases additional suitable cryptographic mechanisms must be in place to provide an appropriate level of protection for integrity and confidentiality, as well as authentication of the data origin. According to the NATO in-depth cyber defense principles, all layers of the system should have such protection measures implemented. However, when using public cloud infrastructure, it is imperative to provide end-to-end security at the application layer. To protect the confidentiality and integrity of the information, the information must be encrypted at the point of origin, that is, the sensor node itself or a sink/gateway node owned by NATO, and stored and transported in this form as long as possible. This protection could also include using homomorphic encryption mechanisms that enable the processing of encrypted data; however, this particular aspect is beyond the scope of our current work and is for future research endeavors. In addition, traditional measures such as various types of digital signatures can provide authentication of the data source, although these techniques are also beyond the scope of our current investigation.

It is unrealistic to apply military grade security (including COMSEC, INFOSEC, and COMPUSEC) to IoT devices due to their low cost, low power, limited connectivity, and limited computational power. Moreover, many IoT scenarios use a decentralized trust model, where trust does not rely on a single entity or artifact. Instead, nodes can compare their values against those around them to see if their neighbors have been compromised. The analysis of data provided by IoT sensors can do the same. However, a consensus is needed before taking action rather than trusting all inputs.

6.6 Security Measures

As discussed in Sect. 6.1, several open and proprietary mechanisms for securing IoT have been proposed by both academic researchers and industry. In the following, we discuss some of the security measures that have been implemented and validated during the demonstrations and experiments described in Chap. 8.

6.6.1 Physical Layer Mechanisms

Physical layer security techniques can effectively complement traditional higher layer cryptography by providing additional protection for the network's data flows and additional resilience to the overall system. In particular, since the signal is made intelligible only within the intended service area and is unusable elsewhere [61], removing a node from such an area results in the immediate and inherent rejection of the node's transmission at the antenna stage. This yields some beneficial effects:

- The network is made *inherently resilient* to malicious signals sent by an attacker or rogue sensor located outside the intended service area.
- *Physical* node tampering activities are made more difficult; nodes that are removed from their assigned location, for example, to be reconfigured or hacked, disappear from the network. If they are subsequently returned to their original locations, they can be marked as untrusted by the upper OSI layers based on their black-out time statistics.
- *Digital* node tampering requires an attacker to physically enter the target service area to access the network.

Using Physically Unclonable Functions (PUFs) is a possible solution to the security of low-cost physical devices, employing the physical characteristics of simple semiconductors to provide elements of security functions. Although not computationally complex, such functions are difficult to replicate and add a level of security that is not achievable using conventional security techniques on low-cost hardware. Another way to protect IoT nodes against unauthorized interference is to use the Platform Configuration Registers (PCR) offered by the TPM in each node.

6.6.2 Establishing Trust

In traditional Internet, the trust structure is usually based on the use of Public Key Infrastructure (PKI) and Certification Authority (CA). However, in the case of IoT nodes with relatively limited resources, such an approach is not always appropriate. A more suitable approach focuses on creating local trust structures for several dozens of IoT nodes (i.e., a cluster) based on TPM. If all IoT nodes are equipped with TPM, the local trust structure can be based on trust in the certified asymmetric Endorsement Key (EK) of each TPM. This key is part of each TPM and is signed by the TPM manufacturer's private key and is accompanied by the appropriate certificate. Based on this trust structure, asymmetric keys can be generated, which can then be used to build a trust relationship between individual IoT nodes and to build a secure distribution system for symmetric keys inside the cluster. Symmetric keys, together with the use of hash functions and the Hash-based Message Authentication Code (HMAC) function, support storing sensitive data in TPM, storing secure data outside the module and the exchange of data between cluster nodes. The cost of using TPM is the increase in energy expenditure necessary for TPM. However, TPM power consumption is relatively low, and the module offers significant security benefits. An example of a solution that implements the described mechanisms is presented in [24, 25].

In the context of military IoT applications it is recommended to use TPM modules in version 2.0, which provides the possibility of using Advanced Encryption Standard (AES), Elliptic Curve Cryptography (ECC), and Secure Hash Algorithm 2–256 bit (SHA-256) algorithms that are compliant with NIST guidance for the strength of cryptographic algorithms. On the other hand, TPM version 1.2 has lower requirements for the node's RAM resources, capacity, and energy demand. Therefore, one must consider whether it is worth using TPM v.1.2 for nodes that are merely a source of simple sensor data, at the expense of a slight deviation from the NIST requirements, but with the benefit of the node's power supply.

6.6.3 Confidentiality and Integrity Protection

Ensuring confidentiality concerns data stored in the sensor nodes and data sent via communication links. Symmetric cryptography is commonly used for this purpose. When using cryptography, it is necessary to apply secure procedures for generating and securely storing cryptographic keys. Furthermore, if cryptography is to be used to protect transmitted data, a secure key distribution procedure is required between the parties exchanging data.

These requirements can be met using the built-in random number generator of the TPM and the internal Non-Volatile Random-Access Memory (NVRAM) memory protected by TPM. Thus, using asymmetric cryptography offered by TPM, it is possible to develop a secure procedure to distribute symmetric keys. Although ensuring integrity usually relates to the integrity of data stored in the sensor node resources and data transmitted to other nodes, integrity should also apply to the hardware and software components of the node. Data integrity is often validated using hash functions, and the TPM offers hardware support for, for example, Secure Hash Algorithm 2–256 bit (SHA-256) and HMAC computation.

6.6.4 Sensor Monitoring and Detection of Sensor Tampering

Sensor nodes are often located in an environment where it is challenging to organize the physical protection of such nodes. Therefore, each node is exposed to unauthorized tampering of the hardware and software configuration.

The PCR registers of TPM can be used to detect unauthorized modifications to the node hardware and software. These registers contain hashes of metrics representing the system and hardware configuration, software components, or other sensitive data stored inside the TPM or in the sensor node resources.

One can read these registers without authorization, but the modification of these registers is not straightforward. While it is possible to modify a particular register, the modification is not a simple overwriting of its previous value, but a combination of the current value of the register and the new value. Consequently, each PCR change affects the future values of the registry. This is one of the safeguards that prevents unauthorized modification of the hardware or software configuration, circumventing TPM's detection.

Upon detecting tampering with the sensor node, various actions are possible, for example, informing other nodes about the situation, blocking sensor node operation, resetting sensitive data stored in the node's memory, or even physically destroying the node.

Remote attestation is relatively easy to implement when sensor nodes cooperate within a secure cluster. If each sensor node is TPM-equipped, then, using the local trust structures created based on these modules, it is possible to build a secure domain of sensor nodes. Each node of this domain undergoes a preregistration procedure, during which it is authenticated and equipped with the appropriate cryptographic material necessary for further secure data exchange within the domain.

In addition, protected data exchange between clusters can be configured using similar mechanisms. Within a cluster of sensor nodes, the safety of individual nodes and the entire cluster should be continuously monitored. For this purpose, system diagnosis procedures that rely on diagnostic communications can be used. These diagnostic communications may focus not only on the security of sensor nodes but also on assessing the quality and correctness of the operation of individual nodes.

6.7 Object-Level Protection

The concept of OLP [58] was developed to support NATO Network Enabled Capability (NNEC) [16]. OLP is a system-wide standard approach to data protection, built on two fundamental pillars:

1. Protection is applied to individual data objects (or portions thereof) rather than a collection of data objects and systems.
2. Metadata are bound to data objects and are used by protection enforcement mechanisms to determine the protection requirements for a data object.

OLP also complements the cyber defense component of a cybersecurity architecture by supporting data leakage prevention and defense in depth. In particular, OLP capability strengthens resilience to unforeseen threats because if perimeter protection turns out to be insufficient, then there is a second layer of protection at the object level, especially when Cryptographic Access Control (CAC) is deployed [57].

6.7.1 OLP Dimensions

The following three dimensions define OLP space:

1. *Detail of content description:* the level of detail in the description (i.e., metadata) of the content of an information object
2. *Granularity of access control:* the granularity of the information about the actor (i.e., a human user or automated process that requests access to a data object) and the environment that the protection policy enforcement capability can support
3. *Level of object protection:* the extent to which it is possible to protect an object regardless of location and time

There may be multiple ways to implement mechanisms. Therefore, we discuss the individual dimensions in more detail in the following.

6.7.1.1 Detail of Content Description

Metadata can describe an information object. An example of metadata compliant with NATO Core Metadata Specification (NCMS) [54] is shown in Fig. 6.1. Protection mechanisms use metadata to determine the protection requirements for a data object. The level of detail that describes an information object determines the

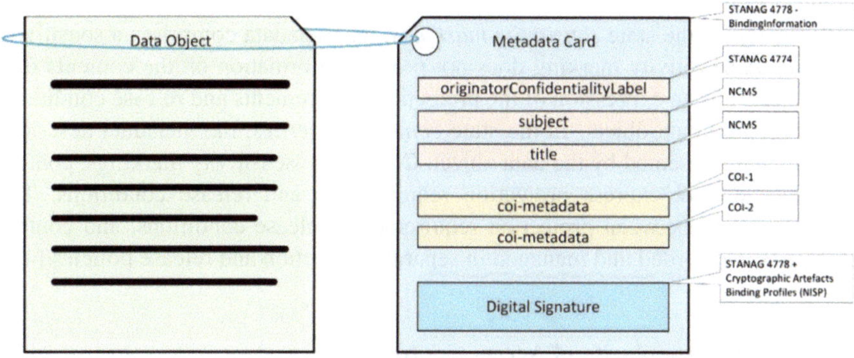

Fig. 6.1 Example of a metadata card compliant with NATO Core Metadata Specification STANAG 5636 [54]

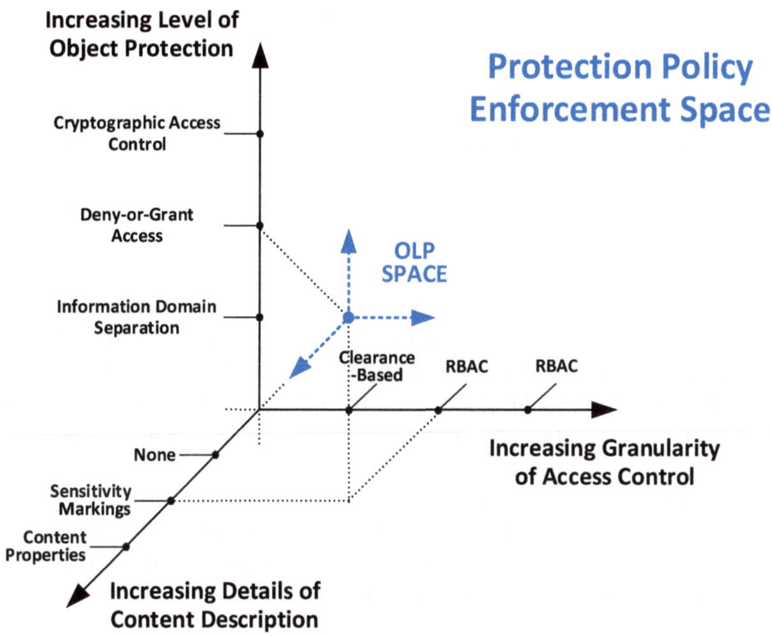

Fig. 6.2 Three-dimensional space of object-level protection

precision in formulating and enforcing the protection requirements. Note that it also depends on the type of metadata used, as shown in Fig. 6.2.

In the state *none*, protection mechanisms do not rely on any metadata. Instead, a protection policy is enforced at the system (or network) level and applied to all data objects (e.g., as in the *system high* approach). There is no protection at the object level in this state, so it is not part of the OLP Space. However, *none* is included in the discussion to clearly establish the distinguishing characteristics of OLP clearly. In the state *sensitivity markings*, the metadata comprises a sensitivity marking. A sensitivity marking does not provide information on the contents of a data object; it is an expression of the protection requirements and release conditions that apply to a data object. In the state *content properties*, the metadata describes the content represented by the data object. Contrary to sensitivity markings, content properties do not express protection requirements and release conditions. The correspondence between protection requirements, release conditions, and content properties is recorded and managed in separate protection and release policies [4].

6.7.1.2 The Granularity of Access Control

The information about an actor and its environment can differ in granularity, depending on the targeted state. For example, in the state *clearance-based* [39], the information about an actor and its environment is limited to the actor's clearance level or the system (or network) classification from which the actor requests access to a data object. The clearance-based model relies on Mandatory Access Control (MAC), also known as Lattice-Based Access Control (LBAC) or multilevel security. It is based on mandated regulations determined by a central authority. This model is usually used to protect information stored and processed by government and military organizations. In MAC, subjects and objects have sensitivity labels assigned to them. The label of a subject, called *clearance*, specifies the level of trust placed on the subject by the organization. The object label, called *sensitivity*, specifies the level of trust required for access. Labels form an ordered set of values, for example, *secret>restricted>unclassified*. To obtain access to a given object, the subject must have a clearance equal to or higher than the requested object's sensitivity. Policies in the "core" Bell-La Padula model (i.e., the one assuming that user and object labels do not change, called the "tranquility condition" in the literature) describe a set of access control rules that use security labels on objects and clearances for subjects. Security labels range from the most sensitive (e.g., "Top Secret") to the least sensitive (e.g., "Unclassified"). The main idea underlying the Bell-La Padula model is that high-level information cannot flow into objects of a low level. This requirement is captured in the following two properties:

1. The *simple security property:* A subject at a given security level may not read an object at a higher security level (*no read-up*).
2. The **-property* (or *confinement*) *property:* A subject at a given security level must not write to any object at a lower security level (*no write-down*).

In this way, users can create content only at or above their own security level. In contrast, users can view content only at or below their own security level.

In *Role-Based Access Control (RBAC)* [20], the clearance level of an actor is expanded with the concept of roles, enabling a more fine-grained expression of the authorizations of an actor. Role-Based Access Control (RBAC) policies control access depending on the roles users have within the system and on rules stating what accesses are allowed to users in given roles. RBAC is currently the most commonly used access control model for commercial applications. RBAC policies can be mapped to MAC policies. For example, [56] proposes a method to check whether an RBAC policy has an equivalent information flow policy in MAC and if this policy is consistent with the security model. In RBAC a role determines a set of permissions that can include both complex operations, such as an e-commerce transaction, and simple actions, such as read or write. The process of defining roles is based on a careful analysis of how an organization operates. RBAC decisions are based on the roles that individual users have as part of an organization.

In the state *Attribute-Based Access (ABAC)* [33], clearance levels (and system/network classifications) and roles are expanded to include more detailed sets of attributes, which describe the actors involved in accessing data objects and the technical capabilities of the systems used for access. *Attribute-Based Access Control (ABAC)* has been proposed as a solution to the challenges related to RBAC [35]:

- Implementing fine-grained authorization can lead to explosive growth of number of roles.
- Role design and engineering are difficult and expensive.
- Adjustment based on local/global situational factors is difficult.

In ABAC, access is not granted based on subject permissions associated with its role, but based on user attributes. An attribute is expressed as a pair *(name, value)* associated with any entity, such as a user, object, or environment. To support the proper processing of access control requests, the user must prove to the access control mechanism the so-called *claims* about the attributes he possesses. An ABAC policy specifies the claims that must be satisfied to grant access to an object. For example, the claim could be *I am a watch officer*, and any user who can prove this claim is granted access. If claims can be proved anonymously, for example, by means of anonymous credentials, users can remain anonymous, and authentication and identification are not required. Examples of attributes are user identity, role, and military rank of users; identifier and sensitivity of resources; and, for context, time of day, and threat level. Under ABAC, appropriately defined attributes can represent security labels, clearances, and classifications (for the encoding of MAC, identities and access control lists (for Discretionary Access Control (DAC)) and roles (for RBAC). *eXtensible Access Control Markup Language (XACML)* [55] is the most widely used policy specification language based on ABAC.

6.7.1.3 Level of Object Protection

As mentioned above, enforcing a protection policy at the data object level requires the ability to apply a protection policy to an individual data object regardless of its location and time. The extent to which this can be achieved is called *level of object protection*, for which three general states are distinguished.

The state *information domain separation* takes its name from the enforcement of a protection policy at the level of the information domain, where all the systems that make up the information domain inherit the policy. In this state, a data object is protected based on its information domain membership and therefore must not be transferred to an information domain under a different (i.e., less stringent) protection policy. Currently, prevention is commonly accomplished by separating systems or networks (where systems or networks contain information and form an information domain). As mentioned above, there is no protection at the object level in this state, so it is not part of the OLP Space. However, the state is included here to clearly distinguish the characteristics of OLP (and the OLP Space).

The state Deny or Grant Access (DOGA) introduces protection mechanisms that can enforce a protection policy at the object level. The enforcement can be coarse-grained, that is, access control is enforced on the data object as a whole, or fine-grained in the sense that it is possible to enforce a protection policy on portions of a data object. Therefore, if an actor is allowed to access a data object, access is granted by releasing the data object to that actor (i.e., the data object is transferred from its information domain to the actor's information domain). An example is the release of data objects from one domain to another through the use of a cross-domain guard, such as NATO Medium Assurance (MA) XML Labeling Guard (XLG) [78], which makes release decisions for (portions of) data objects. In the state Cryptographic Access Control (CAC) [37], the protection policy is not enforced according to the DOGA principle but by encrypting data objects (or portions thereof). Using CAC increases the level of object protection because an encrypted data object is protected regardless of location. In contrast, in the case of DOGA, once the data object has been released, the protection policy of the originating information domain cannot be enforced.

When CAC is used, the access control decision is not enforced directly in response to a request for a data object. Instead, the data objects are encrypted, and the access control decision is delayed until an actor makes a decryption attempt. When the actor successfully decrypts the data object, this implies the usage of the correct key material, and it is concluded that the actor is authorized to access the data object. When decryption fails, this implies that the actor is not authorized. Note that the decision to release decryption key material to an actor depends on whether the actor has authorization. However, when using CAC, the request for key material can, in principle, be made independently of the attempt to decrypt (i.e., access) the data object.

6.7.2 Applying OLP to IoT

OLP with CAC provides an attractive solution for protecting data in MIoT systems. The advantage of using cryptographic access control at the information object level, instead of network layer confidentiality mechanisms, is that objects can be protected end to end, including during possible interim storage. The disadvantage of CAC is the need for an efficient and scalable key management mechanism that can support a potentially very large number of individual encryption keys involved.

Furthermore, the fine-grained access control offered for OLP facilitates the sharing of information generated by the IoT systems with external partners, such as local civilian authorities, nongovernmental organizations, the International Committee of the Red Cross (ICRC), and the United Nations (UN). Implementing a flexible information sharing capability within IoT from the beginning is crucial because IoT systems can generate a large amount of data, which would be impossible to analyze and release using the classical (manual) process.

In public-key encryption schemes, every entity holds both a private and a public key according to predetermined global parameters. To securely transmit a message to a recipient, a sender first encrypts it using the recipient's public key. The recipient can then decrypt the message using his private key. Public keys are usually distributed through public key certificates that contain a copy of a public key, some owner identity information, and a signature on the certificate provided by the central authority for the public key infrastructure.

Identity-based encryption is a variant of public-key encryption that does not use public-key certificates. Instead, the public key for a recipient can be determined directly from the global parameters of the scheme and the public identifier associated with a recipient. Unlike regular public-key infrastructure cases in which every entity can generate its private key, the central authority typically generates and distributes private keys in identity-based encryption schemes.

Attribute-Based Encryption (ABE) is an extension of identity-based encryption, in which decryption can be enabled based on the result of a predicate evaluation on several attribute values associated with the recipient. This is accomplished by assigning a priori a number of private keys corresponding to these attributes. Attribute-Based Encryption (ABE) allows fine-grained control of the decryption capability for a message; the sender does not necessarily need to know the recipients' precise identities but can instead define an attribute-based policy for the recipients.

Attribute-Based Encryption (ABE) can be further abstracted into predicate encryption and functional encryption. Hidden index predicate encryption schemes are useful when the access policy or individual attributes consist of sensitive information and, therefore, should also remain hidden from unauthorized parties. ABE and predicate encryption are special cases of a higher level functionality called functional encryption. Functional encryption schemes allow messages to be encrypted so that decryption of the resulting ciphertext produces an evaluation of the message with respect to a key-dependent function.

We have investigated the use of ABE schemes for object-level protection [57]. These schemes can be combined with classical symmetric encryption schemes, similar to the approach commonly used with traditional public-key cryptography to construct a hybrid encryption scheme. This hybrid encryption scheme limits the performance overhead introduced by using ABE. Several researchers have investigated the use of ABE in the context of IoT, for example, [32, 81]. A recent survey of ABE schemes suitable for IoT can be found in [64].

6.8 Data-Centric Security Architecture for IoT

A secure architecture has been developed that takes into account a complete chain of security: from a trusted sensing platform through a path of secure communication links to the secure processing of information. This architecture is depicted in Fig. 6.3.

Where possible, trusted computing environments can be leveraged to ensure the integrity and authentication of IoT devices, including purposely deployed sensors, edge devices, and end user terminals used to access situational awareness systems. The wireless links between sensors and devices are secured with technology-specific cryptographic mechanisms, suitable for use on computationally constrained platforms and over bandwidth-restricted wireless communication channels. The trusted edge devices perform authentication of the sensor devices and verify the integrity and confidentiality of data received from these devices. Finally, edge devices apply OLP to the information obtained from the sensors, that is, each data object is individually labeled and encrypted, enabling end-to-end confidentiality and integrity protection of its content at the application level.

The hybrid encryption scheme used for OLP combines several state-of-the-art cryptographic mechanisms. First, ABE is used to enforce CAC based on ABAC policies. Employing ABAC policies can enforce fine-grained control of access to the data, taking into account the attributes of an object or subject and the context of access, such as location or time. In addition, a simple secret-sharing scheme ensures that authorized users can access the data only when using an authorized terminal. For example, while an authorized user can access some of the data from any device, including a smartphone, more sensitive data can require access from a secure terminal located in a command center. Finally, using partially or fully

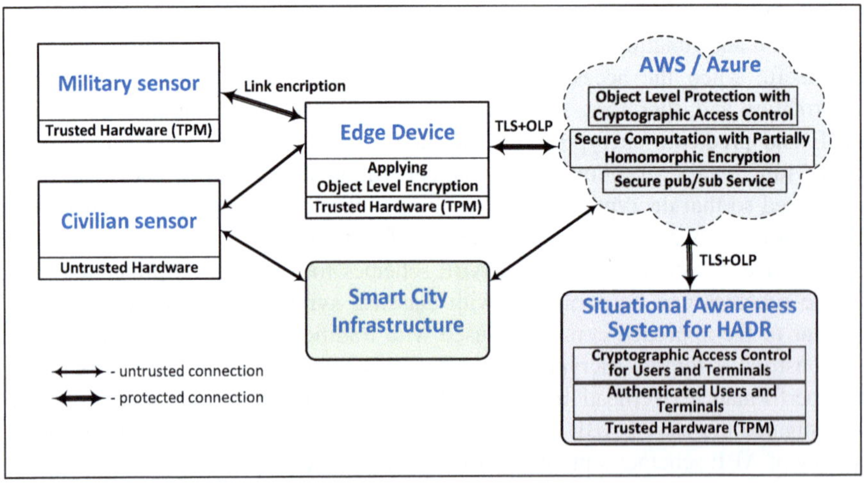

Fig. 6.3 Security mechanisms implemented within IST-147 proof-of-concept demonstrator

homomorphic encryption enables secure processing of the encrypted data, reducing the need for trust into processing infrastructure, and thus supporting new use cases, such as processing of the data in a public cloud. The existing fully homomorphic encryption schemes are still rather computationally inefficient [46]. However, some partially homomorphic schemes, such as Paillier encryption [60], can effectively calculate simple statistical measures on sensor data, such as average or total value. A separate class of encryption schemes can be used for data where the main objective of processing is to perform a search, enabling efficient queries on encrypted data stored in a public cloud [62].

A carefully planned coordination between military and civilian organizations might allow one to implement an a priori federation of identity and access management, enabling emergency response teams to leverage smart city services and assets when needed according to a predefined security policy. An approach to meeting legal requirements and facilitating public acceptance is partial data anonymization and/or perturbation to preserve the privacy of citizens [26]. Alternatively (or in complement), smart city platforms could be designed to support a *breaking glass* policy and enter an emergency mode when needed. Breaking glass security policies are designed to handle dire emergency situations and enable a controlled override of standard security measures. When operating in a breaking glass mode, smart city platforms should enforce strict auditing and logging measures, enabling an a posteriori analysis of operations performed during emergencies and their attribution, consequently ensuring appropriate accountability and enforcement of responsibility.

Our approach is to apply the Data-Centric Security (DCS) paradigm to the IoT environment. DCS has recently been proposed for use within military systems and is the subject of an ongoing investigation within the ACT TIDE Community. The core concept behind DCS is that protection is provided at the level of the data object, as opposed to the lower layers, such as the network or session, and that the type of protected data determines the applied security measures.

We aimed to experiment with a flexible DCS architecture for IoT devices. In our architecture, low-cost and low-energy sensors based on commercial technology collect relevant information (e.g., movement data, temperature, and seismic activity) from their vicinity. These sensors then communicate their information to a local gateway/sink node, which relays it to a commercial cloud storage service. The sink node can be static or mobile, as described in the proposed use cases.

The concept of Content-based Protection and Release (CPR), developed at the NATO Communications and Information Agency (NCIA) [5], offers a specific instantiation of DCS. The CPR policies are a special case of ABAC policies. In particular, CPR defines two ABAC policies: a *release policy*, determining who can access the content, and a *protection policy*, prescribing how the content needs to be protected, for example, specifying the minimum security requirements on devices used to process and transport the content. An example of a naive release policy formulated in plain English would be a policy stating that a logistics officer involved in the Resolute Support mission has access to information about the movement and temperature of containers with supplies at the Kandahar Air Field base. In this case, the user's attributes are rank (officer), discipline (logistics), and assignment

(Operation Resolute Support). Possible content attributes are type of content (food supplies), precision (containers), and type of metadata (location and temperature). Accordingly, a naive protection policy could state that the above information can only be processed on a terminal with particular attributes, for example, equipped with hard drive encryption and biometric authentication.

In our initial architectural pattern, the sensors use only symmetric encryption to communicate the detected information to the gateway securely. The gateway then re-encrypts the information in a way that provides support for CAC enforcement. Several design patterns are possible here to ensure the appropriate integration of CAC. First, a fully symmetric key-based solution can be implemented, as proposed, for example, in [1] or [13]. Second, a traditional hybrid solution could be used, employing a symmetric algorithm for content encryption and public-key cryptography for key distribution. This design pattern has the advantage of relying on well-known and currently deployed technologies at the cost of potentially cumbersome key management required to enforce the DCS access control policies. Finally, attribute-based or functional encryption could be used to enforce access control policies directly within the encryption process, thus simplifying the required key management processes. These design patterns align with the OLP integration levels, as defined in [58].

Our initial solution does not remove the need to pre-distribute the symmetric keys to the sensing nodes. This is due to a resource limitation of the Arduino nodes used in our current prototype and could be reconsidered in the future when some of the functionality currently implemented in the sink node could be moved to the more performant individual sensor nodes. Nevertheless, the issue of symmetric key pre-distribution is limited in our scenario to the local sensor network, formed by the sensor nodes and the sink node; a survey of possible solutions to this problem is provided in [70, 80, 82]. Finally, note that our solution does remove the need to pre-distribute symmetric keys to the information consumers. This offers an advantage, as the population of the information customers is potentially highly heterogeneous and dynamic.

6.9 Security in Publish-Subscribe Services

Publish-subscribe approaches applicable to IoT systems have been introduced and discussed in detail in Sect. 4.6.2. In this section, we discuss various approaches for securing publish-subscribe services in federated environments.

In the context of DCS architecture introduced in Sect. 6.8, an interesting security approach based on Policy Attribute-Based Encryption (CP-ABE) is proposed in Secure MQTT [71]. This solution can be used in both the MQTT and MQTT-SN environments. It is attractive from the point of view of DCS, as it allows the sensor node to perform an encryption procedure once based on a set of attributes that accompany the encrypted data. Data secured in this way can be transmitted over the network unchanged and decrypted by multiple recipients, but only by those who meet the requirements specified in the attributes of the encrypted data. The shortcomings of this solution are that it uses the MQTT frame of the "0000" type, the use of which the MQTT protocol specification does not describe, and additionally that the MQTT broker has a cryptographic key generator function. As a consequence, the standard MQTT software available in the market might not support this solution. Furthermore, such a solution is burdened with the problem of revocating compromised keys, a feature of Ciphertext Policy Attribute Based Encryption (CP-ABE). There are indeed solutions in which a time parameter (validity time) is added to the list of protected data attributes. However, this approach only partially solves the problem because there is no practical possibility of revoking the key before the validity time expires. Shortening the validity time, in turn, causes a large transmission overhead after setting a new validity time and generating new keys.

Another solution based on the MQTT protocol is Key Generating and Renewing (KGR) [22]. The KGR system generates, renews, and securely distributes symmetric cryptographic keys to sensor nodes. Sensor nodes only use this system to generate or renew keys and can normally function in other systems that use KGR-generated keys to secure data storage and exchange. The structure of this system is shown in Fig. 6.4.

Clients in the KGR system can be sensor nodes (N_i) or other nodes that require secure data exchange with N_i sensor nodes. The Key Server (KS) node acts as a cryptographic key generator. It is responsible for authenticating the nodes that request key generation or renewal and distributing the generated keys to the N_i nodes. The KS node stores the generated keys only until the distribution of these keys is completed. The security mechanisms of the TPM v. 2.0 protect all procedures related to the operation of the KS node and N_i nodes. The Authorization Center (AC) node is responsible for preparing the credentials for the nodes using the KGR system. The system also uses another node that acts as an MQTT broker. However, this node is only an intermediary for data exchange between other system nodes

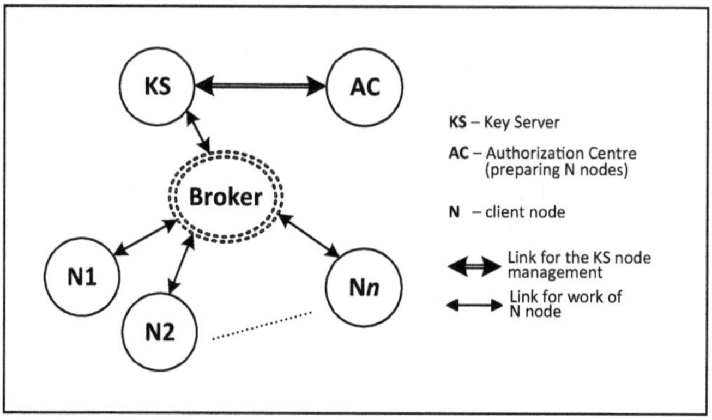

Fig. 6.4 The structure of Key Generating and Renewing (KGR) system

and therefore does not require access to the generated keys and the meaning of the topics used in the KGR system. As the topics used during individual phases of the system's operation have random values, the approach is suitable for sensor nodes operating in the IoT environment with high confidentiality requirements.

6.10 Related Work and Open Issues

There is a wide choice of middleware solutions proposed for IoT (and earlier wireless sensor network) applications [65]. Nevertheless, most of these solutions are proprietary and either do not use commercially available components or do not account for NATO's multitude of security requirements [76]. In addition, only a handful of solutions have been designed with end-to-end security in mind. For example, an approach similar to ours, offering end-to-end security using ABE for Constrained Application Protocol (CoAP)-based IoT applications, was presented in [12]. However, this solution utilized ABE to wrap the symmetric keys used for exchanging CoAP messages between IoT nodes. In contrast, we use ABE to cryptographically enforce access control policies for consumers of the data gathered from the sensing nodes. More generally, the performance impact of enforcing ABAC policies on IoT applications was investigated in [43].

An interesting aspect of using public cloud services for security-sensitive applications is the potential to process encrypted content. Several cryptographic mechanisms have been proposed that provide the capability for some form of encrypted data processing without the need for decryption. For a recent survey of these mechanisms, see, for example, [41]. In general, these schemes decrease in efficiency with increasing richness of the processing functions supported by a scheme. Our future plans include investigating various lightweight methods for aggregating encrypted content at the gateway layer and using more robust processing of encrypted data in the public cloud, including support for verifiable computation.

The performance of symmetric encryption on an Arduino platform was investigated in [8] for AES block cipher and in [36] for Rivest Cipher 4 (RC4) stream cipher. These results confirmed that state-of-the-art symmetric algorithms can be implemented effectively in Arduino. Similar results for Raspberry Pi were presented in [69] and [6]. In addition, the use of bilinear pairing on the Raspberry Pi platform was investigated in [11] in the context of Boneh–Lynn–Shacham signature scheme (BLS) and in [38] in the context of noninteractive zero-knowledge proofs, indicating some performance limitations of single-board computers in the context of implementation of these more advanced cryptographic primitives.

Few recent publications focus on evaluating the performance of ABE schemes. In [40] a performance evaluation of ABE was attempted in the context of IoT applications; however, performance measurements were made in a relatively powerful virtual machine environment. The results of a comparative analysis of CP-ABE and Key Policy Attribute Based Encryption (KP-ABE) in the context of IoT were presented in [75], and similar results were obtained for an Android-based implementation of ABE in [3].

An essential objective of DCS is to enforce information-specific security policies at all IT system layers, including the application and network layers [79], and with respect to all security dimensions, that is, C-I-A [50]. Our future work aims to address these cross-layer objectives.

6.11 Conclusion

We have discussed several use cases and related threats applicable to the use of IoT systems in military applications. Furthermore, we presented various aspects of IT network security in military applications, the places to apply hardware TPM to build security mechanisms, and how to use services based on the pub/sub mechanism. Moreover, we paid particular attention to the MQTT service, which can enable secure data exchange between the IoT network and federated C2 systems.

Finally, we presented the concepts of Object-Level Protection (OLP) and Cryptographic Access Control, which are currently considered possible security models for future NATO operations. We concluded that OLP and CAC can be applied to IoT to provide end-to-end data protection in transit and storage.

The deployment of CAC in IoT introduces several challenges, including key management and implementation of cryptographic algorithms in constrained environments. The computational overhead introduced by many modern cryptographic mechanisms is too cumbersome for many of today's IoT devices. However, experience has shown that device performance improves much faster than expected and that devices can, in a short time, improve sufficiently to perform even complex cryptographic algorithms. Although there will always be a market space for low-end IoT devices, for example, nanothings [2], the majority of IoT devices will be powerful enough to support state-of-the-art encryption mechanisms such as ABE before the regular life cycle of military systems results in the widespread adoption of IoT technology.

An interesting open research question concerns the extent to which data can be kept encrypted throughout its entire life cycle, that is, not only in storage and transit but also during processing [42].

Our work aimed to investigate the feasibility of implementing DCS mechanisms for MIoT applications while relying on commercially or openly available software and hardware components and services to the greatest extent possible. Our proof of concept is a first step toward building a working prototype of a secure IoT system based on commercial components, which could be used for larger-scale experimentation and as a use case for the validation of NATO DCS concept.

The initial evaluation of available hardware and software libraries and public cloud services shows that existing commercial and open-source components offer sufficient functionality to implement a wide variety of MIoT applications at a lower cost compared to fully customized solutions. In addition, the reduced reliability and environmental resistance of the components could be mitigated by using redundant devices and services. Finally, the architectures relying on commonly used IoT components and platforms can enable better Civil–Military Cooperation (CIMIC) and the reuse of civilian IoT resources available in the conflict area for military purposes.

References

1. Akl, S.G., Taylor, P.D.: Cryptographic solution to a problem of access control in a hierarchy. ACM Trans. Comput. Syst. **1**(3), 239–248 (1983)
2. Akyldiz, I., Jornet, J.M.: The Internet of Nano-Things. IEEE Wirel. Commun. **17**(6), 58–63 (2013)
3. Ambrosin, M., Conti, M., Dargahi, T.: On the feasibility of attribute-based encryption on smartphone devices. In: Proceedings of IoT-System, pp. 49–54 (2015)
4. Armando, A., Grasso, M., Oudkerk, S., Ranise, S., Wrona, K.: Content-based information protection and release in NATO operations. In: Proceedings of the 18th ACM Symposium on Access Control Models and Technologies (SACMAT), p. 261–264. Amsterdam, The Netherlands (2013)
5. Armando, A., Oudkerk, S., Ranise, S., Wrona, K.: Formal modelling of content-based protection and release for access control in NATO operations. In: Proceedings of the Foundation of Practical Security, vol. 8352, pp. 227–244 (2013)
6. Banerjee, S., Sethia, D., Mittal, T., Arora, U., Chauhan, A.: Secure sensor node with Raspberry Pi. In: Proceedings of the International Conference on Multimedia Signal Processing and Communication Technologies (IMPACT), pp. 26–30 (2013)
7. Barcena, M.B., Wueest, C.: Insecurity in the Internet of Things. Whitepaper, Symantec (2015)
8. Biswas, K., Muthukkumarasamy, V., Wu, X.w., Singh, K.: Performance evaluation of block ciphers for wireless sensor networks. In: Advanced Computing and Communication Technologies, vol. 452, pp. 443–452. Springer, Berlin (2016)
9. Buriro, A., Crispo, B., Del Frari, F., Wrona, K.: Touchstroke: smartphone user authentication based on touch-typing biometrics. In: New Trends in Image Analysis and Processing–ICIAP 2015 Workshops, pp. 27–34. Springer, Berlin (2015)
10. Checkoway, S., Mccoy, D., Kantor, B., Anderson, D., Shacham, H., Savage, S., Koscher, K., Czeskis, A., Roesner, F., Kohno, T.: Comprehensive experimental analyses of automotive attack surfaces. In: SEC'11 Proceedings of the 20th USENIX conference on Security (2011)
11. Chifor, B.C., Bica, I., Patriciu, V.V.: Sensing service architecture for smart cities using social network platforms. Soft Comput. **21**(16), 4513–4522 (2017)
12. Choi, J., In, Y., Park, C., Seok, S., Seo, H., Kim, H.: Secure IoT framework and 2D architecture for End-To-End security. J. Supercomput. **2**(i), 1–15 (2016)
13. Crampton, J.: Practical and efficient cryptographic enforcement of interval-based access control policies. ACM Trans. Inf. Syst. Secur. **14**(1), 1–30 (2011)
14. DARPA Microsystems Technology Office: N-ZERO Envisions 'Asleep-yet-Aware' Electronics That Could Revolutionize Remote Wireless Sensors. DARPA (13 Apr 2015). https://www.darpa.mil/news/2015/n-zero-aware-electronics. Accessed 28 Sept 2025
15. Division, N.I.T.L.A.C.: NIST cybersecurity for IoT program. https://www.nist.gov/itl/applied-cybersecurity/nist-cybersecurity-iot-program. Accessed 01 July 2023
16. Domingo, A., Wietgrefe, H.: A NNEC-compliant approach for a future mission network. In: Proceedings of the Military Communications Conference (MILCOM) (2012)
17. ENISA: Good Practices for Security of IoT: Secure Software Development Lifecycle. Technical report, November, ENISA (2019). www.enisa.europa.eu

18. ENISA: Internet of things (IoT). https://www.enisa.europa.eu/topics/iot-and-smart-infrastructures/iot/. Accessed 01 July 2023
19. Fagan, M., Marron, J., Watrobski, P., Souppaya, M., Mulugeta, B., Symington, S., Harkins, D., Barker, W., Richardson, M.: Trusted Internet of Things (IoT) device network- layer onboarding and lifecycle management. Special Publication December, NIST (2022)
20. Ferraiolo, D.F., Sandhu, R., Gavrila, S., Kuhn, D.R., Chandramouli, R.: Proposed NIST standard for role-based access control. ACM Trans. Inf. Syst. Secur. **4**(3), 224–274 (2001)
21. Fisher, K.: HACMS: high assurance cyber military systems. In: HILT'12 Proceedings of the ACM SIGAda Annual Conference on High Integrity Language Technology (2012)
22. Furtak, J.: Cryptographic keys generating and renewing system for IoT network nodes—a concept. Sensors **20**(17), 5012 (2020)
23. Furtak, J., Zielinski, Z., Chudzikiewicz, J.: Security techniques for the WSN link layer within military IoT. In: Proceedings of the WF-IoT, pp. 233–238 (2016)
24. Furtak, J., Zieliński, Z., Chudzikiewicz, J.: A framework for constructing a secure domain of sensor nodes. Sensors **19**(12), 2797 (2019)
25. Furtak, J., Zieliński, Z., Chudzikiewicz, J.: Procedures for sensor nodes operation in the secured domain. Concurrency Comput. Pract. Exper. **32**(13), e5183 (2020)
26. Garfinkel, S., Abowd, J.M., Martindale, C.: Understanding database reconstruction attacks on public data. Commun. ACM **62**(3), 46–53 (2019)
27. GlobalPlatform Inc.: Global platform device technology TEE internal API specification—version 1.0. Technical report, GlobalPlatform (2011)
28. Greenberg, A.: Hackers can disable a sniper rifle—or change its target. Wired Magazine (2015). https://www.wired.com/2015/07/hackers-can-disable-sniper-rifleor-change-target/
29. Halperin, D., Heydt-Benjamin, T.S., Fu, K., Kohno, T., Maisel, W.H.: Security and privacy for implantable medical devices. IEEE Pervasive Comput. **7**(1), 30–39 (2008)
30. Halperin, D., Heydt-Benjamin, T., Ransford, B., Clark, S.S., Defend, B., Morgan, W., Fu, K., Kohno, T., Maisel, W.H.: Pacemakers and implantable cardiac defibrillators: software radio attacks and zero-power defenses. In: Symposium on Security and Privacy (2008)
31. Hartmann, K., Steup, C.: The vulnerability of UAVs to cyber attacks—an approach to the risk assessment. In: 5th International Conference on Cyber Conflict (CyCon), 2013, pp. 1–23 (2013)
32. Hernández-Ramos, J., Bernabe, J., Moreno, M., Skarmeta, A.: Preserving smart objects privacy through anonymous and accountable access control for a M2M-enabled Internet of Things. Sensors **15**(7), 15611 (2015)
33. Hu, V.C., Friedman, A.R., Lang, A.J., Cogdell, M.M., Scarfone, K., Kuhn, R., Ferraiolo, D., Schnitzer, A., Sandlin, K., Miller, R.: Guide to attribute based access control (ABAC) definition and considerations (Draft). Technical report, National Institute of Standards and Technology, NIST (2013)
34. IERC: European research cluster on the Internet of Things. http://www.internet-of-things-research.eu/index.html. Accessed 01 July 2023
35. Jin, X., Krishnan, R., Sandhu, R.: A unified attribute-based access control model covering DAC, MAC and RBAC. In: DBSec'12: Data and Applications Security and Privacy XXVI, no. 7371 in LNCS, pp. 41–55 (2012)
36. Kabir, A., Sultanul, F.M., Shorif, M.A., Li, H., Yu, Q.: A study of secured wireless sensor networks with XBee and Arduino. In: 2nd International Conference on Systems and Informatics, ICSAI, pp. 492–496 (2014)
37. Kiviharju, M.: Towards pervasive cryptographic access control models. In: SECRYPT (2012)
38. Lajoie-Mazenc, P., Anceaume, E., Guette, G., Sirvent, T., Tong, V.V.T.: Privacy-preserving reputation mechanism: A usable solution handling negative ratings. In: IFIP Advances in Information and Communication Technology, vol. 454, pp. 13–29. Springer, Berlin (2015)
39. LaPadula, L.J., Bell, D.E.: Secure computer systems: mathematical foundations. Technical report, MITRE Technical Report 2547, vol. II (1973)
40. Li, D., Aung, Z., Williams, J., Sanchez, A.: P3: Privacy preservation protocol for automatic appliance control application in smart grid. IEEE Internet Things J. **1**(5), 414–429 (2014)

41. Li, S., Gao, J.: Security and privacy for big data. In: Yu, S., Guo, S. (eds.) Big Data Concepts, Theories, and Applications, pp. 281–313. Springer, Berlin (2016)
42. Libicki, M.C., Tkacheva, O., Feng, C., Hemenway, B.: Ramifications of DARPA's programming computation on encrypted data program. Technical report, RAND National Defense Research Institute (2014)
43. Lim, L., Marie, P., Conan, D., Chabridon, S., Desprats, T., Manzoor, A.: Enhancing context data distribution for the Internet of Things using QoC-awareness and attribute-based access control. Ann. Telecommun. **71**(3–4), 121–132 (2016)
44. Luscombe, R.: Dick Cheney feared assassination by shock to implanted heart defibrillator. The Guardian **10**(19) (2013). https://www.theguardian.com/world/2013/oct/19/dick-cheney-heart-assassination-fear. Accessed 28 Sept 2025
45. Maisel, W.H., Kohno, T.: Improving the security and privacy of implantable medical devices. N. Engl. J. Med. **362**(13), 1164–1166 (2010)
46. Marcolla, C., Sucasas, V., Manzano, M., Bassoli, R., Fitzek, F.H.P., Aaraj, N.: Survey on fully homomorphic encryption, theory, and applications. Proc. IEEE **110**(10), 1572–1609 (2022). https://doi.org/10.1109/jproc.2022.3205665
47. Markey, E.: Tracking & hacking: Security & privacy gaps put American drivers at risk. Technical report, US Senator for Massachusetts (2015)
48. Marty, J.A.: Vulnerability analysis of the MAVLink protocol. Ph.D. thesis, Air Force Institute of Technology, Wright-Patterson Air Force Base, Ohio (2014)
49. McCarthy, C., Harnett, K., Carter, A.: Characterization of potential security threats in modern automobiles: a composite modeling approach. Technical report, October, National Highway Traffic Safety Administration, Washington (2014)
50. Melrose, J., Wrona, K., Guenther, T., Haakseth, R., Nordbotten, N., Westerdahl, L.: Labelling for integrity and availability. In: Proceedings of the International Conference on Military Communication and Information Systems (ICMCIS) (2016)
51. Miessler, D.: Securing the Internet of Things : Mapping attack surface areas using the OWASP IoT Top 10. In: RSA Conference (2015)
52. Miller, C., Valasek, C.: A survey of remote automotive attack surfaces. Technical report, IOActive (2014)
53. National Cybersecurity Center of Excellence (NCCoE): Internet of things (IoT). https://www.nccoe.nist.gov/iot. Accessed 01 July 2023
54. NATO Standardization Office: NATO core metadata specification. NATO Standardization Agreement (STANAG) 5636, NATO Standardization Office (2019)
55. OASIS: eXtensible Access Control Markup Language (XACML) version 3.0. Standard, OASIS (2013)
56. Osborn, S., Sandhu, R., Munawer, Q.: Configuring role-based access control to enforce mandatory and discretionary access control policies. In: ACM Transactions on Information and System Security (TISSEC), pp. 85–106 (2000)
57. Oudkerk, S., Wrona, K.: Cryptographic access control in support of object level protection. In: Proceedings of the Military Communications and Information Systems Conference (MCC), St. Malo (2013)
58. Oudkerk, S., Wrona, K.: A common approach to the integration of object level protection in NATO. In: Proceedings of the STO Symposium on Cyber Security Science & Engineering, Tallinn (2014)
59. OWASP: OWASP IoT Top 10 2018. https://owasp.org/www-pdf-archive/OWASP-IoT-Top-10-2018-final.pdf
60. Paillier, P.: Public-key cryptosystems based on composite degree residuosity classes. In: Eurocrypt, pp. 223–238 (1999)
61. Pellegrini, V., Principe, F., Guidi, R., Demauro, G., V.Martorelli, Cioni, R.: Cryptographically secure radios based on directional modulation. In: 2018 IEEE—ICASSP 2014, pp. 1–8 (2014)
62. Pham, H., Woodworth, J., Amini Salehi, M.: Survey on secure search over encrypted data on the cloud. Concurrency Comput. Pract. Exper. **31**(17), e5284 (2019). https://doi.org/10.1002/cpe.5284, https://onlinelibrary.wiley.com/doi/abs/10.1002/cpe.5284. E5284 cpe.5284

63. Pike, L., Hickey, P., Bielman, J., Elliott, T., Hamberg, E., Dubuisson, T.: Building a high-assurance unpiloted air vehicle. In: MEMOCODE'13 Eleventh ACM-IEEE International Conference on Formal Methods and Models for Codesign (2013)
64. Rasori, M., Manna, M.L., Perazzo, P., Dini, G.: A survey on attribute-based encryption schemes suitable for the Internet of Things. IEEE Internet Things J. **9**(11), 8269–8290 (2022). https://doi.org/10.1109/JIOT.2022.3154039
65. Razzaque, M.A., Milojevic-Jevric, M., Palade, A., Clarke, S.: Middleware for Internet of Things. IEEE Internet Things J. **3**(1), 70–95 (2016)
66. Rieback, M., Crispo, B., Tanenbaum, A.: Is your cat infected with a computer virus? In: Proceedings of the Fourth Annual IEEE International Conference on Pervasive Computing and Communications (PERCOM), pp. 169–179. Washington, DC (2006)
67. Rose, S., Borchert, O., Mitchell, S., Connelly, S.: Zero trust architecture. Special Publication 800-207, NIST (2020)
68. Roufa, I., Miller, R., Mustafa, H., Taylor, T., Oh, S., Xu, W., Grutser, M., Trappe, W., Seskar, I.: Security and privacy vulnerabilities of in-car wireless networks: A tire pressure monitoring system case study. In: Proceedings of the 19th USENIX Conference on Security (2010)
69. Shah, D.K., Bharadi, V.A., Kaul, V.J., Amrutia, S.: End-to-end encryption based biometric SaaS: using raspberry pi as a remote authentication node. In: Proceedings of 1st International Conference on Computing, Communication, Control and Automation (ICCUBEA), pp. 52–59 (2015)
70. Simplício, M.A., Barreto, P.S., Margi, C.B., Carvalho, T.C.: A survey on key management mechanisms for distributed wireless sensor networks. Comput. Netw. **54**(15), 2591–2612 (2010)
71. Singh, M., Rajan, M., Shivraj, V., Balamuralidhar, P.: Secure MQTT for Internet of Things (IoT). In: Fifth International Conference on Communication Systems and Network Technologies, pp. 746–751. IEEE, Piscataway (2015)
72. Sorniotti, A., Gomez, L., Wrona, K.: Secure and trusted in-network data processing in wireless sensor networks: a survey. J. Inf. Assur. Secur. **2**(4), 191–207 (2007)
73. Transformation, N.A.C.: Data-centric security vision and strategy for the alliance federation, including the NATO enterprise. Technical report, NATO (2019)
74. van den Berg, M., de Graaf, P., Kwant, P.O., Slewe, T.: Mass Surveillance—Part 2: Technology Foresight. Report, European Parliament (2015)
75. Wang, X., Zhang, J., Schooler, E.M., Ion, M.: Performance evaluation of attribute-based encryption: toward data privacy in the IoT. In: Proceedings of the IEEE ICC, pp. 725–730 (2014)
76. Wrona, K.: Securing the Internet of Things: A military perspective. In: Proceedings of the IEEE World Forum on Internet of Things (WF-IoT), pp. 502–507 (2015)
77. Wrona, K., Oudkerk, S.: Content-based Protection and release architecture for future NATO networks. In: Military Communications Conference (MILCOM), San Diego, CA (2013)
78. Wrona, K., Oudkerk, S., Hallingstad, G.: Designing medium assurance XML-labelling guards for NATO. In: Proceedings of the Military Communications Conference (MILCOM), pp. 1794–1799. IEEE, San Jose (2010)
79. Wrona, K., Oudkerk, S., Szwaczyk, S.: Content-based security and protected core networking with software-defined networks. IEEE Commun. Mag. **54**(10) (2016)
80. Xiao, Y., Rayi, V.K., Sun, B., Du, X., Hu, F., Galloway, M.: A survey of key management schemes in wireless sensor networks. Comput. Commun. **30**(11–12), 2314–2341 (2007)
81. Yao, X., Chen, Z., Tian, Y.: A lightweight attribute-based encryption scheme for the Internet of Things. Future Gener. Comput. Syst. **49**, 104–112 (2015)
82. Zhang, J., Varadharajan, V.: Wireless sensor network key management survey and taxonomy. J. Netw. Comput. Appl. **33**(2), 63–75 (2010)
83. Zheng, D., Carter, W.: Leveraging the Internet of Things for a more efficient and effective military. Technical report, Center for Strategic & International Studies, Washington, DC (2015)

Chapter 7
Reliability and Fault Tolerance

Zbigniew Zieliński, Janusz Furtak, and Konrad Wrona

7.1 Introduction

The adaptation of the Internet of Things paradigm in the military domain (e.g., Internet of Battlefield Things (IoBT) [23]) emphasizes the connectivity of military things/objects (e.g., military vehicles; weapons; materials; plenty of military sensors; soldier devices; and plenty of embedded, tiny, in many cases even invisible, low-energy consumption devices). These things/objects can exchange information and take contextually dependent actions/decisions even without human (soldiers) participation [21–23]. Thus the MIoT is the interconnection of uniquely identifiable objects within the existing military communication infrastructure. Such a large number and variety of objects/devices will significantly impact the reliability of services in the MIoT system. The MIoT deploys a multitude of network-aware devices in a dynamic, error-prone, and unpredictable environment and is expected to run for a long time (during all military missions) without failure. To commission such a system and to keep it operational, it is therefore essential that the system's design is reliable. Furthermore, a harsh or even hostile physical environment, for example, radiation, fire on the battlefield, or falling of hard objects on sensor hardware, can also completely damage some devices. Hence, faults and failures should be treated as "normal" facts in the MIoT system. In short, the MIoT system has to deal with faults arising from unreliable hardware, limited energy, connectivity interruption, and harsh or hostile physical environments. These findings highlight

Z. Zieliński · J. Furtak
Military University of Technology, Warsaw, Poland

K. Wrona (✉)
NATO Communications and Information Agency (NCIA), The Hague, The Netherlands

Military University of Technology, Warsaw, Poland
e-mail: konrad.wrona@ncia.nato.int

N. Suri et al. (eds.), *Military Applications of Internet of Things*, Internet of Things,
https://doi.org/10.1007/978-3-032-03378-9_7

the need to consider several factors necessary to ensure the reliability and security of MIoT services. Therefore, the main goal is to achieve robustness of the MIoT system against attacks and resilience against failures.

Building a highly reliable computing system has always been an essential requirement for the military. Undoubtedly, when designing a MIoT system, one must consider its basic requirement, that is, safety-critical real-time applications. Therefore, systems used in critical military applications must be highly reliable. An Internet of Things (IoT)system's dependability (reliability) generally means certainty of the operation of the system, which allows users to have reasonable confidence in the results of services (processing) provided by this system.

Reliability refers to the probability that a system provides its services as required throughout the specified period. Along with reliability, other closely related quality attributes include *availability*, *safety*, and *performability* [18]. Informally, availability is the fraction of time a system can be used for its intended purpose (i.e., operational to the user) within a specified period. Note that availability is also one of the essential security attributes. Safety refers to the probability that a system does not fail in such a manner to cause significant damage to its own and/or allied side. Performability refers to the performance level of the MIoT system in the presence of failures. In this chapter, we will focus on reliability and availability, as without them, it is impossible to meet other objectives, such as safety or performability. It is important to stress that reliability and availability objectives are closely related to security requirements—this is particularly true in military systems that are, by design, exposed to a high risk of malicious interference by highly skilled and motivated threat actors.

Assuming that the system is binary in terms of the reliability state: fit—"up" or unsuitable/unfit—"down," $f(t)$ is the failure probability density function (continuous random variable), and T denotes the time for the system to fail, we can formally define the reliability of a system $R(t)$ as the probability that the system will not fail until time t [19]. It can be expressed as $R(t)=\Pr\{T > t\}=1 - F(t)$, where $F(t)$ is the failure function: $F(t)=\Pr\{T \leq t\}=\int_0^t f(\tau)\,d\tau$.

Similarly to the reliability measure (the average system uptime), the Mean Time to Failure (MTTF) for the system is the expected value of the failure density function $= \int_0^\infty tf(t)dt$, which can be ultimately represented as $\int_0^\infty R(t)dt$.

Availability $A(t)$ means the ratio of the time during which the system is fit for the range $[0, t]$ to the length of this range. Availability at time t, $A_p(t)$, is the probability that the system is fit for the moment t. The long-term availability $A=\lim_{t\to\infty} A(t)=\lim_{t\to\infty} A_p(t)$. This measure is used in recovery/repair systems. Knowing the MTTF and Mean Time to Repair (MTTR) values, we can express the availability as $A=\frac{MTTF}{MTTF+MTTR}$.

From the above definition, we can conclude two main observations:

1. It is possible to design a highly available MIoT system even with components with relatively small MTTF (which is common in the case of using some COTS sensors and systems).

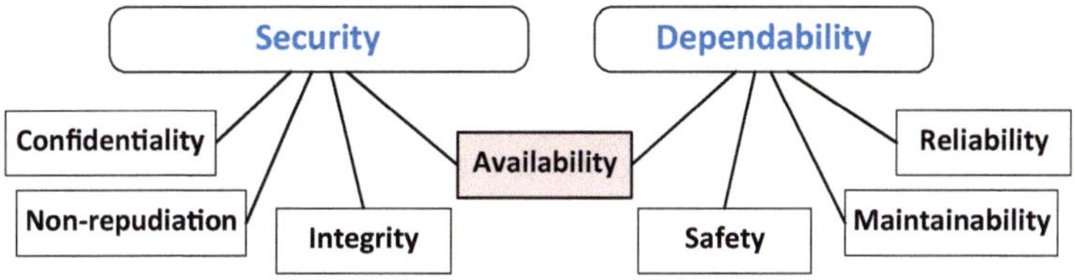

Fig. 7.1 Relationships between basic security and dependability attributes

2. It becomes quite interesting for the MIoT system to adopt the Recovery-Oriented Computing (ROC) [17] paradigm.

In the context of MIoT, the use of the traditional measures of reliability or availability mentioned above is often insufficient. Therefore, an integrated approach is necessary to provide a holistic perspective that includes all the dependability and security attributes. In particular, there is a need to capture the impact of random faults, as well as nonrandom, deliberate adversarial actions aimed at degrading systems. The relationship between fundamental dependability and security attributes is presented in Fig. 7.1. These basic attributes do not include all possible features relevant to MIoT, for example, resilience [14] and survivability [6] are other useful measures in the context of military systems.

7.2 Characteristics of the MIoT and Reliability Challenges

The MIoT architecture depends on a variety of military application scenarios and, in general, should synthesize the characteristics and requirements of all of them. In particular, it should describe the general services of IoT and a structured representation of the function, behavior, and role of entities. In [22], a five-layer architecture of IoT for military purposes was proposed, which included the following layers: sense, access, network, service, and application.

We will use the concept of a layered MIoT architecture presented in Chap. 3 to identify the main reliability challenges and robustness.

We believe that the Perception and Actuation and Communication Layers are where key reliability issues in the MIoT can occur. Furthermore, the Perception Layer is more vulnerable to an external attack, such as an external attack against a battery of WSNs, which can also influence MIoT reliability. Therefore, establishing lightweight and effective intrusion detection and recovery mechanisms to improve system robustness is another important security problem. Meanwhile, establishing a trust and reputation model to evaluate suspicious node behavior can significantly reduce the influence of malicious nodes. Conventional intrusion detection methods need to be modified and improved for application to the IoT due to certain limitations, including resource-constrained devices, the limited memory and battery capacity of nodes, and specific protocol stacks. For example, a lightweight IDS for anomaly detection in the IoT proposed in [7] aimed at detecting a common type of DDoS attack. This work proposed a lightweight attack detection strategy that uses a supervised machine learning-based SVM to detect an adversary that tries to inject unnecessary data into the IoT network. The proposed IDS focused on two prominent issues: the attribute of the receiving data to classify the signal and the machine learning-based classifier. An SVM-based classifier was used for classification purposes, with input as two or three simple features.

7.2.1 Categories of the Military Applications

Based on [19], we can classify MIoT applications into the following categories from a reliability and availability perspective:

1. *Mission-critical applications* (no failure tolerance): This includes MIoT systems placed in a mission-critical scenario, particularly during health care and evacuation of wounded soldiers from an area at risk during a military operation. Another vital mission-critical scenario is building a precise image/map of threats on the battlefield to develop commanders' decisions on the rapid impact and destruction of targets. In these cases, any network-aware devices monitoring the soldiers' health or system components recognizing threats that interact with a middleware platform do not tolerate any failure during their mission time (i.e., when they are actively working).
2. *Recoverable*: In this category, it is acceptable from the end user's point of view to recover the fitness of applications after a system (or its component) failure while meeting the MTTR requirements. For example, a MIoT system for logistics (supply chains) with components embedded in vehicles can afford to restart if the embedded component fails.
3. *Error tolerant* (acceptance of some errors): Erroneous behavior of some system parts can be tolerated within the safety limits defined by the end user as long as the malfunctioning parts return to correct functioning within a predefined time. For example, an unmanned reconnaissance system that provides routine information about CBRN threats in a given area in real time can afford to send bad/incorrect video data before correcting it.

7.2.2 Failure Scenarios

Employing a vast number of smart objects, some of which will be civilian devices with relatively low reliability (e.g., from COTS) and their dependencies on wireless communication technologies, makes MIoT more vulnerable to faults. These faults may pose daunting challenges to experts as managing the participating MIoT components becomes increasingly important.

By the term fault (defect), we mean the root cause of an error. Note that defects can arise due to design and implementation errors, material defects, physical damage, etc. An error is a case of incorrect system operation that causes an unwanted change in the system state, for example, a data transmission error or arithmetic operation error, which can result in a system failure. Failure is the system's or its component's visible inability to perform the required function within specified limits. A fault (or failure) can be a hardware defect or software/programming error (bug). In contrast, an error is a manifestation of fault/failure/bug [11].

Hardware faults can be classified according to several aspects. With respect to their duration, hardware faults can be classified as permanent, transient, or intermittent. A permanent fault reflects the permanent going out of commission of a component. A transient fault causes a component to malfunction for some time, but the fault disappears after that time and the functionality of the component is fully restored. An intermittent fault never disappears entirely; it oscillates between dormant and active. Hardware faults can also be classified as benign or malicious. A fault that causes a unit to fail (silent) is called benign. Such faults are the easiest to manage. Far more tricky are the faults that cause a unit to produce a reasonable but incorrect output or make a component "act maliciously" and send, for example, differently valued output to different receivers. These are called malicious or Byzantine faults [11].

Many researchers and practitioners have extensively studied various faults and remedial actions to keep a system operational in mission-critical scenario systems. In general, these faults are categorized [19] as (1) development faults, (2) infrastructure faults, and (3) interaction faults. Development faults are caused by incorrect software or hardware design and implementation, whereas unanticipated hardware faults cause infrastructure faults. Interaction faults occur due to interaction with other software modules/components or incorrect data format. In addition, the external environment can provide unexpected input to IoT entities, resulting in a computation failure in the MIoT devices.

The cluster of networked devices in a MIoT-based system is expected to operate in unanticipated scenarios. However, some of these scenarios can lead to infrastructure failures; for example:

1. In a given IoT military application scenario, network-enabled devices are embedded in a specific environment (e.g., an unmanned vehicle) to gather and process the data stream. These devices can fail due to their physical condition (e.g., radiation exposure) and interference with the environment in which they operate.
2. The processors/microcontrollers of IoT sensor nodes, usually with limited computational and communication resources, have been designed considering a small form factor rather than making them highly reliable. This applies in particular to a *sensing unit* and significantly impacts the reliability of the entire sensor node. As a result, the IoT sensor nodes can be much more fault-prone than ordinary computers [20].

3. Many MIoT devices run on batteries, which can severely limit their computing time and can cause unexpected termination of a computation.
4. Faults in the network communication layer (Fig. 3.15), where the entire network or the communication components can fail

Consider the architecture reference model shown in Fig. 3.15, where the service platform acts as the base of the Data Management and Intelligent Service Layers. The platform will be built from different third-party components and integrate many third-party products. Even if we assume that this middleware has been thoroughly tested for its functionality, many transient faults can occur in individual components. In some cases, the reliability of third-party components can be questionable and can often cause a main system failure. Furthermore, external systems connected to the IoT application middleware can fail or provide incorrect data, resulting in middleware failure. We treat these failures caused by platforms built with various third-party components as infrastructure errors.

The causes of interaction faults in MIoT can be:

1. Interoperability errors of heterogeneous devices, the causes of which may lie in the irregularities, misunderstanding, or misinterpretation of the protocols and data formats used
2. Interaction faults caused by unexpected workload coming from various MIoT components, which different suppliers often produce

7.2.3 The Fault Model

The fault model should define the behavior of IoT units/components when they become unfit. Failure can lead to various consequences for which system failure classes are determined: fail-stop, fail-crash, omission fault, time-stop fault, incorrect computation, and Byzantine (malicious) fault.

The set of allowable types of fault determines a fault model. At each level of abstraction of the MIoT, we can distinguish different types of faults. In general, system faults can be grouped into several categories: sensor faults, actuator faults, microcontroller/processor faults, communication links faults, and system faults caused by software or hardware faults.

In the sensing layer of MIoT, the components of the nodes can be divided into two groups. The first group consists of nodes with a processor/microcontroller, storage subsystem, and power supply infrastructure. The second group includes sensors and actuators. All components of the first group are much more reliable than those of the second group, which means that they have a much lower failure rate [3]. However, we cannot exclude the possible simultaneous occurrence of microprocessor and sensor faults. Some examples of sensor faults to consider might include [12] systematic calibration error, random noise error, and complete malfunctioning. Systematic errors could be addressed through time redundancy, whereas complete malfunctioning faults should be resolved using hardware redundancy. Faulty nodes and their sensors should be diagnosed and isolated from the system. At the network layer of MIoT, we assume communication links faults and communication node failures (base station, gateway faults, etc.).

7.2.4 *Reliability Challenges for MIoT*

To guarantee the quality of service and performance of IoT applications, it is essential to detect faults and perform something akin to healing and recovery from events that might cause faults or node misbehavior in the network. Moreover, some military applications are used in real-time mode (or near real time) and require high reliability and data processing efficiency throughout a mission. Consequently, in the design phase of a MIoT application, several factors that influence reliability must be seriously considered, including self-organization, self-diagnosis, fault tolerance, and security.

Making Service Available to the User

Most MIoT applications aim to provide an immersive service experience through a tightly coupled human-device interaction in real time. Therefore, the user's perspective is crucial when judging the system's availability. This is known as "user-perceived availability' [5], where perceived availability is about providing the service to the user, not just surviving through failure. The IoT service platform has an essential role to play in improving user perception of availability by ensuring that user service requests always receive a response, even in extreme circumstances.

It is quite likely that a MIoT system will have a set of devices that can dynamically enter and join the system. In such a case, ensuring serviceability without disrupting ongoing activities is more difficult in the case of IoT devices. In a traditional high-availability system, the nodes and software running on them go through scheduled maintenance when the software and patches are updated to prevent any impending failure. However, this traditional type of maintenance may not be feasible for an IoT system.

Reliability at the Communication Layer

The MIoT networks are not only mostly heterogeneous but can also be unmanaged (e.g., logistic applications). This implies that the network must be able to configure itself as environmental conditions or components in the network itself change so that information can always be transmitted reliably from one application to another. Moreover, there can be additional complexities when using sensor devices. The links utilized in most sensor networks today use completely unregulated frequency bands. As a result, signals from one sensor device can very easily interfere with those from another, making the links unreliable. For example, if a newly deployed IoT sensor network starts to use the same channel as someone's existing WLAN access point, interference can disable critical sensor data reporting. Links in sensor networks are often more unreliable than the Internet because of a lack of regulation.

Reliability at the Device (Perception) Level

Even when the network is reliable, there are scenarios in which the applications running on these devices may generate poor-quality data, making the entire computation unreliable. For example, consider a scenario where a device needs to gather images from the physical world where it is embedded. Due to environmental conditions, the quality of the captured images can be below the acceptable level; as a result, the associated inferences drawn from the captured images become unreliable. For sensor devices, the environmental condition can result in bit error.

7.3 Fault-Tolerant Techniques for the MIoT

The foundation of all Fault-Tolerance Techniques (FTT) is redundancy (hardware, information, or time). "Classical" fault tolerance [11] considers three main types of concern: fault models, fault detection and diagnosis, and resilience mechanisms (fault recovery). As mentioned previously, the set of allowable types of faults determines a fault model. In addition, each level of abstraction has its types of faults. Fault detection is a phase that recognizes that an unexpected event has occurred. For MIoT applications, online detection is the most crucial to identify fault in real time and is performed simultaneously during normal network operation. Fault tolerance in distributed systems (also in WSNs) has been the subject of active research in recent decades. In [12], a framework was designed to create highly autonomous fault-tolerant distributed sensor networks with plug-and-play capabilities to diagnose faulty sensors and reconfigure the manufacturing process control systems network.

The field of fault tolerance already comprises a considerable body of work. However, applying FT techniques to MIoT requires a comprehensive approach that accounts for several specific characteristics of the military domain and ensures the high availability and security of services within MIoT. Although it is crucial to make the MIoT systems reliable and secure by design, the possibility of faults can never be excluded. This relates to economic factors, as designing high-assurance systems is often prohibitively expensive and time consuming. Another factor is the unpredictability of a military operational environment. Therefore, MIoT must be able to handle faults that affect it during the execution of the mission. Figure 7.2 shows the relationships between the proposed dependability techniques in MIoT. The MIoTcan use two basic dependability techniques: fault prevention, which is a proactive technique, and fault tolerance, which is a reactive fault management technique. In addition, hybrid approaches are possible, such as Virtual Fault Tolerance and using self-healing, preemptive migration, and replication policies [16].

Fault prevention implies that faults are prevented from occurring in the run-time system. Fault prevention is achievable by applying appropriate quality assurance techniques when designing and implementing a MIoT system or forecasting and removing the part of the system that can potentially fail (or replacing parts with suitable spares). Applying appropriate quality assurance techniques during the

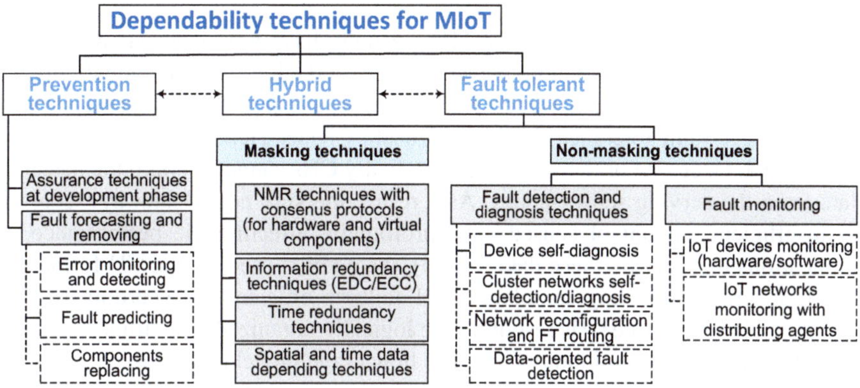

Fig. 7.2 Dependability techniques in MIoT

design process and implementation usually means decomposing and dividing the system into components/modules to reduce the likelihood of making a mistake. In the case of the MIoT-based application, prevention becomes quite difficult, especially in the data perception layer, as there are many unanticipated external and transient faults due to the harsh or even hostile environment, often beyond our control. Another approach to preventing failures in high-availability systems is to perform scheduled maintenance and timely software updates. However, such a technique is not always applicable to MIoT devices.

The fault tolerance mechanism enables the system to provide the user with correct data processing results even when certain classes of defects/ errors in the system components are present. Fault tolerance can be achieved mainly by using error masking or non-masking techniques. Masking techniques are based on redundancy (hardware, software, information, and timing). They can include error-correcting codes and N Modular Redundancy (NMR), which could be hardware and/or software. NMR techniques also require the use of an appropriate consensus mechanism, that is, one that avoids the so-called single point of failure. Consensus refers to how a set of distributed processes reconciles one value from the set of values initially proposed by these processes. In contrast, non-masking techniques are based mainly on fault monitoring or diagnosis during processing and system recovery. In the case of MIoT, there are several possibilities, for example:

- The use of "natural" redundancy of the distributed MIoT system and the ability of processing units with communication capabilities (e.g., processors/microcontrollers) to test other networked units
- Application of self-diagnostic IoT devices (e.g., Built-In Self-Test (BIST) circuits, watchdog monitors)
- The use of other monitoring mechanisms such as checkpoints, watchdog processors, etc.

Different methods can be introduced to increase fault tolerance in MIoT. Most physically close sensor nodes in the MIoT usually have homogeneous properties. Still, some nodes may have more power (energy) and additional functionality (e.g., data aggregation) than others. Therefore, we believe that a cluster-based approach could achieve fault tolerance. In this approach, Cluster Heads (CHs) forward the sensed data. Once one or more CHs fail, the faulty CHs cannot forward the detected data from the serving sensor nodes. As a result, the sink node (gateway) receives an insufficient amount of sensed data, profoundly affecting the performance of applications that rely on these data. In [13], the authors used virtual CH formation and flow graph modeling to tolerate CH failures efficiently. According to this proposal, all failure-free CHs resources are logically organized as a virtual CH to be the common backup of all faulty CHs. The flow graph model is then used to achieve fault tolerance with the minimum total energy consumption of all failure-free CHs. Virtualization in wireless sensor networks has received significant attention due to the growing service domain for IoT, and some FTT have recently been proposed in virtualized networks [10]. This work proposed a framework for optimizing fault tolerance in virtualization in WSNs, focusing on heterogeneous networks for service-oriented IoT applications. A fault-tolerant solution for tolerating physical sensor failures through virtualization in WSNs is presented; a four-layer network architecture is considered, the lowest physical layer of which is represented by real sensor nodes capable of performing different types of tasks (acquiring different data from the environment). Task-based sensor virtualization is performed in the next layer, sensor virtualization, resulting in more virtual sensors than actual physical ones. In the next higher layer, the access layer, various wireless virtual networks are created based on fault-tolerant task-oriented sensor embedding. Each embedded network is a kind of access agent for physical sensors, enabling sensor fault tolerance for various IoT applications. Thus, the fault tolerance mechanisms of the sensors themselves are implemented mainly at the access layer.

7.3.1 Fault Masking Techniques

Fault masking is typically based on various redundancy methods, such as NMR or active-active/active-standby. These methods improve the reliability of a system by relying on a spare component of the system equivalent to the one used to provide service and using it as a replacement in the event of a failure. An active-active solution typically comprises at least two nodes actively running the same kind of service simultaneously. In turn, an active-passive cluster also consists of at least two nodes. However, not all nodes are going to be active. For example, in the case of two nodes, if the first node is already active, the second node must be on standby. The standby node will become active in the event of failure detection in the active node.

These methods ensure correct service delivery within the specified limits (e.g., up to a predetermined number of tolerated errors). However, they have the disadvantage of being very expensive. For example, NMR usually requires more than 300 % overhead in resources. This significant overhead is often due to the need to implement a reliable distributed consensus mechanism to avoid a single point of failure. For example, to mask erroneous data captured by sensor nodes in the Perception and Actuation Layer (PAL) (Chap. 3), the sensor nodes send their data to the voting system, which, based on an appropriate consensus protocol (e.g., according to the majority rule), agrees on common values of sensory data considered correct. The consensus should be achieved in near real time to support the time-critical applications of MIoT.

We believe that building this type of system in a higher layer of the MIoT (Chap. 3) (i.e., Data Management Layer (DML), Intelligent Service Layer (ISL) functioning in the cloud environment) will be more appropriate due to the flexibility and effectiveness of such a solution in the cloud environment. However, the deployment of these solutions in PAL (mainly applicable to WSN) would be significantly less effective. First, it would cause greater energy consumption for communication between sensor nodes to obtain consensus, which, in turn, can accelerate the occurrence of faults due to battery depletion. Second, it could result in longer delays in reaching a consensus. However, applying redundancy would be viable when using cheap COTS devices in PAL. This solution requires the use of cross-layer masking techniques in PAL and higher layers (DML, ISL). Moreover, leveraging the temporal and spatial dependence of the data gathered in PAL could compensate for PAL's limited energy and computational and storage capacity.

Providing high availability for MIoT services requires multiple monitoring and diagnosis functions for infrastructure and applications. However, most architectures proposed in the FTT field of cloud computing have implemented another detection method in their fault detection phase. It is essential to note that all the architectures studied so far are reactive or proactive. However, the architecture described in [4] is hybrid, termed Virtualization and Fault Tolerance (VFT) . The VFT architecture uses self-healing, preemptive migration, and replication policies.

To conclude, the shortcomings of existing solutions are as follows:

- The highest efficiency of the fault tolerance architecture is obtained through hybrid architectures. However, among all the investigated architectures, the VFT architecture is the only hybrid one, so the lack of a hybrid architecture is widely felt in this scope.
- VFT architecture enjoys integrative and proactive policies. However, it has only used the replication method among reactive policies and has not utilized all the methods integratively. Thus, it cannot have maximum reliability.
- The VFT architecture uses a different fault detection method, considered average among the detection methods.
- The fault recovery method of the VFT architecture has been the only previous hybrid architecture at the system level where fault refinement has not occurred at the level of the Virtual Machines (VMs). Moreover, the fault masking method has not enjoyed this method in the fault recovery phase of architecture.

7.3.2 Fault Detection and Diagnosis

7.3.2.1 Fault Monitoring

The two main approaches to fault monitoring in MIoT are device and network fault monitoring, which can be combined for better results. Regarding device fault monitoring, traditional methods to detect the failure of any node in the real-time IoT system are based on the use of heartbeats and watchdogs [11]. A watchdog (often implemented on the hardware) detects errors by monitoring the behavior of a particular component. It performs concurrent system-level error detection by monitoring the system buses that connect the processor and memory. This monitoring focuses primarily on control flow checking, ensuring that the main processor executes the correct code blocks in the correct order. Such monitoring can detect hardware and software faults (mistakes/bugs) that cause either an erroneous instruction(s) to be executed or a wrong program path. Watchdog timers can provide fault monitoring for both individual devices and services located in higher layers, for example, DML and ISL. The recovery procedure can (1) restart the process or the system from its checkpointed state and (2) invoke an appropriate recovery routine. In a heartbeat approach, which mainly applies to CL, a network node periodically sends a heartbeat message to indicate that it is functioning. A properly designed monitoring scheme for objects and services, using watchdogs and heartbeats, could also be used for many other purposes, for example, security.

On the other hand, network fault monitoring involves monitoring for faults in communication links and nodes in CL. It is often based on the agent/manager model involving Distributed Intelligent Agent (DIA) monitoring network flow in separate diagnostic domains, consisting of a set of DIAs and a Network Management Station (NMS). Each DIA monitors the reliability and availability status of the assigned nodes and communication links. DIA relies on a local database of collected monitoring information and a local decision module, which the NMS could update. DIA should be lightweight enough to be installed on edge nodes with moderate communication and computing capabilities. DIAs can perform periodic tests of assigned nodes using a polling-based protocol to improve the diagnostic efficiency of unobservable faults. Individual DIA and NMS exchange notifications about the discovered symptoms of a potential failure.

7.3.2.2 Fault Detection and Diagnosis at the Perception Layer

We distinguish between fault detection and diagnosis activities. Fault detection aims to determine whether a given node or a subset of nodes is suitable. Detection approaches can utilize different techniques, such as error detection codes, time redundancy, fault monitoring, or self-testing. In contrast, fault diagnosis consists of determining the state of components, that is, determining which components are suitable for purpose and which are not. Usually, external testers or self-diagnosing tests perform the diagnosis. MIoT should also have self-healing properties; therefore, we focus on self-detection and self-diagnosing techniques. Diagnosis is a stage in which the exact occurrence of a fault is attributed to a specific atomic piece of the IoT subsystem (which will usually be a node or a subset of nodes). Most devices in the perception layer of the MIoT system are power- and process-constrained infrastructures that incorporate sensors/actuators, processors/controllers, memory/storage facilities, and communication units, all embedded in a Sensor/actuator Node (SN) or a Collect Sensor Node (CSN). These MIoT devices are used mostly for information acquisition and monitoring or surveillance to build precise situational awareness in different military scenarios. However, the greatest challenge is to identify faulty sensor nodes that generate erroneous data, which can break the global situational awareness picture of the MIoT subsystem. Thus, we should have mechanisms that prevent erroneous data transmission to the higher levels of data processing where data fusion is performed (fusion centers), or we should have mechanisms in the data fusion nodes that will detect incorrect data obtained from damaged sensors and reject them.

The potential for simultaneous sensor and microprocessor fault occurrences can be assumed, and diagnosis should be performed during two phases: at the processing nodes (microcontrollers) phase, where the reliability state of the processing elements is diagnosed, and at the subsystem level, where the health of all sensors/actuators is determined. Recently, fault tolerance and management in WSNs have attracted the attention of researchers. Some proposed approaches include energy-efficient detection schemes and algorithms for the detection of failures and identification of faulty sensors [2, 8, 9, 12]. The mutual testing method was adopted at the processor level, where each processing element (sensor node) can test its neighbors [3]. In addition, several approaches for interconnected processors have been developed to diagnose faulty processors among themselves. The MM method, a major approach first proposed by Malek and Maeng [15], performs the diagnosis by sending the same input to pairs of processors and comparing their responses. Based on the collective results of the comparisons, one may be able to claim the faulty/fault-free status of the processors in the system.

The communication between sensors depends on their positions and the transmission range. Communication is implemented directly via a single-hop or a multi-hop network, a sequence of links using intermediate nodes. Wireless multi-hop networks are formed by a group of nodes (clusters) that can communicate with each other over a wireless channel. However, identifying the faulty nodes using the notion of t-diagnosability in the global sense may be ineffective in the MIoT network due to the existing challenges:

- Diagnosing faulty sensors in a centralized manner may be too expensive for the base station, and fault detection should be conducted in a real-time mode with low latency and high throughput.
- Potential substantial excess comparison tests lead to unnecessary energy consumption in sensor networks.

We propose a method of fault detection in the sensor network based on the network's logical structure clustering and determining relatively small so-called td-detectable subnets (clusters).

7.3.2.3 Fault Diagnosis at the Communication Layer

A diagnosis of communication nodes and communication links must be made at the communication layer. At the system level, we propose communication link fault diagnosis and communication node fault diagnosis based on DIA similar to SNMP agents. The MIoT network is divided into hierarchical clusters, which constitute diagnostic domains. Each diagnostic domain has some DIAs and NMS (network management station). A DIA is responsible for determining the reliability status of the assigned nodes (gateways, base station, etc.) and communication links. A process on the Simple Network Management Protocol (SNMP) agent internally generates periodic tests using a poll-based protocol. Distributed Diagnostic Agents (DDAs) send diagnostic information to the NMS station. In the proposed solution, we can integrate the system-level diagnostics method based on mutual testing between DIAs within the diagnostic domain and the method based on the monitoring of the network flow by an agent. DIAs also collect diagnostic information from the base station (fault detection status) within a given cluster of sensors. One of the possible ways to achieve fault tolerance is to reconfigure the network into a smaller system after diagnosing the fault. The new (i.e., degraded) network continues its work after reassigning resources and under the condition that it meets special requirements.

Several problems must be considered when designing a fault-tolerant system:

1. One must first select a fault model.
2. The chosen system must be able to detect each fault in the model and isolate each fault from the functioning portion of the system to prevent the spread of the faulty behavior.
3. As a fault detection mechanism can detect more than one possible fault, the system must also address the fault diagnosis process (or localization), which narrows the set of possible faults and allows more efficient fault isolation techniques to be employed.

7.4 Integrating Fault Tolerance and Security Techniques in the MIoT

7.4.1 Reliability of the Security Mechanisms

Security mechanisms also ensure the availability of services and their failure can significantly reduce the reliability of the system. Therefore, security mechanisms must be reliable and resilient. For example, where possible, they should be designed to avoid a single point of failure and support distributed and federated modes of operation. Distributed ledgers are an example of a recent paradigm that, after a period of initial hype, has evolved into a practical approach to implementing reliable distributed trust management and authorization infrastructure, supporting a dynamic federation of mission participants and operational redundancy. In the context of MIoT, permissioned distributed ledger solutions, such as Hyperledger Fabric [1], are especially interesting. Distributed ledger technology could bring multiple benefits with respect to reliability. For example, it could enable the implementation of immutable security logs, decentralized key generation, and access control, as well as more complex security functions in the form of programmable transactions (so-called smart contracts).

Cyber resilience offers a life cycle approach to ensuring system security and reliability based on four phases: Plan, Absorb, Recover, and Adapt [14]. The concept of resilience recognizes that although preparation is vital, ensuring that the system remains fit for purpose during mission execution also requires the ability to deal with impactful attacks. MIoT must provide the ability to absorb ongoing attacks without compromising minimum military operational requirements and must be able to recover from the impact of an attack efficiently and effectively. Finally, all layers of MIoT must provide the flexibility required to adapt to new threats and attack vectors and support any reconfiguration required to maintain sufficient resistance in the face of evolving threats and attack vectors. Such flexibility can be achieved by increasing the use of software-defined components, including software-defined radio, storage, and local and wide area networks, as well as virtualized and software-defined computing infrastructure, including (when appropriate) the cloud.

7.4.2 Security of the Reliability Mechanisms

Reliability mechanisms must be integrated by design with the appropriate security measures to provide adequately resilient and trustworthy reliability mechanisms in an environment exposed to adversarial attacks. It is critical to preserve the integrity of fault monitoring and diagnostic information and appropriately authenticate these information sources. Otherwise, fault tolerance techniques, designed to improve the system's reliability, could open a new attack surface on the system's availability, enabling the attacker to inject fake error reports and control information. Strong authentication and message authentication codes are crucial for communication between all nodes and system components. Although the confidentiality of fault diagnostic information is not critical from the perspective of operating a fault tolerance mechanism, it can be necessary for some deployment scenarios, as the content of fault diagnostic messages and related communication patterns could disclose potentially sensitive configuration and system status information to an adversary. For example, intercepting fault-related traffic might inform the adversary that certain capabilities are not fully functional and thus provide an opportunity to plan and perform a successful attack.

In a federated mission environment like the NATO Alliance Federation, a distributed trusted computing platform could be implemented through a combination of node- and network-level capabilities. Individual nodes could rely on Trusted Platform Module (TPM) or secure enclaves offered by modern Intel SGX and AMD SEV architectures [1] to provide enhanced authentication, integrity, and confidentiality protection. As described earlier, a distributed ledger could provide additional protection against fake reports and compromised nodes and a trusted repository of reliability-relevant information at the federated mission network level.

7.5 Conclusions

Achieving an appropriate level of dependability of MIoT requires a comprehensive approach to reliability and security across all MIoT layers and phases of the life cycle. Among other things, fault tolerance techniques should be applied consistently at each layer. Moreover, building related mechanisms into several layers, for example, data-oriented cross-layer masking techniques, is necessary. Our assessment of the applicability of selected fault-tolerant techniques to MIoT is summarized in Table 7.1, which shows that most techniques are not applicable to real-time scenarios.

Table 7.1 Applicability assessment of fault-tolerant techniques

Techniques	Examples of solutions	Layer	Applications
Fault masking	N-Modular Redundancy: HW/virtual component replication, SDN distributed controllers	DML, ISL	Real-time, mission-critical
Data fusion-driven fault detection	Fault detection with defense-in-depth principles		
Device fault monitoring	Watchdog timers, built-in self-test techniques, fault recovery	PAL	Restartable, time-dependent
Network fault monitoring	Heart beats, agent/manager model with DIA, network reconfiguration	PAL, CL	
Fault detection within clusters	td-detectable clusters, FT routing		
Faulty nodes diagnosing within clusters	Self-diagnosing MM methods, agent/manager model with DIA, FT routing		

Finding the best solution to make MIoT immune to different types of fault is a great challenge, especially in mission critical scenarios. In particular, fault-tolerant techniques should be integrated with security solutions used within MIoT, as mutual reinforcement is highly beneficial.

We have indicated that NMR techniques are very useful for the ISL. This is particularly relevant for time-critical applications of MIoT, where it is necessary to agree on a common result delivered by several (n) modules (i.e., sensor nodes) to the voting system in real time or slightly delayed. Masking techniques that use a voting system should be built as distributed and have an effective consensus protocol to avoid a single point of failure. Moreover, the consensus process should work in near real-time mode to reach agreement among the common results of time-critical MIoT applications. We believe that building such a system in the higher layers of the MIoT (i.e., DML and ISL functioning in the cloud environment) will be more appropriate due to the flexibility and effectiveness of such a solution in the cloud environment. For example, each voting system component (i.e., virtual machines) could be located on different physical nodes of the cloud environment and use fast communication links to reduce the delay of the consensus process. However, as discussed above, applying these solutions in the PAL layer (mainly applicable for WSN networks) will be significantly less effective due to the following factors:

- Greater energy consumption for communication between the sensor nodes to obtain consensus with agreement among common captured data, which, in turn, can accelerate the occurrence of faults due to battery depletion
- Longer delays in reaching consensus, which may exclude such solutions from real-time MIoT applications

However, the redundancy of cheap sensors and devices in PAL makes the use of cheap devices in MIoT a viable option. However, this requires the use of different data-oriented cross-layer masking techniques, mainly in higher MIoT (i.e., DML) layers. In this case, due to the limited energy, computational, and storage capacities in the PAL layer, a processing scheme is needed in the DML layer that can use the temporal and spatial dependence of the data. Such a scheme should detect most of the faulty data and even increase the error correction capacity of the data aggregation process.

References

1. Brandenburger, M., Cachin, C.: Challenges for combining smart contracts with trusted computing. In: Proceedings of the 3rd Workshop on System Software for Trusted Execution (2018)
2. Chen, J., Kher, S., Somani, A.: Distributed fault detection of wireless sensor networks. In: Workshop on Dependability Issues in Wireless Ad Hoc Networks and Sensor Networks (2006)
3. Chudzikiewicz, J., Furtak, J., Zielinski, Z.: Fault-tolerant techniques for the Internet of Military Things. In: World Forum on Internet of Things (2015)

4. Das, P., Khilar, P.M.: VFT: A virtualization and fault tolerance approach for cloud computing. In: 2013 IEEE Conference on Information Communication Technologies, pp. 473–478 (2013). https://doi.org/10.1109/CICT.2013.6558142
5. Diwan, A.S., Ardelean, D., Lunney, J.J., Hoffmann, P.E., Hauer, T.: Meaningful availability. In: 17th USENIX Symposium on Networked Systems Design and Implementation (NSDI'20) (2020)
6. Ghosh, A.K., Voas, J.M.: Inoculating software for survivability. Commun. ACM **42**(7), 38–44 (1999)
7. Jan, S.U., Ahmed, S., Shakhov, V., Koo, I.: Toward a lightweight intrusion detection system for the Internet of Things. IEEE Access **7**, 42,450–42,471 (2019). https://doi.org/10.1109/ACCESS.2019.2907965
8. Jiang, P.: A new method for node fault detection in wireless sensor networks. Sensors **9**(2), 1282–1294 (2009)
9. Joshi, B., et al.: Hierarchical plug-and-play self-diagnosable intelligent sensor networks for process control. In: International Conference on Electro/Information Technology (2008)
10. Kaiwartya, O., Abdullah, A.H., Cao, Y., Lloret, J., Kumar, S., Shah, R.R., Prasad, M., Prakash, S.: Virtualization in wireless sensor networks: fault tolerant embedding for Internet of Things. IEEE Internet Things J. **5**(2), 571–580 (2018). https://doi.org/10.1109/JIOT.2017.2717704
11. Koren, I., Krishna, C.M.: Fault-Tolerant Systems. Elsevier, Amsterdam (2020)
12. Koushanfar, F., Potkonjak, M., Sangiovanni-Vincentelli, A.: Fault tolerance in wireless sensor networks. In: Handbook of Sensor Networks. CRC, Boca Raton (2004)
13. Lin, J.W., Chelliah, P.R., Hsu, M.C., Hou, J.X.: Efficient fault-tolerant routing in IoT wireless sensor networks based on bipartite-flow graph modeling. IEEE Access **7**, 14,022–14,034 (2019). https://doi.org/10.1109/ACCESS.2019.2894002
14. Linkov, I., Kott, A.: Fundamental concepts of cyber resilience: introduction and overview. In: Cyber Resilience of Systems and Networks. Springer, Berlin (2019)
15. Maeng, J., Malek, M.: A comparison connection assignment for diagnosis of multiprocessor systems. In: IEEE International Symposium on Fault-Tolerant Computing (1981)
16. Nabi, M., Toeroe, M., Khendek, F.: Availability in the cloud: State of the art. J. Netw. Comput. **60**, 54–67 (2016)
17. Patterson, D., Brown, A., Broadwell, P., Candea, G., Chen, M., Cutler, J., Enriquez, P., Fox, A., Kiciman, E., David Oppenheimer, M.M., Sastry, N., Tetzlaff, W., Traupman, J., Treuhaft, N.: Recovery oriented computing (ROC): motivation, definition, techniques. Technical report, University of California at Berkeley, Berkeley, CA (2002)
18. Qian, Y., Joshi, J., Tipper, D., Krishnamurthy, P.: Information Assurance. Dependability and Security in Networked Systems. Elsevier, Amsterdam (2008)
19. Sarkar, S.: Internet of Things – Robustness and Reliability. In: Internet of Things: Principles and Paradigms. Elsevier, Amsterdam (2016)
20. Siewiorek, D.P., Chillarege, R., Kalbarczyk, Z.: Reflections on industry trends and experimental research in dependability. IEEE Trans. Dependable Secure Comput. **1**(2), 109–127 (2004)
21. Suri, N., Tortonesi, M., Michaelis, J., Budulas, P., Benincasa, G., Russell, S., Stefanelli, C., Winkler, R.: Analyzing the applicability of Internet of Things to the battlefield environment. In: International Conference on Military Communications and Information Systems (ICMCIS), pp. 1–8. IEEE, Piscataway (2016)
22. Yishi, L., Fei, J., Hui, Y.: Study on application modes of Military Internet of Things (MIoT). In: International Conference on Advanced Information Systems Engineering, vol. 3, pp. 25–27. Proceedings of 2012 IEEE International Conference on Computer Science and Automation Engineering (CSAE) (2012)
23. Zheng, D., Carter, W.: Leveraging the Internet of Things for a more efficient and effective military. Technical report, Center for Strategic & International Studies, Washington, DC (2015)

Part III
Experimental Results and Recommendations

Chapter 8
Experimental Validation

Manas Pradhan, Niranjan Suri, Konrad Wrona, James R. Michaelis, Frank T. Johnsen, Janusz Furtak, and Michal Dyk

8.1 Introduction

This chapter summarizes the experiments that were conducted as part of the group to test and validate IoT technologies and their application to the military domain. Five different categories of experiments were conducted:

- LoRa communications technology for IoT devices
- Security techniques and solutions for IoT devices and information generated from IoT devices
- IoT body-worn sensors for physiological monitoring

M. Pradhan
DKSR, Berlin, Germany

N. Suri (✉)
US Army DEVCOM Army Research Laboratory (ARL), Adelphi, MD, USA

Florida Institute for Human and Machine Cognition (IHMC), Pensacola, FL, USA
e-mail: nsuri@ihmc.org

K. Wrona
NATO Communications and Information Agency (NCIA), The Hague, The Netherlands

Military University of Technology, Warsaw, Poland
e-mail: konrad.wrona@ncia.nato.int

J. R. Michaelis
US Army DEVCOM Army Research Laboratory (ARL), Adelphi, MD, USA

F. T. Johnsen
Norwegian Defence Research Establishment (FFI), Kjeller, Norway

J. Furtak · M. Dyk
Military University of Technology, Warsaw, Poland

N. Suri et al. (eds.), *Military Applications of Internet of Things*, Internet of Things,
https://doi.org/10.1007/978-3-032-03378-9_8

- Military operations that can leverage IoT in the context of smart cities, including:
  - MQTT as middleware for exchanging data
  - Sensor integration
  - Solution architecture and ontology

8.2 IoT Communications Protocol Testing in Urban Environments

8.2.1 Motivation and Problem Statement

The Long-Range Wide-Area Network (LoRaWAN) protocol (introduced in Sect. 4.4.1.2) has gained significant adoption through combined support for low-power, long-range transmission [1], thereby supporting Low-Power Wide-Area Networking (LPWAN) within IoT infrastructures. LoRaWAN's growing usage in urban environments, mainly due to low deployment costs paired to good receiver sensitivity, has prompted recent investigation on its potential usage within military systems for supporting urban operations, e.g., [10, 11, 13, 17, 18].

This section reviews work conducted in [17, 18], which aimed to conduct an initial investigation on the viability of LoRaWAN to support IoT communications in dense urban environments through assessment of signal coverage (i.e., the portion of transmitted messages received by a gateway) under varying conditions. Key research questions for this work included the following:

1. How would distance between a transmitting LoRaWAN device and receiving gateway impact signal coverage in varying urban conditions, including (i) outdoor settings featuring dense urban infrastructure, such as central business districts and (ii) indoor infrastructure, such as subway platforms?
2. How does variation in LoRaWAN data rate impact signal coverage under varying urban conditions?
3. How does placement configuration of a LoRaWAN gateway (e.g., elevation from ground level) impact signal coverage under varying urban conditions?

Toward addressing these questions, a set of LoRaWAN data collection tests were conducted in the cities of Montreal [17] and New York [18] aimed at testing device coverage in and around dense urban infrastructure (e.g., skyscrapers, narrow streets, and indoor/subterranean environments). This work was undertaken as a key initial step to better understand suitability and applicability of civilian IoT communication protocols within tactical C2 systems for urban deployment. These efforts focused on testing LoRaWAN from two perspectives:

1. Indoor Transmission Coverage: assessing the impact of both indoor structures and lack of line of sight on LoRa signal coverage between transmitting devices and receiving gateways
2. Outdoor Transmission Coverage: assessing the impact of gateway elevation on ability to receive LoRa signals from moving vehicles, accounting for variations in data rate and payload size

Data collection tests conducted in Montreal focused exclusively on outdoor transmission coverage, while those in New York included both indoor and outdoor transmission coverage. In both locations, LoRaWAN performance was compared for data rates on the North American Industrial, Scientific, and Medical (ISM) 915 MHz band.

8.2.2 Supporting Architecture

To support LoRaWAN experimentation in Montreal [17] and New York [18], a COTS IoT architecture for tactical C2 support was utilized, based on work from [10]. Details on the hardware and software setup used in the supporting architecture are described below.

8.2.2.1 Supporting Hardware

LoRa Nodes A set of LoRa nodes was assembled using PyCom LoPy4 microcontrollers[1] equipped with sensors for transmitting varying forms of information, including GPS coordinates, passive infrared readings, and ambient environmental conditions (as featured on the PySense expansion board). For purposes of data collection efforts described in this section, two classes of LoRa nodes were utilized (as depicted in Fig. 8.1): GPS-equipped nodes, and nodes equipped with environmental sensors for measuring temperature, humidity, ambient light levels, and barometric pressure. Across experimentation, GPS nodes were utilized during outdoor transmission coverage, while environmental nodes were utilized for indoor transmission coverage.

LoRaWAN Gateway A MultiConnect Conduit IP67 Base Station[2] was utilized, which was configured to support processing of incoming LoRa messages through the preinstalled Node-RED development framework.[3]

[1] https://pycom.io/ [Accessed March 6, 2023].

[2] https://www.multitech.com/brands/multiconnect-conduit-ip67 [Accessed March 6, 2023].

[3] https://nodered.org [Accessed March 6, 2023].

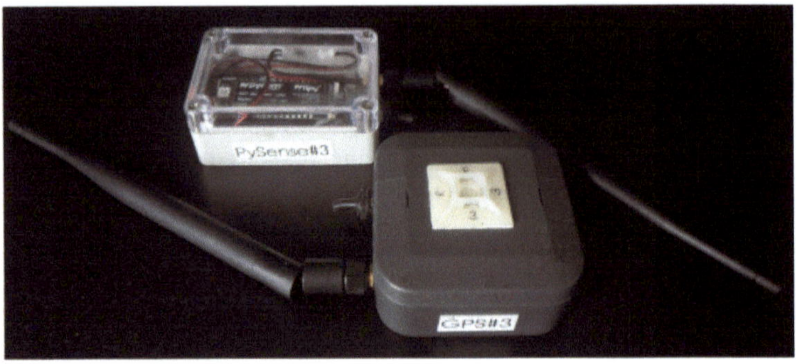

Fig. 8.1 LoRa nodes built with LoPy4 microcontrollers. Top: LoRa node with environmental sensors. Bottom: LoRa node with GPS sensor

8.2.2.2 Supporting Software

Route Logging To support logging of routes covered during data collection, GPS tracking software[4] was enabled on an Android smartphone, which logged a position for the vehicle once every 3 s. During outdoor transmission coverage testing, GPS nodes were configured to send dummy GPS values (latitude: 0, longitude: 0) if the GPS signal were lost. In such cases, the position of the message could be determined by cross-referencing transmission time with route logs collected separately on the smartphone.

LoRa Node PyCom Code Each LoRa node was configured with the PyCom code based on the code intended to transmit messages under varying data rates on the ISM 915 MHz band, defined based on particular combinations of Spreading Factors and Bandwidth such as those listed in Table 8.1. This code was configured to run with LoRaWAN operating in Class A mode, Adaptive Data Rate (ADR) being disabled, and transmissions being performed with one attempt for each message (missing acknowledgments would not trigger retransmissions). Two versions of the PyCom code were developed: one for the GPS LoRa nodes and another for the Environmental LoRa nodes. This version of this code intended for GPS nodes was configured to transmit latitude-longitude coordinates, while the environmental nodes were configured to transmit a mix of measures including temperature, humidity, ambient light levels, and barometric pressure. Under normal operating conditions, each node sent one message every 9–10 s.

[4] https://play.google.com/store/apps/details?id=com.ilyabogdanovich.geotracker [Accessed March 6, 2023].

Table 8.1 LoRaWAN data rate specification for uplink transmissions, North American ISM 915 MHz band

Data rate	Spreading factor (SF)	Bandwidth (BW)	Data rate (bits/s)	Maximum payload size (bytes)
DR0	SF10	125 KHz	980	11
DR1	SF9	125 KHz	1760	53
DR2	SF8	125 KHz	3125	129
DR3	SF7	125 KHz	5470	242
DR4	SF8	500 KHz	12,500	242

8.2.3 Montreal Experimentation

Two sets of experimentation were conducted in Montreal, one aimed at testing coverage on Data Rate (DR) 0 (Experiment 1) and another focused on testing Data Rates 1 and 2 (Experiment 2). A series of eight routes were driven: four for Experiment 1 and four for Experiment 2. As part of our route selection, we attempted to obtain as comprehensive a degree of coverage as possible for streets in and around the downtown area, as well as around the downtown periphery. For these tests, a gateway was placed on a building on a hill overlooking downtown Montreal, with an elevation approximately 150 m above street level in the central business district.

Figure 8.2 shows the tracks (in blue) followed, depicting an aggregation of all eight routes. The farthest distance driven from the LoRa gateway was approximately 8 km. However, most of the driving took place within a 5-km radius of the LoRa gateway.

Tables 8.2 and 8.3 provide statistics for Experiments 1 and 2, respectively, on messages sent by each LoRa node versus the amount received by the LoRa gateway, grouped by transmission distance. Figure 8.3 provides a visual plot of the results of Experiment 1, showing the locations of the DR0 messages received by the LoRa gateway. Likewise, Fig. 8.4 provides visual graphs of Experiment 2 results, depicting the locations of the DR1 and DR2 messages received.

Based on the findings of Experiment 1, messages could be received up to 5.5 km from the gateway. However, the percentage of messages received after 1 km was pretty low and would likely be unacceptable as configured. Note that the percentage of messages received was similar in the 1–5 km range. Retransmission was not enabled in this experiment, which could potentially improve performance beyond the 1 km range.

Experiment 2 was intended to compare the relative performance of two different Data Rates for LoRa. Note that Experiment 2 followed a different route through the city, which implies that a comparison cannot be made between DR0 used in Experiment 1 and DR1/DR2 used in Experiment 2. For Experiment 2, DR1 consistently outperformed DR2 in all ranges, with messages for both rates received up to 4.8 km from the gateway. However, the percentage of messages received after 1 km was

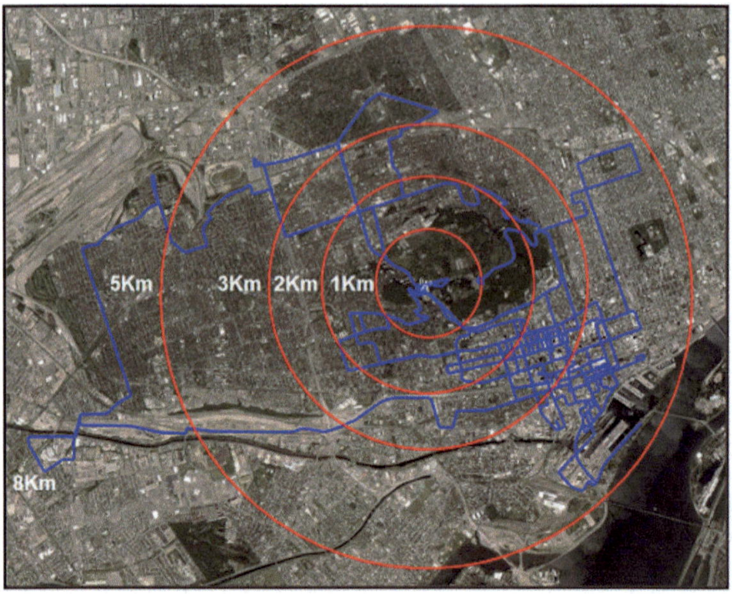

Fig. 8.2 Tracks logged during data collection (in blue), corresponding to the eight routes driven in Montreal. The circles (in red) denote the following distances from the LoRa gateway: 1 km, 2 km, 3 km, and 5 km

Table 8.2 Experiment 1: messages sent and received for LoRa data rate 0

Location	Msg. sent	Msg. received	Percent received
<1 km	407	230	56.51
1–2 km	1061	81	7.63
2–3 km	1974	180	9.12
3–5 km	870	65	7.47
>5 km	311	1	<0.01
Total	4623	557	12.05

pretty low in Experiment 2 just as in Experiment 1. In both Experiments 1 and 2, LoRaWAN coverage could be established in sections of downtown Montreal with significant amounts of potentially obstructing infrastructure (located on average 2.5 km from the LoRa gateway). Differences in performance between Experiments 1 and 2 indicate potential impact of urban infrastructure (e.g., skyscrapers) on signal coverage, to which further insights may be obtained through analysis of supplemental datasets providing the height of infrastructure along the routes (e.g., LIDAR-based 3D models of urban terrain [25]).

Table 8.3 Experiment 2: messages sent and received for LoRa data rate 1 and 2

Location	Data rate	Msg. sent	Msg. received	Percent received
<1 km	DR1	385	266	69.09
	DR2	386	230	59.59
1–2 km	DR1	1004	188	18.73
	DR2	1008	80	7.94
2–3 km	DR1	1867	292	15.64
	DR2	1875	111	5.92
3–5 km	DR1	1241	84	0.07
	DR2	1247	13	0.01
>5 km	DR1	295	0	0
	DR2	296	0	0
Total	DR1	4792	830	17.321
	DR2	4812	454	9.435

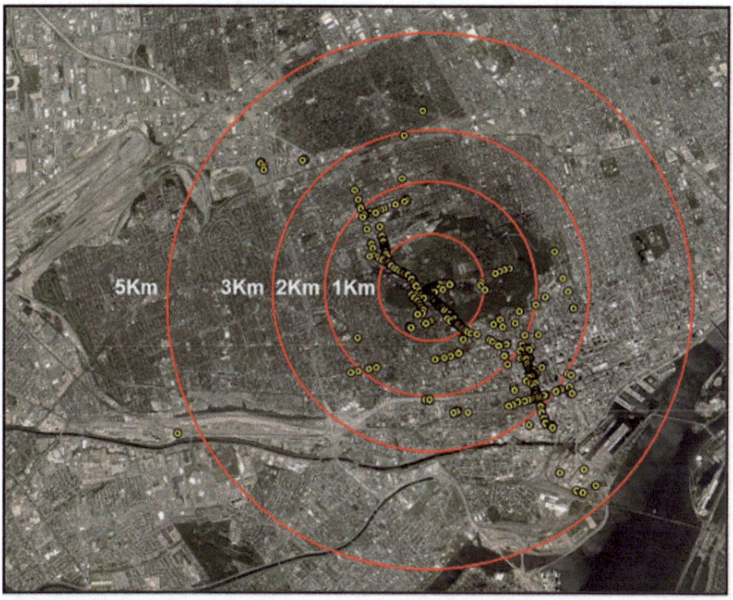

Fig. 8.3 Visual plot of Montreal Experiment 1 results, depicting locations of DR0 messages received by LoRa gateway

8.2.4 New York Experimentation

Three sets of experimentation were conducted in New York: one focused on indoor coverage testing and two on testing of outdoor coverage.

Fig. 8.4 Visual plots of Montreal Experiment 2 results. Top: locations of DR1 messages received by LoRa gateway. Bottom: locations of DR2 messages received by LoRa gateway

8.2.4.1 Experiment 1: Subway Platform Testing

The objective of this experiment was to assess the impact of both indoor structures and lack of line of sight on LoRa signal coverage between transmitters and receiving gateways.

Testing for this experiment was conducted in a decommissioned subway station. For this testing, ten LoRa environmental sensor nodes were placed along the track of a platform 16 m below street level, each transmitting 11-byte payloads on data rate DR0. In turn, signal coverage was compared between two LoRa gateway configurations:

- Configuration 1: with a gateway located on the subway platform 3 m from the track surface
- Configuration 2: with a gateway mounted on a vehicle outside the subway entrance, 4 m above street level

Figure 8.5 shows both side-view and top-down diagrams of the subway station, indicating both LoRa sensor and gateway placement locations. Testing was conducted over two trials, where Trial 1 was conducted with Configuration 1 and Trial 2 was conducted with Configuration 2.

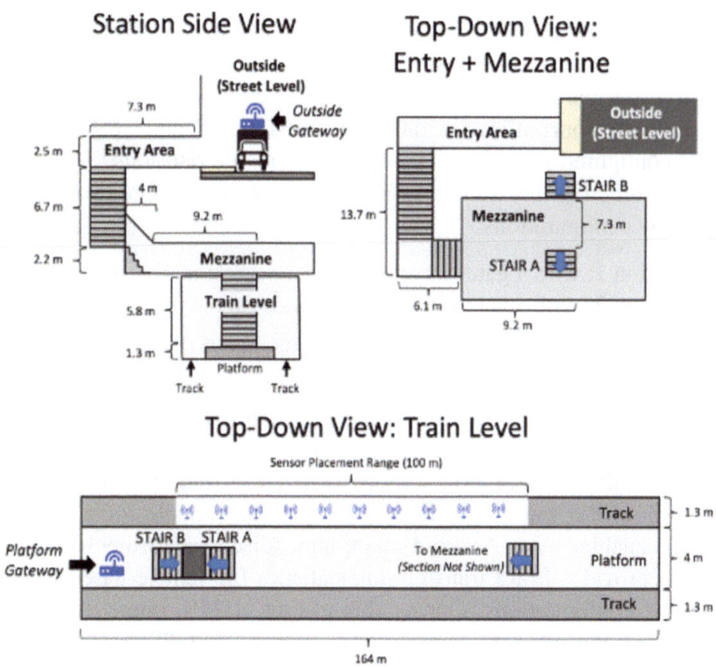

Fig. 8.5 Side-view and top-down diagrams of the subway station, indicating both LoRa sensor and gateway placement locations

Table 8.4 Experiment 1 results

	Gateway placement	Msg. sent	Msg. received	Percent received
Trial 1	Configuration 1 (train platform)	8296	7225	87.09
Trial 2	Configuration 2 (on vehicle)	6310	4463	70.73

Table 8.4 provides LoRa transmission statistics for Experiment 1 across Trials 1 and 2, corresponding to aggregate totals across the ten LoRa sensors. Percentages of messages received by the gateway in Trials 1 and 2 were approximately 87% and 71%, respectively, indicating an approximate 16% drop in received messages for Configuration 2 versus Configuration 1. Here, the performance for the outside gateway placement indicates LoRa's viability in handling transmissions through subterranean urban structures, specifically in cases where line of sight is obstructed between transmitting sensors and receiving gateways.

8.2.4.2 Experiment 2: Gateway Elevation Testing

In contrast to Experiment 1, which involved coverage testing with fixed-position LoRa sensors, Experiment 2 involved coverage testing using sensors attached to a moving vehicle. The objective of this experiment was to assess the impact of both LoRa gateway elevation and mobile sensors on signal coverage in outdoor urban environments.

For this experiment, LoRa testing was conducted using three of the vehicle-mounted LoRa sensors outlined in Table 8.5: GPS1, GPS2, and GPS3. These sensors were each configured to transmit 11-byte payloads on data rates DR0, DR1, and DR2, respectively. For this testing, LoRa signal reception was compared across two LoRa gateway configurations:

- Configuration 1: with a gateway attached to a vehicle at 4 m above street level
- Configuration 2: with a gateway located outside on the roof of a 52-m building

Separate drives were conducted for both trials at distances up to 2.5 km from the gateways, in continuous proximity to buildings between 40 m and 100 m in height (see Fig. 8.6). Both drives attempted to follow predetermined routes, each intended to last approximately 1 hour. However, variations in road accessibility, due to factors including construction activity, resulted in route variation concentrated within 0.5 km of the gateway locations. To track the vehicle route, GPS tracking software was enabled on an Android smartphone (discussed further in Sect. 8.2.2.2).

Table 8.5 provides LoRa transmission statistics for vehicle-based testing using both gateway configurations. During this testing, a 24% improvement in signal reception across the full vehicle route was observed on GPS 3 (DR2) for Configuration 2 versus Configuration 1, while an 18% improvement could similarly be observed for GPS 2 (DR1). Under the testing conditions, no noticeable difference could be observed in coverage for GPS 1 (DR0). These findings appear to indicate

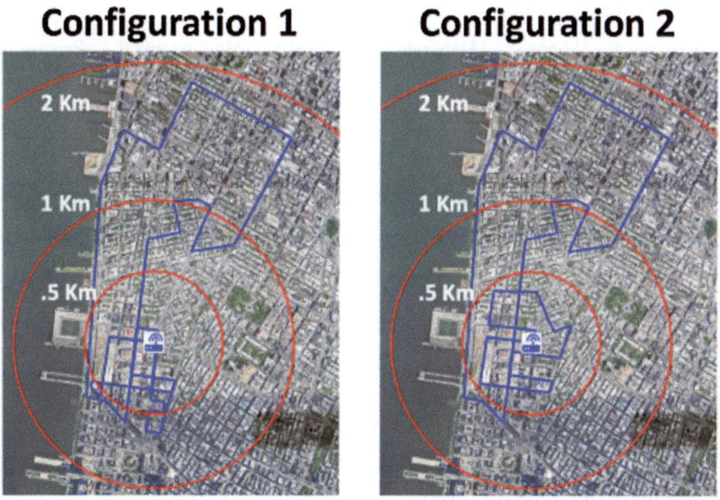

Fig. 8.6 Visualizations of vehicle routes taken during New York Experiment 2

the utility of gateway placement at higher elevations in urban environments for supporting LoRa transmissions, particularly as data rate increases.

The data collection efforts in Montreal and New York only used one gateway at a time. As such, additional data collection efforts featuring multiple gateways would presumably be required to make a more complete assessment regarding the performance of LoRa in urban environments for military use.

8.2.4.3 Experiment 3: LoRa Payload Testing

Extending upon Experiment 2, the objective of this experiment was to assess the impact of variation in both the LoRa data rate and the size of the message payload on the signal coverage received by an elevated gateway, accounting for the distance to the receiving gateway. For this experiment, LoRa testing was carried out using each of the four vehicle mounted LoRa sensors outlined in Table 8.6: GPS1, GPS2, GPS3, and GPS4. A single driving test was carried out, with a gateway located on the roof of a 172-m building (3.5 times the height of the building used in Experiment 2). This 2.5-hour drive covered a route with distances up to 4 km from the gateway, in continuous proximity to buildings ranging from 40 to over 100 m in height. To track the vehicle route, GPS tracking software was enabled on an Android smartphone (discussed further in Section 8.2.2.2).

Table 8.6 provides LoRa transmission statistics for vehicle-based testing in Experiment 3 for GPS 1, 2, 3, and 4, accounting for vehicle distance to the gateway. In turn, Fig. 8.7 provides map-based visualizations that compare LoRa signal coverage per GPS, where green dots indicate the locations of the LoRa

Table 8.5 Experiment 2 results

Distance to gateway	Sensor	Data rate	Msg. sent	Msg. received	Percent received
Gateway configuration 1 (street level)					
<0.5 km	GPS1	DR0	141	109	77.30
	GPS2	DR1	137	99	72.26
	GPS3	DR2	140	101	72.14
0.5–1 km	GPS1	DR0	71	36	50.70
	GPS2	DR1	71	17	23.94
	GPS3	DR2	70	9	12.89
1–2 km	GPS1	DR0	114	57	50.00
	GPS2	DR1	109	18	16.51
	GPS3	DR2	108	15	13.89
Total	GPS1	DR0	326	202	61.96
	GPS2	DR1	317	134	42.27
	GPS3	DR2	318	125	39.31
Gateway configuration 2 (elevated level)					
<0.5 km	GPS1	DR0	204	165	80.88
	GPS2	DR1	199	153	76.88
	GPS3	DR2	208	159	76.44
0.5–1 km	GPS1	DR0	99	49	49.49
	GPS2	DR1	87	47	54.02
	GPS3	DR2	87	49	56.32
1–2 km	GPS1	DR0	107	45	42.05
	GPS2	DR1	116	43	37.07
	GPS3	DR2	117	53	45.30
Total	GPS1	DR0	410	259	63.17
	GPS2	DR1	402	243	60.44
	GPS3	DR2	412	261	63.35

messages received, blue tracks indicate the route followed by the vehicle, and red circles denote varying distances from the gateway. Along the vehicle route, greater signal coverage could be observed for GPS 1 (DR0, 11-byte payload) and GPS 2 (DR1, 11-byte payload) versus GPS 3 (DR2, 11-byte payload). Additionally, an 18% decrease in signal coverage could be observed between GPS 2 (DR1, 11-byte payload) and GPS 4 (DR1, 53-byte payload) along the route. This finding appears to indicate a noticeable impact of message payload size on LoRa signal coverage under conditions where the data rate is kept constant.

Note that the performance of DR1 seemed to be better overall than DR0. This could have been related to a calibration or other issue with the device that was configured to use DR0. Unfortunately, there was insufficient time to repeat the experiments with different hardware to further analyze the performance results.

As in the Montreal experiments, the LoRaWAN transmitting devices were configured to operate in Class A mode, Adaptive Data Rate (ADR) was disabled,

Table 8.6 Experiment 3 results

Distance to gateway	Sensor	Data rate	Msg. sent	Msg. received	Percent received
<1 km	GPS1	DR0	254	131	51.57
	GPS2	DR1	253	148	58.50
	GPS3	DR2	253	107	42.29
	GPS4	DR1	253	101	39.92
1–2 km	GPS1	DR0	274	88	32.12
	GPS2	DR1	266	120	45.11
	GPS3	DR2	266	51	19.17
	GPS4	DR1	262	58	22.14
2–3 km	GPS1	DR0	124	28	22.58
	GPS2	DR1	123	38	30.89
	GPS3	DR2	123	7	5.69
	GPS4	DR1	126	13	10.32
>3 km	GPS1	DR0	156	5	0.03
	GPS2	DR1	155	9	0.06
	GPS3	DR2	158	0	0.0
	GPS4	DR1	155	2	0.01
Total	GPS1	DR0	808	252	31.19
	GPS2	DR1	797	315	39.52
	GPS3	DR2	800	165	20.63
	GPS4	DR1	796	174	21.86

and transmissions were performed with one attempt for each message (missing acknowledgments would not trigger retransmissions).

8.2.5 Concluding Remarks

Based on initial findings from the Montreal and New York data collect, additional directions for follow-on research present themselves.

Expanded Dataset Analysis Focus for this data collection effort was placed on driving around downtown Montreal, in the presence of significant urban infrastructure (including several buildings over 100 m tall in both cities). To date, limited research has been conducted on LoRaWAN's coverage capabilities in Central Business Districts [19], including identification of areas where coverage gaps may likely occur. As such, follow-on analysis of this data collect is envisioned which would account for both infrastructure concentration and building heights, featuring supplemental datasets detailing the height of infrastructure around collection routes (e.g., LIDAR-based 3D models of urban terrain [25]).

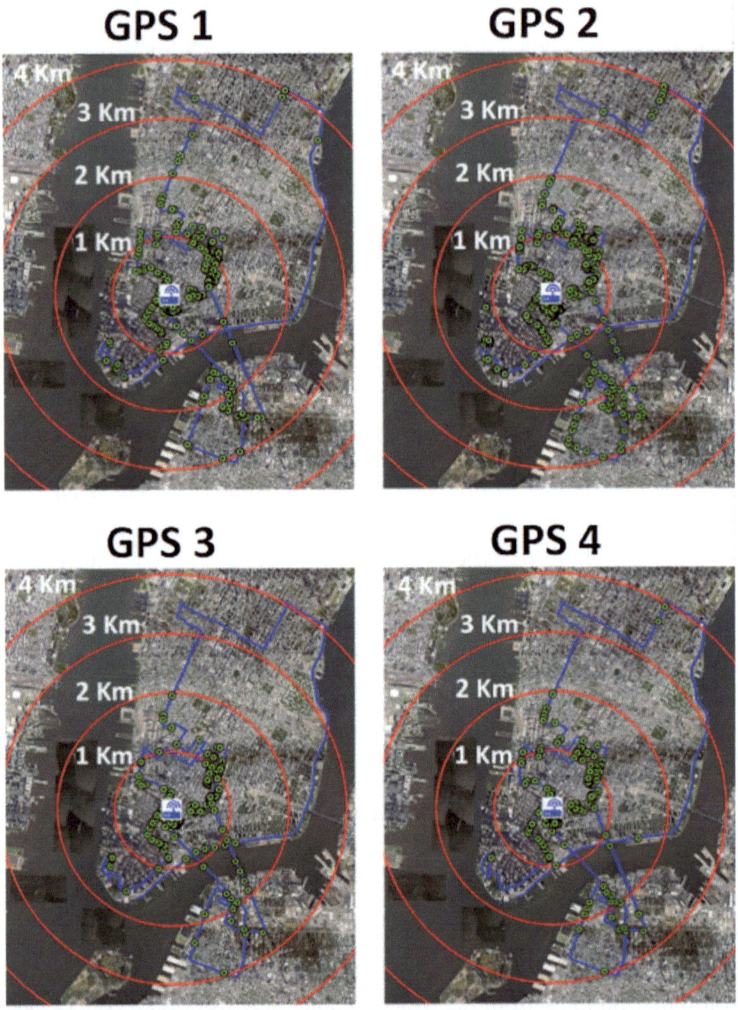

Fig. 8.7 Visualizations of vehicle routes taken during New York Experiment 2

Comparison of LoRaWAN with Commercial Alternatives Toward incorporating COTS IoT protocols into tactical C2 systems, a comparison of their coverage and capabilities in dense urban settings becomes desirable. Alternative commercial protocols to LoRaWAN include Narrowband IoT (NB-IoT) [19] and Sigfox [4], which both provide long-range wireless communications with low throughput and low energy consumption. Toward comparing both the coverage and reliability of these protocols, future experimentation will be planned in urban settings to compare performance for indoor and outdoor settings, as well as for vehicle tracking tasks.

8.3 Security Techniques and Solutions for IoT

The main aim of our security-related experiments was to evaluate the ability of COTS components to provide the adequate level of protection required for unclassified military applications and Civil–Military Cooperation (CIMIC). In particular, our objectives were to:

- Evaluate encryption capabilities of the Arduino sensor boards while using existing software libraries
- Demonstrate the ability to implement cryptographic access control, relying on a hybrid cryptographic solution and enabling enforcement of CPR access control policies
- Evaluate possible options for symmetric key management and trust establishment to build a secure domain of cooperating sensor nodes
- Validate the use of modern hardware security components, such as Trusted Platform Module (TPM) and Quantum Random Number Generator (QRNG) in IoT scenarios

8.3.1 Use Cases

We aimed to investigate security solutions and techniques that are suitable for a large set of different military IoT use cases. In the following, we briefly present two examples of concrete use cases that could be used to evaluate the functionality and performance of the proposed security approach.

8.3.1.1 Securing a Critical Facility

In [9], a specific use case vignette has been defined to secure a critical facility. This vignette could be expanded to provide additional perimeter monitoring using low-cost commodity sensors. Specific cases of such vignettes, for example, related to harbor protection during military logistic operations or a smart military base, are topics of further work performed within the IST-147 group on Military Applications of IoT.

8.3.1.2 Collecting Outpost Data

In this scenario, during a typical operation, a patrol passes several deployed outposts equipped with sensors monitoring the environment. Sensors deployed at the outposts have no long-haul network connection to the base. Instead, they offer a short-range (e.g., WLAN, Bluetooth, NFC) or medium-range (e.g., LoRa) communication technology to read out the collected data. The collected data is read and stored when the patrol passes an outpost and detects connectivity with a sensor network. The data is forwarded to the smart base's networking infrastructure when the patrol returns to the base.

8.3.2 Design

We have implemented a simple proof-of-concept (PoC) sensing system composed of commercially available, low-cost Arduino and Raspberry Pi devices. The information collected by the low-energy Arduino sensor nodes is collected by a Raspberry Pi 2 (RP2) sink node and transmitted to a public cloud IoT service offered by Amazon. Wireless interfaces used for transmission are Wi-Fi and low-energy Bluetooth for short-range communications and LoRa for mid-range communications. Our PoC configuration is depicted in Fig. 8.8.

In our PoC, the CAC is enforced via ABE. The ABE is applied at an RP2, which acts as a sink node and an ABE encrypting device for information collected from several Arduino devices equipped with motion and temperature sensors. The Amazon Web Services (AWS) IoT platform is used as a cloud-based repository for all data collected from the sensors. The data is stored by the AWS IoT in an encrypted form and is decrypted first at the end user's terminal.

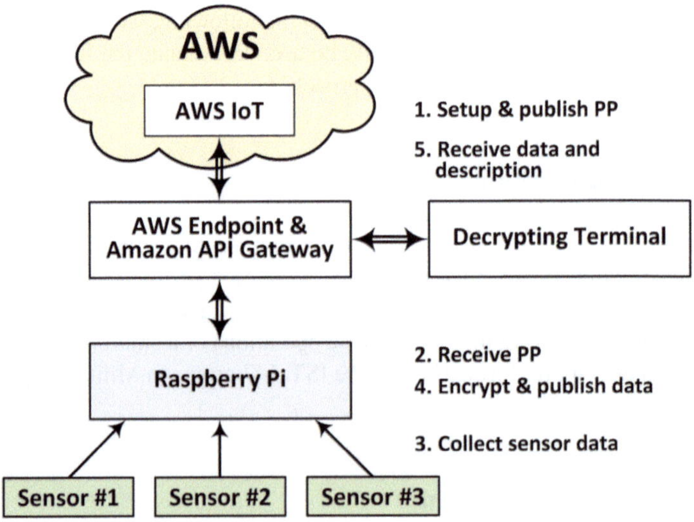

Fig. 8.8 Proof-of-concept configuration

8.3.2.1 Cloud Services

The AWS IoT platform is a service offered by AWS to help connect Amazon cloud services with Internet-connected devices, such as sensors, actuators, and other embedded devices. This allows users to collect and store data from a vast array of devices and create applications that control the behavior of these devices. The AWS IoT platform opens a bidirectional channel between the IoT devices (so-called *Things*) and the applications and services in the cloud. A device with the correct access permissions can connect to an AWS end point and subscribe or publish messages to a topic of their choosing. The topics' names are formed by following a hierarchy, such as house $>$ garage $>$ door. The message broker then forwards each message published to a topic to all the clients subscribed to the said topic. AWS uses the MQTT communications protocol to send JSON objects in its payload. MQTT is an M2M communication protocol, which uses the publish/subscribe method of transporting messages. It is designed to be lightweight and is especially useful in environments where a small code footprint is required, or the network bandwidth is limited. The Python scripts use the *multiple messages* method to reduce the network traffic generated when connecting to the MQTT server.

8.3.2.2 Sink and Sensor Nodes

Several development boards are readily available for implementing IoT applications. We chose two options for our PoC due to their wide availability, high community support, large software libraries, and easiness of use. In particular, we chose the Arduino UNO as a platform for sensor nodes and the Raspberry Pi 2 as a platform for the data aggregation gateway, that is, the sink node.

8.3.2.3 Physical Sensors

There are several types of sensors for detecting movement: passive infrared (PIR), microwave, ultrasonic, tomographic motion detector, and cameras (with video processing). We opted for a two-layer movement sensing system where the PIR functions continuously while the microwave (Doppler) sensor is only activated when the PIR is triggered. This dual system allows for lowering the number of false positives individual sensors produce while limiting the current consumption of the entire sensor system. The disadvantage of PIR-based triggering is that circumventing the PIR sensor will compromise both sensors attached to the node.

PIR sensors are passive electronic devices capable of sensing infrared fluctuations. The PIRs are widely used, inexpensive, easy to maintain, and have a low current consumption (ca. 100 μA), making them a good option for triggering a secondary higher precision (but also higher current consumption) sensor. False triggers to a PIR can be caused by unstable voltage at the detector, sudden heat changes, white light, direct wind, and electromagnetic/radio interference.

Microwave sensors detect movement through the Doppler effect: A signal is transmitted by the device and *multiplied* with the received signal reflected from a target to detect movement. The object's speed relative to the antenna causes a frequency shift detected by the sensor. The advantages of the microwave motion detection sensors include:

- High sensitivity to motion
- The ability of the emitted radiation to penetrate objects (e.g., walls)
- Applicability to harsh environments with a wide range of temperatures

The disadvantages of these types of sensors lie in the fact that they:

- Are easily triggered by small movements, for example, leaves blowing in the wind, therefore causing false alarms
- Can be susceptible to electromagnetic and radio interference
- Have a relatively high current consumption (ca. 40 mA)

8.3.2.4 Cryptographic Mechanisms

In our PoC, we use a combination of symmetric and ABE. For performance reasons, we use symmetric AES encryption to secure communication between sensor nodes and the sink node. The sink node splits the session key into two shares if a fresh session key is used at the UNO for encrypting every data object. If a static key is used at the sensor node, the sink node generates two new random key shares and the resulting encryption key. Then, the key shares are encrypted using ABE, as described in Sect. 8.3.2.5.

8.3.2.5 Separation of Release and Protection Policies

When cryptographically enforcing CPR policies, our objective is to ensure that only the specific combination of authorized user AND authorized terminal is granted access to the information [20]. We can achieve this by requiring both the terminal and the user to contribute to data object decryption. One of the solutions providing such capability is a simple secret sharing scenario described below.

The RP2 sink node generates two random AES keys: a terminal key K^T_{AES} and a user key K^U_{AES}. These keys act as shares of the data encryption key K_{AES} and are securely distributed in an ABE envelope to the authorized terminals and users, respectively. The data encryption key K_{AES} is calculated as an XOR of the terminal key K^T_{AES} and the user key K^U_{AES}: $K_{AES} = K^T_{AES} \otimes K^U_{AES}$. The data received from a sensor is then encrypted using the AES-256 algorithm operating in the Galois/Counter Mode (GCM) and K_{AES}. In turn, the keys K^T_{AES} and K^U_{AES} are encrypted using the ABE algorithm and the keys K^T_{ABE} and K^U_{ABE}, respectively. Finally, the RP2 sink node sends both the encrypted data object and enveloped keys to the AWS IoT platform, where the user can obtain data via the pull or push (i.e., publish/subscribe) mechanism.

8.3.2.6 Cryptographic Access Control with ABE

ABE enables decryption based on the outcome of a predicate evaluation of several attributes associated with the recipient's private key. An advantage of the ABE is that the sender does not necessarily need to know the precise identities of the appropriate recipients but can instead define an access control structure for the recipients based on their attributes, such as rank, affiliation, or role within an organization. However, a specific challenge is related to the computation of access structures required for ABE in order to enforce the organizational CPR policies. In particular, the CP-ABE access structure can be seen as an inverse to a typical ABAC policy. That is, instead of defining who can see certain information, it defines attributes of all types of information that a particular user or terminal can see.

In our PoC, we have implemented an ABE scheme defined in [15] using Python and the Charm library [2]. Charm is a framework that enables developers to create advanced cryptosystems in Python rapidly. It is based on a hybrid design: The lower level mathematical operations are implemented in C modules to increase their performance, and the higher level cryptosystem operations are written in Python to reduce the programming effort.

Our ABE scheme is formed by two setup scripts: one executed on the sink node and the other executed on the Decrypting Terminal. The setup scripts need to be executed only once. In addition, an encrypting script runs on the sink node, and a decrypting script is executed at the Decrypting Terminal.

The setup script executed on the Decrypting Terminal interacts with the user to create and store locally the secret ABE keys SK_{ABE}^{T} and SK_{ABE}^{U} needed to decrypt information for accessing which terminal and user are authorized. In the case of SK_{ABE}^{U}, the key can be stored in a tamper-resistant module, such as a smart card, which is separate from the terminal and owned by the user. The setup script also publishes the Public Parameters (PPs) corresponding to the SK_{ABE}^{T} and SK_{ABE}^{U} to the AWS IoT topic `MSK/exchange`, to which the RP2 setup script is subscribed. Once the setup is complete, the RP2 receives the relevant PP and stores them locally.

After receiving measurement data from a sensor node, the sink node generates three AES keys, K_{AES}, K_{AES}^{U}, and K_{AES}^{T}, and performs an encryption process, as discussed in Sect. 8.3.2.5. Both the ABE encrypted keys K_{AES}^{T} and K_{AES}^{U} and the encrypted data are published to a specific AWS IoT topic.

The terminals, which subscribe to the same AWS IoT topic, will receive the published message. The Decrypting Terminal will decrypt the K_{AES}^{T} with the SK_{ABE}^{T} and ask the user to decrypt the K_{AES}^{U} with the SK_{AES}^{U} known to the user. After receiving both decrypted shares, the terminal reconstructs the K_{AES} and decrypts the data provided by the sensor nodes.

8.3.2.7 Performance Analysis

We have evaluated the execution times of the implemented cryptographic mechanisms, which were executed at the RP2 (the sink node) and Arduino UNO devices (the sensor nodes), as well as the overall size of the exchanged messages. In addition, since the sensor nodes are battery-powered, we also evaluated power consumption at the Arduino nodes (Fig. 8.9).

The resources consumed at the sink node were measured in Python using the Timeit and Psutil libraries. The measurements were logged to a file using a test script that executes the encrypting script multiple times. The obtained results are provided in Tables 8.7 and 8.8. The sensor measurements transmitted in our PoC are 4 bytes long and padded to 16 bytes to enable encryption with AES256-GCM. The total size of payloads, including both ABE encrypted AES keys and the AES encrypted sensor data, is 2518 bytes. The payload related to a single ABE encrypted AES key is 1166 bytes long—this is expanded to 1214 bytes due to JSON encoding.

Fig. 8.9 Arduino and Raspberry Pi 2 boards with LoRa modules used in performance evaluation

Table 8.7 Time spent in different execution phases

Phase	Time
Generation of three AES keys	0.15 ms
ABE public parameters: deserialization and load	5.81 ms
Conversion of two payloads to JSON format	1.48 ms
Sending messages and setting up communications	1.25 s
Script execution (total)	3 s
CPU (user)	1.87 s
CPU (system)	0.06 s

Table 8.8 Performance of different cryptographic algorithms on Raspberry Pi 2

Algorithm	Encryption	Decryption	Setup time
ABE-LW12	19.98 ms	14.76 ms	593.99 ms
RSA-1024	1.97 ms	11.50 ms	266.60 ms
AES-256	0.02 ms	0.02 ms	0.05 ms
3DES-128	0.02 ms	0.02 ms	1.88 ms

Table 8.9 Performance of different variants of AES encryption on Arduino UNO

Algorithm	Mode	Encryption	Decryption
AES-256	ECB	47 μs/byte	96 μs/byte
AES-128	ECB	33 μs/byte	67 μs/byte
AES-256	GCM	125 μs/byte	125 μs/byte
AES-128	GCM	112 μs/byte	111 μs/byte

We have evaluated two block ciphers, AES256 and AES128, in both Electronic Codebook (ECB) and Galois/Counter Mode (GCM), which are implemented in an openly available `Crypto.h` library. Table 8.9 presents the results of tests performed on a 16-MHz 8-bit ATMega328P UNO board. The implementation of encryption algorithms occupies 23% of the FLASH memory and 17% of the dynamic memory (SRAM), leaving enough space to implement data acquisition and transmission applications.

We have also focused on testing the power consumption of the Arduino board—this is critical for ensuring an acceptable duration of maintenance-free operation of the sensor nodes. Several tests were performed, and we concluded that the board has a current draw of 49 mA at 5 V input in a normal mode when no peripherals are attached. In addition, two sleep modes can be implemented, reducing the current consumption to 36 mA. Moreover, consumption can be lowered to 12 mA if the board is powered at 3.3 V. At 5 V, and attaching a LoRaWAN Dragino board in continuous transmission mode increases the current consumption to 61 mA. The waking-up procedure depends on the algorithm implemented on the board: One option allows the board to sleep for a determined period of time, while the other leaves a couple of digital pins on, making it possible to wake up the board with the output from a digital sensor, for example, PIR.

8.3.3 The Secure Domain for Sensor Nodes

The sensor node is usually built based on a constrained device and has limited resources. In particular, we are referring to the limited size of the available RAM, relatively low computing power, limited power supply, and the use of primarily wireless connectivity with a short range. Because of these limitations, security issues in IoT networks must be considered within a network of sensor nodes and organized locally in small clusters of sensor nodes [8], which will be called the sensor node domain. Mechanisms offered by the Trusted Platform Module (TPM), which should

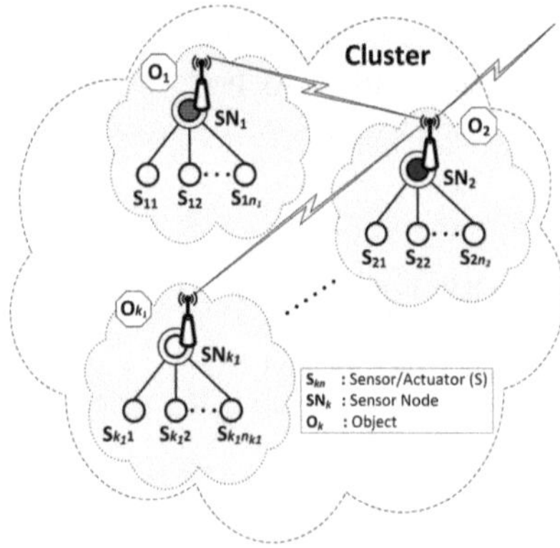

Fig. 8.10 Structure of sensor nodes security domain

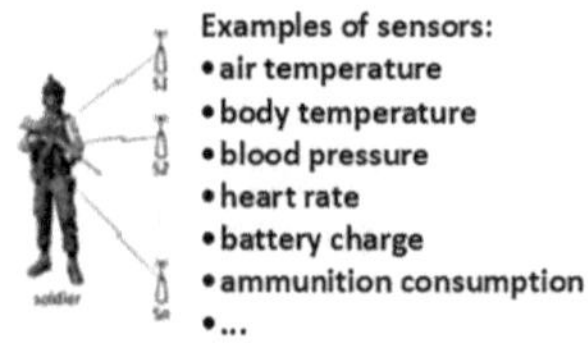

Fig. 8.11 An example of one cluster object—soldier equipment

be a mandatory hardware component of each sensor node, can be used to build a local trust structure in the domain of sensor nodes.

One domain of sensor nodes can represent one group of cooperating objects (sensor nodes). The structure of such a domain is shown in Fig. 8.10. An example of a group of cooperating objects could be one person (a patient, rescuer, soldier, etc.) equipped with various actuators (e.g., body sensors) or a vehicle (a mobile robot, drone, etc.) equipped with onboard sensors. Figure 8.11 shows an example of such a group of objects.

In the domain of sensor nodes, one node acts as the *Gateway* for all the cooperating nodes. This node mediates the data exchange between the domain nodes and the surrounding domain. To support security mechanisms within the domain of sensor nodes, one node acts as a security authority in the domain and plays the role of the *Master*.

Two models of security domains for sensor nodes were prepared:

1. Model I: A simple model that assumes that each sensor node is a source of data. In the domain, there is only one sensor node, which is simultaneously the recipient of data generated by domain nodes (i.e., it acts as a *Gateway*) and a security authority for the domain (i.e., it concurrently acts as a *Master*).
2. Model II: A model in which each sensor node can act in the *Gateway* and *Master* roles. Different domain nodes can perform these roles at a given time, but it is possible that the same sensor node will play both roles simultaneously. It was assumed that diagnostic procedures are performed periodically in the domain to determine the suitability of the domain nodes and that each sensor node in the domain can take over the role of *Master* or *Gateway* if the node currently

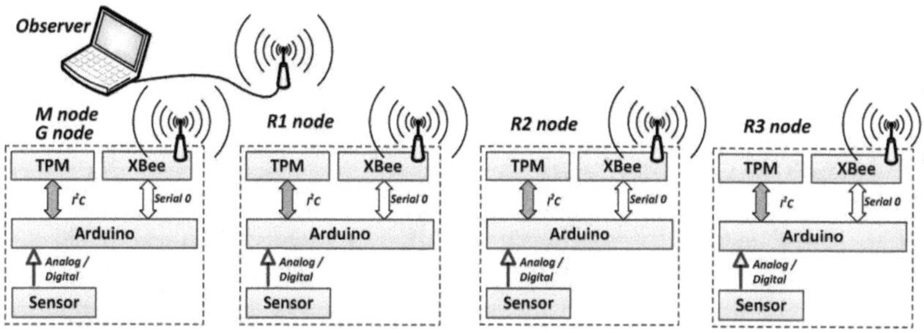

Fig. 8.12 Structure of demonstrator of the security domain for sensor nodes

fulfilling one of these roles is deemed unfit. In addition, to increase the level of security, it was assumed that each sensor node must be preprepared and registered in the domain before starting normal operation. The setup procedure takes place in a safe and controlled environment outside the domain's normal operating area. Finally, the source of authentication data for prepared nodes is a special node that acts as a *Base*.

8.3.3.1 The Security Domain for Sensor Nodes: Model I

Figure 8.12 shows a prototype of a security domain for the sensor nodes [6]. Each sensor node can perform additional functions necessary to:

- Gather and preliminarily process the data deriving from sensors located in each of the objects
- Perform a protected transfer of data collected by objects of one cluster to other clusters
- Ensure transmission security within the cluster and authentication of objects in the cluster

Clusters of sensor nodes consist of two interpenetrating domains, which include only the SN_k nodes:

- **Security domain:** The nodes will exchange data used to authenticate the nodes. The diagnostic procedures will also be implemented in this domain. The authority in this domain is the node, which plays the role of the *Master*.
- **Domain of protected transmission** of the sensory data from SN_k nodes to one of the SN_k nodes, which plays the *Gateway* role in the domain. The gateway node is also responsible for protected data transfer to other clusters.

Properties of the security domain include:

- One node, which plays a role of a *Master* (the *M* node), and the others that play the role of *Replica of Master* (the *R* node)

Fig. 8.13 The prototype of security domain for sensor nodes (Model I)

- A unicast transmission and the symmetric cryptography (AES in CBC mode with individual NSK (Node Security Key) key of CSN_k node), always used during the transmission of the data between CSN_k nodes and M node.

Properties of the domain of protected transmission of the sensory data are:

- The central node and the sink node for the sensory data of the domain is the gateway node (the G node), and the transmission of the sensory data in the domain is cryptographically protected using a common symmetric key (NTK—Node Transfer data Key).
- The gateway node is also responsible for the protected data transfer to other clusters.

The technical implementation of the demonstrator of a security domain for the sensor nodes with strong authentication is shown in Fig. 8.13.

8.3.3.2 The Security Domain for Sensor Nodes: Model II

The main assumptions behind Model II development include:

- Cryptographic protection of data stored in the resources of sensor nodes and data sent from sensor nodes.
- Symmetric algorithms as the base of cryptographic security—asymmetric algorithms are used only during the registration procedures of nodes in the security domain of sensor nodes.
- All cryptographic security processes use mechanisms offered by TPM —TPM is an obligatory hardware component of each sensor node.
- The domain can execute diagnostic procedures to verify the correct operation of both individual nodes and the entire domain to construct a self-organizing

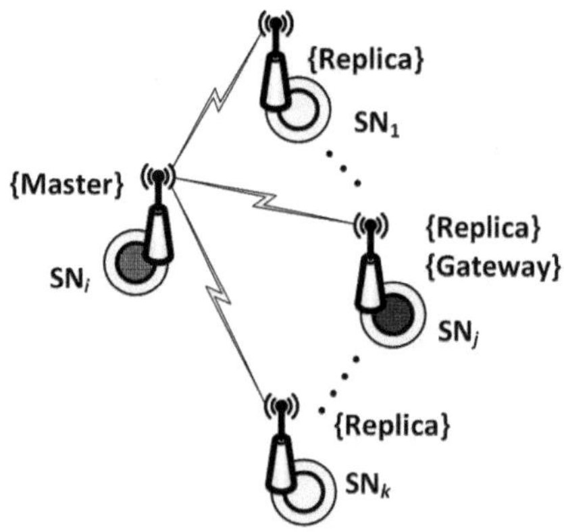

Fig. 8.14 Model II—logical network topology in the security domain

network of sensor nodes immune to failure, compromise, or hostile takeover of individual sensor nodes.

In the security domain of sensor nodes [7], all domain nodes belong to two overlapping domains. As described above, these are the security and sensor data transmission domains.

In the security domain, the nodes exchange data to authenticate the sensor nodes. The authority in this domain is a single sensor node, known as the *M* node, which acts as the *Master*. The remaining domain nodes, referred to as *R* nodes, can take over the role of the *Master* if the node currently performing this role malfunctions. The data transfer in the security domain is cryptographically secured using the NSK symmetric key (*Node Security Key*). In the security domain, the network has a logical star topology. The structure of such a network is shown in Fig. 8.14.

Diagnostic procedures are also performed in the security domain to detect failed sensor nodes. In the event of detecting the malfunction of the sensor node currently playing the role of *Master* or *Gateway* in the domain, the procedure of selecting the node that will take over the failed node's role is started.

In the domain of sensor data transmission, the receiving node is one of the domain nodes, referred to as the *G* node, which currently acts as a *Gateway* in the domain. The transmission of sensor data is cryptographically secured using a symmetric NTK key (*Node Transfer Data Key*), a common key for all domain nodes. The *G* node is also responsible for securely transferring data from the domain to recipients outside the domain. The logical topology of the sensor data transfer domain is shown in Fig. 8.15.

The sensor node is a mobile device. The XBee module is used to exchange data between domain nodes. The LoRa interface, enabling data transmission over longer distances, is intended for data exchange with other domains in a situation where a given sensor node acts as a *gateway* in the domain. The serial interface is used only while preparing the sensor node to work in the domain, and once this procedure is completed, it is removed from the node. An Arduino 2560 microcontroller can be an element that processes data in the sensor node. The TPM module is an obligatory

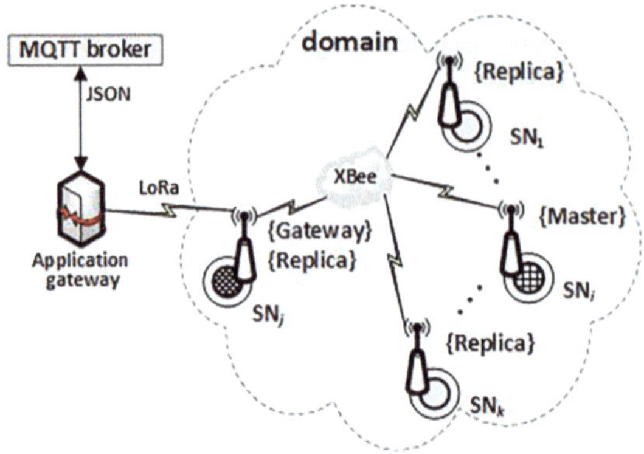

Fig. 8.15 Model II—logical topology in the domain of sensor data transfer

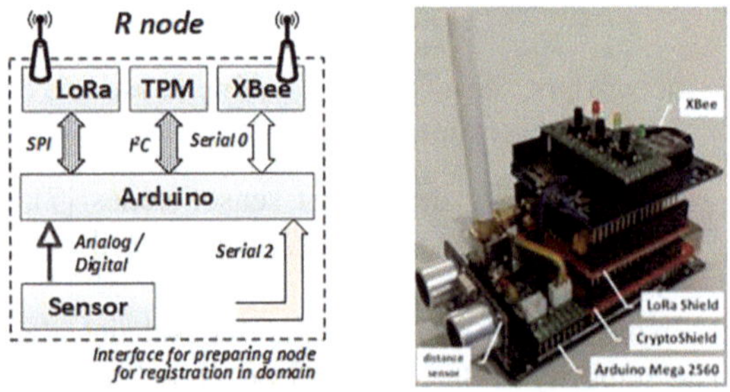

Fig. 8.16 The block diagram of the sensor node and the sensor node's hardware implementation

Fig. 8.17 The demonstrator of security domain for sensor nodes (Model II)

component of the sensor node. The block diagram of the sensor node and the sensor node's hardware implementation are both shown in Fig. 8.16.

The technical implementation of the demonstrator of a security domain for the Model II sensor nodes is shown in Fig. 8.17.

Fig. 8.18 Quantum random number generator shield for Arduino

8.3.3.3 Quantum Random Number Generator

A random number generator is a crucial element when creating cryptographic keys in all cryptographic security solutions. Only a source of random numbers with very good entropy makes it possible to create secure cryptographic solutions. In many experiments to study the security domain for sensor nodes, it has been convenient to use a random number generator, which is a component of TPM.

For this project we used a quantum random number generator with excellent entropy and very high efficiency, namely a Quantis OEM component generating a random bit stream of 4 Mbits/s. In addition, we developed a shield compatible with the Arduino and Raspberry Pi platforms and a software library for this component. The sensor node using the developed shield prepared for the Arduino platform, shown in Fig. 8.18, has wholly fulfilled its tasks in the presented security domain for the sensor nodes.

8.3.3.4 Security Domain: Conclusions

Demonstrators were developed for both domain models, and their performance was tested. The biggest challenge for software development was the size of the available RAM on the Arduino 2596 boards, which was only 8 KB. This memory had to accommodate the sensor node software, which occupied about 4.8 KB, and the software data. One data structure dedicated to one RSA-2048 asymmetric key required about 850 bytes. The Electrically Erasable Programmable Read-Only Memory (EEPROM) of the Arduino board was used to store data for which there was insufficient space in RAM.

The following procedures were tested in the Model I experiments:

- The gateway node is also responsible for the protected transfer of the data to other clusters.

- The sensor domain initiation procedure including *M* node initiation
- The procedure for registering the *R* node in the domain
- Transfer of sensor data from *R* node to *M* node
- Removal of the *R* node from the domain
- Procedure for mutual authentication verification by domain nodes
- Test of sensor nodes domain integration
- Renewal of node cryptographic keys

All procedures tested worked correctly. The experiments revealed the following shortcomings of Model I:

- *M* node simultaneously acts as a receiver of data from all sensor nodes in the domain and is, therefore, the single critical point of failure in the domain.
- Due to the size of the resources, the *M* node is not able to handle the actuators of this node.
- The procedure for initializing and registering the node provides the opportunity to take over the public key part of the domain, allowing a fake node to be registered in the domain.
- The overly complicated and resource-intensive way of securing frames to send sensor data to the *M* node
- Due to the large size of the asymmetric keys, the resources of the *M* node limited the domain size to only 11 sensor nodes.

The solutions used in Model II were intended to either eliminate or reduce the shortcomings of Model I. Following the development of the Model II demonstrator, similar experiments were performed as for Model I. The results confirmed the correctness of the developed solutions.

The diagnostic procedures implemented in the system detect unfit nodes in the domain, e.g., *M* node. If an unfit *M* node or *G* node is detected, a procedure is triggered to elect a new node to perform the corresponding role, thus eliminating a single point of failure in the domain.

The use of symmetric encryption to secure the data reduced the resource requirements of the node and allowed each node to be the data source regardless of the role it performed in the domain. The same resources allowed a domain containing 62 sensor nodes to be supported.

The parameters of the wireless link used are stored in the secure resources of the node and are relatively easy to reproduce. This approach eliminates the possibility of interception of the transmission link.

In some experiments, power banks with a capacity of approximately 6600 mAh were used to power the nodes. Such a power supply allowed experiments to be performed that lasted between 4 and 6 hours. This is not a long time, but during the experiments, many routines were used to monitor the nodes' behavior, which strained the power sources—these routines will not be used in the regular operation of the nodes.

8.4 Body Area Network

A Body Area Network (BAN) is a class of short-range wireless networks dedicated to monitoring an individual's state and nearest surroundings. BANs utilize mainly low-power, short-range technologies such as Bluetooth Low Energy, ZigBee, ANT+, and RFID. There are other proprietary solutions; however, they are primarily based on the 2.4 GHz ISM band.

In IST-147, we evaluated the BAN class network as a base concept for real-time monitoring of soldiers' condition, health state, and equipment. The crucial consideration was the *combat triage* process. Generally, in healthcare, *triage* is the process of categorizing the criticality of a patient's condition. The person responsible for triage (usually a first responder) performs a brief, focused assessment and assigns the patient a triage acuity level, which is a proxy measure of how long an individual patient can safely wait for a medical screening examination and treatment.

Civilian and combat triage have a lot in common. Both refer to situational crises and use similar methods. For instance, in the USA, The National Association of Emergency Medical Technicians (National Association of Emergency Medical Technicians (NAEMT)) adopts a military Tactical Combat Casualty Care (Tactical Combat Casualty Care (TCCC)) course to train civilian Emergency Medical Services (EMS). However, some factors make civilian and combat triage different. According to TCCC, those are:

- Hostile fire
- Darkness
- Environmental extremes
- Different wounding epidemiology
- Limited equipment
- Need for tactical maneuver
- Long delays to hospital care
- Different medic training and experience

In many cases, these factors cause combat triage to be more challenging than civilian triage, especially when first responders, medics, and patients themselves are under constant threat. It is also worth pointing out that the patient is only part of the mission during combat operations, whereas in a civilian setting, the patient is the mission. That is why combat triage has, in fact, three goals:

1. Treat the casualty.
2. Prevent further casualties.
3. Complete the mission.

Triage is a process where health state categorization is relatively simple and can be supported by wearable sensors connected in a BAN network. As shown in Fig. 8.19, multiple health parameters and conditions can be monitored and identified by noninvasive sensors, including:

- Blood pressure: a variety of Photoplethysmography (PPG) sensors
- Blood saturation: a variety of PPG sensors

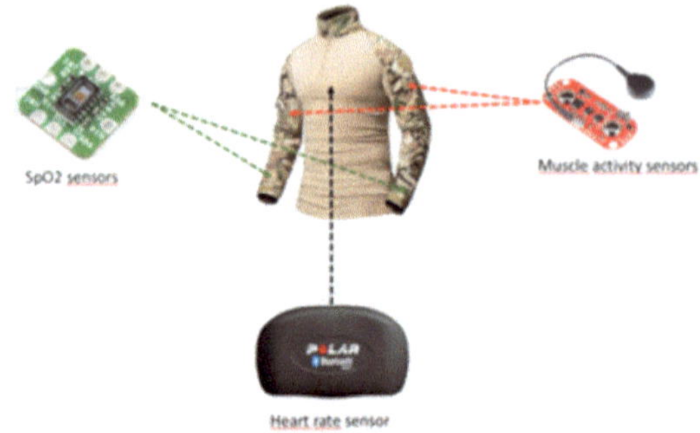

Fig. 8.19 Wearable triage sensors

Fig. 8.20 EMG sensors setup with Genuino 101 microcontroller

- Heart rate: a variety of PPG sensors or ECG sensors in the form of chest straps
- Heart condition: Elektrokardiogramm (German for ECG) (EKG) sensors like Comarch PMA, AD8232 Heart Rate Monitor
- Head injury: dedicated head impact sensors like Triax Smart Impact Monitor
- Muscle activity: a variety of EMG sensors

There is an obvious need for automated triage, especially in military scenarios. Since wearable sensors and BAN technologies are widely available in IST-147, we decided to implement a PoC of a triage solution. According to this idea, a soldier will be equipped with a personal sensor system for health monitoring. It consists of:

- EMG sensor for muscle activity monitoring: SparkFun MyoWare
- Heart rate sensor: Polar WearLink
- Microcontroller: Genuino 101
- Android-compatible smartphone

EMG sensors are connected to a microcontroller board (see Fig. 8.20), which transmits muscle activity signals via BLE to a smartphone device. Similarly, the heart rate sensor connects with the Android device via BLE. The smartphone runs an application that assesses the soldier's health state. As mentioned in Sect. 8.5.1.2, a soldier can be classified into one of four categories: HEALTHY, EXHAUSTED,

NEEDS TREATMENT, and CRITICAL CONDITION, based on sensed vital signs. Before a soldier begins a mission, the application is calibrated. The type of mission is selected, and the soldier needs to register. The soldiers' smartphones create a MANET network on top of BLE, and information about their health state and equipment is sent to the team leader.

In order to perform triage classification, we have designed a problem-solving ontology that represents some of the sensor and medical domain concepts. The Sensor Data processing path delivers seven-stage combat triage utilization of ontology model forms terminology and model constraints on which a triage knowledge base is formulated. The knowledge base supported by the reasoning mechanisms delivers model consistency checks of the instance base (data), ensuring valid relations between data instances, performing classification tasks for instance data, and executing rules to infer new facts in the knowledge base. These tasks have been used as tools for solving the problem of classifying an individual's health state based on the sensor data measuring stimuli. Implemented in the ontology concept definitions and rules is the performance of data classification tasks evaluating sensor data introduced within the system. The evaluation process takes the preliminary data package and confronts the data with "evaluation rules," which analyze specific data characteristics (discrete or continuous in nature) to aggregate and produce information about the inspected (monitored) object. In order to perform the analysis, the rules contain evaluation or decision predicates. The predicates determine if a calculated characteristic contains useful information in the context of the evaluated object and environment. For example, PPG sensor data containing a photoplethysmogram can be processed to evaluate heart rate, which, depending on the subject's age and physiological stamina, can determine the stress level, exhaustion, and health state. The ontology, in that case, offers a set of rules that interpret sensor data (inertial, biomedical) to evaluate an object's characteristics in the context of the current health state.

The Combat Triage Ontology delivers a means to produce more than one predicate based on one instance of a SensorDataPackage, which can help to evaluate sensor data in the context of health state and other monitored object's characteristics. The next processing stage aims to evaluate the ObjectState based on the series of associated SensorDataPackages. The ObjectState concept holds several Description Logic (DL) constructors, which assign an ObjectStateDomain value to each and every ObjectState. The ObjectStateDomain is a nominal concept (enumeration) that holds evaluation statuses for any monitored object. The evaluation status can be understood as an outcome of the evaluation process produced by the reasoning mechanism. The classification mechanism in the knowledge base iteratively performs sensor data analysis, after which instances of ObjectState are evaluated and linked with an adequate ObjectStateDomain value. The ObjectStateDomain concept provides detailed taxonomy leading toward HumanStateDomain and TriageHumanStateDomain and is further specialized according to the AJP-4.10.A standard (AJP-4.10A-TriageHumanStateDomain) and S.M.A.R.T. (S.M.A.R.T-TriageHumanStateDomain). Depending on the object type and aim of the object's state evaluation, the reasoner can evaluate a particular ObjectState with status taken from the AJP-4.10A combat triage standard. With this result, the reasoner can evaluate instances of SensorDataPackages, engage a set of rules that produce information about a given monitored object, and confront sensor information within the environmental context in which the object is found (combat mission, medical treatment, etc.).

The knowledge base also utilizes built-in semantic mapping between enumerated values. This feature is useful for mapping purposes, in which a well-defined concept or instance has a corresponding entity—a synonym or entity of very similar semantics. In the case of Combat Triage Ontology, semantic mapping has been used to map various triage standards and approaches to each other. Using the owl axiom owl:sameAs (for individuals) and owl:equivalentTo (for concepts), we have been able to map AJP-4.10.A triage statuses with the S.M.A.R.T. methodology statuses and more found in crisis management methodologies. We made the decision to restrain the concept list to ensure classification efficiency. Note that this also ensures model readability and supports future modularization. The ontology has been developed using Web Ontology Language (OWL) 2.0 RL, merging OWL and SWRL language capabilities. The final form of the ontology has been developed using SHOIN(D) Description Logic dialect, which demonstrates the complexity of ontology definitions, containing over 20 SWRL rules, 50 concepts, 20 object properties, and over 20 datatype properties. One of the important characteristics of the ontology is that it contains over 29 defined concepts implementing both value and cardinality restrictions. Moreover, the model has been modularized to consist of three parts: the base terminology, defined concepts terminology, and instance base. Such construction supports further extensions of terminology. Most of all, however, it separates the meta-model from the data instances and prepares the knowledge base to be fed with data from the real sensor system or simulation software (Fig. 8.21).

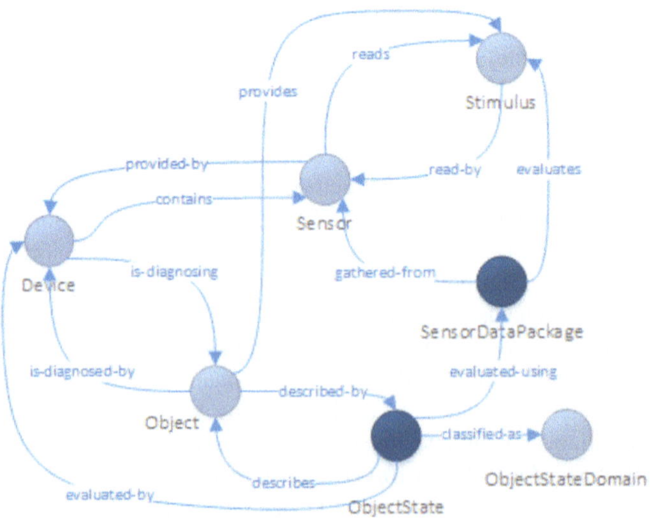

Fig. 8.21 Triage ontology main concepts

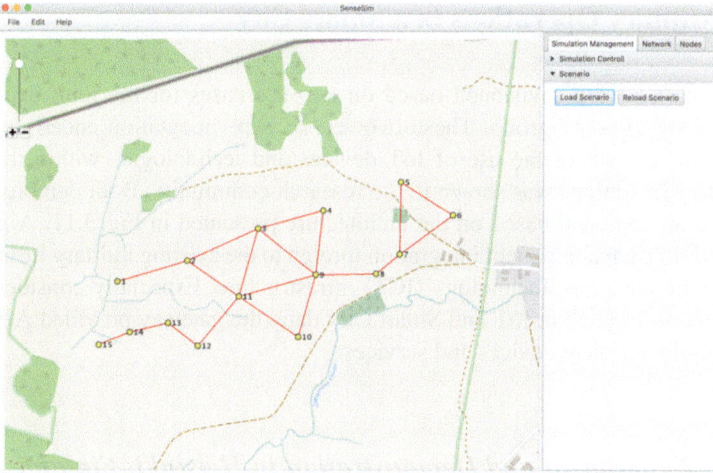

Fig. 8.22 SenseSim scenario

For demonstration purposes, part of the network is simulated in the SenseSim simulator with real hardware in the loop. In this scenario, a military patrol of 15 soldiers is deployed in the city. Each soldier is equipped with the health monitoring system as described above. Still, only one set is real hardware, and the rest is simulated using real health data from the PhysioBankATM site. Overall, there are 15 entities that take part in the scenario; 14 of them are simulated, and 1 is hardware in the loop.

Figure 8.22 shows the initial state of the scenario simulated in SenseSim. The network has 15 devices that connect to one another using a wireless channel. It is assumed that each device has the same communication capabilities, and all messages are sent peer-to-peer. The network's topology is not strict and may change in time due to device movement. Devices can communicate with each other as long as they stay within radio range. In this scenario, the maximum range is 200 m. Devices use a flooding routing algorithm for distributing messages. However, flooding is not an efficient algorithm: It generates a lot of network traffic and may cause redundant messages. On the other hand, it enables a high probability of message delivery, especially when the network topology may change, and devices have little or no knowledge about it.

Thanks to ontology reasoning, expert-level knowledge is not required from first responders to properly carry out the triage process. Some vital signs registered by sensors, such as ECG, may be hard to interpret without a medical background, but there are methods for automatically processing and extracting valuable information. Moreover, using IoT devices for triage enables this process to become continuous with minimal manual effort.

8.5 Military Operations in a Smart City

Two scenarios were envisioned based on two use cases for the joint experimentation of the IST-147 group. These two sets of experimentation ended up as live demonstrations where the use of IoT devices and technologies within the realm of military operations was shown to the research community. Both demonstrations followed an approach based on the architecture presented in Fig. 3.11. A *gateway* was used for each component integration foreign to the existing military Information and Communications Technology (ICT) infrastructure. Especially considering the consumption of civilian IoT and Smart City data, the gateway provided API-based access to the physical devices and services.

8.5.1 Experiment and Demonstration in Helsinki, Finland

In 2017, during the International Conference on Military Communications and Information Systems (ICMCIS) in Oulu, Finland, the IST-147 group held the first demonstration of IoT systems in use with existing military systems [13, 24]. A complex scenario was envisioned to show the real-time usability of such a system. It involved multiple military ICT systems from various NATO partner countries plugged in for joint SA during ISR ops.

The city of Helsinki was the basis for the smart city scenario where the military had to conduct an operation. The aim was to showcase a convoy's passage through the city streets from the airport to the US Embassy in Helsinki. During transit, along with the traditional means and methods used by the military to secure the pathways and enable ground-level intelligence, the convoy also used the support of the smart city assets.

The core idea of the whole experiment was to find out how to use modern ICT domain technologies in conjunction with the ICT systems traditionally used. The IST group implemented support for collecting data from Helsinki's sensors and pushing them into the military's C2 systems. Apart from the smart city's sensors, a range of military-owned IoT-based assets were used to collect data about the Area of Operations (AOOs), and (simulated) military units leveraged this information. The data was consumed in widely used formats and processed on a mobile cloud running on the mobile Tactical Operations Center (TOC) deployed on a base convoy vehicle. The ground sensors and actuators monitored the city's pathways as the convoy traversed the designated path to keep the military's SA up to date, enabling evasive actions based on the collected and analyzed data and risks identified. Additionally, the soldiers' on-body sensors monitored their health and status in case of an accident or attack on the convoy. Secure data transmission methods using IoT communication technologies prevented data manipulation and tapping by enemy forces.

8.5.1.1 Solution Architecture

Figure 8.23 shows an overview of the systems used during the demonstration.

As seen in Fig. 8.23 and from the scenario described, the primary component for gathering SA data was the smart city domain. In this experiment, the Finnish Transport Authority (FTA) provided the ICT assets in the form of sensors and access to the range of cameras deployed across Helsinki through the Application Programming Interfaces (APIs). The FTA used OpenAPI Specification (OAS) to describe the Digitraffic API, which provides access to different kinds of sensors deployed across Finland. By leveraging metadata information provided by the APIs, it is possible to query sensors for their types, such as traffic cameras, weather stations, precipitation meters, wind sensors, visibility meters, road surface condition sensors, etc. In addition, metadata provides sensor information about their geographical location or municipality, the time of the last update or the frequency of updates, current status, and so on.

For our implemented use case, APIs provided access to 40 live streams from the cameras. Additionally, the Android Assault Toolkit (ATAK) application allowed the cameras to be visible and accessible. The ATAK is a handheld C2 application designed for use by mobile or ground soldiers and any other military entity, according to the requirements. The toolkit provides customization and actuation options and displays a range of SA data on embedded maps using various military symbols. Figure 8.24 shows the cameras available, shown as symbols on the ATAK map. Clicking on any of these symbols enabled access to the physical cameras through the respective Digitraffic APIs. Furthermore, it enabled live video streaming to show the status at the respective geographic location. In addition to the cameras, clicking on the yellow symbols shown in the figure (i.e., the suns) enabled access to live weather data from the FTA weather sensors.

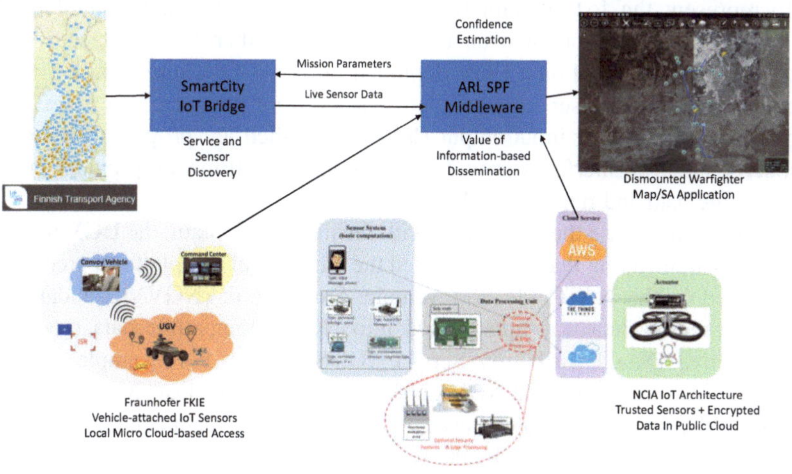

Fig. 8.23 Experiment setup at ICMCIS 2017

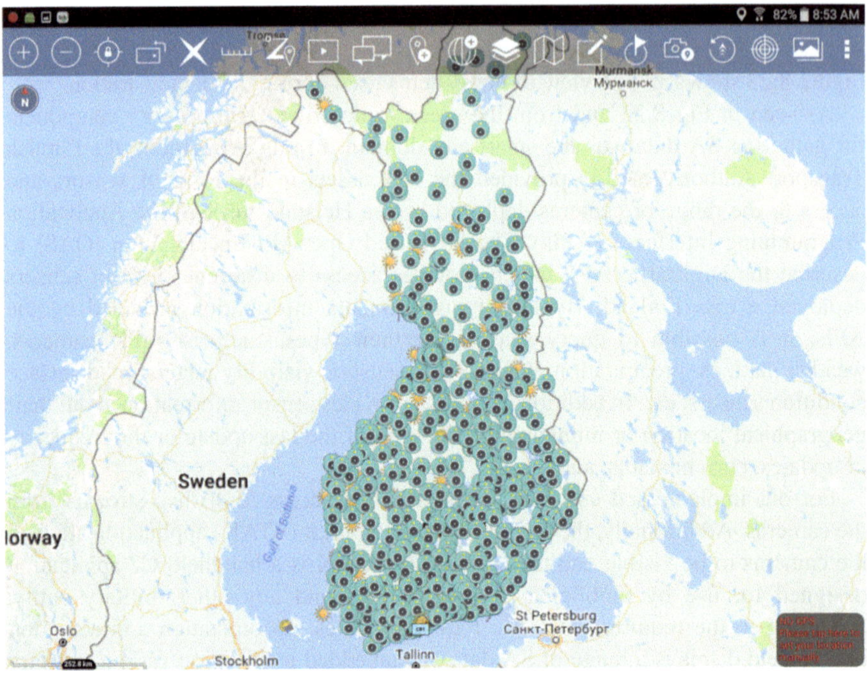

Fig. 8.24 Available cameras provided by FTA

For the FTA APIs to be accessible on the ATAK platform, we implemented the Smart City IoT Bridge, as shown in Fig. 8.23. These streams provided the ISR data necessary to combine with other deployed military ICT assets. Figure 8.25 shows the architecture of the components deployed by ARL for the demo.

To represent the IoT domain from the military end point, Fraunhofer FKIE deployed IoT sensors mounted on IoT platforms. Further, these IoT sensors and platforms were mounted on a UGV deployed in the AOO. These acoustic, motion detection, GPS, and laser range finder sensors were connected to a Raspberry Pi platform through the input/output (I/O) pins, which are programmable on the Raspberry Pi. Java interfaces provided by the sensors' firmware provider enabled access to the data and function of the sensors.

To further showcase the integration with the military domain, the UGV used was NGVA-based, meaning that the UGV used the data model and platform architecture as specified in the NATO Generic Vehicle Architecture (NGVA) standard. The individual sensor data was transformed into NGVA-compatible data models for the respective Data Distribution Service (DDS) topics. For example, the GPS sensor data was mapped to the NGVA Navigation Reference Data Model (NRDM). The NRDM provides the topics used to represent GPS parameters, such as the platform's latitude and longitude, the platform's identifier, the device ID, etc. The NGVA-formatted data can be exchanged interoperably with any other NGVA-

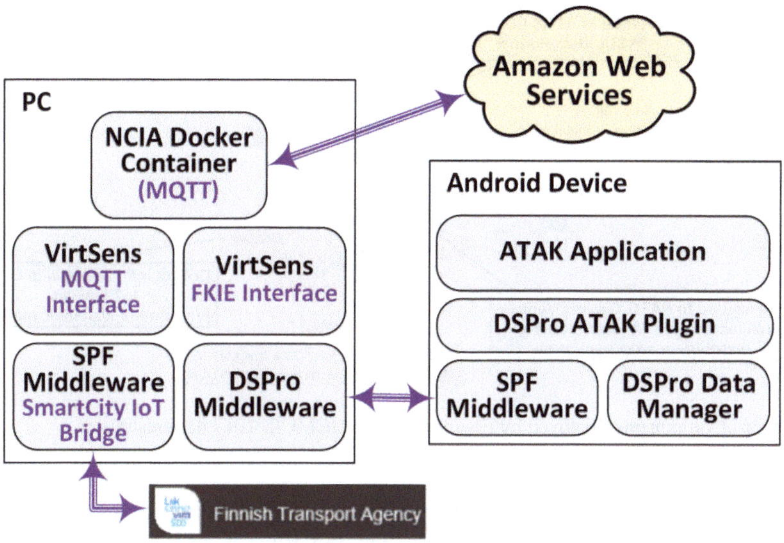

Fig. 8.25 ARL components for ICMCIS 2017 demonstration

platform compliant subsystem or NGVA-compliant vehicle using the DDS data exchange protocol. The sensors mounted on the UGV represented an independent UGV conducting ISR operations on behalf of the NATO forces and reporting any identified abnormality from the ground. For example, when the convoy passes through an AOO, there should be complete sanitation performed by NATO forces to provide safe passage to the convoy and assure the ground units. Thus, the UGV performed this ISR sweep in parallel to the convoy route reporting any unwanted activity. This activity report was sent to an NGVA-compliant vehicle associated with the UGV, which then passed the filtered activity report to the higher echelons, such as a Tactical Operations Center (TOC).

Because it was a simulated demonstration and not an actual convoy with military units deployed on the ground, a GPX file showed the route traversal by the convoy and the UGV. However, the events reported by the sensors were authentic, that is, the sensors reported physical events such as motion detected by the sensor platform in an area that was supposed to be empty. As mentioned above, the data from NGVA to be sent to the higher echelons must be transformed into a format that can be understood by the corresponding system. For the demo use case, the C2 system ATAK was the higher echelon system that consumed this NGVA data from the NATO ground support vehicle. Since NGVA and ATAK are not interoperable, a gateway function was needed at both ends—on the NGVA side to convert from NGVA into the common representation defined by the group and then on the ATAK side to consume the common representation and convert it into the format supported by ATAK.

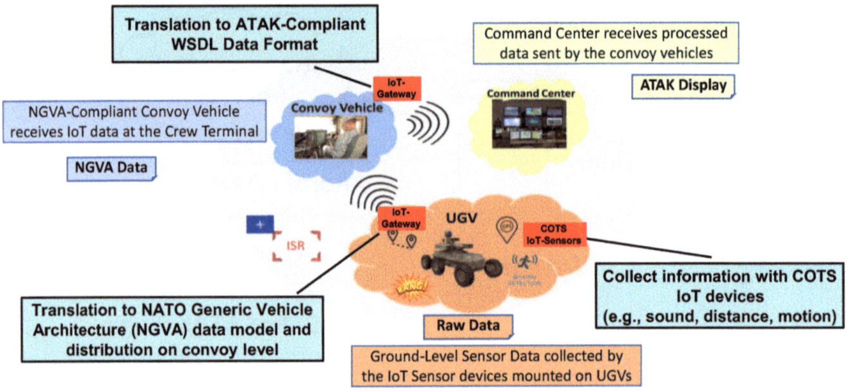

Fig. 8.26 Components deployed by Fraunhofer at the ICMCIS 2017 demonstration

Web services based on SOAP enabled data exchange between the two systems. WSDL-based XML-defined data contained the filtered data about the reported events. This event data in XML format over SOAP could be sent to ATAK, which consumed and transformed the required data. Figure 8.26 shows the architecture of the components Fraunhofer deployed for the demo.

Figure 8.27 presents the SA data and route shown on the ATAK User Interface (UI). The SA gathered from the various sources is analyzed and therefore shown as events on the UI. The route originated from the port of Helsinki and continued to the US embassy, shown as a route overlay in the UI. As the convoy passed through, the neighborhood FTA cameras and weather sensor station symbols were overlaid, showing the camera streams and weather data when clicked. In addition, the FKIE IoT platforms sent out the IoT data as warning events, shown as yellow exclamation marks around the route track. Based on SA data received, the respective cameras in the locations presented by the FTA were accessed to see live video streams to gain ISR preciseness.

8.5.1.2 Real-Time Monitoring of Soldier's Capabilities

In order to demonstrate how MIoT can support commanders in the field, a simulated scenario was presented. The scenario's course is as follows: Two military patrols are deployed in the city with a peacekeeping mission. The situation is unstable, and asymmetric threats such as IEDs, sniper fire, or hostile citizens are highly possible. In such a dynamic environment, a commander needs real-time information about subordinate soldiers to quickly and adequately respond and make decisions. Therefore, each soldier has a smart device with connected sensors for:

- Health monitoring:
  - ECG sensor

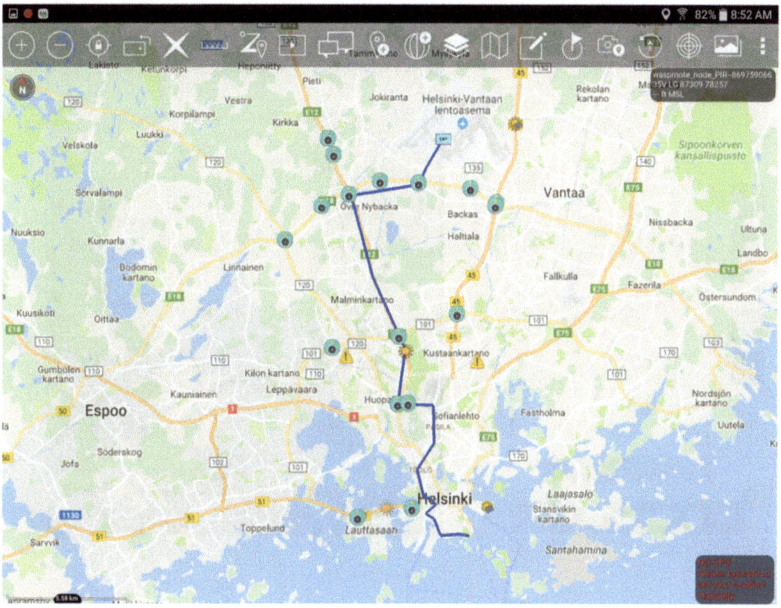

Fig. 8.27 ATAK user interface showing planned mission route and available IoT assets at ICMCIS 2017 demonstration

  - Pulse oximeter sensor
  - Thermometer

- Equipment monitoring:

  - RFID tag reader
  - Each important equipment element has a unique RFID tag

The role of each device is to assess a soldier's state in terms of health condition and available equipment. Based on sensed vital signs, a soldier can be classified into one of four categories: HEALTHY, EXHAUSTED, NEEDS TREATMENT, and CRITICAL CONDITION. Based on RFID tags recognized by a reader, data about currently available equipment such as the primary weapon, secondary weapon, ammunition, or medicines is collected and quantified.

The devices in each squad form a network cluster with M2M wireless communication. The device owned by a squad leader acts like a cluster head (CH), which communicates with the platoon leader over a tactical network. The CH collects reports from subordinate devices and sends them to a mobile Common Operational Picture (mCOP) application (see Fig. 8.28). The capabilities of mCOP include providing the platoon leader real-time information concerning each subordinate soldier's position, health status, and available equipment and supplies.

mCOP is an Android-based decision support tool developed at the Military University of Technology, Warsaw, Poland. The application deeply integrates

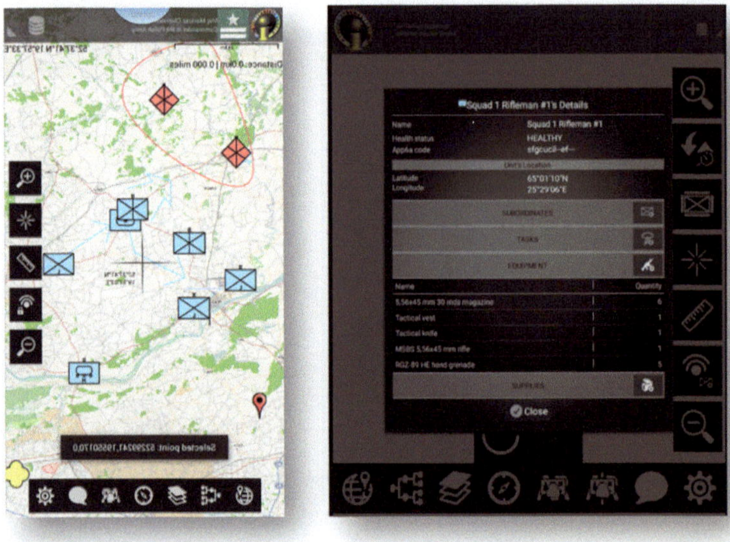

Fig. 8.28 mCOP application

with any android handheld (smartphone and tablet) to deliver Common Tactical Picture (CTP) and Common Operational Picture (COP) products for lower level commanders and group leaders. mCOP delivers military and civilian operational pictures with standardized GIS data sources and operational symbology standards (STD2525, APP6A, and crisis management). Therefore, mCOP supports the implementation of detailed battlespace information: units, equipment, and installations, and their affiliation and status. As a result, the command can gather specific tactical information on equipment, warfare, and unit supplies. Furthermore, the detailed combat information contains static data and a real-time description of the unit's behavior (tasks).

SenseSim, an IoT and WSN simulator focused on cognitive and behavioral aspects of smart devices embedded into a changeable environment, simulated the devices and M2M communication (see Fig. 8.29). Thanks to its flexible design and mathematical model, SenseSim can simulate various phenomena. These include weather (rain, wind, high/low humidity, etc.), situational crises (fire, contamination, etc.), vital signs (ECG, blood saturation, temperature, etc.), or urban specific (e.g., car or pedestrian traffic), which are observed by IoT devices. The devices in SenseSim can build a heterogeneous network in which some may be simple sensors with basic perceptual and communication capabilities, and others may be smart devices with some intelligence and cognitive capabilities. This is possible thanks to the devices' layered architecture, which consists of:

- Hardware and Firmware layer: It is responsible for sensing and communication

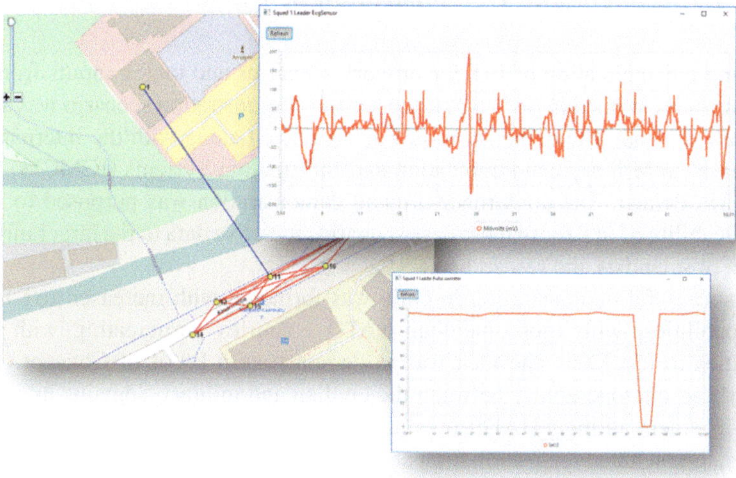

Fig. 8.29 SenseSim IoT simulator

- Device API layer: It provides basic functions that allow programmatic manipulation of the device.
- Device Middleware: It provides services and required resources for IoT applications that run on the device.

Finally, SenseSim enables the possibility to freely redesign and re-implement each layer depending on the needs of the scenario.

The presented scenario showed how MIoT could support commanders in the field, where soldiers' health and physiological information are crucial for mission success and soldiers' lives. Wearable devices can support C2 solutions by supplying them with more sophisticated data than just the unit's position, which is the typical information presented currently. Indeed, this could be further enhanced by logistics information (e.g., ammunition supplies, water, fuel, food rations) that could also be automatically determined via MIoT.

Thus, for future research, work related to creating ontologies commonly agreeable to the participating countries and an IoT exchange standard must be identified. Furthermore, these ontologies and standards must be based on Open Standards since the IoT domain is still evolving. Creating a closed-loop military-specific mechanism would not allow for extensibility and scalability.

8.5.2 *Experiment and Demonstration in Warsaw, Poland*

Furthering the application of IoT for military scenarios and taking inputs from the experiments mentioned in Sect. 8.5.1, the second demonstration scenario was set up in Warsaw, Poland. The demonstration was conducted along with the International Conference on Military Communication and Information Systems (ICMCIS) 2018 in Warsaw, Poland. The conference-aligned demonstration was purposed to show the applicability of IoT for real use case scenarios with live data to the larger military diaspora of scientists and researchers.

The successful demonstration in Oulu was furthered with increased IoT assets and capabilities, with more member NATO countries participating with their capabilities to test [22]. The idea was to demonstrate a Proof of Concept (PoC) for federated interoperability between the civilian and military domains in a smart city environment during a HADR operation.

The setup included the information and sensors coming from the city of Warsaw, showcasing the utilization of smart city assets and devices in a disaster scenario to help increase the SA of the soldiers and civilians by coupling the data collected to the assets from the military domain. The military assets consisted of IoT sensors, actuators, C2 systems, static and dynamic assets, edge computing assets, gateways deploying various IoT communication technologies, security-related implementations, etc. The scenario required the civilians and military to assist each other by fusing their available assets for a faster recovery process. The tasks included ISR missions, logistics and transportation of resources, CBRN responsiveness, and building ad hoc communication networks. The networks replicated the communication means that would be set up for providing connectivity when traditional network means fail.

Figure 8.30 shows the architecture for the experiment setup at ICMCIS 2018 by the NATO IST-147 group. The components from the previous demo were carried forward with the existing ATAK C2 system and the NGVA-based UGV platform. In addition, there were other integrated C2 systems, including an MIP-based C2 system and a generic OpenStreet maps-based map to demonstrate a civilian C2 system.

8.5.2.1 MQTT as the Middleware for Exchanging IoT Data

In IST-147, we chose to use MQTT as the publish/subscribe protocol since it is already widely used for civilian IoT applications. Through our PoC, we performed a comparative evaluation of the protocol between three civilian standards: AMQP, MQTT, and WS-Notification. We measured the end-to-end delay, that is, the time the message was sent from the publisher until the consumer received it via the broker. We chose this metric because we wanted to identify the protocol that would induce the least delay from data production to consumption, since timeliness is essential to achieve SA in many cases. Table 8.10 shows the comparison of the protocols. In experiments by the Norwegian Defence Research Establishment (FFI)

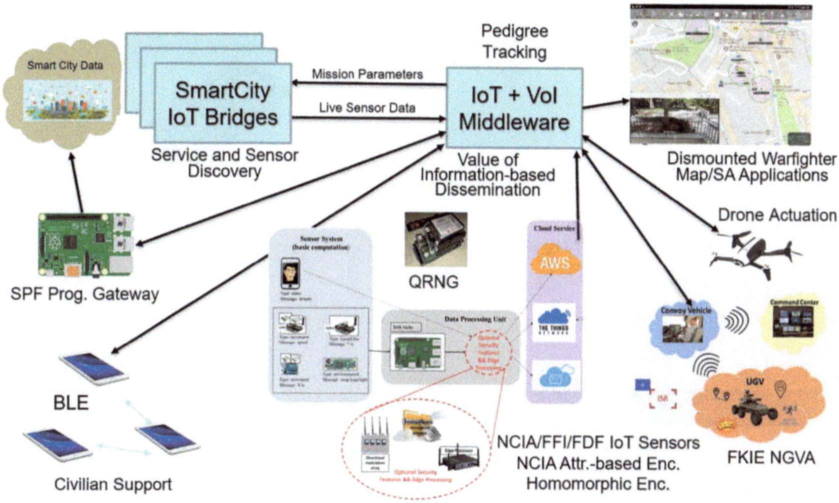

Fig. 8.30 Architecture for demo setup at ICMCIS 2018

Table 8.10 Comparison of end-to-end delay between AMQP, MQTT, and WS-notification. Time is measured in seconds and constitutes the duration from the start until the end of a burst of 100 messages sent and received

AMQP	MQTT	WS-notification
3.103124	2.576307	11.085122

we used KDA WM600 radios setup back-to-back in our lab using the actual radio module, but communicating via coaxial cables through a topology emulator with configurable dampening to create a controlled environment. This setup allowed us to execute tests in the radios' sweet spot, i.e., a good, stable connection across the ad hoc network. Also, by configuring the topology emulator, we were able to conduct tests at the very limit of the radios' reach, hence being able to test performance during very poor networking conditions. Here, we see that for our purpose (lowest delay of sensor data), we should opt for MQTT (findings that are also supported in the more elaborate study here [12]).

8.5.2.2 Sensor Integration

This section details the contribution from Norwegian Defence Research Establishment (FFI) to the demo, where, as shown in Fig. 8.31, we used Raspberry Pi 3 single-board computers as the hardware platform for our PoC. We chose this platform because Raspberry Pi is a cheap, capable, and representative device for IoT applications. In addition, many sensors are also available, making it a good platform for rapid IoT prototyping. Our setup consists of three Raspberry Pi 3s:

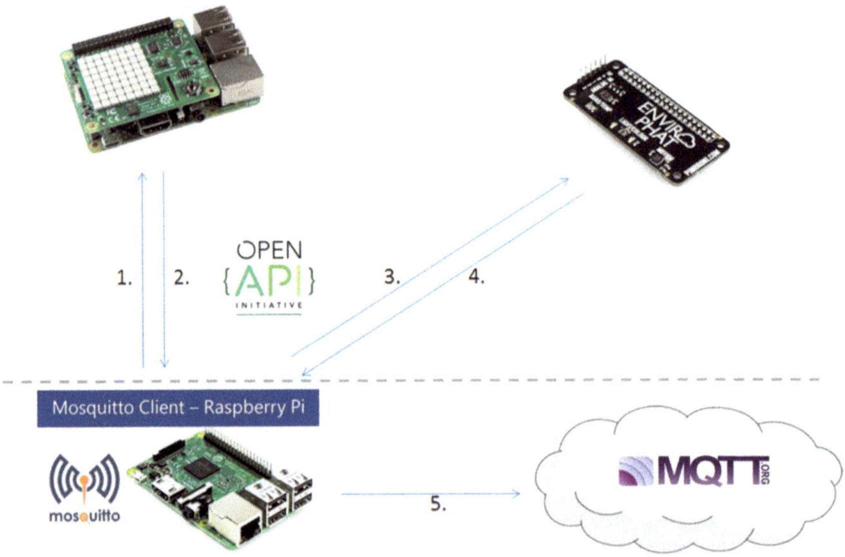

Fig. 8.31 Proof of concept: implementation overview. Our two sensors provide data according to the open API specification (steps 1–2 and 3–4) to the MQTT gateway, which publishes the data to a common MQTT bus (step 5)

The first is equipped with a Sense Hat [23], and the second is equipped with an Enviro pHat [21]. These two provide the sensing capabilities of our PoC. We have exposed the sensor capabilities with Swagger and OAS. The third unit does not have sensors but serves as a gateway accessed through the Raspberry Pi3 Wi-Fi module to a common MQTT bus using the Mosquitto [5] MQTT client implementation. Consequently, this unit also serves as the gateway between the request/response and the publish/subscribe protocols. In key with TIDE specifications, the Raspberry Pi 3 devices are discoverable using mDNS. The components work together in the following manner:

1. The Mosquitto Client (i.e., the Raspberry Pi 3 functioning as the gateway to MQTT) initiates a REST request to the service exposing the Sense Hat API.
2. The REST service providing the Sense Hat API processes the request and responds with a payload of JSON-formatted discrete sensor data.
3. The Mosquitto Client initiates a REST request to the Raspberry Pi 3, exposing the Enviro pHat API.
4. The REST service providing the Enviro pHat API processes the request and, just like the Sense Hat service, returns a JSON payload.
5. The Mosquitto Client publishes the data from the sensors to the MQTT broker, splitting the data into different sensor values and corresponding MQTT topics.

As a result of this process, the information from the sensors is available through the MQTT broker. This information can then be forwarded into the military domain

```
{
    "lat": latitude,
    "UTC": timestamp,
    "msg_id": uuid,
    "lon": longitude,
    "value": value
}
```

Listing 1: JSON message template: The latitude and longitude are decimal values, the timestamp and uuid are strings, and the value differs with the sensor type

by a gateway, which can perform a transformation from MQTT to the protocols supported in the military domain [3] if needed. Note that steps 1–2 and 3–4 are pull-pattern request/response, whereas step 5 is push-pattern publish/subscribe. That said, the pull steps do not need to go across the network, as the REST to MQTT functionality can co-reside with the REST service on the actual device so that only the messages published with MQTT traverse the network. This is a matter of how and where the Mosquitto Client software is deployed.

It is important to note that due to the current lack of agreement on the "best" ontology to use together with SensorML, we chose a slightly different scheme for identifying sensor data in our PoC. Here, we let the topic under which we publish the data indicate the type of data to expect in the corresponding payload. This solution works where this is the agreed-upon approach (i.e., IST-147 bespoke topics and data format), but in the long run NATO, as a community, needs to find a viable approach to describing sensors, as discussed earlier in Chap. 4.

In IST-147, we arranged the topic names to carry meaning beyond merely being a name for a data dissemination channel. Lacking a standardized approach, we chose a scheme where the topic indicated how to interpret the data received. Therefore, the JSON data we sent was always structured the same way for these discrete sensors, but the meaning of the "value" changed from topic to topic. Also, the sensor owner was indicated in the topic, so a topic containing "NOR" implied a Norwegian sensor. This approach to structuring topics is further discussed in [16]. The JSON message template is shown in Listing 1, and an example of a temperature reading from the Enviro pHat is shown in Listing 2. There we can see that the message conveyed as JSON contains a "lat" and a "lon" value to give the position of the sensor reading. A timestamp ("UTC"-field denotes that Universal Time Coordinated (UTC) is the time zone used) gives the time of the sensor reading, whereas the "msg_id" gives a unique identifier for this particular message. The identifier is generated as per uuid4 described in RFC4122 [14]. Finally, the "value" is the actual sensor reading. Here, and also for our Sense Hat, it should be interpreted as Celsius, but another sensor type might indicate a different interpretation. Table 8.11 summarizes the different topic suffixes provided in our PoC.

```
{
    "lat": 52.245621,
    "UTC": "2018-02-01 10:17:30.227125",
    "msg_id": "4427ccee-ab06-4614-92b9-af209edd2b1c",
    "lon": 21.012371,
    "value": 27.4
}
```

Listing 2: This example is taken from a message exchange where the topic suffix is "sensor/enviro/temp." This indicates that this is a sensor of type Enviro pHat and that the reading is a temperature, implying that the value in the JSON above should be interpreted as Celsius

Table 8.11 Sensors, topic suffixes used, and how to interpret the JSON "Value" field

Origin sensor	Topic suffix	Value interpretation
Enviro pHat	Sensor/enviro/temp	Degrees Celsius
Enviro pHat	Sensor/enviro/light	Lumen
Sense Hat	Sensor/astro/temp	Degrees Celsius
Sense Hat	Sensor/astro/humidity	%rH (relative humidity)
Sense Hat	Sensor/astro/pressure	Millibars

This approach was used for all sensor data integration in the demo, as further explained in Sect. 8.5.2.6 on ontology.

8.5.2.3 Solution Architecture

Bluetooth low-energy (BLE)-based information exchange was demonstrated using smartphones when the HADR responders were sent on ground missions and did not have access to traditional network connectivity such as Long-Term Evolution (LTE).

Drones from the US ARL and FKIE were integrated to showcase the use of UAVs based on COTS IoT products. FKIE mapped the drone controls based on the underlying Robot Operating System (ROS) platforms, and the US used the drone API to be mapped to corresponding service calls from ATAK.

Smart city systems from Warsaw were engaged to showcase the integration of a smart city into the HADR scenario. Various agencies are responsible for separate applications for weather, traffic, pollution, street cameras, etc., and prior authentication and permissions were established with the corresponding city authorities to access these systems. A *Smart City IoT Bridge* consumed and disseminated all the Smart City data. The individual countries deployed their own gateways to interface with their consumption sources. These gateways enabled asset discovery, access, and information storage at the service end points, which were further passed on to specific military systems.

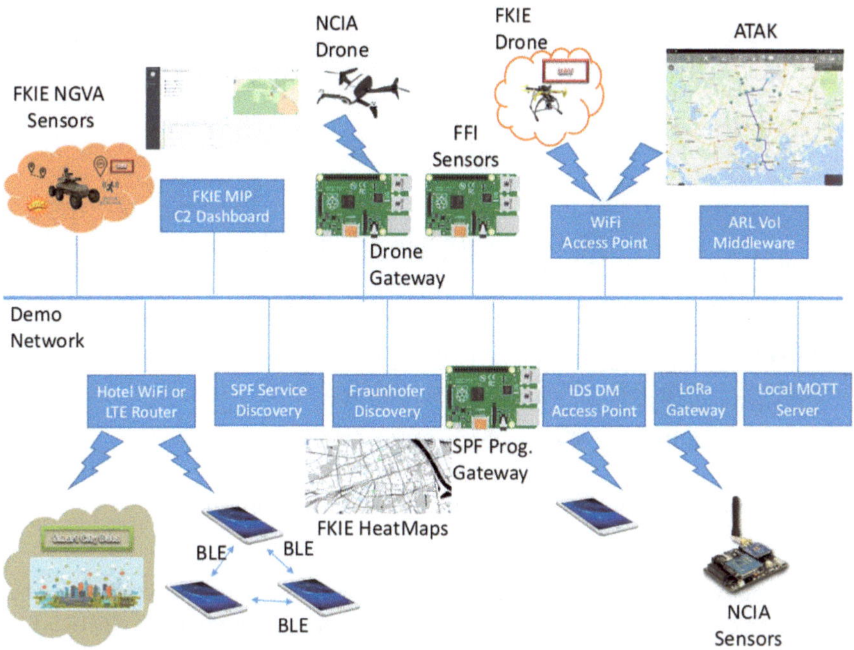

Fig. 8.32 Demo setup at ICMCIS 2018

FKIE exploited the use of smart home devices by end users to showcase the concept of using civilian "smart devices" for HADR operations. For this purpose, a prearranged device with prior permissions from the civilian was obtained. The civilian allowed access to certain kinds of sensor data during the experiment, such as temperature and humidity. The idea was to show that in case of disaster, Smart Home data can be used to determine civilian safety. For example, such use cases include fire detection, that is, temperature readings from the smart home sensors.

The Norwegian Defence Research Establishment (FFI) specifically investigated sensor integration from the Raspberry Pi 3. This sensor setup is indicated as "FFI Sensors" in Fig. 8.32 and was further explained in Sect. 8.5.2.2. All communication components used the network setup as explained in Sect. 8.5.2.4.

8.5.2.4 Networking

A common demo network was used. All the systems communicated using a wireless Wi-Fi switch and an ad hoc LTE network setup provided by ARL. The LTE network enabled sending requests and connecting to the smart city services and the AWS service. Member nations connected to a secured external public domain network using this LTE network, allowing them to perform service discovery and access civilian assets. Further, the individual member nations connected over a secured

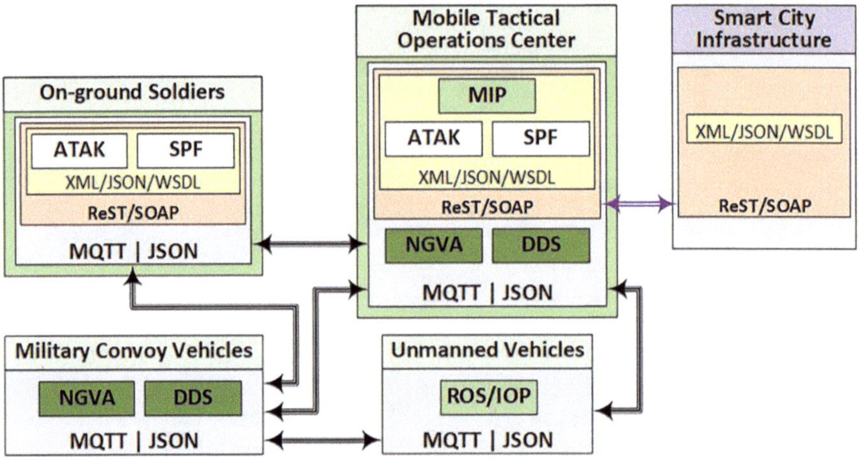

Fig. 8.33 Exchange of data between components with interoperable ontology

Ethernet network via the Wi-Fi switch. This showed the network separation when consuming data from civilian networks while still on a secured military network.

8.5.2.5 Demo Setup

Figure 8.32 shows the components used in the experiment setup at ICMCIS 2018 by the NATO IST-147 group. An interaction, as presented in Fig. 8.33, would be needed for the demo setup involving the components individually using specific gateways. These national systems will ideally have federation interfaces that follow the agreed FMN standards for data models and system interfaces. Furthermore, smart cities implement many commercial and widely used data models and formats. These data formats vary widely from city to city and between various services and applications provided by the city.

- A Mobile Tactical Operations Center (MTOC) is a mobile node that can be quickly moved into position and established as a C2 node to oversee HADR operations. The Multilateral Interoperability Program (MIP) was devised for automated report generation and data exchange on a command center level to support decision-making. It uses the MIP Information Exchange Specification (IES) to classify and define several exchange patterns based on the MIP Information Model (MIM). In addition, it uses open and standard technologies such as ReST and SOAP, according to exchange pattern technology (e.g., the publish/subscribe mechanism), to map rules documented as MIP4 design principles, rules, and decisions. It defines standardized data formats for describing events and patterns for effective and reliable information exchange. Finally, it can ingest data and services from heterogeneous applications that conform to the

MIP data format. The MTOC runs an MIP C2 application designed for higher echelons for better reporting and establishing better SA. For MIP to interact with other participating applications in the coalition, customized mapping is required to allow the message exchange. C2 applications would connect to the specific web-based, exposed ReST or SOAP end points to access raw or processed data. Based on the web technologies these applications use, the data formats can be either XML- or JSON-based.

- For military support vehicles, the NGVA standard is used to exchange data. The NGVA data model (DM) is used to define the message set for subsystem data exchange, using Data Distribution Service (DDS) middleware. For interaction with unmanned vehicles, it uses the "NGVA/Interoperability Profiles (IOP)" bridge as the external gateway to map between messages. Similarly, for interaction with the web-based application for dismounted soldiers, it uses the gateway for mapping NGVA to SOAP messages. Finally, it uses NGVA to MIP message (based on ReST) mapping for interaction with the web-based MIP application.
- Dismounted soldiers are equipped with mobile devices (e.g., a mobile phone or a tablet) that run applications for soldiers to send and receive SA data, reports, and logistics information. The ATAK Android application was used for our demo to show the map display. ATAK communicates with various back-end systems running on the military support vehicles and the MTOC to send and receive up-to-date SA data. In the HADR and IoT context, ATAK can receive up-to-date SA data from smart city services and allow soldiers to generate reports based on their observations, which are then made available to the MTOC and other dismounted soldiers using ATAK.
- To enable interoperability between various types of Unmanned Vehicles (UxVs), the US Army devised a bridge between Robotic Operating System (ROS) and the Joint Architecture for Unmanned Systems (JAUS) profiles. Further, ROS/JAUS is mapped to the IOP profiles for IOP standardization requirements, allowing for seamless integration of legacy and future sensors and subsystem modules in UxVs. It defines various capabilities and requirements for supporting the capabilities related to the employment and usage of UxVs to perform robotic missions. The devised IOP standard allows for JAUS messages to support interoperability on a variety of missions and objectives, vehicle classes/types, controller classes/types, payload classes/types, physical/software architectures, and interaction with external systems.
- For integrating smart city data, primarily web-based end points based on ReST or SOAP are used to deliver data in XML or JSON format. As per the policy set between the military and the city, the city can set various access levels to let the military access data from the end points. Further, the city might allow the military to access the data directly from crowdsourcing applications and commercial entities, which can also be web service end points.

Figure 8.33 shows the complete map of data exchange between the demonstrated systems and their corresponding data models, formats, and mechanisms. It enables data exchange using IoT gateways by wrapping internal data exchange mechanisms with the MQTT protocol and custom JSON messages. On the other hand, NATO proposes the use of a different publish/subscribe mechanism, called WS-Notification, for use between mission partners in the FMN context. However, the WS-Notification protocol is more complex than MQTT and supports only XML data, and it is therefore not very suitable for use in IoT contexts.

As discussed in the improvements needed for federation and interoperability in the Helsinki demo and considering the heterogeneous gateways needed at each level for the demo use cases and components, an open standard IoT protocol was needed to show the IoT use case for military data exchange.

8.5.2.6 Demo Ontology

A common ontology was devised based on open IoT ontology specifications. Figure 8.34 shows an example of data exchange based on this ontology pre-agreed between the member nations. The ontology was represented as a topic structure where the topic fields or qualifiers represented each level of capability or ownership. The common topic structure ensured interoperability between information producers (sensor owners) and information consumers (e.g., military C2 applications). The data exchange using IoT gateways was achieved by wrapping data based on the topic structure with the MQTT protocol and custom JSON messages. The levels of the tree-based topic structure and the corresponding JSON message are as follows:

1. <Org_id> describes the organization identifier that exchanges the messages.
2. <Country_id> describes the country that is the origin of the message.
3. <ownership_type> describes the ownership type of the message, such as public, private, military, etc.
4. <device_class> describes the device class sending out the message, such as sensor, actuator, or microcontroller.
5. <device_type> describes the type of device sending the message, such as environmental or astronomical sensor.
6. <message_type> describes the message type, including status, command, etc.
7. <data_type> describes the type of data the message contains, such as temperature or acoustic information, etc.

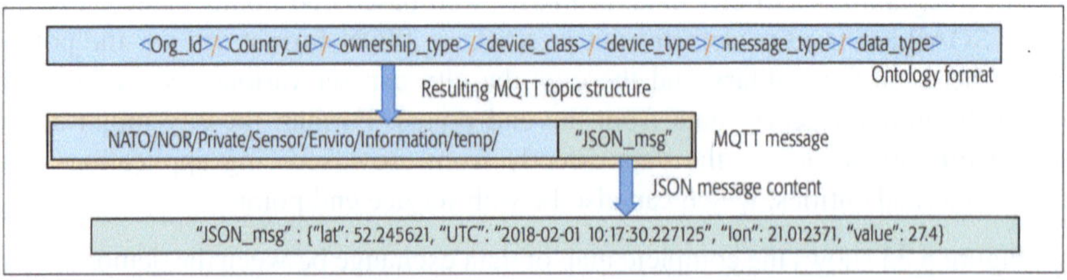

Fig. 8.34 Ontology for experimentation

A string-based representation of this topic structure was attached to the MQTT message as metadata when the message was published. In this case, the actual message content was carried in the message payload as a JSON string. Figure 8.34 shows an actual message with the MQTT message topics in the metadata and the JSON string as the message payload, resulting from the ontology format. The JSON string consists of the following fields: `Obj_id` stating the unique identifier for the resource, latitude (`lat`) and longitude (`lon`) stating the position of the temperature sensor, `UTC` stating the timestamp of the information, and `value` stating the actual temperature value recorded by the sensor. These MQTT messages wrapped around the messages being exchanged between the applications running on MTOC, UxVs, vehicles, and soldiers. Thus, it enabled the heterogeneous systems to communicate with each other without having to write individual gateways to interact with other applications, thus providing the required data interoperability. The MQTT wrappers could be applied to military systems but not to the smart city systems since each city has its own format for data dissemination. Also, a city might provide access to its assets, such as sensors and actuators, which have their own ways of disseminating data, and wrapping each dataset at the source (i.e., from the city ICT systems, individual hardware manufacturers, and service providers) is not practically feasible.

8.6 Conclusions

In our experiments, we have used several architectural building blocks and middleware solutions that we identified during our research as suitable to support the effective implementation of military applications of IoT. In particular, we have demonstrated how the data from smart city sensors could be obtained and used by military C2 systems. We identified four challenges that we have addressed in our initial Proof of Concept (PoC) using open standards and best practices. Based on our investigation, we believe that fog computing is the most suitable approach to military IoT applications in Tactical Network (TN). To provide an effective mechanism for sensor discovery, we suggest that mDNS specifications for service discovery, proposed by the TIDE community, should be experimentally investigated for their effectiveness and suitability. Sensors can be described using the SensorML standard as the overarching metadata for all smart city sensors. An associated ontology to be used along with SensorML must be evaluated by NATO. We have experimented with different communication paradigms, i.e., request-response and publish-subscribe, that can be used to retrieve sensor data. Our experiments confirmed that both paradigms can be integrated to provide IoT data to military systems. However, it is essential to use APIs and open standards to ensure interoperability.

The demonstration showed how MIoT could be leveraged while combined with traditional military systems and scenarios. Although military standards could talk to each other, they were very limited. For example, German systems, based on

NGVA and web services, could only talk to US systems. Furthermore, the Polish and NCIA systems involved were not integrated into the overall demonstration setup and were standalone components showing their respective use cases. As a result, participating systems did not fully realize the concept of *interoperability*, mainly due to the lack of an IoT ontology and a data exchange mechanism that could be commonly agreed upon for federation. Such a mechanism must also be suitable for IoT-specific applications as envisioned for the group's goal. NATO Federated Mission Networking (FMN) spirals propose using web services for exchange but are not always suitable for IoT applications due to the constraints of the devices and the network. The existing FMN specification proposes using WS-Notification as the publish/subscribe mechanism for data exchange. However, such a mechanism is not suitable for IoT implementations requiring less complexity and a low footprint of code and resource usage. Furthermore, only one C2 application was integrated for the demonstration, which is hardly a real-world situation. Multiple countries involve multiple C2 systems, thereby increasing the complexity of interoperable and federated operations.

We plan to continue our work, with the objective of performing larger interoperability experiments involving several military and civilian partners from NATO nations and developing a strategy for future DCS implementation in the IoT domain.

References

1. Adelantado, F., Vilajosana, X., Tuset-Peiro, P., Martinez, B., Melia-Segui, J., Watteyne, T.: Understanding the limits of LoRaWAN. IEEE Commun. Mag. **55**(9), 34–40 (2017)
2. Akinyele, J., Garman, C., Miers, I., Pagano, M.W., Rushanan, M., Green, M., Rubin, A.D.: Charm: a framework for rapidly prototyping cryptosystems. J. Cryptogr. Eng. **3**(2), 111–128 (2013)
3. Bertelsen, E., et al.: Federated publish/subscribe services. In: 9th IFIP International Conference on New Technologies, Mobility & Security 26 to 28 February 2018, Paris (2018)
4. Centenaro, M., Vangelista, L., Zanella, A., Zorzi, M.: Long-range communications in unlicensed bands: the rising stars in the IoT and smart city scenarios. IEEE Wirel. Commun. **23**(5), 60–67 (2016)
5. Eclipse Foundation: Eclipse Mosquitto™. https://mosquitto.org/
6. Furtak, J., Chudzikiewicz, J.: Secure transmission in wireless sensors' domain supported by the TPM. In: Advances in Network Systems. iNetSApp 2015. Advances in Intelligent Systems and Computing, vol. 461, pp. 129–148. Springer, Berlin (2017)
7. Furtak, J., Zieliński, Z., Chudzikiewicz, J.: A framework for constructing a secure domain of sensor nodes. Sensors **19**(12), 2797 (2019)
8. Furtak, J., Zieliński, Z., Chudzikiewicz, J.: Procedures for sensor nodes operation in the secured domain. Concurrency Comput. Pract. Exp. **32**(13), e5183 (2020)
9. Gierłowski, K., Gierszewski, T., Fongen, A., Fuchs, C., Furtak, J., Jarmakiewicz, J., Krygier, J., Marks, M., Niewiadomska-Szynkiewicz, E., Zieliński, Z.: Internet of military things: framework for the application of Internet of things in the military domain. Technical report, NATO STO (2014)
10. Jalaian, B., Gregory, T., Suri, N., Russell, S., Sadler, L., Lee, M.: Evaluating LoRaWAN-based IoT devices for the tactical military environment. In: 2018 IEEE 4th World Forum on Internet of Things (WF-IoT), pp. 124–128. IEEE, Singapore (2018)

11. Jalaian, B., Suri, N., et al.: Investigating LoRa for the Internet of battlefield things: a cyber perspective. In: MILCOM 2018–2018 IEEE Military Communications Conference (MILCOM), pp. 749–756. IEEE, Piscataway (2018)
12. Johnsen, F.T.: Using publish/subscribe for short-lived IoT data. In: Proceedings of the IoT ECAW 2018, Poznan, Poland (2018)
13. Johnsen, F.T., Zieliński, Z., Wrona, K., Suri, N., Fuchs, C., Pradhan, M., Furtak, J., Vasilache, B., Pellegrini, V., Dyk, M., Marks, M., Krzysztoń, M.: Application of IoT in military operations in a smart city. In: 2018 International Conference on Military Communications and Information Systems (ICMCIS), pp. 1–8 (2018)
14. Leach, P., Mealling, M., Salz, R.: A Universally Unique IDentifier (UUID) URN namespace. In: Request for Comments (RFC) 4122. IETF (2005)
15. Lewko, A., Waters, B.: New proof methods for attribute-based encryption: achieving full security through selective techniques. In: Advances in Cryptology – CRYPTO, pp. 180–198 (2012)
16. Manso, M., Johnsen, F.T., Brannsten, M.R.: A smart devices concept for future soldier systems. In: ICCRTS (2017)
17. Michaelis, J., Morelli, A., Raglin, A., James, D., Suri, N.: Leveraging LoRaWAN to support IoBT in urban environments. In: 2019 IEEE 5th World Forum on Internet of Things (WF-IoT), pp. 207–212. IEEE, Piscataway (2019)
18. Michaelis, J., Morelli, A., Hernandez, L., James, D., Freeman, J., Suri, N.: Lorawan testing for military communications in urban environments. In: 2021 IEEE 7th World Forum on Internet of Things (WF-IoT), pp. 885–890. IEEE, Piscataway (2021)
19. Mwakwata, C.B., Malik, H., Mahtab Alam, M., Le Moullec, Y., Parand, S., Mumtaz, S.: Narrowband Internet of Things (NB-IoT): from physical (PHY) and media access control (MAC) layers perspectives. Sensors **19**(11), 2613 (2019)
20. Oudkerk, S., Wrona, K.: Cryptographic access control in support of object level protection. In: Proceedings of the Military Communications and Information Systems Conference (MCC), St. Malo (2013)
21. Pimoroni: Python libraries and examples for the Pimoroni Enviro pHAT. https://github.com/pimoroni/enviro-phat
22. Pradhan, M., Suri, N., Fuchs, C., Bloebaum, T.H., Marks, M.: Toward an architecture and data model to enable interoperability between federated mission networks and IoT-enabled smart city environments. IEEE Commun. Mag. **56**(10), 163–169 (2018)
23. pythonhosted.org: Sense HAT API reference. https://pythonhosted.org/sense-hat/api/
24. Suri, N., Zielinski, Z., Tortonesi, M., Fuchs, C., Pradhan, M., Wrona, K., Furtak, J., Vasilache, D.B., Street, M., Pellegrini, V., Benincasa, G., Morelli A., Stefanelli, C., Casini, E., Dyk, M.: Exploiting smart city IoT for disaster recovery operations. In: 2018 IEEE 4th World Forum on Internet of Things (WF-IoT), pp. 458–463. IEEE, Singapore (2018)
25. Wang, R., Peethambaran, J., Chen, D.: Lidar point clouds to 3-d urban models: a review. IEEE J. Sel. Top. Appl. Earth Obs. Remote Sens. **11**(2), 606–627 (2018)

Chapter 9
Conclusions and Future Directions

Niranjan Suri, Konrad Wrona, and Zbigniew Zieliński

The activities of the NATO IST-147 Research Task Group explored the capabilities of IoT and the roles of IoT in the military domain. In this book, we introduced several different application domains for IoT in the context of military operations, including logistics and sustainment, surveillance and reconnaissance, personal health monitoring, smart military bases, and Humanitarian Assistance and Disaster Relief (HADR). We analyzed the existing architectural frameworks for IoT ecosystems such as oneM2M and highlighted some of the challenges of integrating these types of architectures with existing military infrastructure. We proposed some architectural patterns for achieving this integration of IoT into the military domain, especially from the perspective of the different networks involved. We also reviewed existing IoT-related technologies for sensing, actuation, and communications that are relevant to the military domain.

Military IoT applications must meet some specific requirements, including the ability to process large amounts of heterogeneous data, support for increased automation, robust device and network security, and resilience to electronic warfare measures. Moreover, to allow the use of IoT technology in military operations, it is important to focus on the full spectrum of DOTMLPFI, including training, as well as organizational and doctrinal implications and impediments.

N. Suri (✉)
US Army DEVCOM Army Research Laboratory (ARL), Adelphi, MD, USA

Florida Institute for Human and Machine Cognition (IHMC), Pensacola, FL, USA
e-mail: nsuri@ihmc.org

K. Wrona
NATO Communications and Information Agency (NCIA), The Hague, The Netherlands

Military University of Technology, Warsaw, Poland
e-mail: konrad.wrona@ncia.nato.int

Z. Zieliński
Military University of Technology, Warsaw, Poland

N. Suri et al. (eds.), *Military Applications of Internet of Things*, Internet of Things,
https://doi.org/10.1007/978-3-032-03378-9_9

The broader development and deployment of IoT applications in the military will take time; however, there are areas where the military can generate significant savings from IoT using existing technologies and business practices available in the commercial sector. These areas include condition-based maintenance, real-time fleet management, military base management, and increasing energy efficiency of military operations.

Key enablers for the adoption of IoT technology in military applications include extended connectivity, common security measures suitable for COTS devices and applications, and the development of common standards and protocols. In addition, military organizations must pursue agile ways to access innovation and adopt commercial best practices for technology development and procurement to enable enhanced collaboration with the private sector to quickly field, maintain, and update IoT systems with the latest technologies. NATO, with its member and partner nations, should consider creating a collaborative technology testbed and a live training environment to identify and experiment with technologies that could transform the way the military accomplishes its mission and serve as a link between military personnel and technology developers.

Multiple IoT industrial architectural activities define IoT from different perspectives and try to address different requirements, and there is no leading standard IoT architecture. In a typical deployment, there are one or more IoT device domains that support direct M2M communications between IoT devices. Device domains are sometimes connected to the infrastructure domain using dedicated gateways. The infrastructure domain consists of a core network and additional services, supporting applications using IoT information collected in the device domains.

Most commercial IoT architectures also assume the existence of an enterprise cloud at the back end, as well as a reliable high-bandwidth communications network that connects the gateways (or sometimes the IoT devices themselves) to the enterprise cloud. Although communications within the device domain may be limited (e.g., using low-power long-range standards such as LoRa, NB-IoT, Sigfox, Bluetooth, and Z-Wave), the gateways themselves connect to the enterprise cloud via the commercial Internet or some other high-speed link (e.g., LTE, 5G, and SATCOM). Furthermore, the enterprise cloud is not resource-constrained and is assumed to provide the necessary scalability for data storage and processing to support the desired IoT user-facing applications.

From a military perspective, COTS IoT sensors, actuators, and communications technologies can address a variety of operational needs at significantly lower cost than their military-developed counterparts. By assessing particular mission needs, military users can selectively rely on COTS-based solutions and integrate them into their infrastructures. Assessing the usage of COTS IoT technologies in broader military infrastructures represents a key research objective in the emergent IoT-based IoBT paradigm.

The use of civilian IoT solutions in military operations represents a significant paradigm change relative to the approaches currently practiced. Addressing some of the open issues related to this paradigm change is the objective of NATO IST-176 RTG, which builds on the results obtained by the IST-147 group. It is also important to introduce the military IoT applications to the NATO Federated Mission Networking (FMN) community to be considered for incorporation into the FMN architecture and future FMN spirals.

Many of the military IoT solutions presented in this book are currently being deployed on a distributed NATO testbed with several physical locations for experimentation purposes. For example, CIMIC applications were integrated for demonstration purposes with IoT infrastructures of several cities, including Warsaw (Poland), The Hague (The Netherlands), Pensacola (Florida, USA), and Ferrara (Italy), taking advantage of a heterogeneous set of sensing and computing platforms and interacting with a broad range of smart city services.

The widespread adoption of IoT and the increasing availability of smart city environments present compelling opportunities to increase public safety and security. However, their realization poses several challenges at various levels, including, but not limited to, IT service design, architecture, and integration. We envision that adopting IoT in the military will lead to a new era of dual use IoT solutions and the emergence of innovative and sophisticated civil-military cyber-physical systems. These would incorporate IoT as an extremely valuable, although not entirely reliable, source of information for situational awareness purposes.

We discussed several use cases and related threats applicable to the use of IoT systems in military applications. Furthermore, we presented various aspects of IT network security in military applications, the places to apply hardware TPM to build security mechanisms, and how to use services based on the publish-subscribe mechanism. We paid particular attention to the MQTT service, which can enable secure data exchange between the IoT network and the federated Command and Control (C2) systems.

Finally, we have presented the concept of Data-Centric Security, which is currently considered a possible security model for future NATO operations. We conclude that DCS can be applied to IoT to provide end-to-end data protection in transit and storage; however, its implementation in IoT introduces several challenges, including key management and implementation of cryptographic algorithms in constrained environments. The computational overhead introduced by many modern cryptographic mechanisms is still too high for many of today's IoT devices. However, experience has shown that device performance improves much faster than expected and that devices can, in a short time, improve sufficiently to perform even complex cryptographic algorithms. Although there will always be a market space for low-end IoT devices, the majority of IoT devices are powerful enough to support state-of-the-art security mechanisms, including encryption of data throughout its entire life cycle, that is, not only in storage and transit but also during processing. In our work, we have demonstrated the feasibility of implementing Data-Centric Security (DCS) mechanisms for MIoT applications, while relying on commercially or openly available software and hardware components and services to the greatest extent possible.

Achieving an appropriate level of dependability of MIoT requires a comprehensive approach to reliability and security across all layers and stages of the life cycle of MIoT. Among other things, fault tolerance techniques must be applied consistently at each layer. Furthermore, our assessment of the applicability of selected fault-tolerance techniques to MIoT shows that most techniques are not applicable to real-time scenarios. Finding the best solution to make MIoT immune to different types of fault is a great challenge, especially in mission-critical scenarios. In particular, fault-tolerant techniques should be integrated with the security solutions used within MIoT, as their mutual reinforcement is highly beneficial. The redundancy of cheap sensors and devices makes the use of MIoT a viable option. However, this requires the detection of faulty data and an increase in the error correction capacity of the data aggregation process.

Our initial evaluation of available hardware and software libraries and public cloud services shows that existing commercial and open-source components offer sufficient functionality to implement a wide variety of military applications at a lower cost compared to fully customized solutions. Furthermore, the lower reliability and vulnerability of the components could be mitigated by using redundant devices and services. Finally, the architectures relying on commonly used IoT components and platforms can enable better CIMIC and the reuse of civilian IoT resources available in the conflict area for military purposes. Our proof of concept is a first step toward building a working prototype of a secure IoT system based on commercial components, which could be used for larger-scale experimentation.

In our work, we addressed some but not all the questions and challenges related to the effective implementation of military applications of IoT.

For example, in the context of reliability, allowing hardware redundancy and strong security of everything in an IoT scenario is unrealistic and potentially prohibitively expensive. We assume the possibility of hardware redundancy only to a limited extent and temporal redundancy (which requires repeated data acquisition and processing). However, in practice, the Military Internet of Things application will need to deal with data of various levels of reliability and trustworthiness. Less reliable or trustworthy data may be better than no data if the user understands the reliability and trustworthiness of the obtained information. Therefore, data must be dynamically labeled with associated reliability and trustworthiness indicators, which provide some measure of data quality. These indicators could change over time as we gain or lose faith in a source of data, e.g., a sensor. Applications that use those data must take into account the reliability and trustworthiness of the data during processing and reasoning for decision support. The open problem is how to measure the reliability and trustworthiness of the data. One could, for example, use the results of mutual testing/monitoring and recording of events in edge devices with possible transmission to higher layers. Analyzing the quality of the data itself at the data management layer would also help determine the reliability of the data and its sources. One could introduce a feedback loop; that is, data quality assessed as bad in the data management layer could result in further diagnoses being performed in the device layer and possible exclusion of faulty devices.

Index

N. Suri et al. (eds.), *Military Applications of Internet of Things*, Internet of Things,
https://doi.org/10.1007/978-3-032-03378-9

MIX
Papier aus verantwortungsvollen Quellen
Paper from responsible sources
FSC® C105338

If you have any concerns about our products,
you can contact us on
ProductSafety@springernature.com

In case Publisher is established outside the EU,
the EU authorized representative is:
Springer Nature Customer Service Center GmbH
Europaplatz 3, 69115 Heidelberg, Germany

Printed by Libri Plureos GmbH
in Hamburg, Germany

Evidence-Based Behavioral Health Practices in Pediatric Specialty Settings

Alexandros Maragakis • Mari Janikian
Editors

Evidence-Based Behavioral Health Practices in Pediatric Specialty Settings

Springer

Editors
Alexandros Maragakis
Department of Psychology
The American College of Greece
Athens, Greece

Mari Janikian
Department of Psychology
The American College of Greece
Athens, Greece

ISBN 978-3-031-71003-2 ISBN 978-3-031-71004-9 (eBook)
https://doi.org/10.1007/978-3-031-71004-9

This Springer imprint is published by the registered company Springer Nature Switzerland AG
The registered company address is: Gewerbestrasse 11, 6330 Cham, Switzerland

Contents

Contributors

Remos Armaos, PhD Department of Psychology, The American College of Greece, Athens, Greece

Brigid Behrens, MA Department of Pediatric Psychology, University of Michigan Health, Ann Arbor, MI, USA

Andrew Shaw Benotakeia, MD Department of Psychiatry, Hackensack Meridian Health- Hackensack University Medical Center, Hackensack, NJ, USA

Teryn Bruni, PhD Department of Psychology, Algoma University, Sault Ste. Marie, ON, Canada

Michelle Byrd, PhD Department of Psychiatry, VA Ann Arbor Healthcare System, Howell, MI, USA

Ilona Fishkin, MD, MPH Department of Psychiatry, Hackensack Meridian Health- Hackensack University Medical Center, Hackensack, NJ, USA

Elizabeth Freehling, BA Department of Psychology, Eastern Michigan University, Ypsilanti, MI, USA

Sarah E. Freeman, PhD Department of Psychology, Eastern Michigan University, Ypsilanti, MI, USA

Dana Gadaire, PsyD Department of Pediatric Psychology, University of Michigan Health, Ann Arbor, MI, USA

Jayasudha Gude, MD Department of Psychiatry, Hackensack Meridian Health-Hackensack University Medical Center, Hackensack, NJ, USA

Celeste Hawkins, PhD School of Social Work, Eastern Michigan University, Ypsilanti, MI, USA

Amanda Hicks, MA Department of Pediatric Psychology, University of Michigan Health, Ann Arbor, MI, USA

Brian L. Isakson, PhD Department of Psychiatry and Behavioral Sciences, University of New Mexico, Albuquerque, NM, USA

Maitri Jain, MA Department of Psychology, Eastern Michigan University, Ypsilanti, MI, USA

Mari Janikian, PhD Department of Psychology, The American College of Greece, Athens, Greece

Phoebe Jordan, PhD Department of Pediatric Psychology, University of Michigan Health, Ann Arbor, MI, USA

Elizabeth Koval, MA Department of Pediatrics, Pediatric Psychology, Michigan Medicine, Ann Arbor, MI, USA

Leah LaLonde, PhD Psychiatry and Behavioral Health, Nationwide Children's Hospital, Columbus, OH, USA

Blake Lancaster, PhD Department of Pediatric Psychology, University of Michigan Health, Ann Arbor, MI, USA

Jamie M. Lawler, PhD Department of Psychology, Eastern Michigan University, Ypsilanti, MI, USA

John-Paul Legerski, PhD Department of Psychology, University of North Dakota, Grand Forks, ND, USA

Caroline Lucy, MA Eastern Michigan University, Ypsilanti, MI, USA

Alexandros Maragakis, PhD Department of Psychology, The American College of Greece, Athens, Greece

Stefanos Mastrotheodoros, PhD Department of Youth and Family, Utrecht University, Utrecht, The Netherlands

Department of Psychology, University of Crete, Rethymno, Greece

Lito E. Michalopoulou, PhD School of Psychology, Department of Psychology, Aristotle University, Thessaloniki, Greece

Randy Nedegaard, PhD Department of Social Work Education, California State University, Fresno, CA, USA

Lilian V. Pelios, PhD School of Graduate and Professional Education, The American College of Greece, Athens, Greece

Konstantina Psiachou, MS Department of Psychology, Panteion University, Athens, Greece

Cameron B. Richardson, PhD Clearinghouse for Military Family Readiness at Penn State, Social Science Research Institute, The Pennsylvania State University, University Park, PA, USA

Marie Rodriguez, MS Department of Pediatric Psychology, University of Michigan Health, Ann Arbor, MI, USA

Aidan P. Schmitt, PhD Department of Child & Adolescent Outpatient Psychiatry, Weill Cornell New York-Presbyterian Hospital, New York, NY, USA

Gary Small, MD Department of Psychiatry, Hackensack Meridian Health-Hackensack University Medical Center, Hackensack, NJ, USA

Luke Turnier, PhD Department of Pediatrics, Pediatric Psychology, Michigan Medicine, Ann Arbor, MI, USA

Ulrick Vieux, DO, MS, DFAPA Department of Psychiatry, Hackensack Meridian Health- Hackensack University Medical Center, Hackensack, NJ, USA

Julie Wojtaszek, MHSA, MS Eastern Michigan University, Ypsilanti, MI, USA

Chapter 1
History and Key Concepts of Evidence-Based Practices

Mari Janikian and Alexandros Maragakis

Introduction

The high prevalence of emotional, behavioral, and social difficulties in pediatric populations is extensively recorded, with approximately 68 million children and adolescents suffering from depression and anxiety (UNICEF, 2021). Additionally, following a global pandemic, where an estimated ten and a half million youth experienced the loss of a caregiver or parent (Hillis et al., 2022), a considerable number continue to face the potential for increasing challenges, which could have lasting impacts on communities and society as a whole. As professionals dedicated to promote the well-being of children, psychologists and other mental health clinicians are called to contribute to the identification, development, and dissemination of practices that reflect empirical and scholarly expertise. Within our field, practice and research complement one another, and both benefit the population they serve (American Psychological Association [APA], 2003).

This chapter provides an overview of the history and key concepts of evidence-based practice in pediatric specialty settings. First, we provide a brief overview of the historical development of evidence-based practices (EBP), followed by an examination of the challenges in the application of EBPs with pediatric populations. The next section reviews applications of EBPs including assessment and interventions. Finally, a brief discussion around cultural considerations and the structure of the book are discussed.

M. Janikian (✉) · A. Maragakis
Department of Psychology, The American College of Greece, Athens, Greece
e-mail: mjanikian@acg.edu; amaragakis@acg.edu

A. Maragakis, M. Janikian (eds.), *Evidence-Based Behavioral Health Practices in Pediatric Specialty Settings*, https://doi.org/10.1007/978-3-031-71004-9_1

Defining EBP

By the end of 1990s, Dr. David Sackett (1996) described EBP as "…the conscientious, explicit and judicious use of current best evidence in making decisions about the care of the individual patient. It means integrating individual clinical expertise with the best available external clinical evidence from systematic research" (Sackett et al., 1996, p. 71). EBP was further redefined in 2000 as "the integration of best available evidence, clinical expertise, and patient preferences and values" (Sackett et al., 2000, p. 1). EBP has been commonly depicted as a three-legged stool where all three factors are considered fundamental in health care practice (Fig. 1.1; APA, 2021): (a) The "best available evidence" is obtained by a systematic and critical evaluation of relevant scientific literature, with an emphasis on studies that employ rigorous research methods (i.e., randomized controlled trials (RCTs) and meta-analyses); (b) "clinical expertise" is the capacity to integrate information from accessible scientific data with personal knowledge to make relevant recommendations that are contextually appropriate for patients or clients; and (c) "patient preferences and values" involves patient in the decision-making process and considers their worldviews, expectations, and cultural background (Sackett et al., 2000).

In 2005, APA endorsed the following definition: "Evidence-based practice in psychology (EBPP) is the integration of the best available research with clinical expertise in the context of patient characteristics, culture, and preferences" (APA Presidential Task Force on Evidence-Based Practice, 2006, p. 273). In 2008, the APA Task Force on EBP with children and adolescents continued to emphasize the importance of the science-practitioner model, adopting a pertinent developmental and contextual lens.

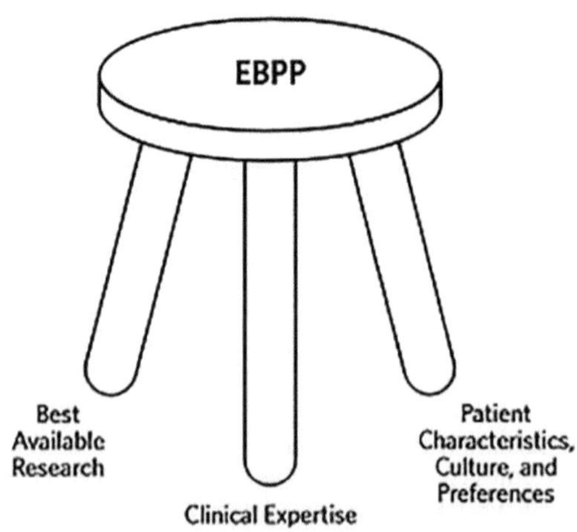

Fig. 1.1 Components of evidence-based practice in psychology. (EBPP; APA, 2021)

Historical Development

The evolution of EBP in child and adolescent mental health parallels a larger shift in the integration of research findings into clinical decision-making. EBP originated in the medical field, where pioneers, like Archie Cochrane, advocated for the use of systematic reviews to guide medical decisions and laid the groundwork for the founding of organizations like the Cochrane Collaboration. With the aim of providing robust scientific evidence for the efficacy of interventions, experimental research—in particular randomized controlled trials (RCTs)—was given greater importance in the second half of the twentieth century.

In the late twentieth century, evidence-based principles were applied to pediatric populations. Researchers and clinicians recognized the need for empirically supported treatments specifically targeted to the developmental needs of youth. In the 1990s and early 2000s, professional organizations, including the APA, began creating clinical practice guidelines (CPGs) and protocols with the purpose of standardizing and guiding practitioners in the selection and administration of EBPs. Recently, the APA Clinical Practice Guideline Panel published CPGs for the treatment of obesity and overweight (2018), as well as for depression (2019) in children and adolescents.

Whereas the Diagnostic and Statistical Manual for Mental Disorders (DSM; APA, 2013) continues to take a prominent role in the organization of disease entities, broader contextual aspects of emotional and behavioral indicators have been emerging in the child and adolescent mental health field. In response to criticisms regarding the medical model origins of EBP, preventive (WHO, 2020) and recovery-oriented approaches (Dallinger et al., 2023) applied to youth have been highlighted.

To sum up, the evolution of EBP in child and adolescent psychology can be traced back to a shift from a focus on evidence in healthcare to a more sophisticated understanding of how evidence-based principles can be tailored to address the unique needs of pediatric populations. The continued development of EBP in this area demonstrates the commitment to enhancing the quality and efficacy of mental health care provided to children and adolescents.

Key Concepts in Evidence-Based Practices

Sources of Evidence

EBP has traditionally emerged from research-based inquiries (see Table 1.1 for a summary of the main sources utilized), which differ in terms of validity, reliability, and how they influence treatment. For example, case studies are helpful for identifying therapeutic processes but typically have lower external validity; true

Table 1.1 Sources of research evidence

Source of evidence	Description
Research studies	Empirical investigations, including randomized controlled trials (RCTs), longitudinal studies, and experimental designs
Meta-analyses and systematic reviews	Comprehensive reviews that synthesize findings from multiple studies to provide a cumulative understanding of evidence
Clinical guidelines and protocols	Recommendations and protocols developed by professional organizations based on synthesized evidence and expert consensus
Treatment manuals and protocols	Detailed guidelines offering step-by-step instructions for implementing evidence-based interventions
Outcome measures and assessment tools	Valid and reliable instruments used to assess symptoms, monitor progress, and evaluate the outcomes of interventions
Implementation science research	Studies focused on strategies to integrate evidence-based interventions into real-world settings, addressing practical challenges
Consumer and stakeholder feedback	Input from children, adolescents, families, and other stakeholders, providing valuable perspectives on the effectiveness and acceptability of interventions

experimental designs, including RCTs, allow for causal inferences but suffer from a number of limitations. Specifically, many RCTs have the drawback of focusing almost solely on the improvement of primary symptoms of disorders (i.e., depression). Additional research is therefore required to assess other significant facets of children and adolescents' experiences and lives.

Another limitation stems from the design of RCTs. A clinically significant difference between treatment and control groups does not specify whether a patient will have a similar outcome to the treatment group average, given their individual characteristics and past experiences. Outcomes vary greatly across treatment and control groups, especially within Black, Indigenous, and other People of Color populations (Buchanan et al., 2021). The majority of intervention research is grounded in White and Western populations and views, with minimal inclusion of opinions from experts of color (APA, 2021). Also, the insufficient analysis of demographic differences, as well as the understudy of dropout groups and those not responsive to interventions, may further limit research findings. Moving forward, evidence sources can expand further to include community-based participatory research (Freire et al., 2022; Mathur Grunewald & Foley, 2024; Miller et al., 2022) and practice-based evidence (Abe et al., 2018; Leon, et al. 2013) among others, and contribute significantly to the growth of EBPP.

Professional Competence

Research continuously shows that the expertise of individual practitioners is positively correlated with behavioral health outcomes (Lambert et al., 2018). Competent practitioners are able to recognize critical indicators that are pertinent to their decisions (Larrick & Feiler, 2015). Clinical knowledge is essential at every stage of the treatment process, from initial assessment and rapport-building to the final assessment of treatment efficacy. It is also essential for determining and incorporating patient traits, culture, and preferences into treatment, as well as for assessing the best research that is currently available.

However, the role of professional expertise can be limited in a number of ways. For example, common cognitive biases (i.e., fundamental attribution error, confirmation bias, diagnostic overshadowing, and the availability heuristic) can have a negative impact on the accuracy of clinical decisions (Bowes et al., 2020). Indeed, evidence suggests that the accuracy of clinical judgments is frequently affected by biases, even among practitioners who are considered competent, well-informed, and intelligent (Jenkins, & Youngstrom, 2016). Researchers recommend continued monitoring and mitigation of cognitive biases grounded in ethical and evidence-based practices (Bowes et al., 2020) in the context of clinical supervision and consultation.

Patient Preferences

In order to respect patient autonomy and the right to self-determination, and to fulfill basic ethical duties for compassionate care, it is imperative that patient characteristics, culture, and preferences be taken into consideration and integrated into treatment plans. Studies indicate that treating patients according to their preferences and culture reduces the rate of early dropout and improves treatment outcomes (de Haan et al., 2018; de Soet et al., 2023; Simmons et al., 2021).

However, in the case of pediatric populations, the legal authority to decide whether or not they receive treatment rests with a guardian. Also, regardless of patients' preferences, characteristics, or culture, in cases concerning neglect or abuse of children, clinicians are legally mandated to report (Pietrantonio et al., 2013). Still, practitioners should strive to respect patients' choices, traits, and cultural background while engaging in collaborative decision-making, even when alternatives are limited due to ethical, legal, or other factors (Hawley & Morris, 2017). Overall, collaborative decision-making is the recommended way to build cooperative relationships and to increase the likelihood of successful outcomes.

To summarize, in clinical decision-making with patients, clinicians need to be mindful of the advantages and disadvantages of each of the three EBP components and carefully assess factors unique to each patient. Next, we will describe the application of EBPs in the assessment of children and adolescents.

Applying Evidence-Based Practices in Child and Adolescent Mental Health

Evidence-Based Assessment

Evidence-based assessment (EBA) is essential in providing excellent care for pediatric populations and is an integral part of EBP and a necessary precursor to evidence-based treatment (Hunsley & Mash, 2007; Youngstrom et al., 2017). EBA is defined as "an approach to clinical evaluation that uses research and theory to guide the selection of constructs to be assessed for a specific assessment purpose, the methods and measures to be used in the assessment, and the manner in which the assessment process unfolds" (Hunsley & Mash, 2007, p. 30). What is specified in this definition, is that information provided by assessment procedures or tests must be empirically supported and produce better outcomes for patients.

The aim of EBA is to improve diagnostic or classification utility (*accuracy*) and lead to targeted treatments that enhance the functioning of children and adolescents (*treatment utility*). By taking into consideration empirically supported assessment scores or variables, practitioners would be less likely to employ cognitive biases that influence decision-making and scientific thinking in a negative way (Lilienfeld et al. 2013; Watkins, 2009). To aid in this endeavor, the EBA framework focuses specifically on the "3 Ps" of clinical assessment (Youngstrom, 2013): *prediction* (how well does assessment data relate to a target criterion?), *prescription* (how well does assessment information inform treatment?), and *process* (is the assessment data useful for identifying variables that mediate treatment progress?).

Overall, the purpose of EBA is twofold: (a) in the initial assessment, EBA aids practitioners in case formulation and treatment planning. This process includes the recognition of strengths, challenges, and various resources at personal, familial, and community levels while also considering environmental, cultural, and systemic factors (Youngstrom et al., 2015); (b) during treatment, EBA is used to monitor progress and to tailor treatment accordingly for maximizing treatment outcomes. Longer-term impacts are evaluated through posttreatment assessments and follow-up evaluations, with a focus on culturally sensitive measures across symptoms, disorders, and real-life functioning from a variety of informants and settings, including children, parents, teachers, and peers in different contexts like home and school (Ng & Weisz, 2016). EBA has been linked with increased child treatment engagement and response (Scott & Lewis, 2015). Thus, a truly evidence-based approach to assessment would improve the clinical outcomes for the patient being assessed.

Within the assessment domain, the use of tests and procedures to aid in psychodiagnosis or classification and treatment recommendations are guided by specific principles prescribed by the Standards for Educational and Psychological Testing (American Educational Research Association, American Psychological Association, & National Council on Measurement in Education, 2014). Although child and adolescent psychological instruments have been traditionally used in assessment and refined over the years, gaps still remain in the evidence regarding their appropriate use, especially with populations which differ significantly from the original norm groups. In a recent systematic review, Becker-Haimes et al. (2020) provided a list of short measures that can be used in EBA with children and adolescents. The review identified measures with excellent psychometric properties that are available for practitioners freely (e.g., Behavior and Feelings Survey by Weisz et al., 2019; PROMIS Depression by DeWalt et al., 2015). Future research should continue evaluating existing measures for use with routine progress monitoring and ease of implementation in pediatric settings.

Further, there are general and specific ethical principles (APA, 2017; NASP, 2020) that address test score use and interpretations that guide assessment practice. EBA is implicitly guided by these principles because applications of information derived from tests or procedures must be empirically supported to be included, and only information that has been shown to substantively reduce uncertainty in clinical decisions is considered useful.

Evidence-Based Intervention Strategies

To keep up with the rapidly increasing demand of mental health services, in the context of limited mental health providers (MHPs), EBP services have evolved in multiple ways to best meet the needs of the pediatric population. Specifically, there has been a rapid expansion of EBPs focused on preventative services and addressing systemic issues from a holistic perspective to address difficulties beyond mental health diagnoses (Wu et al., 2015). These EBPs shift away from the traditional "siloed" model of mental healthcare delivery (i.e., delivering individual psychotherapy within an MHPs private practice without any collaboration with other healthcare providers) and include a multidisciplinary team within settings where there are high mental health needs (Maragakis, et al., 2018), such as the primary care setting (see Chap. 8) and school settings (see Chap. 6). Furthermore, innovative strategies such as stepped-care approach (see Chap. 4) and the use of new technologies to expand the dissemination of services (e.g., telehealth; see Chap. 3) highlights an important paradigm shift in psychological services (Maragakis & Hatzigeorgiou, 2018).

Prevention Strategies

Identifying and creating EBPs within a preventative framework focuses on addressing mental health concerns in children before they become clinical concerns. Specifically, priority on identification of risk factors and "preclinical" concerns, coupled with early, often brief, intervention, is a defining characteristic of this model of care (Carter et al., 2020). Furthermore, these preventative models typically engage in a "high-volume, low touch" approach in settings like primary care clinic and schools, which allows MHPs to identify and address concerns in populations that otherwise may not have accessed providers (Strosahl, 2005). Finally, these preventative models of care not only alleviate the burden of more serious mental health conditions but also contribute to long-term improvements in children's well-being (Stancin & Perrin, 2014).

However, preventative models are not without their challenges. First, funding EBPs from a preventative model raises complex barriers. For example, do preventative services cost less or the same as acute services? If so, what is this difference? Furthermore, does everyone need to "opt in" before being charged for preventative services? A major strength of a preventative model is to be able to screen a broader population than would typically present in traditional mental health settings. What would we do for those who do not wish or more importantly cannot afford these types of preventative services?

Another important challenge with a prevention model is when treatment is successful, it is unclear whether a disorder has been prevented or onset has simply been delayed (Cuijpers et al., 2008). Ideally, prevention would lead to a more permanent prevention of any particular disorder, but long-term data currently does not provide a clear picture of whether or not this occurs.

With preventative services, the proverbial question arises: "How are you sure that what you did (i.e., preventative services) resulted in *something not happening* (i.e., patient never crossed into the clinical threshold)?", as most research around prevention looks at improving theorized protected factors instead of assessing the actual disorder (Gadsby et al., 2023). Given the complexity of some of these mental health issues, it can be difficult to set up scientifically rigorous studies that will capture the data meaningfully.

Finally, EBPs within a preventative framework are significantly different than what most MHPs are typically trained in, thus creating a barrier regarding the number of providers who are adequately trained to work within a preventative framework. Competencies focused on how to adequately train mental health professionals throughout their continuum of training have been provided for specific settings (e.g., primary care settings) (Ham et al., 2022), but more robust shifts in how MHPs are trained in general may be a meaningful path for the field to consider in order to fill this workforce gap (Vriesman et al., 2023).

System-Focused Strategies

A unique aspect of working within pediatric settings is that therapy is rarely solely conducted with the child and that parents are typically the primary agent of therapeutic change (Wu et al., 2015). Therefore, an EBP approach that focuses on addressing system-level (e.g., family structure, school) strengths and barriers typically is needed to ensure that interventions will be successful and sustainable. This systems-focused approach prioritizes the involvement of families and other individuals who come into constant contact with the child (e.g., teachers) in treatment planning and implementation, to various degrees, depending on the age and presenting problem of the child, aiming to address not only the child's individual needs but also the broader context in which those needs arise (Wu et al., 2015).

An important component of a systems-focused model is its ability to leverage the family and other professionals in the child's life to augment the effectiveness and sustainability of interventions. By integrating caregivers and professionals (e.g., teachers) in the therapeutic process, clinicians can assess and intervene in environmental factors that may contribute to or maintain a child's mental health concern. By engaging and intervening in the child's "system," EBPs move beyond the therapy room and increase the probability of meaningful clinical change.

However, a systems model also presents several key challenges that can impact the implementation and success of therapy. First, navigating the dynamics between multiple "systems" (e.g., school, family, and medical providers) in a child's life can add a level of complexity, which can be greatly impacted by cultural and socioeconomic factors. Failure on the clinician's part to be sensitive, and adapting therapy accordingly, to these various dynamics can result in treatment resistance or reluctance by various members within the child's system. Finally, there are times when the parents or caregivers are identified with having an underlying mental health issue that is negatively impacting the well-being of the child. Navigating the role of the MHP and when they should intervene in caregiver mental health issues and when to refer, especially when working from a systems model, can raise important ethical and practical concerns (Maragakis et al., 2021).

Finally, systems models rely heavily on the engagement of multiple individuals (e.g., parents, caregivers, and teachers), which can be very time-consuming and require additional resources. These resources, which may not be easily available depending on socioeconomic status or geographic location, can impact the sustainability of treatment and may limit both the short- and long-term impact of treatment outcomes.

Cultural Considerations in the Implementation of EBP

As society becomes more integrated on a global scale, the effective delivery of EBPs requires MHPs to assure their interventions meet their patients' unique cultural needs. The importance of culturally sensitive providers is augmented by the fact that minority populations have historically been the most ignored when developing EBPs (La Roche & Christopher, 2009). While it is beyond the scope of this chapter to provide an in-depth analysis of cultural factors and considerations that impact the development and implementation of EBPs, a brief overview of key considerations is warranted.

In implementing culturally responsive EBPs, Hayes (2009) recommends several key strategies for MHPs to consider throughout treatment. First, a thorough assessment emphasizing culturally respectful behavior is important to building therapeutic alliance. For example, respecting rapport may not be universally prioritized or may look qualitatively different, particularly among specific cultural groups. This highlights the important adjustment MHPs may need to make when engaging with the patient or family members. It is recommended that MHPs take a hypothesis-testing approach to navigate potential cultural differences, ensuring that assessment processes are sensitive to individual cultural backgrounds and expectations (Hays, 2009).

Another important process of culturally sensitive EBPs is identifying relevant strengths and supports. Recognizing and valuing the patient's cultural identity, spirituality, familial networks, and community resources can play a significant role in developing a system-based intervention in which cultural aspects set the foundation. The explicit consideration of cultural preferences and supports allows practitioners to effectively utilize these resources to adapt EBPs in a meaningful way (Hays, 2009).

Finally, understanding various environmental and cognitive factors, under the framework of culture, can help develop culturally responsive EBPs. For example, cognitive restructuring may be modified due to cultural factors to involve more questions around helpfulness rather than the validity of beliefs. This shift in questioning may be more consistent with the patient's cultural norms and may allow for the patient and the MHP to collaboratively explore alternative perspectives (Hays, 2009).

Overall, much of the literature on cultural responsiveness from an EBP viewpoint does not require radical shifts in how MHPs provide services (Hays, 2009; La Roche & Christopher, 2009). Instead, it requires MHPs to maintain an open mind and curiosity around how cultural factors may influence patient preferences and responsiveness. This in turn allows the provider to make meaningful adjustments to how, when, or where services are provided in order to increase the saliency of treatment and improve the overall treatment experience.

Purpose of This Book

As EBPs and the overall field of pediatric psychology grow, so does the demand of MHPs that specialize in the unique demands of this population. Given that the majority of MHPs are trained as "generalists" (Norcross, et al., 2018) and only 16% identify pediatrics as their primary clinical specialty (Hamp et al., 2016), providing those interested in the field of pediatrics with a "roadmap" of various (not all) settings in which pediatric MHPs engage in meaningful therapeutic work can hopefully help build the much-needed mental health workforce.

With the most salient EBP-related topics written by leaders in their respective fields, this compilation of chapters is dedicated to behavioral health trainees, providers, and supervisors on EBPs in multiple pediatric specialty settings. This text will clarify skills and competencies for effective training and supervision, amounting to specific task lists for a range of settings. The *Evidence-Based Behavioral Health Practices in Pediatric Specialty Settings* book has been organized into two different but equally important sections. The first section looks at broader considerations and approaches to treatment, such as developmental factors or the utilization of stepped-care processes, which can be applicable to most settings. The second section is an overview of specific settings, what EBPs are relevant within that setting, and the unique expectations and practical and training considerations for those interested in working within that field. It is our hope that this book will provide not only a meaningful overview but also significant value in the development of professionals to meet the growing need of pediatric services across these unique settings.

Select Bibliography

American Psychological Association. (2021). *Professional practice guidelines for evidence-based psychological practice in health care*. Retrieved from https://www.apa.org/about/policy/evidence-based-psychological-practice-health-care.pdf

Becker-Haimes, E. M., Tabachnick, A. R., Last, B. S., Stewart, R. E., Hasan-Granier, A., & Beidas, R. S. (2020). Evidence base update for brief, free, and accessible youth mental health measures. *Journal of Clinical Child and Adolescent Psychology*, 1–17. https://doi.org/10.1080/15374416.2019.1689824

Melchert, T. P., Halfond, R. W., Hamdi, N. R., Bufka, L. F., Hollon, S. D., & Cuttler, M. J. (2023). Evidence-based practice in psychology: Context, guidelines, and action. *American Psychologist*. Advance online publication. https://doi.org/10.1037/amp0001253

References

Abe, J., Grills, C., Ghavami, N., Xiong, G., Davis, C., & Johnson, C. (2018). Making the invisible visible: Identifying and articulating culture in practice-based evidence. *American Journal of Community Psychology, 62*(1–2), 121–134. https://doi.org/10.1002/ajcp.12266

American Educational Research Association, American Psychological Association, & National Council on Measurement in Education. (2014). *Standards for educational and psychological testing*. American Educational Research Association.

American Psychiatric Association. (2013). *Diagnostic and statistical manual of mental disorders* (5th ed.). https://doi.org/10.1176/appi.books.9780890425596

American Psychological Association. (2003). Guidelines on multicultural education, training, research, practice, and organizational change for psychologists. *American Psychologist, 58*(5), 377–402. https://doi.org/10.1037/0003-066X.58.5.377

American Psychological Association. (2017). Ethical principles of psychologists and code of conduct (2002, amended effective June 1, 2010, and January 1, 2017). https://www.apa.org/ethics/code

American Psychological Association. (2019). *Clinical practice guideline for the treatment of depression across three age cohorts*. Retrieved from https://www.apa.org/depression-guideline.

American Psychological Association. (2021). *Professional practice guidelines for evidence-based psychological practice in health care*. https://www.apa.org/about/policy/psychological-practice-health-care.pdf

American Psychological Association, Clinical Practice Guideline Panel. (2018). *Clinical practice guideline for multicomponent behavioral treatment of obesity and overweight in children and adolescents: Current state of the evidence and research needs*. Retrieved from http://www.apa.org/obesity-guideline/obesity.pdf

American Psychological Association, Presidential Task Force on Evidence-Based Practice. (2006). Evidence-based practice in psychology. *American Psychologist, 61*(4), 271–285. https://doi.org/10.1037/0003-066X.61.4.271
American Psychological Association Task Force on Evidence-Based Practice for Children and Adolescents. (2008). *Disseminating evidence-based practice for children and adolescents: A systems approach to enhancing care*. American Psychological Association. Retrieved from: https://www.apa.org/practice/resources/evidence/children-report.pdf
Becker-Haimes, E. M., Tabachnick, A. R., Last, B. S., Stewart, R. E., Hasan-Granier, A., & Beidas, R. S. (2020). Evidence base update for brief, free, and accessible youth mental health measures. *Journal of Clinical Child and Adolescent Psychology*, 1–17. https://doi.org/10.1080/15374416.2019.1689824
Bowes, S. M., Ammirati, R. J., Costello, T. H., Basterfield, C., & Lilienfeld, S. O. (2020). Cognitive biases, heuristics, and logical fallacies in clinical practice: A brief field guide for practicing clinicians and supervisors. *Professional Psychology, Research and Practice, 51*(5), 435–445. https://doi.org/10.1037/pro0000309
Buchanan, N. T., Perez, M., Prinstein, M. J., & Thurston, I. B. (2021). Upending racism in psychological science: Strategies to change how science is conducted, reported, reviewed, and disseminated. *The American Psychologist, 76*(7), 1097–1112. https://doi.org/10.1037/amp0000905
Carter, B. D., Tsang, K. K., Brady, C. E., & Kullgren, K. A. (2020). Pediatric consultation-liaison: Models and roles in pediatric psychology. In *Clinical handbook of psychological consultation in pediatric medical settings* (pp. 11–24).
Cuijpers, P., Van Straten, A., Smit, F., Mihalopoulos, C., & Beekman, A. (2008). Preventing the onset of depressive disorders: A meta-analytic review of psychological interventions. *American Journal of Psychiatry, 165*(10), 1272–1280.
Dallinger, V. C., Krishnamoorthy, G., du Plessis, C., Pillai-Sasidharan, A., Ayres, A., Waters, L., Gloom, Y., Sweeney, K., Anderson, L., Rees, B., & Burton, L. (2023). Conceptualisation of personal recovery and recovery-oriented care for youth: Multisystemic perspectives. *International Journal of Mental Health and Addiction*. Advance online publication. https://doi.org/10.1007/s11469-023-01170-3
de Haan, A. M., Boon, A. E., de Jong, J. T., & Vermeiren, R. R. (2018). A review of mental health treatment dropout by ethnic minority youth. *Transcultural Psychiatry, 55*(1), 3–30.
de Soet, R., Vermeiren, R. R. J. M., Bansema, C. H., van Ewijk, H., Nijland, L., & Nooteboom, L. A. (2023). Drop-out and ineffective treatment in youth with severe and enduring mental health problems: A systematic review. *European Child & Adolescent Psychiatry*. https://doi.org/10.1007/s00787-023-02182-z. Advance online publication.
DeWalt, D. A., Gross, H. E., Gipson, D. S., Selewski, D. T., DeWitt, E. M., Dampier, C. D., Hinds, P. S., Huang, I. C., Thissen, D., & Varni, J. W. (2015). PROMIS(®) pediatric self-report scales distinguish subgroups of children within and across six common pediatric chronic health conditions. *Quality of Life Research, 24*(9), 2195–2208. https://doi.org/10.1007/s11136-015-0953-3
Freire, K., Pope, R., Jeffrey, K., Andrews, K., Nott, M., & Bowman, T. (2022). Engaging with children and adolescents: A systematic review of participatory methods and approaches in research informing the development of health resources and interventions. *Adolescent Research Review, 7*(3), 335–354. https://doi.org/10.1007/s40894-022-00181-w
Gadsby, E. W., Hotham, S., & Merritt, R. (2023). The theoretical and practical difficulties of evaluating a community-based 'whole systems' obesity prevention intervention: A research team's critical reflection. *Perspectives in Public Health, 143*(6), 337–346.
Ham, H., Lancaster, B., Maragakis, A., Harris, S., Turnier, L., & Bruni, T. (2022). Competencies for the practice of pediatric integrated primary care: A continuum of training. *Training and Education in Professional Psychology, 16*(2), 190–196.
Hamp, A., Stamm, K., Lin, L., & Christidis, P. (2016). *2015 APA survey of psychology health service providers*. http://www.apa.org/workforce/publications/15-health-service-providers/index.aspx
Hawley, S. T., & Morris, A. M. (2017). Cultural challenges to engaging patients in shared decision making. *Patient Education and Counseling, 100*(1), 18–24. https://doi.org/10.1016/j.pec.2016.07.008

Hays, P. A. (2009). Integrating evidence-based practice, cognitive–behavior therapy, and multicultural therapy: Ten steps for culturally competent practice. *Professional Psychology: Research and Practice, 40*(4), 354–360. https://doi.org/10.1037/a0016250

Hillis, S., N'konzi, J. P. N., Msemburi, W., Cluver, L., Villaveces, A., Flaxman, S., & Unwin, H. J. T. (2022). Orphanhood and caregiver loss among children based on new global excess COVID-19 death estimates. *JAMA Pediatrics, 176*(11), 1145–1148.

Hunsley, J., & Mash, E. J. (2007). Evidence-based assessment. *Annual Review of Clinical Psychology, 3*, 29–51.

Jenkins, M. M., & Youngstrom, E. A. (2016). A randomized controlled trial of cognitive debiasing improves assessment and treatment selection for pediatric bipolar disorder. *Journal of Consulting and Clinical Psychology, 84*, 323–333. https://doi.org/10.1037/ccp0000070

La Roche, M. J., & Christopher, M. S. (2009). Changing paradigms from empirically supported treatment to evidence-based practice: A cultural perspective. *Professional Psychology: Research and Practice, 40*(4), 396.

Lambert, M. J., Whipple, J. L., & Kleinstäuber, M. (2018). Collecting and delivering progress feedback: A meta-analysis of routine outcome monitoring. *Psychotherapy, 55*(4), 520–537. https://doi.org/10.1037/pst0000167

Larrick, R. P., & Feiler, D. C. (2015). Expertise in decision-making. In G. Keren & G. Wu (Eds.), *The Wiley Blackwell handbook of judgment and decision-making* (1st ed., pp. 696–721). Wiley. https://doi.org/10.1002/9781118468333.ch24

Leon, S. C., Stoner, A. M., Lyons Usher, A. M., & Carey, D. (2013). Measuring children's response to inpatient treatment: Use of practice-based evidence. *Psychiatric Services, 64*(3), 252–256. https://doi.org/10.1176/appi.ps.201100448

Lilienfeld, S. O., Ritschel, L. A., Lynn, S. J., Cautin, R. L., & Latzman, R. D. (2013). Why many clinical psychologists are resistant to evidence-based practice: Root causes and constructive remedies. *Clinical Psychology Review, 33*(7), 883–900. https://doi.org/10.1016/j.cpr.2012.09.008

Maragakis, A., & Hatzigeorgiou, M. N. (2018). The transformation of the healthcare system: Integrated primary care and the role of stepped care interventions for behavioral health providers. In A. Maragakis & W. T. O'Donohue (Eds.), *Principle-based stepped care and brief psychotherapy for integrated care settings* (pp. 9–17). Springer International Publishing Company.

Maragakis, A., Lindeman, S., & Nolan, J. (2018). Evidence based and intensity specific services in the integrated care setting: Ethical considerations for a developing field. *Behavior Analysis: Research and Practice, 18*(4), 425–435.

Maragakis, A., Ham, H., Bruni, T., Byrd, M. R., LaLonde, L., Neenan, A., Caserta, A., & Turiner, L. (2021). Promoting a holistic family care model: Ethical and practical considerations for treating adult caregivers in integrated pediatric settings. *Clinical Practice in Pediatric Psychology, 9*(2), 209–218.

Mathur Grunewald, N., & Foley, N. M. (2024). Exploring the utility of community-based participatory research to investigate mindfulness-based interventions in schools. *Psychology in the Schools, 61*(1), 398–412. https://doi.org/10.1002/pits.23059

Miller, A. B., Davis, S. H., Mulder, L. A., Winer, J. P., Issa, O. M., Cardeli, E., & Ellis, B. H. (2022). Leveraging community-based mental health services to reduce inequities for children and families living in United States who have experienced migration-related trauma. *Psychological Trauma: Theory, Research, Practice, and Policy*. https://doi.org/10.1037/tra0001392

National Association of School Psychologists. (2020). The professional standards of the National Association of School Psychologists. https://www.nasponline.org/x55315.xml

Ng, M. Y., & Weisz, J. R. (2016). Annual research review: Building a science of personalized intervention for youth mental health. *Journal of Child Psychology and Psychiatry, 57*(3), 216–236. https://doi.org/10.1111/jcpp.12470

Norcross, J. C., Sayette, M. A., & Pomerantz, A. M. (2018). Doctoral training in clinical psychology across 23 years: Continuity and change. *Journal of Clinical Psychology, 74*(3), 385–397. https://doi.org/10.1002/jclp.22517

Pietrantonio, A. M., Wright, E., Gibson, K. N., Alldred, T., Jacobson, D., & Niec, A. (2013). Mandatory reporting of child abuse and neglect: Crafting a positive process for health professionals and caregivers. *Child Abuse & Neglect, 37*(2–3), 102–109. https://doi.org/10.1016/j.chiabu.2012.12.007

Sackett, D. L., Rosenberg, W. M., Gray, J. A., Haynes, R. B., & Richardson, W. S. (1996). Evidence based medicine: What it is and what it isn't. *BMJ (Clinical Research Ed.), 312*(7023), 71–72. https://doi.org/10.1136/bmj.312.7023.71

Sackett, D. L., Straus, S. E., Richardson, W. S., Rosenberg, W., & Haynes, R. B. (2000). *Evidence-based medicine: How to practice and teach EBM*. Churchill Livingstone.

Scott, K., & Lewis, C. C. (2015). Using measurement-based care to enhance any treatment. *Cognitive and Behavioral Practice, 22*(1), 49–59. https://doi.org/10.1016/j.cbpra.2014.01.010

Simmons, C., Meiser-Stedman, R., Baily, H., & Beazley, P. (2021). A meta-analysis of dropout from evidence-based psychological treatment for post-traumatic stress disorder (PTSD) in children and young people. *European Journal of Psychotraumatology, 12*(1), 1947570. https://doi.org/10.1080/20008198.2021.1947570

Stancin, T., & Perrin, E. C. (2014). Psychologists and pediatricians: Opportunities for collaboration in primary care. *American Psychologist, 69*(4), 332.

Strosahl, K. (2005). Training behavioral health and primary care providers for integrated care: A core competencies approach. In *Behavioral integrative care: Treatments that work in the primary care setting* (pp. 15–52).

United Nations Children's Fund. (2021). *Seen, counted, included: Using data to shed light on the well-being of children with disabilities*. UNICEF. https://data.unicef.org/resources/children-with-disabilities-report-2021/

Vriesman, M., Dhuga, J., LaLonde, L., Orkopoulou, E., Tepple, T., Good, J., Lucy, C., & Maragakis, A. (2023). Clinical psychologists as T-shaped professionals. *Perspectives on Psychological Science, 18*(5), 996–1008.

Watkins, M. W. (2009). Errors in diagnostic decision making and clinical judgment. In T. B. Gutkin & C. R. Reynolds (Eds.), *Handbook of school psychology* (4th ed., pp. 210–229). Hoboken, NJ: Wiley.

Weisz, J. R., Vaughn-Coaxum, R. A., Evans, S. C., Thomassin, K., Hersh, J., Ng, M. Y., et al. (2019). Efficient monitoring of treatment response during youth psychotherapy: The behavior and feelings survey. *Journal of Clinical Child & Adolescent Psychology, 1–15*. https://doi.org/10.1080/15374416.2018.1547973

World Health Organization. (2020). *Guidelines on mental health promotive and preventive interventions for adolescents: Helping adolescents thrive*. World Health Organization. https://iris.who.int/handle/10665/336864

Wu, M. S., Hamblin, R. J., & Storch, E. A. (2015). Evidence-based psychological treatments of pediatric mental disorders. *Advances in Pediatrics, 62*(1), 165–184. https://doi.org/10.1016/j.yapd.2015.04.007

Youngstrom, E. A. (2013). Future directions in psychological assessment: Combining evidence-based medicine innovations with psychology's historical strengths to enhance utility. *Journal of Clinical Child & Adolescent Psychology, 42*, 139–159. https://doi.org/10.1080/15374416.2012.736358

Youngstrom, E. A., Choukas-Bradley, S., Calhoun, C. D., & Jensen-Doss, A. (2015). Clinical guide to the evidence-based assessment approach to diagnosis and treatment. *Cognitive and Behavioral Practice, 22*(1), 20–35. https://doi.org/10.1016/j.cbpra.2013.12.005

Youngstrom, E. A., Van Meter, A., Frazier, T. W., Hunsley, J., Prinstein, M. J., Ong, M. L., & Youngstrom, J. K. (2017). Evidence based assessment as an integrative model for applying psychological science to guide the voyage of treatment. *Clinical Psychology: Science and Practice, 24*(4), 331–363. https://doi.org/10.1111/cpsp.12207

Part I
General Considerations and Novel Models of Care

Chapter 2
Developmental Considerations Across the Pediatric Population

Lito E. Michalopoulou, Cameron B. Richardson, and Stefanos Mastrotheodoros

Humans develop in different ways; we have different starting points, and our life trajectories differ based on several factors associated with person-level characteristics, context, and life events. Developmental issues and health, including behavioral health, are intertwined in countless ways. This chapter allows behavioral health providers and trainees to gain a foundational understanding of the theoretical and empirical background of children's and adolescents' development. In specific, major theories, concepts, and research that can be of particular interest to a pediatric context are reviewed. Last, implications for practice and training are presented to allow trainers, trainees, and practitioners to consider applications of these theories and concepts into practice.

Theoretical Background of Children's Development

In the 1970s, Uri Bronfenbrenner proposed an ecological approach focusing on direct and indirect contextual and reciprocal influences on child development, initially known as the ecological systems theory or the social-ecological model (Tudge et al., 2022). His theory of human development grew for more than 20 years, and in

L. E. Michalopoulou (✉)
School of Psychology, Aristotle University of Thessaloniki, Thessaloniki, Greece

C. B. Richardson
Clearinghouse for Military Family Readiness at Penn State, Social Science Research Institute, The Pennsylvania State University, University Park, PA, USA

S. Mastrotheodoros
Department of Youth and Family, Utrecht University, Utrecht, The Netherlands

Department of Psychology, University of Crete, Rethymno, Greece

A. Maragakis, M. Janikian (eds.), *Evidence-Based Behavioral Health Practices in Pediatric Specialty Settings*, https://doi.org/10.1007/978-3-031-71004-9_2

its "mature" version of the mid-1990s, the theory transformed into the bioecological or the Process-Person-Context-Time model of development (Bronfenbrenner, 1994; Bronfenbrenner & Morris, 2006; Fiese et al., 2008; Tudge et al., 2009). Bronfenbrenner's evolving theory is considered contextualist because the different factors that influence one's development are thought to function interdependently and synergistically rather than in isolation (Tudge et al., 2022). A child's environment is therefore conceptualized as an interconnected arrangement of the following contextual structures:

1. The *microsystem* is any setting in which the child has direct interactions with other people, objects, and symbols; it is characterized by reciprocity and defined by a functional social system (e.g., home, school, and extracurriculars).
2. The *mesosystem* encompasses the interrelations and linkages among major settings containing the developing person at a specific time in their life, and it is a system of microsystems (e.g., home interactions with the school, home interactions with a place of worship).
3. The *exosystem* refers to formal and informal social structures in which the developing person does not actively participate, but which affect them (e.g., neighborhood activities, parents'/guardians' work environment, and parents'/guardians' social activities).
4. The *macrosystem* encompasses institutional systems (e.g., economic, social, educational, legal, and political) of a society or subculture group that impact a developing person's micro-, meso-, and exosystems.

Another model that enhances the ecological perspective and has explicit applications to pediatric psychology is Sameroff and Chandler's transactional model (Fiese & Sameroff, 1989). In this model, a child's development is conceptualized as the developing person's continuous dynamic transactions and the experiences obtained in their environment over time (Sameroff, 2009). As with Bronfenbrenner's bioecological model, the transactional model emphasizes the bidirectional and interdependent effects between children (e.g., a child's temperament) and their context (e.g., high resource environment typically associated with higher socioeconomic status). Although transactional models have been traditionally applied to parent–child transactions, such occurrences happen in various social units (e.g., parent, school, and culture) across a person's systems. Another important aspect of the transactional model is that the environment is characterized by plasticity and is paired with the idea that the developing person has an active role in their own growth (Sameroff, 2009). As such, the experiences provided by the environment are not viewed as independent of the child (Sameroff & MacKenzie, 2003).

This brief overview of these two models helps us understand the developmental context of children in pediatric settings. Before we examine child-level and family-level developmental considerations along with structural and social ecologies, let us discuss how a case study may be approached using the bioecological and transactional models.

The Case of Natalie

Natalie was born with hip dysplasia, a condition that seems to run on her mother's side of the family. She was raised in a rural setting with limited access to health care. At the age of three, due to her imbalance associated with hip dysplasia, she fell down a flight of stairs, breaking and dislocating both her legs. In order to receive treatment and rehabilitation, her parents and Natalie moved to a large city. Given that both her parents needed to work to afford her care and could not provide home-based services nor could they provide daily transportation to receive outpatient treatment, she remained hospitalized in the pediatric orthopedics wing for approximately 6 months. Her parents would visit her daily and spend almost 2 hours with her. Upon discharge, and given her continuing needs for rehabilitation, the family decided to remain in the city. She started attending the local preschool, and by the end of kindergarten, she was diagnosed with a speech and language impairment for which she started receiving speech and language therapy at a child center based in a local public hospital.

When adopting the bioecological model, practitioners need to consider the different contextual systems Natalie influences and that influence her development, such as "What have been Natalie's interactions with her parents before and after the accident?" "What have been Natalie's interactions with her healthcare providers in terms of exposure to language?" or "How did Natalie spend her time in the hospital—alone, in conversation with healthcare providers or other patients if colocated in a shared room?" Next, practitioners can move toward clarifying the linkages between Natalie's microsystem to the meso-, exo-, and macrosystems with questions such as the following:

- How long does it take for the family to commute to the hospital for language therapy and continuing rehabilitation?
- What types of foods does the family have ready access to; are there any safety concerns in the neighborhoods?
- Does the family's joint income cover the costs of renting/owning a place?
- How will such a diagnosis and therapy responsiveness impact her later schooling experiences?
- Might Natalie's need for services confirm stereotypes of her "group" by those in the dominant culture?

From a transactional model perspective, additional questions to ask may pertain to which of Natalie's characteristics are serving to invite or discourage interactions/transactions with others and especially her caregivers (e.g., temperament/personality, irritability, persistence, and responsiveness) and what are her personal resources/assets (e.g., language ability level, prior experiences, and prior access to developmental activities and interactions). By posing and trying to find answers to such questions, practitioners can better understand the developing person and their context along with a child's developmental outcomes at any point and, thus, provide more sensitive and tailored behavioral health services in a pediatric setting.

The next section details some of the major developmental concepts and research associated with childhood and adolescence in an attempt to further refine our understanding of the pediatric population.

Developmental Considerations: From Childhood to Adolescence

Typically developing children experience changes in cognitive abilities, reasoning skills, including moral reasoning, communication skills, emotional capabilities, and independence from birth throughout young adulthood, impacting the expression and manifestation of behavioral health concerns. At the same time, each child progresses at different rates depending on the idiosyncrasies of their genetic and environmental context and life experiences. As such, one useful distinction to draw is a child's developmental age, indicating how well a child performs on a standardized task in comparison to other children, versus a child's chronological age, indicating how old a child is.

Childhood

Childhood is the developmental period roughly from ages 2 to 11, with early childhood comprising ages 2 to 6, and middle childhood from 6 to 11. Major developmental milestones in childhood center on social skills and developing meaningful friendships (Palermo et al., 2014a). These tasks follow the more basic developmental tasks of infancy and toddlerhood such as learning about one's environment through monitoring of basic cause and effect relations (e.g., "when I push this object it rolls away from me; when my caretaker's voice turns playfully high pitched, she's going to tickle me" [Zaadnoordijk et al., 2020]), learning that they can trust their environment (e.g., "my caretaker provides for me" [Koenig & Woodward, 2010]), learning how to communicate (e.g., "this utterance is often paired with a finger point or an eye gaze in the direction of this object and then back to my eyes" [Donnelly & Kidd, 2021]), and learning about societal and/or cultural expectations for appropriate behavior (e.g., "they clap when I share things with others" [Boe & Woods, 2018]).

Research on early patterns of development supports the premise that the early years (the first 3 years of a person's life) provide a foundation for subsequent growth trajectories (Peterson, 1987). Individual differences coupled with the high vulnerability among young children to environmental and biological factors create a learning pattern and a rate of development that is likely to be sustained unless major environmental changes occur, such as the changes introduced by early intervention programs (Guralnick, 1997; Peterson, 1987). For this age group, developmental outcomes hinge on numerous parent/caregiver influences, such as parent–child transactions, family-oriented child experiences, and family-provided child health and safety (Guralnick, 2020).

With school-age children, parents/caregivers help children through their growing need for autonomy and their need for adult guidance and oversight. Parenting behaviors play an immense role in the success or failure of many of the developmental tasks as well as in their health (Frosch et al., 2021; Pierce et al., 2020). For instance, parents with emotional health concerns may attend to their children less than in normative parenting contexts, which may result in delays in communication and literacy (Brookman et al., 2020). So too, a lack of trust in the adults in one's environment may stem from exposure to inconsistent caregiving behaviors, which may manifest as disorganized attachment behavior (i.e., inconsistent approach toward and avoidance of the caregiver in the presence of strangers [Van Ijzendoorn et al., 1999]).

Success in connecting to others is an important developmental task, with those children who share and are less aggressive being seen as more likable (Denham & Holt, 1993). Empirical research has indicated that children (early and middle childhood) with medical conditions, such as chronic pain, however, may experience poor social functioning, evidenced by fewer friendships, greater victimization and isolation, and less positive peer ratings compared to children without such conditions (Forgeron et al., 2010). And while direct causal evidence has yet to be documented (Goodacre et al., 2023; Slomkowski & Dunn, 1996), the child's budding abilities to take the perspective of others is expected to be one of many precursors (e.g., self-regulatory capacity, empathy, and information processing capacity) for the development of robust connections to others.

Adolescence: Autonomy and Identity Development

Adolescence is the developmental period roughly from ages 12 to 18. Puberty typically signifies the start of this period, whereas the gradual emergence of more-or-less adult roles signifies the end of adolescence and the transition to emerging adulthood (Arnett, 2007). Adolescence is a challenging developmental period because many fast-paced changes happen in the physical, cognitive, social, and interpersonal spheres, and young people need to adjust to these changes. Resulting from these changes, adolescents need to gradually assume more autonomy (Soenens et al., 2017) and start forming a sense of personal identity (Luyckx et al., 2011), including plans for their future. Developing autonomy and a sense of personal identity are two of the core developmental tasks during adolescence that need to be considered in health care in pediatric settings. According to both Bronfenbrenner's model (Bronfenbrenner & Morris, 2006) and the transactional model of psychosocial development (Sameroff, 2010), these two developmental tasks happen within the context of family relationships.

Adolescent autonomy is often described as a key issue during development. Even though autonomy and independence are often used interchangeably, theory (Ryan & Deci, 2017) and research (Van Petegem et al., 2013) suggest that they differ from each other. Independence implies a more interpersonal construct (independent decision-making, i.e., without the help of others), whereas autonomy implies a more intrapersonal construct (decision-making based on the core inner wishes and values). Both aspects are key to healthy development during adolescence and are associated with better adolescent adjustment (Ryan & Deci, 2017). A related concept is autonomous motivation (internal motivation), which describes deeply held, personal reasons for which one remains engaged in action. Increased autonomy is expected to be associated with increased autonomous (internal) motivation, which then is associated with sustained behavior. These ideas have important implications for pediatric care during adolescence.

The importance of autonomy and independence during adolescence does not mean that adolescents are capable of and should be solely responsible for health-related decisions (DiMatteo & Miller, 2013). Whereas adolescents should not be making serious health-related decisions independently, enhancing autonomy and autonomous motivation are expected to improve health outcomes (Ntoumanis et al., 2021). For example, allowing space for adolescents to express their inner wishes and preferences (more autonomy) can be expected to increase treatment adherence via increased internal motivation (Goethals et al., 2019).

A second key developmental task during adolescence is the development of identity. Identity is a broad concept that has been approached from several different angles (Schwartz et al., 2011), with terminology such as *ego identity, personal identity, national identity, ethnic identity, and racial identity* often used in the empirical literature to describe its different aspects. Personal identity development describes the extent to which adolescents engage in exploration of and commitment to alternatives regarding who they are and what makes them unique (e.g., Årseth et al., 2009; Crocetti et al., 2008; Marcia, 1993). Partly based on the advances in the cognitive domain (new cognitive skills, abstract thinking, etc.), as well as in the changes in the social domain (e.g., new school contexts and more varied peer groups), adolescents face several different options for what could make them unique, and who they want to become. Theory (Erikson, 1968) and research (e.g., Becht et al., 2019, 2021) suggest that adolescents who engage at least in some form of exploration and commitment making are better adjusted, compared to adolescents who avoid issues pertaining to their identity.

Identity development has also been found to be an important resource for adolescents who face illness, such as diabetes (Luyckx et al., 2008). For example, a stronger sense of identity, characterized by higher exploration and stronger commitments, was associated with more adaptive coping strategies and fewer illness-related problems (Luyckx et al., 2008). Therefore, providing opportunities for youth to explore and shape their personal identities might have positive implications for their pediatric issues.

Family Relationships During Childhood and Adolescence

While Locke's concept of tabula rasa (blank slate) has been called to task for its oversimplification of infancy and childhood, it can serve as a useful starting point for differentiating family relationship dynamics across childhood and adolescence. For instance, family relationships in childhood are marked by a need for both warmth (i.e., positive affect, dedication, closeness, and acceptance) and control (i.e., parental interventions intended to modify behavior, emotions, or thoughts), which may look different for different family systems, cultures, and for different child characteristics (Deater-Deckard et al., 2011). Control that is punitive in nature (i.e., rigid rules, hostility, and penalization of errors) can stifle a child (Zubizarreta et al., 2019). In contrast, control that uses logical (e.g., a child spills cereal on the floor and is asked to pick it up) and natural consequences (e.g., failing to wear a jacket leads to being cold) is considered more tolerable to children (Mageau et al., 2018; Smetana et al., 2021; Tisak, 1986). As the child grows in experience, efforts to modify the child's behavior that fail to take into account the child's growth may become less effectual.

Family relationships change during adolescence, as adolescents strive for increased independent decision-making, while they navigate several developmental challenges, whereas parents often face their own developmental—midlife—challenges while they still are responsible for their adolescent child's development (Branje et al., 2021). Parents and adolescents hold significantly different expectations about when certain milestones should be achieved (Deković et al., 1997), which often contributes to parent–adolescent conflict (Mastrotheodoros et al., 2020).

To understand how the parent–adolescent relationship evolves over time during adolescence, two core aspects of change need to be considered (Branje et al., 2021). On the one hand, mean-level stability or change captures the degree to which a concept increases or decreases, on average across people of a certain group. On the other hand, rank-order stability or change captures the similarity among two scores (e.g., a parent report and an adolescent report), regardless of the level of each variable. Together, these two aspects provide a better understanding of parent–adolescent relationships during adolescence (Branje et al., 2021).

Overall, several aspects of parenting and parent–adolescent relationships show significant mean-level changes during adolescence. Warmth, which describes the extent to which parents openly express their positive emotions toward adolescents, and behavioral control, which describes the extent of clear rules and limit setting by parents, generally decrease throughout adolescence (Mastrotheodoros et al., 2019). At the same time, conflict intensity shows some increase, but only in the eyes of adolescents (Mastrotheodoros et al., 2020). While these mean-level changes imply important changes in parent–adolescent relationships, other studies focusing on rank-order stability show significant relationship continuity (Branje et al., 2021). Furthermore, the relationship quality before adolescence will partly determine how parent–adolescent relationship quality will develop during adolescence. Specifically, those who had relatively worse relationship quality during childhood tended to have relatively worse relationship quality during adolescence (Laursen et al., 2010). Therefore, notwithstanding the mean-level changes that, on average, happen throughout adolescence, there is also relative continuity in parent–adolescent relationships (Branje et al., 2021).

The relationship quality between parents and adolescents has implications for pediatric populations. Specifically, better parent–adolescent relationship quality has been associated with better treatment adherence (DiMatteo & Miller, 2013). Therefore, pediatric healthcare specialists should also pay attention to the relationships between parents and adolescents in order to fine-tune the effectiveness of pediatric health care.

Child Health Inequalities and Development

Around the world, health inequities impacting children's health and development across the lifespan are created and maintained through complex interactions occurring in the ecological system (Link & Phelan, 1995; Spencer et al., 2019). Such inequities are more prominent and widespread in the world's poorest nations, but even in wealthy countries, children coming from poor and disadvantaged environments have the highest risk of poor health and developmental outcomes (Moore et al., 2015; Spencer et al., 2019). Determinants of child health inequities stem from social and structural ecologies (Shahidullah et al., 2023). The World Health Organization (WHO) defines these social drivers of health as "the conditions in which people are born, grow, work, live, and age, and the wider set of forces and systems shaping the conditions of daily life" (WHO, n.d.). The related ecologies are typically manifested in the form of racism, systemic discriminatory practices, oppression, and so forth (Shahidullah et al., 2023). Among these facets, racism is particularly salient. As per the American Academy of Pediatrics, a child's, youth's, and adolescent's social environment shapes their development, and, within this environment, the experience of racism can function detrimentally as a social determinant of health that not only influences children per se but also their families (Trent et al., 2019). Further, racial and ethnic discrimination manifests in patterns of inequitable treatment across pediatric care services, such as neonatology, primary care, emergency medicine, inpatient and critical care, surgery, developmental disabilities, mental health care, endocrinology, and palliative care (Slopen et al., 2024). Child health inequalities can be considered avoidable and can be reduced; a major role in this shift is placed on pediatric health organizations and professionals who are called at a national and international level to promote and advocate for child health equality (Spencer et al., 2019; Trent et al., 2019).

Referencing back to Natalie's case, let us consider the factors that may have affected her health and development at this macro-level. She was conceived, born, and initially raised in an area with limited access to health services to a low socioeconomic status family, conditions that placed her in a higher risk group for poor health outcomes. In addition, let us imagine that she did not only move from a rural area to an urban area, but she also had to move to a neighboring, wealthier country. It has been established that for immigrant families, social determinants can impact short-term psychological adaptation as well as long-term developmental trajectories (Suárez-Orozco et al., 2018). Consequently, as professionals, we may wonder how Natalie's lived experiences of potential discrimination, xenophobia, and lack of access to health and other opportunities throughout her childhood and adolescent years, experiences characteristically shared among youth from minoritized and oppressed groups, may have affected her socio-emotional development along with her educational and vocational attainment, especially in the absence of protective factors (e.g., supportive social network, sense of heritage, and ethnic identity). The question, then, becomes, how can professionals working in behavioral health and pediatric care help and support children like Natalie, and overall promote child health equity?

Implications for Training and Practice

When it comes to developmental considerations and their implications for practitioners working with children and adolescents in pediatric care settings, first and foremost is that one needs to develop and utilize not only age-appropriate (i.e., based on a child's chronological age) but primarily developmentally appropriate (i.e., based on a child's developmental age) strategies, techniques, and interventions. It has been indicated that a child's developmental level has often been overlooked in some pediatric populations, such as interventions for children with chronic pain (Palermo et al., 2014a). A direct implication is that a clear understanding of a child's developmental age across domains (cognitive, socio-emotional, physical, and ethical/moral) is warranted. By understanding through formal and informal assessment methods, providers can tailor services to a child's developmental level, needs, and strengths. This is perhaps the most critical aspect of developing and implementing interventions, as doing so increases the potential for positive outcomes and better-quality services for professionals. Thus, all providers and trainees in such settings need to become familiar with developmental theories, such as Bronfenbrenner's and Sameroff's, along with other principles and determinants of child and adolescent development that facilitate their services from case conceptualization to intervention planning and service delivery (Murphy & Christner, 2012; Palermo et al., 2014b). Nonetheless, having a sound understanding of typical, adaptive, and maladaptive cognitive, socio-emotional, behavioral, and physical development in the broader context of developmental expectations, parent/caregiver behavior, and systems influence (Madan-Swain et al., 2012; Palermo et al., 2014b) is considered a core competency for training in pediatric psychology. An underlying hypothesis for considering the knowledge and understanding of development considerations as a core competence and more broadly considering the research/theory-practice gap is that scientifically literate practitioners will facilitate effective psychological services (Cook & Coyne, 2005; Jones & Mehr, 2007).

Another implication arises from developmentally grounded practice: beyond differences across individual children, intrapersonal (within-child) differences across the several developmental domains also need to be taken into account (Murphy & Christner, 2012). In other words, each child's physical, cognitive, socio-emotional, and moral development may not be progressing at the same rate based on genetic and environmental factors. This means that a professional may be working with a child whose cognitive development aligns with their chronological age and who can eloquently express themselves, but their socio-emotional development may be that of a younger child and, consequently, they may experience challenges expressing their feelings and adjusting to their treatment protocol.

Next, providers and trainees in healthcare settings need to recognize that children, unlike adults, develop rapidly and their health depends largely on their cognitive, emotional, and physical growth (Forrest et al., 1997). This dimension of providing healthcare services to children is associated with the scope that pediatric care functions habilitatively (i.e., the goal of the provided services is to help patients learn or improve skills or functions necessary for daily life, which they are not able to develop naturally on their own) rather than rehabilitatively (i.e., the treatment goal is to aid patients recover skills or functions they have lost due to injury or illness) and as a resource that boosts children's developmental trajectories (Stille et al., 2010). Thus, providers and trainees need to be knowledgeable and aware of how biological, cognitive, social, affective, sociocultural, and life span developmental factors impact children's health and medical conditions, including interfacing mechanisms and mediational pathways (Palermo et al., 2014b). For example, the prevalence, severity, and symptoms of chronic illnesses in childhood may change over time making the provision of appropriate services a dynamic process.

Last, across childhood and adolescence, access to care is provided via an adult "gatekeeper," most often a parent. For young children who may have not yet developed language and communication skills, a major challenge is for parents and caregivers to recognize symptoms and seek help (Rapoff et al., 2023), and consequently seek services. Nonetheless, from a child's birth onward and beyond parent awareness and identification issues, one of the greatest challenges to meeting children's health, including mental health, needs, relates to parents'/caregivers' empowerment, willingness, or ability to engage in seeking and receiving care (Ashcraft et al., 2019; Hoagwood, 2005; Mian, 2014). Hence, for practitioners working with families, parent empowerment and engagement strategies may be beneficial in improving collaboration with the families along with treatment adherence and outcomes.

Multiple strategies that can enhance parent and family engagement have been identified in the published literature, including the following: (a) practitioner reinforcement practices (e.g., monetary rewards and verbal praise) used to increase desired behaviors related to intervention engagement (e.g., attendance and homework completion); (b) expectation setting, problem-solving, and parent coping achieved through exercises and strategies designed to enhance the parents/caregivers' ability to handle stressful situations that may impede intervention participation; (c) accessibility promotion, meaning any strategy used to make services convenient and accessible in order to proactively encourage and increase engagement; (d) use of change talk (e.g., using motivational-interviewing strategies, such as probing reading to change though examining disadvantages of the current situation and advantages of change in terms of increased engagement); (e) psychoeducation and peer pairing (i.e., pairing the child/adolescent and/or parent(s)/caregiver(s) with another child/adolescent and/or parent(s)/caregiver(s) with the goal to provide support around issues of seeking help, encouraging intervention participation, enhancing skill development, and sharing information); and (f) practitioner–family rapport building, meaning strategies aiming to increase the quality of the relationship between the child/parent(s)/caregiver(s) and the practitioner so as to increase youth/family engagement in the services (Desai, 2019; Haine-Schlagel & Walsh, 2015; Lindsey et al., 2014).

Communication Tailored to Children's and Youth's Abilities and Prior Experience

A crucial aspect of engaging families as well as designing and using developmentally appropriate interventions and strategies pertains to how a provider talks to children and youth. Good clinical and interviewing competence implies a sensitivity to a child's communication and cognitive/socio-emotional development abilities by adapting one's communication techniques to the child's level (McConaughy, 2013). In addition, practitioners need to consider how the language of children coupled with adult assumptions about the meaning of words can create the perception that children are less capable or thoughtful than they are. McCabe (1996) describes this situation well by talking through the potential meanings of "I don't want to…" type of comments. One interpretation is that the child is refusing treatment, whether that treatment plan be a shot (to inoculate against disease), bedtime (to inoculate against immune compromise due to exhaustion), or school (to inoculate against ignorance), while another interpretation is that the "want" language is expressing an emotional reaction to hearing about a course of action that the child will go along with, even if grudgingly.

Of special interest to the pediatric population service is how to communicate with, involve, and engage children and adolescents in decision-making processes about their own health and treatment. For instance, talking with them about medical or health issues not only demonstrates respect for them as persons but also allows them an opportunity to think about the future while listening and responding to their concerns further engage them in the decision-making process. The implication from the empirical literature pertains to how to involve children in the context of medical decision-making and allow them to bring their fears, desires, understanding of consequences of different courses of action, and other considerations to bear on decisions. Determining when a child is ready to make a medical or treatment decision for themselves is a complex business that requires assessing a child's ability to understand the consequences of different courses of action (Larcher & Hutchinson,

2010). Further, a child's involvement can occur along a continuum, depending on their *developmental age and experience* (Larcher et al., 2015). As children grow older and mature, they may seek more autonomy in the treatment decision-making process (Rapoff et al., 2023). Even making the final decision while explicitly taking into account the child's considerations represents a form of autonomy granting. Finally, inviting continued discussion about the situation they find themselves in will serve to allow them to practice taking control of how they think about what happens to their body (Larcher et al., 2015). Verbal and nonverbal communication techniques for talking to children of varying developmental ages, including how to better engage them in a treatment setting, are provided in Table 2.1.

Beyond a child's developmental age, a practitioner should also take into account a child's prior experiences. In the context of pediatric care, a patient who has little experience with the medical system may react differently to the realization that a medical appointment is coming up than a patient who has had numerous encounters as a result of a challenging chronic condition. Considering the perspective-taking concept, a medical provider may be able to help their patient who is struggling to regulate their fear of medical procedures by taking the perspective of what the child can process about the end goal of medical tests and interventions. Using analogies to situations they have experienced in the past can help them better grasp what is being considered. Further, visualizing the process of medical decision-making using a toy analogy may help calm a child's fears and/or lack of understanding of the logic behind medical decision-making (e.g., Have you ever heard of or played with dominos [if not, describe it or show a video]? Have you ever set up multiple domino tracks? Well, that's sort of how medical professionals think—we hear about symptom[s] [the end-most domino[s], and start knocking down different dominos [running different tests, testing different interventions] to figure out which track [the underlying disease process] has caused the symptoms.").

Age as an Identity Aspect

In theoretical models conceptualizing individuals' multidimensional identities for health service psychologists and counselors, chronological and biological ages are considered separate aspects of a person's identity. For example, in the ADDRESSING model (Hays, 2008)—standing for *A*ge and generation, *D*evelopmental or other *D*isability, *R*eligion and spirituality, *E*thnic and racial identity, *S*ocioeconomic status, *S*exual Orientation, *I*ndigenous heritage, *N*ational Origin, and *G*ender—the dimension of age is coupled with generational influences. Within this model, providers are asked to consider not only a person's chronological age but also cultural influences that have affected members of each generation (e.g., Generation Z is the first generation to have grown up with access to the Internet and digital technology from a young age) and generational roles that have shaped one's identity, such as being the youngest sibling out of five in a family (Hays, 2012).

Table 2.1 Communication techniques

Age	Indicated communication techniques
<6 months	1. As a behavioral health clinician, model appropriate ways of speaking to babies for the parents.
6 months to 3 years	1. Reassure the child by allowing them to stay close to the parent(s)/guardian(s). 2. Avoid constant, direct eye contact with the child younger than 2 years as this may be perceived as threatening. 3. Approach the child gently and gradually. Judge their acceptance by their body language (e.g., willingness to leave a parent's/guardian's side). 4. Use play (peekaboo, keys, flashlight, or toy) to capture the child's attention and ease their anxiety. 5. Prepare the child for any physical examination or other procedures by modeling/imitating with the parent/guardian or a doll/stuffed animal.
3 to 6 years	1. Use simple language (e.g., limit the length and complexity of questions, avoid embedded phrases or clauses). Remember, a child's receptive language is more advanced than their expressive language. 2. Encourage the child to ask questions (e.g., the use of extenders can encourage more child talk). 3. Allow them to participate in their examination or talk to them about their health care.
6 years to 12 years	1. Talk about familiar settings and activities. Children in this age group enjoy talking about family, friends, school, and other life events. Their interests and strengths are easily determined. 2. Avoid abstract questions, too many factual questions, and too much direct questioning. Rephrase or simplify questions when the child has misunderstood or not responded. 3. In your discussions, an explanation of procedures, assessments, and intervention/treatment plans becomes important and helpful in eliciting the preadolescent's cooperation.
>12 years	1. As a sign of respect for independence and individuality, interview the adolescent separately from the parent(s)/guardian(s) and allow time to cultivate the therapeutic alliance. 2. Be clear about limits and issues of confidentiality. 3. Elicit and address the adolescent's concerns. 4. Do not pressure the adolescent to talk. Be patient and respect the adolescent's privacy and limits. 5. Discuss sensitive topics (e.g., drugs, sex, depression, anxiety, and eating disorders) in a nonjudgmental manner after rapport has been developed. Consider using indirect questions and statements as well as asking first about the experience of the adolescent's friends. 6. Acknowledge that the adolescent is responsible for personal health care and advocate for the adolescent with parents/guardians.

Adapted from Liu and Stein (2018) and McConaughy (2013)

Taking into account that age is just one aspect of a person's individuality, it is important to adopt a more holistic approach to developing children and their parents/guardians. Utilizing a multidimensional lens in practice means being aware that a person's interlocking identities (e.g., age, race, ethnicity, and gender sexuality) overlap and interface with psychological needs, attitudes, values, and worldviews and are significantly impacted by how these and other dimensions intersect with each other (D'Andrea & Daniels, 2001). For children and families coming from minoritized backgrounds, it is also important to acknowledge the intersectionality in their identities, that is, the merging or intersection of multiple marginalized identities (e.g., belonging to a racial or ethnic minority, low SES, and member of the LGBTQI+ community) that reflects macro-level forms of biases and discrimination, such as racism, classicism, and heteronormativity.

For example, as aforementioned, for adolescents, personal identity development is a major developmental task (Crocetti et al., 2008; Marcia, 1993). Youth of color, specifically, experience another major developmental task during this period, that of ethnic and racial identity (ERI) development. ERI has been conceptualized as a multidimensional construct that encompasses the beliefs and attitudes that people have about their ethnic/racial group memberships, as well as the mechanisms enacted for these beliefs and attitudes to develop (Umaña-Taylor et al., 2014). ERI is an identity aspect that emerges in early childhood, and its formation is associated with adolescents' cognitive milestones, the onset of puberty, and social and environmental contexts, such as family, peers, and media (Rivas-Drake et al., 2014; Umaña-Taylor et al., 2014). It has been demonstrated that in minoritized adolescent samples, different ERI dimensions tend to be positively associated with psychosocial functioning both directly and indirectly in adolescence among ethnic minority youth, while for some ethnic and racial groups, more positive ERI attitudes are associated with decreased health risk outcomes, as group members are less likely to be contracting HIV or abusing substances (Rivas-Drake et al., 2014). So, practitioners working with adolescents from minoritized backgrounds need to understand that personal identity formation not only is of the utmost importance but also pertains to ERI aspects and is impacted by the health conditions these adolescents may experience.

Conclusion

Children's health and development are intertwined with each other and associated with their context, especially, their family system. In the context of pediatric populations, this chapter presented core developmental theories and concepts regarding childhood to adolescence. Further, it highlighted the applications and implications of these theories and concepts in daily practice for professionals, trainers, and trainees alongside recommendations on how to incorporate them into evidence-based behavioral health practices in pediatric specialty settings.

Select Bibliography

Guralnick, M. J. (2011). Why early intervention works: A systems perspective. *Infants & Young Children, 24*(1), 6–28. https://doi.org/10.1097/IYC.0b013e3182002cfe. Early intervention is crucial to the developmental trajectories of at-risk children and children with disabilities. Guralnick's conceptual and empirically grounded approach to service delivery via developmental and prevention science is presented in this article. Emphasis is placed on family-centered early intervention practices.

Larcher, V., & Hutchinson, A. (2010). How should paediatricians assess Gillick competence? *Archives of Disease in Childhood, 95*(4), 307–311. Developmentally appropriate autonomy granting is an important consideration for adults charged with the care of children and/or adolescents. This article presents some practical techniques and considerations surrounding the documentation of a minor's ability to make or contribute to medical treatment decisions.

Palermo, T. M., Janicke, D. M., McQuaid, E. L., Mullins, L. L., Robins, P. M., & Wu, Y. P. (2014). Recommendations for training in pediatric psychology: Defining core competencies across training levels. *Journal of Pediatric Psychology, 39*(9), 965–984. https://doi.org/10.1093/jpepsy/jsu015. This article presents the core competencies of pediatric psychologists. Its content can provide readers with a broad understanding of how developmental considerations fit within a competency-based training approach along with implications for practice.

Rapoff, M. A., Duncan, C., & Karlson, C. (2023). Developmental considerations in assessing and improving adherence. In *Adherence to Pediatric Medical Regimens* (pp. 81–114). Springer International Publishing. https://doi.org/10.1007/978-3-031-27484-8_4. Adherence to recommended treatments in pediatric settings is a key component associated with treatment effectiveness. Utilizing literature from several pediatric conditions (chronic and acute) and adopting a developmental approach, this chapter provides a foundation for conceptualizing and promoting treatment adherence among children and youth.

Sameroff, A. (Ed.). (2009). *The transactional model of development: How children and contexts shape each other*. American Psychological Association. https://doi.org/10.1037/11877-000. The transactional model has explicit applications to pediatric psychology and care. This book, written by renowned scientists in the field of child developmental psychology, offers an overview of the model and its history, presents empirical support and related challenges alongside the studies that have utilized it. The book has been recommended as a graduate textbook and a valuable book for developmental researchers.

References

Arnett, J. J. (2007). Emerging adulthood: What is it, and what is it good for? *Child Development Perspectives, 1*(2), 68–73. https://doi.org/10.1111/j.1750-8606.2007.00016.x

Årseth, A. K., Kroger, J., Martinussen, M., & Marcia, J. E. (2009). Meta-analytic studies of identity status and the relational issues of attachment and intimacy. *Identity, 9*(1), 1–32. https://doi.org/10.1080/15283480802579532

Ashcraft, L. E., Asato, M., Houtrow, A. J., Kavalieratos, D., Miller, E., & Ray, K. N. (2019). Parent empowerment in pediatric healthcare settings: A systematic review of observational studies. *The Patient-Patient-Centered Outcomes Research, 12*, 199–212. https://doi.org/10.1007/s40271-018-0336-2

Becht, A. I., Luyckx, K., Nelemans, S. A., Goossens, L., Branje, S. J. T., Vollebergh, W. A. M., & Meeus, W. H. J. (2019). Linking identity and depressive symptoms across adolescence:

A multisample longitudinal study testing within-person effects. *Developmental Psychology, 55*(8), 1733–1742. https://doi.org/10.1037/dev0000742

Becht, A. I., Nelemans, S. A., Branje, S. J. T., Vollebergh, W. A. M., & Meeus, W. H. J. (2021). Daily identity dynamics in adolescence shaping identity in emerging adulthood: An 11-year longitudinal study on continuity in development. *Journal of Youth and Adolescence, 50*(8), 1616–1633. https://doi.org/10.1007/s10964-020-01370-3

Boe, J. L., & Woods, R. J. (2018). Parents' influence on infants' gender-typed toy preferences. *Sex Roles, 79*, 358–373. https://doi.org/10.1007/s11199-017-0858-4

Branje, S., Mastrotheodoros, S., & Laursen, B. (2021). Family relationships during adolescence. In I. A. L. Vangelisti (Ed.), *The Routledge handbook of family communication* (3rd ed., pp. 247–261). Routledge. https://doi.org/10.4324/9781003043423-22

Bronfenbrenner, U. (1994). Ecological models of human development. *International Encyclopedia of Education, 3*(2), 37–43.

Bronfenbrenner, U., & Morris, P. A. (2006). The bioecological model of human development. In W. Damon (Series Ed.) & R. M. Lerner (Vol. Ed.), *Handbook of child psychology: Theoretical models of human development* (pp. 793–828). Wiley.

Brookman, R., Kalashnikova, M., Conti, J., Xu Rattanasone, N., Grant, K. A., Demuth, K., & Burnham, D. (2020). Depression and anxiety in the postnatal period: An examination of infants' home language environment, vocalizations, and expressive language abilities. *Child Development, 91*(6), e1211–e1230. https://doi.org/10.1111/cdev.13421

Cook, J. M., & Coyne, J. C. (2005). Re-envisioning the training and practice of clinical psychologists: Preserving science and research orientations in the face of change. *Journal of Clinical Psychology, 61*(9), 1191–1196. https://doi.org/10.1002/jclp.20163

Crocetti, E., Rubini, M., & Meeus, W. (2008). Capturing the dynamics of identity formation in various ethnic groups: Development and validation of a three-dimensional model. *Journal of Adolescence, 31*(2), 207–222. https://doi.org/10.1016/j.adolescence.2007.09.002

D'Andrea, M., & Daniels, J. (2001). Respectful counseling: An integrative multidimensional model for counselors. In D. B. Pope-Davis & L. K. Coleman (Eds.), *The intersection of race, class, and gender in multicultural counseling* (pp. 417–466). Sage. https://doi.org/10.4135/9781452231846.n17

Deater-Deckard, K., Lansford, J. E., Malone, P. S., Alampay, L. P., Sorbring, E., Bacchini, D., Bombi, A. S., Bornstein, M. H., Chang, L., Di Giunta, L., Dodge, K. A., Oburu, P., Pastorelli, C., Skinner, A. T., Tapanya, S., Tirado, L. M. U., Zelli, A., & Al-Hassan, S. M. (2011). The association between parental warmth and control in thirteen cultural groups. *Journal of Family Psychology, 25*(5), 790–794. https://doi.org/10.1037/a0025120

Deković, M., Noom, M. J., & Meeus, W. (1997). Expectations regarding development during adolescence: Parental and adolescent perceptions. *Journal of Youth and Adolescence, 26*(3), 253–272. https://doi.org/10.1007/s10964-005-0001-7

Denham, S. A., & Holt, R. W. (1993). Preschoolers' likability as cause or consequence of their social behavior. *Developmental Psychology, 29*(2), 271–275. https://doi.org/10.1037/0012-1649.29.2.271

Desai, N. (2019). The role of motivational interviewing in children and adolescents in pediatric care. *Pediatric Annals, 48*(9), e376–e379. https://doi.org/10.3928/19382359-20190816-01

DiMatteo, M. R., & Miller, T. A. (2013). Treatment adherence in adolescence. In W. T. O'Donohue, L. T. Benuto, & L. Woodward Tolle (Eds.), *Handbook of adolescent health psychology* (pp. 373–386). Springer. https://doi.org/10.1007/978-1-4614-6633-8_24

Donnelly, S., & Kidd, E. (2021). The longitudinal relationship between conversational turn-taking and vocabulary growth in early language development. *Child Development, 92*(2), 609–625. https://doi.org/10.1111/cdev.13511

Erikson, E. H. (1968). Identity and identity diffusion. *The Self in Social Interaction, 1*, 197–205.

Fiese, B. H., & Sameroff, A. J. (1989). Family context in pediatric psychology: A transactional perspective. *Journal of Pediatric Psychology, 14*(2), 293–314.

Fiese, B. H., Spagnola, M. A., & Everhart, R. (2008). Family context in developmental and behavioral pediatrics. In M. Wolraich, D. Drotar, P. H. Dworkin, & E. Perrin (Eds.), *Handbook of developmental and behavioral pediatrics* (pp. 79–108). Mosby. https://doi.org/10.1016/B978-0-323-04025-9.50008-8

Forrest, C. B., Simpson, L., & Clancy, C. (1997). Child health services research: challenges and opportunities. *JAMA, 277*(22), 1787–1793.

Forgeron, P. A., King, S., Stinson, J. N., McGrath, P. J., MacDonald, A. J., & Chambers, C. T. (2010). Social functioning and peer relationships in children and adolescents with chronic pain: A systematic review. *Pain Research and Management, 15*, 27–41.

Frosch, C. A., Schoppe-Sullivan, S. J., & O'Banion, D. D. (2021). Parenting and child development: A relational health perspective. *American Journal of Lifestyle Medicine, 15*(1), 45–59.

Goethals, E. R., Soenens, B., de Wit, M., Vansteenkiste, M., Laffel, L. M., Casteels, K., & Luyckx, K. (2019). "Let's talk about it" the role of parental communication in adolescents' motivation to adhere to treatment recommendations for Type 1 diabetes. *Pediatric Diabetes, 20*(7), 1025–1034. https://doi.org/10.1111/pedi.12901

Goodacre, E. J., Fink, E., Ramchandani, P., & Gibson, J. L. (2023). Building connections through play: Influences on children's connected talk with peers. *British Journal of Developmental Psychology, 41*(3), 203–226. https://doi.org/10.1111/bjdp.12443

Guralnick, M. J. (1997). Effectiveness of early intervention for vulnerable children: A developmental perspective. *American Journal on Intellectual and Developmental Disabilities, 102*(4), 319–345. https://doi.org/10.1352/0895-8017(1998)102<0319:EOEIFV>2.0.CO;2

Guralnick, M. J. (2020). Applying the developmental systems approach to inclusive community-based early intervention programs: Process and practice. *Infants & Young Children, 33*(3), 173–183. https://doi.org/10.1097/IYC.0000000000000167

Haine-Schlagel, R., & Walsh, N. E. (2015). A review of parent participation engagement in child and family mental health treatment. *Clinical Child and Family Psychology Review, 18*, 133–150. https://doi.org/10.1007/s10567-015-0182-x

Hays, P. A. (2008). *Addressing cultural complexities in practice: Assessment, diagnosis, and therapy*. American Psychological Association.

Hays, P. A. (2012). *Connecting across cultures: The helper's toolkit*. Sage Publications.

Hoagwood, K. (2005). Family-based services in children's mental health: A research review and synthesis. *Journal of Child Psychology and Psychiatry, 46*(7), 690–713. https://doi.org/10.1111/j.1469-7610.2005.01451.x

Jones, J. L., & Mehr, S. L. (2007). Foundations and assumptions of the scientist-practitioner model. *American Behavioral Scientist, 50*(6), 766–771. https://doi.org/10.1177/0002764206296454

Koenig, M. A., & Woodward, A. L. (2010). 24-month-olds' sensitivity to the prior inaccuracy of the source: Possible mechanisms. *Developmental Psychology, 46*(4), 815–826.

Larcher, V., & Hutchinson, A. (2010). How should paediatricians assess Gillick competence? *Archives of Disease in Childhood, 95*(4), 307–311.

Larcher, V., Craig, F., Bhogal, K., Wilkinson, D., & Brierley, J. (2015). Making decisions to limit treatment in life-limiting and life-threatening conditions in children: A framework for practice. *Archives of Disease in Childhood, 100*(Suppl 2), s1–s23. https://doi.org/10.1136/archdischild-2014-306666

Laursen, B., DeLay, D., & Adams, R. E. (2010). Trajectories of perceived support in mother–adolescent relationships: The poor (quality) get poorer. *Developmental Psychology, 46*(6), 1792–1798. https://doi.org/10.1037/a0020679

Lindsey, M. A., Brandt, N. E., Becker, K. D., Lee, B. R., Barth, R. P., Daleiden, E. L., & Chorpita, B. F. (2014). Identifying the common elements of treatment engagement interventions in children's mental health services. *Clinical Child and Family Psychology Review, 17*, 283–298. https://doi.org/10.1007/s10567-013-0163-x

Link, B. G., & Phelan, J. (1995). Social conditions as fundamental causes of disease. *Journal of Health and Social Behavior*, 80–94. https://doi.org/10.2307/2626958

Liu, Y. H., & Stein, M. T. (2018). Talking with children. In M. Augustyn & B. Zuckerman (Eds.), *Zuckerman Parker handbook of developmental and behavioral pediatrics for primary care* (4th ed.). Wolters Kluwer.

Luyckx, K., Seiffge-Krenke, I., Schwartz, S. J., Goossens, L., Weets, I., Hendrieckx, C., & Groven, C. (2008). Identity development, coping, and adjustment in emerging adults with a chronic illness: The sample case of Type 1 diabetes. *Journal of Adolescent Health, 43*(5), 451–458. https://doi.org/10.1016/j.jadohealth.2008.04.005

Luyckx, K., Schwartz, S. J., Goossens, L., Beyers, W., & Missotten, L. (2011). Processes of personal identity formation and evaluation. In *Handbook of Identity Theory and Research* (Vol. 1, pp. 77–98). https://doi.org/10.1007/978-1-4419-7988-9

Madan-Swain, A., Hankins, S. L., Gilliam, M. B., Ross, K., Reynolds, N., Milby, J., & Schwebel, D. C. (2012). Applying the cube model to pediatric psychology: Development of research competency skills at the doctoral level. *Journal of Pediatric Psychology, 37*(2), 136–148. https://doi.org/10.1093/jpepsy/jsr096

Mageau, G. A., Lessard, J., Carpentier, J., Robichaud, J. M., Joussemet, M., & Koestner, R. (2018). Effectiveness and acceptability beliefs regarding logical consequences and mild punishments. *Journal of Applied Developmental Psychology, 54*, 12–22.

Marcia, J. E. (1993). The ego identity status approach to ego identity. In *Ego identity* (pp. 3–21). Springer.

Mastrotheodoros, S., Van der Graaff, J., Deković, M., Meeus, W. H. J., & Branje, S. J. T. (2019). Coming closer in adolescence: Convergence in mother, father, and adolescent reports of parenting. *Journal of Research on Adolescence, 29*(4), 846–862. https://doi.org/10.1111/jora.12417

Mastrotheodoros, S., Van der Graaff, J., Deković, M., Meeus, W. H. J., & Branje, S. (2020). Parent–adolescent conflict across adolescence: Trajectories of informant discrepancies and associations with personality types. *Journal of Youth and Adolescence, 49*(1), 119–135. https://doi.org/10.1007/s10964-019-01054-7

McCabe, M. A. (1996). Involving children and adolescents in medical decision making: Developmental and clinical considerations. *Journal of Pediatric Psychology, 21*(4), 505–516.

McConaughy, S. H. (2013). *Clinical interviews for children and adolescents* (2nd ed.). The Guilford Press.

Mian, N. D. (2014). Little children with big worries: Addressing the needs of young, anxious children and the problem of parent engagement. *Clinical Child and Family Psychology Review, 17*, 85–96. https://doi.org/10.1007/s10567-013-0152-0

Moore, T. G., McDonald, M., Carlon, L., & O'Rourke, K. (2015). Early childhood development and the social determinants of health inequities. *Health Promotion International, 30*(suppl_2), ii102–ii115. https://doi.org/10.1093/heapro/dav031

Murphy, V. B., & Christner, R. W. (2012). A cognitive-behavioral case conceptualization approach for working with children and adolescents. In R. B. Mennuti, R. W. Christner, & A. Freeman (Eds.), *Cognitive-behavioral interventions in educational settings: A handbook for practice* (2nd ed., pp. 81–114). Routledge.

Ntoumanis, N., Ng, J. Y. Y., Prestwich, A., Quested, E., Hancox, J. E., Thøgersen-Ntoumani, C., Deci, E. L., Ryan, R. M., Lonsdale, C., & Williams, G. C. (2021). A meta-analysis of self-determination theory-informed intervention studies in the health domain: Effects on motivation, health behavior, physical, and psychological health. *Health Psychology Review, 15*(2), 214–244. https://doi.org/10.1080/17437199.2020.1718529

Palermo, T. M., Valrie, C. R., & Karlson, C. W. (2014a). Family and parent influences on pediatric chronic pain: A developmental perspective. *American Psychologist, 69*(2), 142–152.

Palermo, T. M., Janicke, D. M., McQuaid, E. L., Mullins, L. L., Robins, P. M., & Wu, Y. P. (2014b). Recommendations for training in pediatric psychology: Defining core competencies across training levels. *Journal of Pediatric Psychology, 39*(9), 965–984.

Peterson, N. L. (1987). *Early intervention for handicapped and at-risk children*. Love Publishing Company.

Pierce, M., Hope, H. F., Kolade, A., Gellatly, J., Osam, C. S., Perchard, R., Kosidou, K., Dalman, C., Morgan, V., Di Prinzio, P., & Abel, K. M. (2020). Effects of parental mental illness on children's physical health: Systematic review and meta-analysis. *The British Journal of Psychiatry, 217*(1), 354–363. https://doi.org/10.1192/bjp.2019.216

Rapoff, M. A., Duncan, C., & Karlson, C. (2023). Developmental considerations in assessing and improving adherence. In *Adherence to pediatric medical regimens* (pp. 81–114). Springer International Publishing. https://doi.org/10.1007/978-3-031-27484-8_4

Rivas-Drake, D., Seaton, E. K., Markstrom, C., Quintana, S., Syed, M., Lee, R. M., Schwartz, S. J., Umaña-Taylor, A. J., French, S., Yip, T., & Ethnic and Racial Identity in the 21st Century Study Group. (2014). Ethnic and racial identity in adolescence: Implications for psychosocial, academic, and health outcomes. *Child Development, 85*(1), 40–57. https://doi.org/10.1111/cdev.12200

Ryan, R. M., & Deci, E. L. (2017). *Self-determination theory: Basic psychological needs in motivation, development, and wellness*. Guilford Press.

Sameroff, A. (2009). The transactional model. In A. Sameroff (Ed.), *The transactional model of development: How children and contexts shape each other* (pp. 3–21). American Psychological Association. https://doi.org/10.1037/11877-001

Sameroff, A. (2010). A unified theory of development: A dialectic integration of nature and nurture. *Child Development, 81*(1), 6–22. https://doi.org/10.1111/j.1467-8624.2009.01378.x

Sameroff, A. J., & Mackenzie, M. J. (2003). A quarter-century of the transactional model: How have things changed? *Zero to Three, 24*(1), 14–22.

Schwartz, S. J., Luyckx, K., & Vignoles, V. (2011). *Handbook of identity theory and research*. Springer. https://doi.org/10.1007/978-1-4419-7988-9

Shahidullah, J. D., Hostutler, C. A., Coker, T. R., Allmon Dixson, A., Okoroji, C., & Mautone, J. A. (2023). Child health equity and primary care. *American Psychologist, 78*(2), 93–106. https://doi.org/10.1037/amp0001064

Slomkowski, C., & Dunn, J. (1996). Young children's understanding of other people's beliefs and feelings and their connected communication with friends. *Developmental Psychology, 32*(3), 442–447. https://doi.org/10.1037/0012-1649.32.3.442

Slopen, N., Chang, A. R., Johnson, T. J., Anderson, A. T., Bate, A. M., Clark, S., Cohen, A., Jindal, M., Karbeah, J., Pachter, L. M., Priest, N., Suglia, S. F., Bryce, N., Fawcett, A., & Heard-Garris, N. (2024). Racial and ethnic inequities in the quality of paediatric care in the USA: A review of quantitative evidence. *The Lancet Child & Adolescent Health, 8*(2), 147–158. https://doi.org/10.1016/S2352-4642(23)00251-1

Smetana, J. G., Yoo, H. N., Nguyen, N., & Ball, C. L. (2021). Moms know best?: Children's evaluations of mothers' unfair responses to peer conflicts. *Journal of Experimental Child Psychology, 201*, 104993. https://doi.org/10.1016/j.jecp.2020.104993

Soenens, B., Vansteenkiste, M., Van Petegem, S., Beyers, W., & Ryan, R. (2017). How to solve the conundrum of adolescent autonomy? On the importance of distinguishing between independence and volitional functioning. In *Autonomy in adolescent development* (pp. 1–32). Psychology Press. https://www.taylorfrancis.com/chapters/edit/10.4324/9781315636511-1/solve-conundrum-adolescent-autonomy-bart-soenens-maarten-vansteenkiste-stijn-van-petegem-wim-beyers-richard-ryan

Spencer, N., Raman, S., O'Hare, B., & Tamburlini, G. (2019). Addressing inequities in child health and development: Towards social justice. *BMJ Paediatrics Open, 3*(1), e000503. https://doi.org/10.1136/bmjpo-2019-000503

Stille, C., Turchi, R. M., Antonelli, R., Cabana, M. D., Cheng, T. L., Laraque, D., & Perrin, J. (2010). The familycentered medical home: specific considerations for child health research and policy. *Academic Pediatrics, 10*(4), 211–217. https://doi.org/10.1016/j.acap.2010.05.002

Suárez-Orozco, C., Motti-Stefanidi, F., Marks, A., & Katsiaficas, D. (2018). An integrative risk and resilience model for understanding the adaptation of immigrant-origin children and youth. *American Psychologist, 73*(6), 781–796. https://doi.org/10.1037/amp0000265

Tisak, M. S. (1986). Children's conceptions of parental authority. *Child Development, 57*, 166–176.

Trent, M., Dooley, D. G., Dougé, J., & AAP Section on Adolescent Health, AAP Council on Community Pediatrics, AAP Committee on Adolescence. (2019). The impact of racism on child and adolescent health. *Pediatrics, 144*(2), e20191765. https://doi.org/10.1542/peds.2019-1765

Tudge, J. R., Mokrova, I., Hatfield, B. E., & Karnik, R. B. (2009). Uses and misuses of Bronfenbrenner's bioecological theory of human development. *Journal of Family Theory & Review, 1*(4), 198–210. https://doi.org/10.1111/j.1756-2589.2009.00026.x

Tudge, J. R., Merçon-Vargas, E. A., & Payir, A. (2022). Urie Bronfenbrenner's bioecological theory: Its development, core concepts, and critical issues. In K. Adamsons, A. L. Few-Demo, C. Proulx, & K. Roy (Eds.), *Sourcebook of family theories and methodologies: A dynamic approach* (pp. 235–254). Springer International Publishing.

Umaña-Taylor, A. J., Quintana, S. M., Lee, R. M., Cross, W. E., Jr., Rivas-Drake, D., Schwartz, S. J., Syed, M., Yip, T., Seaton, E., & Ethnic and Racial Identity in the 21st Century Study Group. (2014). Ethnic and racial identity during adolescence and into young adulthood: An integrated conceptualization. *Child Development, 85*(1), 21–39. https://doi.org/10.1111/cdev.12196

Van Ijzendoorn, M. H., Schuengel, C., & Bakermans–Kranenburg, M. J. (1999). Disorganized attachment in early childhood: Meta-analysis of precursors, concomitants, and sequelae. *Development and Psychopathology, 11*(2), 225–250. https://doi.org/10.1017/S0954579499002035

Van Petegem, S., Vansteenkiste, M., & Beyers, W. (2013). The jingle–jangle fallacy in adolescent autonomy in the family: In search of an underlying structure. *Journal of Youth and Adolescence, 42*(7), 994–1014. https://doi.org/10.1007/s10964-012-9847-7

World Health Organization. (n.d.). *Social determinants of health*. Available at https://www.who.int/health-topics/social-determinants-of-health#tab=tab_1

Zaadnoordijk, L., Meyer, M., Zaharieva, M., Kemalasari, F., van Pelt, S., & Hunnius, S. (2020). From movement to action: An EEG study into the emerging sense of agency in early infancy. *Developmental Cognitive Neuroscience, 42*, 100760.

Zubizarreta, A., Calvete, E., & Hankin, B. L. (2019). Punitive parenting style and psychological. problems in childhood: The moderating role of warmth and temperament. *Journal of Child and Family Studies, 28*, 233–244. https://doi.org/10.1007/s10826-018-1258-2

Chapter 3
Evidence-Based Practice via Telehealth in Pediatric Psychology

Teryn Bruni, Elizabeth Koval, and Luke Turnier

Overview

The global pandemic expanded the use of telehealth modalities in medicine across the United States and the world (Curfman et al., 2021b). Practices and policies evolved rapidly out of necessity across healthcare sectors, including the delivery of mental health care. Pediatric settings where mental health screening, assessment, and interventions are integrated into medical settings were especially adaptable to telehealth (Cahan et al., 2020). Despite today's ubiquitous presence of telehealth within the delivery of pediatric psychology services, the pre-pandemic history of telehealth is important to understand. This chapter provides an overview of telehealth within the field of pediatric psychology, outlining how it has evolved through the global pandemic and how it continues to respond to the needs of pediatric patients and their families. We review the empirical support for telehealth, current best practices for pediatric psychologists, considerations for supervision and training, and the future role of Artificial Intelligence (AI) in telehealth care.

T. Bruni (✉)
Department of Psychology, Algoma University, Sault Ste. Marie, ON, Canada
e-mail: teryn.bruni@algomau.ca

E. Koval · L. Turnier
Department of Pediatrics, Pediatric Psychology, Michigan Medicine, Ann Arbor, MI, USA

A. Maragakis, M. Janikian (eds.), *Evidence-Based Behavioral Health Practices in Pediatric Specialty Settings*, https://doi.org/10.1007/978-3-031-71004-9_3

Empirical Support for Telehealth

The American Telemedicine Association (ATA) broadly defines telemental health as "the practice of mental health specialties at a distance" (Grady et al., 2011, p. 134). Telehealth in the context of mental health services within a pediatric setting includes the assessment and treatment of patients in a physical location that is different from the provider. Patients and parents may be located in a hospital, clinic, school, or, most commonly, their homes, while completing an appointment with a provider who may be conducting care from another clinical setting, office, or home office (Grady et al., 2011). The modality in which telehealth is conducted varies from asynchronous remote monitoring tools, patient portal to provider communication, or parents recording child behavior and sending it to the provider for assessment, to synchronous audiovisual or audio-only (e.g., telephone) platforms. In this chapter, we primarily discuss telehealth in the context of synchronous audio-visual appointments or "video visits." The increased use of teleconference tools to facilitate telehealth sessions has provided patients and providers a more interactive experience that is commensurate with traditional face-to-face interactions and is now common practice in pediatric settings (Chakawa et al., 2021; Ros-DeMarize et al., 2021).

Historically, telehealth has been used to support patient health care and provide education through consultation with specialty providers (Lustig, 2012). Telehealth is considered an important strategy for improving access to pediatric specialty care, including mental health services (Curfman et al., 2021b). Research in this area is continuously and rapidly evolving with advances in technology and changes in healthcare policy. Pre-pandemic, telehealth contributed to increased access to quality care and provided a cost-effective alternative to face-to-face modalities (Bethell et al., 2011; Grady et al., 2011). Telehealth is highly acceptable to patients and provides a convenient and less costly alternative to face-to-face visits (Curfman et al., 2021b). Families who may not have the available resources for traditional mental health visits (e.g., transportation, financial means, and/or time) may particularly benefit from telehealth services. Telehealth is associated with numerous positive health outcomes including fewer hospitalizations (Dayal et al., 2019) and reduced disparities in health care (Bethell et al., 2011). Post-pandemic onset, acceptability and normalization of telehealth among pediatric providers grew substantially, with many providers preferring telehealth or video appointments over in-person visits (Romani et al., 2022).

Telehealth in pediatric settings has been shaped by rapid advances in technology, policy shifts, and the ubiquitous use of technology in the daily lives of most Americans (El-Miedany, 2017). This ever-changing landscape was most profoundly impacted, of course, by the global COVID-19 pandemic in 2020, when the mental health outcomes of children and families reached crisis levels nationwide (Chakawa et al., 2021). Due to the need for social distancing, service providers were forced to pivot to virtual services "almost overnight" (Ros-DeMarize et al., 2021, pp. 2). Medical centers across the country were leaders in the use of telehealth, and many pediatric psychologists were already incorporating virtual care into their daily practice (Aboujaoude & Salame, 2016). This resulted in the agile adoption of telehealth technology, rapid implementation, and swift shifts in policies governing operations and payment (Curfman et al., 2021b). Pediatric psychology services were largely transitioned to virtual formats, not only increasing access to mental health screening, assessment, and intervention at a time when in-person care was not available but also expanding the reach of services to patients who may not have otherwise been able to access in-person care, pre-pandemic (Curfman et al., 2021b). Retrospective analysis of telehealth implementation during the pandemic has provided useful insights on the effectiveness of delivering behavioral health services via telehealth.

Screening and Assessment

Pediatric psychologists play a pivotal role in mental health screening, assessment, and referral within medical settings (Stancin & Perrin, 2014). Recent research has highlighted how telehealth has been leveraged to expand the reach of screening programs, improve referral processes, and conduct valid assessments across commonly diagnosed externalizing and internalizing disorders (Coker et al., 2019; Ros-Demarize et al., 2021). Pre-pandemic, the remote administration and completion of screening measures and rating scales was common practice among many pediatric psychologists (Coker et al., 2019; Komanchuk et al., 2023). The availability of online forms allows for more efficient administration and scoring, makes patient data available to providers prior to appointments, and facilitates referral decisions. Thus, studies were already emerging on how remote digital screening could expand the reach of screening programs (Coker et al., 2019). The use of digital screeners and rating forms has become an acceptable and promising approach to mental health screening for pediatric patients (Talbott et al., 2020), particularly in medical settings where data could be integrated into a patient's electronic health record.

More comprehensive assessments may include remote delivery of rating scales as a component, but many diagnostic decisions require direct assessment aimed to elicit behavior using standardized instruments. Standardized assessment tools have clearly outlined administration requirements and are normed for face-to-face administration (American Psychological Association, 2013). Thus, pre-pandemic, direct assessment using virtual delivery modalities was largely not considered best practice (Dale et al., 2022). Despite the lack of existing guidelines for virtual assessment in practice settings, some pre-pandemic studies were beginning to demonstrate that not only was diagnostic rigor and accuracy comparable to face-to-face assessments but also providers and patients were satisfied with virtual assessment formats (Elford et al., 2000; Myers et al., 2004). These findings are notable due to the inferiority of the available technology coupled with the lack of public exposure and comfort with technology in the early 2000s when these studies were conducted.

Social distancing during the pandemic resulted in a movement toward modifying assessment delivery modalities to ensure patients continued to have access to assessment services. Professional practice standards were rapidly developed and disseminated (Dale et al., 2022; Farmer et al., 2020; Wright et al., 2020). Wright and Raiford (2020) provide a useful practical guide to teleassessment that includes discussions about the administration of a broad range of psychological tests and measures, ethical concerns, and interpretation considerations. In line with this shift, test developers have also begun to offer administration guidance on remote administration. For example, Pearson Assessments (2021) published a telepractice guide for their Wechsler cognitive assessment series, commonly used pediatric assessment tools, which highlighted best practice for digital assessments and outlined their more recently developed digital tools for administration (i.e., Q-global).

Considerations for conducting assessments virtually include the ability to obtain valid behavioral observation data, the need to exercise caution in the interpretation of virtual assessments when deviating from standardization protocols, and the diagnostic complexity of the case (Berger et al., 2022; Pritchard et al., 2020). Developmental assessment and direct assessments for autism spectrum disorder (ASD), for example, involve the evaluation of the presence or absence of highly specific behavioral presentation outlined in diagnostic criteria. Therefore, having enough direct behavioral observation data to make a determination of symptom criteria is essential. Berger et al. (2022) discussed the importance of obtaining synchronous behavioral observations in the context of remote evaluations when determining if a child meets the criteria for ASD.

Additionally, Pritchard et al. (2020) recommend a tiered approach to diagnostic assessment based on case complexity. They discuss their approach in the context of evaluations determining a diagnosis of Attention Deficit-Hyperactivity Disorder (ADHD). "Tier 1" may include third-party rating scales, clinical interview, and a brief virtual behavior observation, and this may be sufficient data to make a determination of whether a child meets the criteria for ADHD. However, for children with comorbid mental health presentations or existing diagnoses, more intensive virtual contact with the child may be necessary to obtain the relevant behavioral data. As the complexity of the case obscures the diagnostic clarity, face-to-face testing may be needed (i.e., "Tier 3").

Intervention

Pediatric psychologists deliver a wide range of interventions, from behavior management training to more traditional psychotherapy (Stancin & Perrin, 2014). They also work on medical teams where they use their intervention skills to help patients adjust to their illness, adhere to medical regimens, or treat common comorbid mental health disorders (e.g., anxiety and depression). Psychological interventions can be highly conducive to remote delivery, especially in the case of parent-mediated interventions and cognitive behavioral therapy. Studies have shown the success of telehealth intervention in pediatric psychology settings, pre-pandemic, including a wide range of modalities, including virtual one-on-one therapy (e.g., cognitive-behavior therapy [CBT]), mobile interventions, and parent training (Aboujaoude & Salame, 2016).

Pre-pandemic reviews on the use of technology in the delivery of pediatric mental health showed promise, yet pointed to a need for more rigorous study designs (Aboujaoude & Salame, 2016). A common question among researchers and practitioners was whether rapport could be developed and maintained if a pediatric psychologist was not face-to-face with their patients/families (Nelson & Sharp, 2016). As technology has advanced, more researchers have evaluated patient outcomes and patient/provider satisfaction in the setting of telehealth. A pre-pandemic study by Tse et al. (2015) found that treatment for ADHD could be delivered using video conferencing technology with comparable outcomes to face-to-face delivery; however, they found caregiver distress was reduced only in the face-to-face condition. It is possible that this difference was attributable to more variable levels of comfort with technology among children and their caregivers, differences in rapport building for caregivers in a virtual vs face-to-face format, or challenges with caregiver engagement during virtual sessions. Despite these findings, many studies conducted both pre- and post-pandemic onset have reported high rates of satisfaction with telehealth service delivery among patients, caregivers, and providers (Bagner et al., 2023; Comer et al., 2017; Dadds et al., 2019; Frye et al., 2022; Nelson & Sharp, 2016). Rapport building continues to be an important concern among pediatric psychologists (Frye et al., 2022), and thus studies have shifted to examine how to best implement telehealth with pediatric populations (Martinez et al., 2022).

As with face-to-face practice, pediatric psychologists must remain within their scope and competencies in the context of telehealth services (American Psychological Association, 2013; Myers et al., 2017). According to *the American Telehealth Association Operating Procedures for Pediatric Telehealth*, a provider should only provide telehealth services to patients in areas where they have received training and/or experience providing in-person or in-previous telehealth work (McSwain et al., 2017). If an organization or patient asks a provider to deliver telehealth services outside her or his scope of practice, the provider should seek out further training or refer the patient to another provider who is competent with the requested service. It is also important to ensure that telehealth providers have professional liability insurance that covers telehealth services (Myers et al., 2017).

In the delivery of psychological interventions, it is important to consider that telehealth may not be the most appropriate option for all pediatric patients. According to Islam et al. (2023), many factors should be considered when weighing the possibility of utilizing telehealth services including symptom presentation, client perception and motivation, logistical barriers, and family comfort. When considering symptom presentation, it is important to determine if symptoms of the disorder would be better treated in-person (American Psychological Association, 2013). For example, if a patient presents with social anxiety, would it be helpful to treat the patient in-person and provide them opportunities to present with exposure to social situations during sessions. This may be more difficult to facilitate in a virtual environment. Patients presenting with high levels of risk (e.g., patients with eating disorders, patients with active suicidal ideation) may require intensive in-person treatment or hospitalization services to ensure the safety of the patient.

Logistical barriers such as lack of transportation, availability, and/or obligations to care for other siblings can limit access to face-to-face services (Curfman et al., 2021a). Thus, some patients or caregivers may perceive telehealth services as more feasible. Additionally, patient perceptions and motivations may play into the choice to engage in one service modality over another. A patient may believe face-to-face services would be more effective and may have less motivation to engage with virtual services (Bagner et al., 2023). Some younger patients may have trouble sitting in front of the camera due to other reinforcing objects in their home environment, so virtual care may be difficult. However, for some children with high levels of aggressive behaviors, virtual care may be preferred to keep clinicians safe and maintain the focus of the parent when explaining behavior management strategies. If family members are sick or immunocompromised, they may prefer virtual services to keep themselves and others safe. Conversely, patients may face home environments where they do not feel they can talk privately or without distraction and thus prefer face-to-face visits (McSwain et al., 2017).

Building Rapport

Rapport and the strength of the therapeutic relationship is a critical factor in the effectiveness of therapeutic interventions (Wampold, 2015). Thus, rapport should continue to be prioritized in the context of telehealth services (Nelson & Sharp, 2016). Telehealth providers may feel that building rapport via telehealth is more difficult and requires more effort than in the context of face-to-face services. Providers should be aware that some pediatric patients may feel uncomfortable on a screen, especially if they have not utilized telehealth services before (Hilty et al., 2019). Some pediatric patients may naturally seek to build rapport with telehealth providers by running up to the camera with a favorite toy, pet, or drawing and talking about it. When appropriately placed during the session, these encounters can be an effective way to build relationships with patients. Providers may also choose to "tour" the patient around their office by tilting the camera, so the patient can see the entire office (McSwain et al., 2017). These short interactions can help providers connect with their patients and learn more about their strengths and interests.

How to Build a Telehealth Practice in Pediatric Settings

Telehealth provides important opportunities for pediatric psychologists to expand access to critical behavioral health services to children and their families. Implementation strategies have been outlined to help pediatric psychologists and institutional leaders employ telehealth services in medical settings (Curfman et al., 2021a). The development of a new telehealth practice relies on institutional and administrative support, evaluation of existing in-person services, assessment of patient needs and potential barriers, and a survey of resources and technology infrastructure.

Needs Assessment and Evaluation

A model of virtual care should meet the needs of patients and practice standards of providers and fit into the clinical workflow of the medical setting. It is important to identify the team members who will likely be involved in virtual care implementation, including medical assistants, information technology staff, clerical staff, and administrators (Curfman et al., 2021a; Myers et al., 2017). Before a pediatric psychology service adopts telehealth, a needs assessment should be conducted in order to discern "site readiness" and "scalability" (Myers et al., 2017, p. 788). The needs assessment should involve a range of institutional and community partners, including medical teams, patients, families, and administration. Including financial partners (e.g., insurance agencies) is critical to establishing expected service coverages and reimbursements for virtual services provided. An organization must consider how behavioral healthcare policies and practices fit within a virtual delivery model, especially when dealing with high-risk situations, such as reports of suicidal ideation or suspected maltreatment of a patient. A needs assessment should strive to identify the target age groups, referral conditions, and potential interventions (e.g., therapy and/or assessment) that will be targeted or utilized in a telehealth format.

Curfman et al. (2021b) also recommend promoting the use of telehealth to relevant stakeholders to communicate clinical feasibility, utility, and assure patients and their families that the services provided will be as effective as in-person services. They emphasize training as an essential step in facilitating comfort with technology, new workflows, and anticipating risk factors that can be associated with telehealth. Implementation of telehealth services should be considered an iterative process. Some aspects of virtual care may be immediately successful, but others may require ongoing evaluation to ensure patient and provider acceptability. Evaluations that include both quantitative and qualitative data allow for ongoing monitoring and adaptation of telehealth services (Curfman et al., 2021b).

Practical Considerations for Providers

Therapeutic Environment

The remote environment of the pediatric psychologist includes the physical space/location, technology used in the context of service delivery, and behaviors on and off camera. The clinical environment should be welcoming, private, and ensure patient confidentiality (McSwain et al., 2017). The provider should be mindful of others in the vicinity of their location hearing or seeing patients during visits. Providers should carefully monitor information displayed on their computer screens when working in a remote environment and also when engaging in screen sharing with patients or other providers (McSwain et al., 2017). The American Academy of Child and Adolescent Psychiatry (AACAP, 2017) also recommends having a sound machine outside of the room to decrease outside noise from coming into the room and ensure maximum patient privacy.

In addition to privacy, the provider should also consider the general ambiance and inviting nature of the clinic space. For example, it is important that the virtual care space has adequate lighting (preferably natural light) and that the provider's face is centered and in camera view for the patient (AACAP, 2017; Myers et al., 2017). The visible treatment space should provide minimal distraction and reflect the professional decor of face-to-face treatment spaces (McSwain et al., 2017). The clinician should keep in mind that virtual backgrounds can be distracting and impact the quality of the interaction. The image of the virtual background should be carefully considered, and if possible virtual backgrounds should be avoided. The clinician should understand the resources and technology available within their care delivery platforms to optimize the quality of care. The use of screen sharing, whiteboards, and chat features can add important interactive elements to therapy sessions. It is also important to think about how nonverbal communication (e.g., eye contact) can be perceived differently in telehealth interactions. For example, for providers to increase rapport, practitioners should "alternate the gaze between the monitor and the camera during the session" (AACAP, 2017, p. 885).

Some telehealth providers may choose to use third-party applications on mobile or computer devices to facilitate treatment. These applications may be an integral part of treatment or something recommended casually during a session. It is important for pediatric psychologists to research the applications they are recommending. Many applications, even those specifically for mental health, are not well regulated and contain security and privacy concerns (Psihogios et al., 2021). Dehling et al. (2015) assessed 24,405 health-related applications, and 95.63% of them posed at least one security or privacy risk to the user. It is especially important for clinicians to screen applications prior to use with pediatric patients as these younger patients may not know what to look for when assessing security or know how to determine if privacy has been breached. Many applications are also written at a higher reading level than many pediatric patients are able to read (Das et al., 2018). Finally, many mental health applications are not well tested on pediatric populations and should be used with caution (Psihogios et al., 2021).

Patient-Centered Care

Strategies to promote patient-centered care should be carefully considered in the setting of telehealth. To accommodate the unique needs of families and provide culturally responsive care, pediatric psychologists must think about how access to telehealth may contribute to mental health disparities among pediatric patients and families (Myers et al., 2017). It is important to identify the patient/caregiver's comfort with technology, need for technology support, and access to reliable Internet. Data has shown that poor access to technology disproportionately impacts underrepresented and marginalized patient groups, patients from low SES households, and those who live in rural regions (Digital Nation Data Explorer, 2022). Strategies to promote equity in telehealth access include offering both face-to-face and virtual options, if possible, to allow patients to choose the modality that best suits their needs. Providers should have access to patient-centered technology support that can be accessed efficiently to avoid appointment failures and develop workflows and processes for technology failures. Using platforms that allow for multiple family members to join visits and creating workflows for interpreters to be used in telehealth sessions will also help promote equitable access to telehealth services.

Provider Wellness

A more recently studied phenomenon is provider burnout in the setting of telehealth (Hilty et al., 2022). Burnout can be more likely due to the potential for increased patient workload, amount of time on screens, workplace isolation, and the need to be trained on new technology and workflows. Ensuring providers are connected and feel rewarded by their telehealth practice is important for reducing the risk of burnout. Managing workload to include breaks and time for engagement with colleagues will help reduce the risk of isolation that can be felt as a result of remote work settings (Hudgins et al., 2023). Having adequate technology, support staff, and efficient documentation workflows also contributes to the experience of delivering telehealth services as well as the health and well-being of providers. Other technical considerations include having multiple screens to allow providers to manage documentation efficiently while engaging in direct patient care. Providers should be mindful of physical symptoms of fatigue when engaging in on-screen activities for lengthy durations. Headaches and vision problems can result if breaks are not scheduled appropriately (Hudgins et al., 2023).

Practical Considerations for the Patient

Patient Expectations

Aligning with patients on their expectations of telehealth appointments prior to the first visit can ensure effective service delivery and avoid difficult conversations that may arise in telehealth contexts (Martinez et al., 2022; Myers et al., 2017; Nelson et al., 2017). Patients should understand what can and cannot be achieved within a virtual appointment. Expectations regarding the physical location of the patient and family members should also be discussed (Nelson et al., 2017). Families may log onto a visit from many different places including their home, school, car, or when they are out in the community. Discussing appropriate space and location for visits is important to ensure that visits can be conducted safely, privately, and without distraction. Providers should always confirm the exact address/location of patients so that they are able to respond in emergency situations if needed and ensure the patient is located in the jurisdiction in which the provider is licensed to practice (Martinez et al., 2022).

It is also important for pediatric psychologists to assess if the patient feels safe in the environment in which they log onto the virtual visit (Myers et al., 2017). Providers must be aware that there may be certain cases where pediatric patients are not being truthful due to fear that individuals in the home could overhear them (Martinez et al., 2022). Using the chat feature and allowing patients to answer questions discretely may help identify if the setting is not private or if the patient feels uncomfortable (Martinez et al., 2022). Parents with young children may prefer to complete the visit in a child's playroom to keep them occupied during the duration of the visit. If a parent of a young child tries to complete the visit in a space that is not comfortable for the child, the provider may need to suggest that the parent move the virtual visit to a different location (Myers et al., 2017). Finally, it is important that providers make sure a capable adult/caregiver is on-site during a visit in order to assist with technical issues, assess potential harm, or intervene if necessary (Myers et al., 2017).

Alternative Locations

If the patient is joining the virtual visit from school, it is important that the provider communicates to the school the requirements for the virtual space (Martinez et al., 2022; Nelson et al., 2017). The patient must be in a private area and have access to staff who can help set up the technology (Myers et al., 2017). Providers should obtain the address of the school in case of an emergency and have a release of information signed by the parent to be able to communicate with school personnel. In the case of completing visits from a vehicle, providers should ensure all motor vehicles are stopped and not in motion during video visits. Parents or adolescent patients may not be aware of this policy, so it may be necessary to remind them immediately during a visit when a provider notices they are in a car. Visits in the car may be difficult for young children who may be strapped in car seats for long periods of time, so a provider may want to inquire if there is a more comfortable place for the child during the visit (Myers et al., 2017). If this is not possible, it may be necessary to suggest the next visit be completed in the home or at a location that is more comfortable for the child.

Occasionally, parents will join visits while in line at the grocery store or with coworkers immediately surrounding them in their place of work. In such cases, the provider can remind the caregiver of the importance of maintaining the confidentiality of their child's health information (Martinez et al., 2022). Additionally, it can be difficult to complete a session when the parent and child are distracted by their environment. A provider may acknowledge that the parent looks busy and inquire if there would be a better time to complete the visit. At the provider's discretion, there may be times when it is appropriate to insist that the visit be rescheduled to another time when the patient and caregiver can more easily attend to the visit and maintain privacy. It can be embarrassing for a child to hear their parents discuss their mental health conditions in public, and it is for this reason that providers should work to establish the expectation that parents be in a private space for the duration of the appointment.

Internet Bandwidth

While there is no evidence that a particular technology is superior to others in the provision of telehealth services, both the telehealth provider and patient must have sufficient bandwidth to support telehealth services (AACAP, 2017). Without adequate Internet speed, it can be difficult to observe the patient and hear their responses. If a patient's wireless Internet is not sufficient, providers could prompt the family to move the phone or computer closer to the Internet router to get a better signal. Occasionally, even with this adjustment, it is still difficult to see or hear the patient. If that is the case, it is necessary to document if technology hindered the provider's ability to provide services (Myers et al., 2017). When seeing a new patient, a telehealth provider should create a contingency plan of what to do if technology fails during a visit (McSwain et al., 2017). For example, the provider and patient could verify appropriate telephone contact information at the beginning of a visit so that the provider can contact the patient via telephone should the video conference fail. A telehealth program may also have a technological support phone number which should be provided to families at the beginning of services. If difficulties with technology are a consistent issue during appointments, it may be necessary to recommend a transition to face-to-face appointments, if possible (Myers et al., 2017).

Ethical and Legal Concerns

It is important to note that all legal and ethical guidelines still apply to all clinicians providing virtual care. *The American Telemedicine Association Guidelines for Telemental Health with Children and Adolescents* states that each clinician is responsible for providing ethical and legally appropriate care while practicing telehealth (Myers et al., 2017; Nelson et al., 2017). Some of these legal and ethical requirements require more forethought when providing care virtually. As the research and practice of telehealth services are continuing to develop, it is important for pediatric psychologists to stay current on how to provide ethically and legally responsive care to all patients (American Psychological Association, 2013; Nelson & Sharp, 2016).

Policies and training are important regarding what to do when safety concerns occur during telehealth appointments (Martinez et al., 2022; Myers et al., 2017).

Some telehealth programs may use echeck-in questions to acquire the address of where the patient is located during the video visits. A triage plan needs to be in place for what to do if telehealth is no longer the most appropriate level of care (McSwain et al., 2017). When emergency situations occur, "the telehealth provider should, if safe and feasible, stay online with the patient until transfer of care is given to the team assuming care" (McSwain et al., 2017, p. 704). Pediatric Psychologists must abide by all federal and state laws regarding child privacy including Children's Online Privacy Protection Act (COPPA), Health Insurance Portability and Accountability Act (HIPAA), Health Information Technology for Economic and Clinical Health (HITECH), and Family Educational Rights and Privacy Acts (FERPA) (McSwain et al., 2017; Nelson et al., 2017).

Occasionally, recording a video visit may be requested by a parent or guardian. Providers should be aware of their institution's policies surrounding recording telehealth visits as well as state and federal laws (McSwain et al., 2017). For example, the recording of minors requires written consent from the parent. It is important to clarify the purpose and intended use of the recording. Parents and legal guardians have the right to have a copy of recordings when requested. Additionally, if images are sent during telehealth services, they must be encrypted (McSwain et al., 2017).

Informed consent for telehealth services must be tailored to the specific services provided and in accordance with jurisdictional regulations (Myers et al., 2017). Parents and legal guardians should be clear during the informed consent process about who is required to be at telehealth appointment, especially if your institutional policy requires a parent or legal adult to be in the home during telehealth services in the case of an emergency (American Psychological Association, 2013; Martinez et al., 2022). All languages must be easily understandable by parents and pediatric patients (American Psychological Assocition, 2013; McSwain et al., 2017). There are certain emergency situations in which informed consent may not be required ahead of time (McSwain et al., 2017), but whenever possible informed consent crafted to telehealth services is important.

Parent Participation and Privacy

Parents or legal guardians may participate in telehealth services unless the patient is legally allowed to oversee their own care (McSwain et al., 2017). At the beginning of a new patient video visit, it is important to ask any adults on the visit to identify themselves and their relationship to a patient (Myers et al., 2017). Telehealth providers should establish "the legal guardianship of the youth as well as custody agreements" (Myers et al., 2017, p. 789). Occasionally, the patient's medical record may be out of date and not reflect the current guardianship agreement. If this is the case, it is essential to work with the patient to get current documentation in the chart.

Providers may want to ask the patient to step away from their parents in their own private space to conduct a risk assessment or discuss other confidential information. It is important for telehealth providers to set the stage for talking privately with the child during the initial session. Providers should discuss with patients and their parents what information is legally mandated to be disclosed (e.g., safety risk), and communicate what information will remain confidential between the provider and the patient (Nelson et al., 2017). If the patient is unable to find a private place in the home to meet for the visit, headphones may be worn by the patient and the provider can ask "yes" or "no" questions to keep responses discrete (Martinez et al., 2022). Another option may be for providers and/or patients to use the chat feature on video conferencing.

Training and Supervision

The COVID-19 pandemic brought about changes in how to effectively deliver adequate supervision of pediatric psychology trainees (Williamson et al., 2022; Phillips et al., 2021), while not losing focus on the core competencies of clinical psychological practice (e.g., research; ethical and legal standards; cultural diversity; professional values, attitudes and behaviors; communication and interpersonal skills; assessment; intervention; supervision, consultation; and interprofessional/interdisciplinary skills) (Ham et al., 2022). Very little research has been conducted on training and supervision in the setting of telehealth for future pediatric psychologists. Thus, the guidelines presented below are based on the author's extensive clinical experience applying training and supervision standards virtually over the last several years.

Telehealth supervision has increased flexibility of scheduling for supervisors and trainees, in addition to making peer consultation more convenient (Jordan & Shearer, 2019). For example, both trainees and supervisors have the ability to reduce or eliminate physical travel time from the workday, providing increased availability for scheduled meetings. However, telehealth supervision may offer benefits beyond flexibility and ease. Telehealth provides an opportunity to diversify supervision experiences with supervisors working in diverse settings (e.g., urban vs. rural) (McCord et al., 2024).

In pediatric psychology training programs where telehealth utilization for patients is high, telecommunication use is also likely highly utilized for other training experiences (e.g., research meetings, didactics, formal supervision meetings, etc.). When the training experience is virtual across multiple domains (e.g., clinic, consultation, supervision and research), it can be difficult to establish working relationships in a more organic way as would happen by working in a shared physical space (Hames et al., 2020). Due to the remote nature of the videoconference format, more attention and organization of the supervisory experience from both the supervisor and the trainee is needed to initially build rapport that establishes a collegial relationship that is also mutually cooperative. In a shared physical space, these types of interactions are able to develop more naturally, so it is likely that more attention needs to be given to establish a working relationship of trust between a trainee and supervisor.

Of course, it is critical that a supervisor establish clear clinical expectations (e.g., when to join virtual visits once a patient has arrived and when to speak up in a co-therapy model of training) (Hames et al., 2020). The flow of clinical visits can feel different for trainees not used to working in a virtual format, so it is important that the structure of the clinic visit be discussed prior to when clinical work begins. In order to help expedite the assimilation of trainees, it is also important that supervisors share their calendar and invite trainees to all relevant research and consultation meetings they would normally attend if clinics were face-to-face. It is easy for supervisors to overlook inviting learners to meetings that could potentially benefit their professional development.

It is also important in a virtual training format to review contingency plans with trainees should technical difficulties arise both in and out of the context of clinical care (Hames et al., 2020). For example, trainees may need to complete a video visit by phone should a patient experience difficulty with connectivity; or, the trainee may need to contact the supervisor by phone should a clinical concern arise that needs immediate attention. These scenarios may present a higher need for trainees and supervisors to share personal contact information to ensure there are always open lines of communication. This increase in accessibility for both supervisor and trainee, while positive, also presents a unique set of challenges that need to be addressed. For example, the lines between hours of work and hours of leisure may be less clear when conducting work mostly virtually. Supervisors should discuss specific days/times trainees need to be available, and also timelines for completing clinical work (e.g., patient charting and consult/patient messages) so as to avoid trainees feeling like work needs to be completed at all hours.

Overall. Supervision in a virtual setting provides many of the same opportunities for personal and professional growth as in-person training (Williamson et al., 2022). However, due to the virtual format relying more on scheduled time, there is less room and opportunity for unanticipated learning opportunities. Virtual supervision can be effective as long as both the supervisor and the trainee make a concerted effort to be deliberate with their expectations, communicate frequently and honestly, and supervisors are willing to plan ahead and be mindful to incorporate trainees into professional experiences that may otherwise easily be overlooked.

Telehealth and Artificial Intelligence

Telehealth within pediatric psychology is ever changing as technology and legislative policies change. One major question for pediatric psychology telehealth providers and researchers is the role that AI will play in the delivery of telehealth care. AI is defined as the capacity of a machine to demonstrate "learned" or "intelligent" behaviors (Matheny et al., 2022, p. 13). AI can work toward demonstrating human intelligence, but it can also use large sets of data "to generate a framework to answer the question(s) of interest" (Matheny et al., 2022, p. 13). AI has the potential to be employed broadly across patient care activities from the referral process, billing, patient interventions, provider documentation, and discharge decisions. Yet, the use of AI comes with major ethical and practical concerns that require further research and guidance.

The use of AI in telehealth excites some clinicians and frightens others. Some clinicians worry that AI will result in human-delivered therapy becoming irrelevant. However, according to Topol (2019), AI rarely replaces human capability, rather it extends their performance in a variety of ways. AI has the potential to assist pediatric psychologists in the completion of more mundane administrative tasks that take up a large portion of a clinician's day, freeing them to engage in more clinically relevant activities. AI also has the potential to help clinicians predict risk, diagnose, and treat pediatric patients (Aylward et al., 2023). It may allow pediatric patients and their families to engage in treatment interventions in between sessions.

AI could also be used to triage and collect background information on patients to allow diagnostic intake sessions to be conducted more efficiently. Also, AI has been suggested as a potential tool to assist with insurance authorization (Matheny et al., 2022), streamlining, and creating efficiency in approval processes. Additionally, AI could be used to enhance telehealth care by employing mobile applications using AI technology which could help patients engage in therapeutic techniques tailored to the individual patient in between sessions with the support of their telehealth psychologist. In areas where there are no specialized providers, AI may be able to bring diagnosis and treatment options to patients (Buch et al., 2018).

While there are many potential benefits of using AI to assist in patient care, it must be approached with caution. Buch et al. (2018) described AI as "a very picky eater" (p. 143). It requires extremely structured and accurate data about an individual or target population to be effective. More research on the use of AI is needed that includes a wide range of patients from diverse backgrounds to avoid exacerbating existing disparities related to access to technology in healthcare settings (Matheny et al., 2022). Additionally, clinicians need to ensure that AI services and technologies are evidence-based, align with ethical practice standards, and maintain the privacy of their patient's health information. Clinicians should use professional judgment when employing AI technology in their treatment plans, particularly in the case of high-risk patients.

Future Directions

The landscape of telehealth in pediatric psychology will continue to evolve with changes in technology, healthcare policy, and public attitudes and comfort with technology. For telehealth services to continue to expand, legislation needs to address reimbursement issues and access to technology infrastructure among pediatric patients and their families. Temporary policies during the COVID-19 pandemic that paid for telehealth or required that telehealth services be paid in parity with in-person services are at risk of not being sustained over time (Curfman et al., 2021b). Pediatric psychologists must continue to advocate for policies that support telehealth services. Despite the practical and ethical considerations outlined in this chapter, telehealth is essential for "creating efficient, innovative, high-value care models in which patients get the right care at the right time, as well as investing in preventative care to reduce costly emergency department and hospital visits" (Curfman et al., 2021a, p. 2). Pediatric psychologists have an opportunity to expand their reach through the virtual delivery of assessment and intervention.

Telehealth expansion in the wake of the COVID-19 pandemic changed how many pediatric psychologists delivered care to patients and their families. Research supports the use of telehealth for the assessment, intervention, and consultation of many common childhood and adolescent disorders. There are many considerations when providing virtual care. While telehealth can be helpful for many, it should not be used in all cases. Clinicians must stay current on emerging research and best practices for telehealth. Many barriers still exist to providing telehealth, and thus further research should include diverse patient groups and continue to evaluate factors in telehealth that will improve patient access to care and behavioral health outcomes.

Select Bibliography

American Psychological Association. (2013, Updated 2023). Guidelines for the practice of telepsychology. *The American Psychologist*, *68*(9), 791–800. https://doi.org/10.1037/a0035001. The American Psychology Association's (APA) guidelines regarding telepsychology. It discusses areas such as informed consent, confidentiality, and testing guidelines. It was written in 2013 and was updated in 2023 to reflect current telehealth research and recommendations.

Cahan, E. M., Mittal, V., Shah, N. R., & Thadaney-Israni, S. (2020). Achieving a quintuple aim for telehealth in pediatrics. *Pediatric Clinics of North America, 67*(4), 683–705. W.B. Saunders. https://doi.org/10.1016/j.pcl.2020.04.015. Details the quintuple aim in pediatric telehealth (ensure equity, improve outcomes, increase access, decrease cost, and support workforce). It also details guiding principles to responsibility integrating telehealth into pediatric care.

Curfman, A. L., Hackell, J. M., Herendeen, N. E., Alexander, J. J., Marcin, J. P., Moskowitz, W. B., Bodnar, C. E. F., Simon, H. K., & McSwain, S. D. (2021). Telehealth: Improving access to and quality of pediatric health care. *Pediatrics*, *148*(3). https://doi.org/10.1542/PEDS.2021-053129/181044. Discusses the importance of telehealth for pediatric patients within a medical home, the impact of COVID-19 on telehealth, and payment for telehealth services. It also provides recommendations for how to increase access and equity in mental health care through pediatric telehealth services.

Myers, K., Nelson, E. L., Rabinowitz, T., Hilty, D., Baker, D., Barnwell, S. S., Boyce, G., Bufka, L. F., Cain, S., Chui, L., Comer, J. S., Cradock, C., Goldstein, F., Johnston, B., Krupinski, E., Lo, K., Luxton, D. D., McSwain, S. D., McWilliams, J., et al. (2017). American telemedicine association practice guidelines for telemental health with children and adolescents. *Telemedicine and e-Health, 23*(10), 779–804. Mary Ann Liebert Inc. https://doi.org/10.1089/tmj.2017.0177. This document provides practice guidelines for practitioners providing telehealth services to pediatric patients. It also details specific telemental health interventions that have been researched and the populations on which they have been shown to be effective.

Ros-DeMarize, R., Chung, P., & Stewart, R. (2021). Pediatric behavioral telehealth in the age of COVID-19: Brief evidence review and practice considerations. *Current Problems in Pediatric and Adolescent Health Care*, *51*(1). https://doi.org/10.1016/j.cppeds.2021.100949. Provides a review of the evidence for telehealth in the assessment and treatment of both internalizing and externalizing disorders. Practice considerations for providing pediatric telehealth are also discussed.

References

Aboujaoude, E., & Salame, W. (2016). Technology at the service of pediatric mental health: Review and assessment. *Journal of Pediatrics, 171*, 20–24. https://doi.org/10.1016/j.jpeds.2015.12.009

American Academy of Child and Adolescent Psychiatry (AACAP) Committee on Telepsychiatry and AACAP Committee on Quality Issues. (2017). Clinical update: Telepsychiatry with children and adolescents. *Journal of the American Academy of Child and Adolescent Psychiatry, 56*(10), 875–893. https://doi.org/10.1016/j.jaac.2017.07.008

American Psychological Association. (2013). Guidelines for the practice of telepsychology. (Updated 2023). *The American Psychologist, 68*(9), 791–800. https://doi.org/10.1037/a0035001

Aylward, B. S., Abbas, H., Taraman, S., Salomon, C., Gal-Szabo, D., Kraft, C., Ehwerhemuepha, L., Chang, A., & Wall, D. P. (2023). An introduction to artificial intelligence in developmental and behavioral pediatrics. *Journal of Developmental and Behavioral Pediatrics, 44*(2), e126–e134. https://doi.org/10.1097/DBP.0000000000001149

Bagner, D. M., Berkovits, M. D., Coxe, S., Frech, N., Garcia, D., Golik, A., Heflin, B. H., Heymann, P., Javadi, N., Sanchez, A. L., Wilson, M. K., & Comer, J. S. (2023). Telehealth treatment of behavior problems in young children with developmental delay: A randomized clinical trial. *JAMA Pediatrics, 177*(3), 231–239. https://doi.org/10.1001/jamapediatrics.2022.5204

Berger, N. I., Wainer, A. L., Kuhn, J., Bearss, K., Attar, S., Carter, A. S., Ibanez, L. V., Ingersoll, B. R., Neiderman, H., Scott, S., & Stone, W. L. (2022). Characterizing available tools for synchronous virtual assessment of toddlers with suspected autism spectrum disorder: A brief report. *Journal of Autism and Developmental Disorders, 52*(1), 423–434. https://doi.org/10.1007/s10803-021-04911-2

Bethell, C. D., Kogan, M. D., Strickland, B. B., Schor, E. L., Robertson, J., & Newacheck, P. W. (2011). A national and state profile of leading health problems and health care quality for US children: Key insurance disparities and across-state variations. *Academic Pediatrics, 11*(3), s22–s33. https://doi.org/10.1016/j.acap.2010.08.011

Buch, V. H., Ahmed, I., & Maruthappu, M. (2018). Artificial intelligence in medicine: Current trends and future possibilities. *British Journal of General Practice, 68*(668), 143–144.

Cahan, E. M., Mittal, V., Shah, N. R., & Thadaney-Israni, S. (2020). Achieving a quintuple aim for telehealth in pediatrics. *Pediatric Clinics of North America, 67*(4), 683–705. https://doi.org/10.1016/j.pcl.2020.04.015

Chakawa, A., Belzer, L. T., Perez-Crawford, T., & Yeh, H. W. (2021). COVID-19, telehealth, and pediatric integrated primary care: Disparities in service use. *Journal of Pediatric Psychology, 46*(9), 1063–1075. https://doi.org/10.1093/jpepsy/jsab077

Coker, T. R., Porras-Javier, L., Zhang, L., Soares, N., Park, C., Patel, A., Tang, L., Chung, P. J., & Zima, B. T. (2019). A telehealth-enhanced referral process in pediatric primary care: A cluster randomized trial. *Pediatrics, 143*(3), e20182738. https://doi.org/10.1542/peds.2018-2738

Comer, J. S., Furr, J. M., Kerns, C. E., Miguel, E., Coxe, S., Meredith Elkins, R., Carpenter, A. L., Cornacchio, D., Cooper-Vince, C. E., DeSerisy, M., Chou, T., Sanchez, A. L., Khanna, M., Franklin, M. E., Garcia, A. M., & Freeman, J. B. (2017). Internet-delivered, family-based treatment for early-onset OCD: A pilot randomized trial. *Journal of Consulting and Clinical Psychology, 85*(2), 178–186. https://doi.org/10.1037/ccp0000155

Curfman, A. L., Hackell, J. M., Herendeen, N. E., Alexander, J. J., Marcin, J. P., Moskowitz, W. B., Bodnar, C. E. F., Simon, H. K., & McSwain, S. D. (2021a). Telehealth: Improving access to and quality of pediatric health care. *Pediatrics, 148*(3), e2021053129. https://doi.org/10.1542/peds.2021-053129

Curfman, A., McSwain, S. D., Chuo, J., Yeager-McSwain, B., Schinasi, D. A., Marcin, J., Herendeen, N., Chung, S. L., Rheuban, K., & Olson, C. A. (2021b). Pediatric telehealth in the COVID-19 pandemic era and beyond. *Pediatrics, 148*(3), e2020047795. https://doi.org/10.1542/peds.2020-047795

Dadds, M. R., Thai, C., Diaz, A. M., Broderick, J., Moul, C., Tully, L. A., Hawes, D. J., Davies, S., Burchfield, K., & Cane, L. (2019). Therapist-assisted online treatment for child conduct problems in rural and urban families: Two randomized controlled trials. *Journal of Consulting and Clinical Psychology, 87*(8), 706–719. https://doi.org/10.1037/ccp0000419

Dale, B. A., McDonald, C., & Neild, R. (2022). Tele-assessment considerations for school psychologists in the Covid-19 era. *Thresholds in Education, 45*(1), 75–84.

Das, G., Cheung, C., Nebeker, C., Bietz, M., & Bloss, C. (2018). Privacy policies for apps targeted toward youth: Descriptive analysis of readability. *JMIR mHealth and uHealth, 6*(1), e7626. https://doi.org/10.2196/mhealth.7626

Dayal, P., Chang, C. H., Benko, W. S., Pollock, B. H., Crossen, S. S., Kissee, J., Ulmer, A. M., Hoch, J. S., Warner, L., & Marcin, J. P. (2019). Hospital utilization among rural children served by pediatric neurology telemedicine clinics. *JAMA Network Open, 2*(8). https://doi.org/10.1001/jamanetworkopen.2019.9364

Dehling, T., Gao, F., Schneider, S., & Sunyaev, A. (2015). Exploring the far side of mobile health: Information security and privacy of mobile health apps on ios and android. *JMIR mHealth and uHealth, 3*(1), e3672. https://doi.org/10.2196/mhealth.3672

Digital Nation Data Explorer, National Telecommunications and Information Administration. (2022, October 5). *Digital nation data explorer*. https://www.ntia.gov/other-publication/2022/digital-nation-data-explorer#sel=internetUser&demo=&pc=prop&disp=chart

Elford, R., White, H., Bowering, R., Ghandi, A., Maddiggan, B., St John, K., Harnett, J., West, R., & Battcock, A. (2000). A randomized, controlled trial of child psychiatric assessments conducted using videoconferencing. *Journal of Telemedicine and Telecare, 6*(2), 73–82. https://doi.org/10.1258/1357633001935086

El-Miedany, Y. (2017). Telehealth and telemedicine: How the digital era is changing standard health care. *Smart Homecare Technology and Telehealth, 4*, 43–51. https://doi.org/10.2147/SHTT.S116009

Farmer, R. L., McGill, R. J., Dombrowski, S. C., McClain, M. B., Harris, B., Lockwood, A. B., et al. (2020). Teleassessment with children and adolescents during the coronavirus (COVID-19) pandemic and beyond: Practice and policy implications. *Professional Psychology: Research and Practice, 51*(5), 477–487. https://doi.org/10.1037/pro0000349

Frye, W. S., Gardner, L., Campbell, J. M., & Katzenstein, J. M. (2022). Implementation of telehealth during COVID-19: Implications for providing behavioral health services to pediatric patients. *Journal of Child Health Care, 26*(2), 172–184.

Grady, B., Myers, K. M., Nelson, E. L., Belz, N., Bennett, L., Carnahan, L., Decker, V. B., Holden, D., Perry, G., Rosenthal, L., Rowe, N., Spaulding, R., Turvey, C. L., White, R., & Voyles, D. (2011). Evidence-based practice for telemental health. *Telemedicine Journal and E-Health, 17*(2), 131–148. https://doi.org/10.1089/tmj.2010.0158

Ham, H. L., Lancaster, B., Maragakis, A., Harris, S. D., Turnier, L., & Bruni, T. (2022). Competencies for the practice of pediatric integrated primary care: A continuum of training. *Training and Education in Professional Psychology, 16*(2), 190–196. https://doi.org/10.1037/tep0000405

Hames, J. L., Bell, D. J., Perez-Lima, L. M., Holm-Denoma, J. M., Rooney, T., Charles, N. E., et al. (2020). Navigating uncharted waters: Considerations for training clinics in the rapid transition to telepsychology and telesupervision during COVID-19. *Journal of Psychotherapy Integration, 30*(2), 348. https://doi.org/10.1037/int0000224

Hilty, D. M., Randhawa, K., Maheu, M. M., McKean, A. J. S., & Pantera, R. (2019). Therapeutic relationship of telepsychiatry and telebehavioral health: Ideas from research on telepresence, virtual reality, and augmented reality. *Psychology and Cognitive Sciences—Open Journal, 5*(1), 14–29. https://doi.org/10.17140/PCSOJ-5-145

Hilty, D. M., Armstrong, C. M., Smout, S. A., Crawford, A., Maheu, M. M., Drude, K. P., Chan, S., Yellowlees, P. M., & Krupinski, E. A. (2022). Findings and guidelines on provider technology, fatigue, and well-being: Scoping review. *Journal of Medical Internet Research, 24*(5). JMIR Publications Inc. https://doi.org/10.2196/34451

Hudgins, A., Garber, K., & Gustin, T. (2023). *Synchronous telehealth fatigue among healthcare providers*. https://doi.org/10.2139/ssrn.4580263

Islam, S., Sanchez, A. L., McDermott, C. L., Clapp, D., Worley, J., & Becker-Haimes, E. M. (2023). To proceed via telehealth or not? Considerations for pediatric anxiety and related disorders beyond COVID-19. *Cognitive and Behavioral Practice*. https://doi.org/10.1016/j.cbpra.2023.01.004

Jordan, S. E., & Shearer, E. M. (2019). An exploration of supervision delivered via clinical video telehealth (CVT). *Training and Education in Professional Psychology, 13*(4), 323–330. https://doi.org/10.1037/tep0000245

Komanchuk, J., Cameron, J. L., Kurbatfinksi, S., Duffet-Leger, L., & Letourneau, N. (2023). A realist review of digitally delivered child development assessment and screening tools: Psychometrics and considerations for future use. *Early Human Development, 183*, 1–9. https://doi.org/10.1016/j.earlhumdev.2023.105818

Lustig, T. A. (2012). *The role of telehealth in an evolving health care environment: Workshop summary*. The National Academies Press. https://doi.org/10.1901/jaba.1972.5-225

Martinez, R. G., Seager van Dyk, I., Kroll, J., Emerson, N. D., & Bursch, B. (2022). Recommendations for building telemental health relationships with youth: A systematic review and resource for clinicians. *Evidence-Based Practice in Child and Adolescent Mental Health, 7*(3), 349–362.

Matheny, M., Israni, S. T., Ahmed, M., & Whicher, D. (Eds.). (2022). *Artificial intelligence in health care: The hope, the hype, the promise, the peril*. National Academy of Medicine.

McCord, C., Garney, W., Garcia, K., Macareno, B., & Williamson, M. (2024). Evaluation of a multi-site health services psychology training program for telehealth and integrated behavioral health. *Frontiers in Psychology, 15*, 1–11. https://doi.org/10.3389/fpsyg.2024.1339319

McSwain, S. D., Bernard, J., Burke, B. L., Cole, S. L., Dharmar, M., Hall-Barrow, J., Herendeen, N., Herendeen, P., Krupinski, E. A., Martin, A., McCafferty, D., Mulligan, D. A., North, S., Ruschman, J., Waller, M., Webster, K., Williams, S., Yamamoto, S., & Yeager, B. (2017). Policy: American telemedicine association operating procedures for pediatric telehealth. *Telemedicine Journal and E-Health, 23*(9), 699–706. https://doi.org/10.1089/tmj.2017.0176

Myers, K. M., Sulzbacher, S., & Melzer, S. M. (2004). Telepsychiatry with children and adolescents: Are patients comparable to those evaluated in usual outpatient care? *Telemedicine Journal and E-Health, 10*(3), 278–285.

Myers, K., Nelson, E. L., Rabinowitz, T., Hilty, D., Baker, D., Barnwell, S. S., Boyce, G., Bufka, L. F., Cain, S., Chui, L., Comer, J. S., Cradock, C., Goldstein, F., Johnston, B., Krupinski, E., Lo, K., Luxton, D. D., McSwain, S. D., McWilliams, J., et al. (2017). American telemedicine association practice guidelines for telemental health with children and adolescents. *Telemedicine and e-Health, 23*(10), 779–804. https://doi.org/10.1089/tmj.2017.0177

Nelson, E. L., & Sharp, S. (2016). A review of pediatric telemental health. *Pediatric Clinics of North America, 63*(5), 913–931. https://doi.org/10.1016/j.pcl.2016.06.011

Nelson, E. L., Cain, S., & Sharp, S. (2017). Considerations for conducting telemental health with children and adolescents. *Child and Adolescent Psychiatric Clinics, 26*(1), 77–91.

Phillips, L. A., Logan, J. N., & Mather, D. B. (2021). COVID-19 and beyond: Telesupervision training within the supervision competency. *Training and Education in Professional Psychology, 15*(4), 284.

Pritchard, A. E., Sweeney, K., Salorio, C. F., & Jacobson, L. A. (2020). Pediatric neuropsychological evaluation via telehealth: Novel models of care. *The Clinical Neuropsychologist, 34*(7–8), 1367–1379. https://doi.org/10.1080/13854046.2020.1806359

Psihogios, A. M., Stiles-Shields, C., & Neary, M. (2021). The needle in the haystack: Identifying credible mobile health apps for pediatric populations during a pandemic and beyond. *Journal of Pediatric Psychology, 45*(10), 1106–1113. https://doi.org/10.1093/JPEPSY/JSAA094

Romani, P. W., Kennedy, S. M., Sheffield, K., Ament, A. M., Schiel, M. A., Hawks, J., & Murphy, J. (2022). Pediatric mental healthcare providers' perceptions of the delivery of partial hospitalization and outpatient services via telehealth during the COVID-19 pandemic. *Evidence-Based Practice in Child and Adolescent Mental Health, 7*(3), 327–340. https://doi.org/10.1080/23794925.2021.1931985

Ros-DeMarize, R., Chung, P., & Stewart, R. (2021). Pediatric behavioral telehealth in the age of COVID-19: Brief evidence review and practice considerations. *Current Problems in Pediatric and Adolescent Health Care, 51*(1), 100949. https://doi.org/10.1016/j.cppeds.2021.100949

Stancin, T., & Perrin, E. C. (2014). Psychologists and pediatricians: Opportunities for collaboration in primary care. *American Psychologist, 69*(4), 332–343. https://doi.org/10.1037/a0036046

Talbott, M. R., Dufek, S., Zwaigenbaum, L., Bryson, S., Brian, J., Smith, I. M., & Rogers, S. J. (2020). Brief report: Preliminary feasibility of the TEDI: A novel parent-administered telehealth assessment for autism spectrum disorder symptoms in the first year of life. *Journal of Autism and Developmental Disorders, 50*(9), 3432–3439. https://doi.org/10.1007/s10803-019-04314-4

Telepractice and the WISC-V. (2021). www.pearsonassessments.com

Topol, E. (2019). *Deep medicine: How artificial intelligence can make healthcare human again*. Basic Books.

Tse, Y. J., McCarty, C. A., Stoep, A. V., & Myers, K. M. (2015). Teletherapy delivery of caregiver behavior training for children with attention-deficit hyperactivity disorder. *Telemedicine Journal and E-Health, 21*(6), 451–458. https://doi.org/10.1089/tmj.2014.0132

Wampold, B. E. (2015). How important are the common factors in psychotherapy? An update. *World Psychiatry, 14*(3), 270–277. https://doi.org/10.1002/wps.20238

Williamson, M., Jackson, K., Wade, L., Lopez, C., Nanavaty, N., Regan, T., et al. (2022). Training health service psychology students in an in-person and telehealth integrated behavioral health service delivery model in primary care. *Families, Systems & Health, 40*(4), 572. https://doi.org/10.1037/fsh0000752

Wright, A. J., & Raiford, S. E. (2020). *Essentials of psychological tele-assessment*. John Wiley & Sons.

Wright, J., Mihura, J. L., Pade, H., & McCord, D. M. (2020). *Guidance on psychological tele-assessment during the COVID-19 crisis*. https://www.apaservices.org/practice/reimbursement/health-codes/testing/ tele-assessment-covid-19

Chapter 4
Stepped-Care Practices in the Pediatric Setting

Caroline Lucy, Julie Wojtaszek, and Leah LaLonde

Overview of Stepped Care

Stepped care is a model of health-care service delivery that aims to more efficiently allocate existing health-care resources and increase the number of individuals connected to effective services. Stepped behavioral health care assumes that a portion of patients will experience the same benefit from a lower-intensity intervention as they would from weekly 50-minute sessions of individual therapy, which is the current default treatment method (Bower & Gilbody, 2005). In stepped care, patients start with the intervention that is the least restrictive in terms of the required cost, effort, or frequency that is still likely to contribute to their clinical benefit. Importantly, not all patients automatically start with the same step; there should be a match between the severity of the patient's presenting concerns and their initial treatment step. If the patient continues to experience symptoms and/or impairment after initial treatment, they are subsequently "stepped up" to more intensive treatment. Patients can also be "stepped down" to less intensive interventions as a way of titrating down services when appropriate (Bower & Gilbody, 2005; Haaga, 2000). Importantly, stepped care relies on both effective intake assessments to inform decisions around which level of treatment is most appropriate for a patient, and on routine progress monitoring throughout treatment to inform whether it is necessary to step up to a higher-intensity treatment (Bower & Gilbody, 2005).

C. Lucy · J. Wojtaszek
Eastern Michigan University, Ypsilanti, MI, USA

L. LaLonde (✉)
Psychiatry and Behavioral Health, Nationwide Children's Hospital, Columbus, OH, USA
e-mail: lmcdiarm@emich.edu

A. Maragakis, M. Janikian (eds.), *Evidence-Based Behavioral Health Practices in Pediatric Specialty Settings*, https://doi.org/10.1007/978-3-031-71004-9_4

In order for stepped-care models to be implemented, treatments of variable intensities must be available. Figure 4.1 presents an example of different treatment steps.

Self-help options include approaches such as bibliotherapy or computer-assisted therapies that involve no or little direct interaction with a provider (Bower & Gilbody, 2005). Guided self-help options include approaches that are partially bibliotherapeutic or computer-assisted, but also include some direct interaction with a provider (Cuijpers et al., 2010). Brief intervention is also referred to as "low-intensity intervention," which has been defined as six hours or less of contact and intervention that can be provided by trained practitioners or nonspecialists (Shafran et al., 2021). Group therapy refers to a time-limited intervention in which multiple patients participate in treatment at the same time with a single provider, often about a common theme or presenting problem (American Psychological Association, 2019). Brief individual therapy is time-limited individual treatment with specific goals, typically involving four to six sessions. For example, integrated behavioral health in primary care often involves four or fewer brief sessions focused on specific problem behaviors (Bryan et al., 2012; Strosahl, 2005). In addition to these lower-intensity steps, watchful waiting is sometimes included as a step prior to active therapeutic intervention in which patients are routinely screened and monitored. If issues persist over time or are complex, patients may be referred to outside providers for a higher-intensity intervention, such as traditional, weekly individual therapy (Strosahl, 2005). In general, medication is considered most appropriate for children and adolescents with more severe symptoms and functional impairment that is not effectively managed with psychological interventions alone (Brewer & Sarvet, 2011). Intensive outpatient, partial hospitalization, and inpatient hospitalization are some of the highest-intensity treatment options, as they require a significant allocation of time and resources for both patients and providers.

Any type of evidence-based psychological treatment (e.g., cognitive behavioral therapy [CBT], dialectical behavior therapy [DBT], behavior modification, acceptance and commitment therapy [ACT], exposure, and response prevention) can be delivered through a stepped-care model, although some types of treatment may be

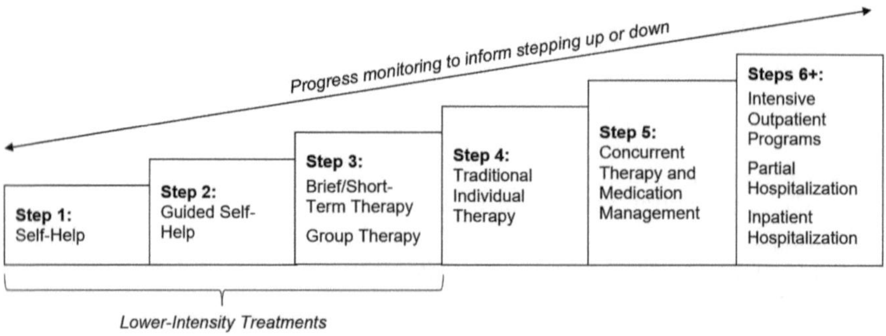

Fig. 4.1 Example of a stepped-care model

more amenable to stepped-care models than others. For example, while a full program of dialectical behavior therapy (DBT) involves individual therapy, group therapy, and therapist consultation, a stepped approach to DBT may involve starting with participation in a DBT group as a stand-alone treatment and then "stepping up" to individual DBT if needed. As another example, web or digital applications based on CBT, ACT, and other empirically supported treatments have been and continue to be developed. These interventions incorporate elements of these therapies but may not involve interaction with a provider. Therefore, stepped-care models do not necessarily inform what treatment is most appropriate, but what "dose" or "how much of the intervention" is likely to be most effective for whom.

Benefits of Stepped Care

Stepped care has a number of benefits, including addressing the unmet need for services, improving treatment outcomes, and reducing the time and cost of services. Multiple factors contribute to poor access to behavioral health care, and particularly relevant to the current chapter is the shortage of pediatric behavioral health providers and long waitlists (Cunningham, 2009; Thomas & Holzer, 2006). The traditional behavioral health service delivery model, which primarily relies on long-term individual therapy, is not a sustainable model to meet the demand (Kazdin & Blase, 2011). Increasing the number of behavioral health providers who can provide evidence-based care is an important part of the solution, but it will not likely solve this problem alone. By emphasizing lower-intensity methods, stepped-care models can help increase the number of individuals participating in empirically supported treatments with existing resources. Importantly, pediatric stepped care has also been shown to be more cost-effective than treatment as usual for a number of clinical presentations, including anxiety (Rapee et al., 2017; Yeguez et al., 2020), depression (Mufson et al., 2018), and trauma (Salloum et al., 2016, 2022).

Next, the emphasis on matching the intensity of the presenting concern and the patient's context to the intensity of treatment makes stepped care a flexible, individualized treatment approach. The progress monitoring within stepped care also facilitates quicker flagging of treatment nonresponse or deterioration. By engaging in measurement-based care and routinely tracking progress throughout treatment, providers can make accurate and appropriate decisions to step treatment up or down (Pettit et al., 2017; Rapee et al., 2017).

Finally, providers, patients, and families are open to stepped care. One study found that 84% of surveyed providers and 71% of surveyed patient participants rated stepped care as a highly acceptable model of treatment (Haugh et al., 2019). Considering specific interventions, patients and families rated stepped interpersonal therapy for depression (Mufson et al., 2018) and trauma-focused cognitive behavioral therapy (Salloum et al., 2014) as highly acceptable.

Risks of Stepped Care

As with any model, there are risks associated with stepped care. The most significant risk is assigning patients to an inappropriate level of care for their needs. Assigning patients to a level of care that is more intensive than their presenting concern or functional impairment, or that requires a level of engagement that exceeds patients' motivation, readiness to engage, context, or ability to participate, may lead to low follow-up and greater dropout, inefficient use of resources, or over-pathologizing.

On the other hand, there are also risks with assigning a patient to a level of care that is less intensive than what would best meet their needs and context. Patient safety and well-being are a primary concern, and limiting access to beneficial treatments is a considerable risk. For example, assigning a patient with severe, persistent suicidality to bibliotherapy or even group therapy alone is likely inappropriate. Further, undertreating can result in the perception that behavioral health care is not helpful altogether (Davison, 2000; Wilson et al., 2000). Patients' willingness to engage with higher levels of care may be diminished due to perceived "failure" of treatment, and the patient may earn the label of "treatment resistant."

Unique Aspects of Stepped Care in Pediatrics

Treatment of pediatric psychological concerns includes unique barriers, which stepped-care models can help to address. One significant barrier is that treatment is still predominantly delivered in person at a mental health clinic and provided by a mental health professional (Kazdin, 2019). As a result, securing childcare and transportation are challenges that interfere with participation in treatment (Fernandez & Eyberg, 2005, 2009). To address these barriers, flexible service delivery models may be implemented in a variety of settings, such as schools and primary care offices, the two systems that most children and adolescents have access to (Abrams, 2022; Burkhart et al., 2020; Society for Adolescent Health and Medicine, 2020; Yonek et al., 2020). While this chapter is not specific to primary care, behavioral health concerns are often first identified within this setting, as parents present to their child's pediatrician, citing concerns related to development, inattention, externalizing behaviors, anxiety, or mood. As a result, stepped-care models within primary care commonly involve assessing needs, delivering psychoeducation, and implementing lower levels of care. Integrated health-care teams, where pediatricians and behavioral health specialists collaborate, can also be an important factor in increasing access and coordinating care to improve outcomes where services are scarce (Burkhart et al., 2020; Snipes et al., 2015; Society for Adolescent Health and Medicine, 2020; Yonek et al., 2020).

Stepped care is also well-suited to meet the often-changing developmental needs of the pediatric population. As children and adolescents develop, their context and needs often change significantly; for example, the needs of a 6-year-old with an ADHD diagnosis will be very different when that child is a preteen and then a teenager. Stepped care is an appropriate model of care to provide lower-intensity interventions at critical points across the developmental trajectory. As stepped care adapts treatment to the individual's needs by stepping treatment up or down in intensity, services can be provided based on what the individual needs at a certain time.

Empirical Support for Stepped Care in Pediatric Setting

Attention-Deficit/Hyperactivity Disorder (ADHD) Assessment

The assessment of ADHD in pediatric populations is an area where stepped care may be a particularly useful model, as the cost, availability of providers, and waitlists for comprehensive assessments can be major barriers to accessing services. The American Academy of Pediatrics (AAP) Clinical Practice Guidelines state that patients between the ages of 4 and 18 with academic or behavior concerns and inattentive, hyperactive, or impulsive symptoms should be evaluated for ADHD (Wolraich et al., 2019). Currently, the majority of ADHD diagnoses are given by pediatricians (Visser et al., 2015). To make an ADHD diagnosis, providers must ensure that DSM-5 criteria are met, assess for any comorbid emotional, behavioral, developmental, or physical conditions, and rule out any alternative causes for the symptoms (Wolraich et al., 2019). According to the AAP guidelines, pediatricians should refer to specialists in cases of more complex presentations, such as those featuring more severe symptoms, comorbid diagnoses, early or late onset of symptoms, and symptoms that do not respond to treatments (Wolraich et al., 2019). While an accurate evaluation and diagnosis can be given in primary care, patients with more complex presentations or those who require a more comprehensive assessment for ADHD differential diagnosis should be referred to psychologists, neurologists, developmental-behavioral pediatricians, psychiatrists, or other specialists (Wolraich et al., 2019). These guidelines fit well within the stepped-care model, as the complexity of the presentation should determine the level of provider expertise needed and the intensity of the assessment process that would most benefit the patient and family.

Alternative approaches to diagnosis that balance efficiency and accuracy have also been explored. The Attention Problems and Related Concerns (ARC) screening clinic undertook a multiyear project aimed at developing and implementing a streamlined, efficient, "fast track" for diagnosing ADHD using a briefer assessment battery and 1 day of clinical interviewing, testing, and feedback (Burrows & Rysdyk, 2022). The clinic had one pediatric psychologist who was able to devote between a quarter to a third of her time to the ARC assessment services (Burrows & Rysdyk, 2022). On average, patients were assessed about 3 months faster than those referred to other sites, and parents reported being highly satisfied with services (Burrows & Rysdyk, 2022). While the question remains for whom these streamlined assessments are most appropriate, such a procedure represents a stepped approach to diagnosis that can more efficiently distribute health care resources and meet family needs.

Autism Spectrum Disorder (ASD) Assessment

Caregivers of children diagnosed with ASD often raise concerns with general practitioners before age 2, yet many families wait over one year for a diagnosis (MacKenzie et al., 2023) and many do not ever complete the evaluations. Additionally, due to the increase in ASD screening and assessment of older children exhibiting symptoms of ASD, who were not evaluated early in development, and an increased awareness of the disorder, there are extensive waitlists at autism assessment clinics.

Screening

Early diagnosis and intervention are critical for neurodevelopmental disorders such as autism. Due to the fact that most children were not being diagnosed until later, regular screening for ASD has increased within pediatric primary care. Screening is the least intensive step, with the goal of quickly identifying toddlers at risk for neurodevelopmental disorders such as ASD. The Modified Checklist for Autism in Toddlers (M-CHAT; Robins et al., 1999) and the Modified Checklist for Autism in Toddlers-Revised with Follow-up (MCHAT-R/F; Robins et al., 2009) are screening tools commonly integrated into the 18-month and 24-month well-child exams (Robins et al., 2014) to detect children who are at risk for ASD and require further assessment. The MCHAT-R/F has high sensitivity and specificity (Robins et al., 2014), and it is recommended that children falling within the at-risk ranges should receive a full ASD evaluation (Lord et al., 2012).

Comprehensive Evaluation

The standard of care for a comprehensive ASD evaluation includes behavioral observation, accurate history taking, clinical interview, parent-report measures, semi-structured interviewing, and/or structured observation tools (e.g., Autism Diagnostic Observation Schedule-2; Lord et al., 2012) from a specialist who is competent to diagnose (Hyman et al., 2020). While many children, particularly older children and adolescents or toddler/young children without severe symptoms, require a full, comprehensive evaluation, toddlers exhibiting consistently severe symptoms may not require a comprehensive evaluation for accurate diagnosis and/

or ongoing monitoring. Streamlined diagnostic assessments integrated within primary care have been shown to reduce wait times and improve the capacity of diagnostic services offered to underserved populations (Hine et al., 2020). By providing a streamlined diagnostic assessment pathway, children are diagnosed at a critical time in development when significant progress can be made, and caregivers can learn about their child's symptoms and how to best support their development (Snider & Lyren, 2023). Future research should implement and evaluate a stepped-care approach to ASD assessment with pathways for briefer evaluations between screening and comprehensive evaluations.

Depression

Screening

Approximately one in five adolescents who meet the criteria for a depressive disorder do not receive any treatment (Merikangas et al., 2011; Mojtabai et al., 2016). As peers, parents, or care providers do not always recognize symptoms, screening is important for treating pediatric depression. Screening increases the identification of subthreshold and moderate symptoms of depression that may not otherwise be identified (Brown & Wissow, 2010), but are responsive to lower-intensity interventions. The Patient Health Questionnaire 9 items (PHQ-9) is a common screener that assesses the symptoms of depression and categorizes scores into none/minimal, mild, moderate, moderately severe, or severe depression to support decision-making related to diagnosis and treatment planning. Similarly, specific screening for suicidality has proliferated over the years as rates of suicide attempts have increased, particularly among adolescents between ages 15 and 17 (Plemmons et al., 2018). Many hospitals screen every pediatric patient for suicide risk (Horowitz et al., 2013).

Lower-Intensity Interventions

Following the initial AAP recommendation for depression screening (Cheung et al., 2007), there are calls for more guidance and research on how to treat children and adolescents after they screen "positive." Psychoeducation is a lower level of intervention that can be utilized by primary care physicians (PCPs) or integrated behavioral health providers who are readily available to provide immediate education regarding symptoms (Richards et al., 2012). Beyond education, single-session interventions can have a positive impact on parent- and youth-reported symptoms of depression (Schleider & Weisz, 2017). Another lower-level intervention is group therapy, which is time-efficient and allows for greater access to care. For example, Acceptance and Commitment Therapy (ACT) groups offered in pediatric primary care showed reduced symptoms of anxiety and depression, increased psychological flexibility, and significant improvements in individually tailored treatment goals within 9 sessions (O'Dell et al., 2020). Similarly, transdiagnostic groups have been found to lead to similar changes in depression scores on the PHQ-9 compared to low-intensity individual interventions (Morris et al., 2023). Broten and colleagues (2011) published a stepped-care review for depression, which outlined a variety of treatment options that range in intensity.

Higher-Intensity Interventions

Clinical practice guidelines include assessing symptoms regularly to determine whether patients show improvement and if not engaging in shared decision-making regarding the next steps, which may include stepping up to a higher level of care (American Psychological Association, 2021a). Adolescents with symptoms that are not improving with lower levels of care or adolescents with more severe, long-standing symptoms of depression or significant impairments in functioning will likely require a higher level of care, such as traditional outpatient therapy. Other intensive forms include school-based and home-based services that also target the systems surrounding the child or adolescent. In some cases, the highest intensity of intervention such as intensive outpatient, partial hospitalization, and inpatient hospitalization is clinically indicated.

Anxiety

Anxiety disorders are one of the most common psychological conditions encountered across pediatrics, often presenting at an early age and with significant functional impacts (Mian et al., 2019). Early onset, ongoing occurrence, and presentation with other disorders lead to a varying array of symptoms, with not all pediatric patients requiring direct treatment by behavioral health providers (Abrams, 2022). Treating symptoms of anxiety using a stepped-care approach intends for patients to receive the care they need aligned with severity level and delivered by the most appropriate providers (Kendall et al., 2016). The stepped approach allows providers to assess needs and disseminate psychoeducation and anticipatory guidance as appropriate, thereby reserving direct behavioral health intervention resources for those who need it most (Kendall et al., 2016).

Screening

Stepped care for anxiety begins with screening. The intent of the screening is to assess stability, duration, and intensity of the anxiety, as well as the level of functional impairment (Kendall et al., 2016). Both the US Preventative Services Task

Force (USPSTF) and the American Academy of Child and Adolescent Psychiatry (AACAP) recommend routine anxiety screening for children and adolescents, beginning at around 8 years of age (US Preventive Services Task Force et al., 2022; Walter et al., 2020). PCPs are in a unique position to detect anxiety disorders among children, as an estimated 90% of children visit their PCP annually (Centers for Disease Control and Prevention, 2023). Additionally, preventative care and symptom monitoring fall within PCP's scope of practice.

Lower-Intensity Interventions

Specific treatment recommendations guide the flexible implementation of a variety of CBT treatment components. Direct intervention should begin when anxiety is interfering with functioning in one or more areas of a child's life (Kendall et al., 2016). Treatment for mild to moderate anxiety includes the delivery of psychoeducation, self-help (print and Internet/app) resources (Kendall et al., 2016), CBT-based Internet applications (Kendall et al., 2016), and individual therapy (Brewer & Sarvet, 2011). While self-help resources for patients and families are recommended, it is important to keep in mind that additional research is needed to confirm the efficacy and specific criteria to use when evaluating when to "step up" from self-help to therapist intervention (Jolstedt et al., 2021; Kendall et al., 2016). In a meta-analysis, single-session interventions were found to have a moderate effect on symptoms of anxiety (Schleider & Weisz, 2017).

Higher-Intensity Interventions

The AACAP Clinical Practice Guidelines recommend CBT, selective serotonergic reuptake inhibitors (SSRIs), and/or a combined approach to treat anxiety in children and adolescents of ages 6–18 (Walter et al., 2020). CBT is typically considered the first-line treatment, with medication as an alternative that is often most appropriate for a higher level of severity. Recommended treatments for moderate anxiety include psychoeducation and individual therapy, as well as routine progress monitoring to determine the need to step up to a more intense approach. If progress is not achieved within 4 to 6 months, it may be appropriate to consider initiating medication to support therapy gains (Brewer & Sarvet, 2011). For those with severe anxiety, it is recommended that treatment start with both medication and therapy for the most effective results. However, medication such as SSRIs should be prescribed with caution as longitudinal data is not yet available on the safety of SSRIs with children, particularly for long-term use (Stutzman, 2021). Taking a care team approach to coordinate care between pediatrician and behavioral health provider(s) and organizations is also recommended (Kendall et al., 2016).

Other Pediatric Conditions

Functional Neurological Disorder

Functional neurological disorder (FND) is a disorder that includes a wide range of symptoms. It requires collaboration between multiple disciplines such as neurology, physical therapy, occupational therapy, and psychology. The most successful treatment programs appear to use a biopsychosocial stepped-care approach in order to address the heterogeneity and complexity of the presentation. The child or adolescent needs are assessed, and the treatment plan is developed based on the severity and pattern of the symptom presentation, as well as the level of functional impairment (Vassilopoulos et al., 2022). For instance, physical and/or occupational therapy can target issues with movement. CBT may be recommended to address maladaptive/unhelpful thinking patterns related to symptoms. Psychoeducation may assist the family in understanding and managing treatment, and school-based interventions may be used to further support the patient's functioning in this setting. While more research is needed on the specifics of how to step up or down, in general, treatment is continuously adjusted over time based on response, which is evaluated through multidisciplinary review and collaboration.

Trauma

Standard trauma-focused CBT (TF-CBT) typically has a 3–6-month duration including 12 weekly visits with a mental health therapist. A stepped-care version of treatment (SC-TF-CBT) has been developed for children (Salloum et al., 2016). Treatment begins with skills-focused strategies, primarily delivered at home by parents and supplemented with three clinician office visits as well as phone- and web-based support. This is followed by a maintenance phase of 6 weeks to practice learned skills, or without sufficient improvement, patients move to step two which includes more in-person office visits with a therapist. In an initial trial, costs were 51% lower with comparable improvements in posttraumatic stress symptoms. A randomized clinical trial again found comparable changes, with the caveat that standard treatment might be more effective for children with externalizing symptoms and parents with high stress (Salloum et al., 2022).

Obsessive Compulsive Disorder

CBT is the first-line treatment for obsessive-compulsive disorder (OCD) in children and adolescents, but requires multiple office visits. An Internet-based version (ICBT) was developed as a guided treatment in which therapists provide support asynchronously, thereby lowering treatment costs and decreasing other barriers to treatment. Symptom reduction was comparable in ICBT to the standard CBT at 6 months (Aspvall et al., 2021) and 2 years posttreatment (Lauri et al., 2023).

Psychoeducation, exposure, and relapse prevention are included in both in-person and Internet-based CBT (ICBT). The key difference is that therapist support is asynchronous on the Internet-based version, and additional telephone support is needed. If ICBT is offered as a first step in treatment, response should be monitored. If a patient does not respond to ICBT after 3 months, then stepping up to in-person CBT is recommended. Medication management with an SSRI could be considered if a patient does not respond to traditional CBT within 6 months (Aspvall et al., 2021).

Transdiagnostic Approaches

Comorbidity of emotional disorders in children and adolescents is common. The Unified Protocols for Transdiagnostic Treatment of Emotional Disorders in Children (UP-C) was developed to target core dysfunction underlying various disorders (Barlow et al., 2010). A stepped-care UP-C was developed to address a wide range of emotional disorders such as depression, anxiety, OCD, and trauma in a more parsimonious and cost-effective manner (Kennedy et al., 2021). The telehealth intervention has two steps containing six sessions each. The first step, containing four 50-minute parent/child sessions and two 30–40-minute parent support sessions, focuses on teaching opposite action to strong emotions and related behaviors. If the patient does not respond to the first six sessions, then they complete an additional six sessions. The content in step one is standardized, but the clinician tailors sessions in step 2 to the patient based on presenting concerns and available interventions. The decision to step up is made based on the review of questionnaire feedback and collaborative discussion between the client, parent, and therapist on perceived progress related to treatment targets. Case studies provided early support for this approach (Kennedy et al., 2021), but further research is needed to determine overall efficacy. Similarly, taking a principle-based approach to treatment allows for a more flexible and idiographic approach to treatment, which can inform ways to take components from evidence-based treatments based on the specific case conceptualization and eliminate unnecessary components (LaLonde et al., 2022).

Translating into Practice

A major barrier to stepped care is the limited guidance on its application and use in complex health-care systems, including where to begin treatment, how to match a patient to the most appropriate starting intervention, and what factors are most important to consider when making these decisions. In both the literature and clinical practice, the number of steps offered, the interventions offered at each step, and the criteria used for stepping up are highly variable (Richards et al., 2012; van Straten et al., 2015). While additional research and clinical guidance are needed, this section reviews a number of important considerations for the effective application of stepped care in pediatric clinical practice.

Deciding on a Treatment Step and Measuring Progress

Decisions regarding which intervention to begin with and when to step up or down are central to effective stepped care. In research, stepped-care models often either begin treatment with the lowest level of care (rather than the most clinically appropriate level of care for pediatric populations) or only include participants that meet strict criteria for a low-intensity intervention. However, this method is susceptible to the risks of undertreating patients described earlier in this chapter. Instead, in clinical practice, a number of unique patient factors need to be considered when deciding on an initial treatment step and when to step treatment up or down.

There is broadly a lack of evidence supporting the administration of screening and/or assessment tools to make decisions regarding the level of care a patient would benefit from. The following tools have been used to assess symptom severity and inform level of care: Anxiety Disorders Interview Schedule for Children and Parents (ADIS-C/P; Silverman & Albano, 1996), Screen for Child Anxiety Related Emotional Disorders (SCARED; US Preventive Services Task Force et al., 2022; Walter et al., 2020), Generalized Anxiety Disorder-7 (GAD-7; Walter et al., 2020), and Patient Health Questionnaire (PHQ-9; Morris et al., 2023).

In addition, other presentation factors such as intensity, stability vs. transience of symptoms, and functional impairment may also impact decision-making in stepped care (Brewer & Sarvet, 2011; Kendall et al., 2016; Vassilopoulos et al., 2022). Cost and other engagement barriers are also important factors to consider when deciding on a level of care (Salloum et al., 2016). Broadly, collaborative decision-making between the patient, parent, and therapist is critical to determining initial treatment steps, whether to step up or down (Kennedy et al., 2021).

Relatedly, stepped care relies on measurement-based care and routine progress monitoring in order to properly assign patients to initial treatment and step the patient up or down as needed. While clinicians tend to rely on their own clinical judgment, their estimates of their patients' progress or deterioration are biased and frequently inaccurate (Hatfield et al., 2010). Given that stepped care relies on the appropriate decisions to step treatment up or down, inaccurate estimates of progress significantly limit the model. Therefore, considering multiple methods and informants for measuring progress is important. To inform decisions, providers may use validated tools such as the GAD-7 or PHQ-9, idiographic tools such as youth top problems (Weisz et al., 2011), and consider a patient's context, while facilitating ongoing discussions with the patient and their families on progress treatment toward goals.

Availability and Coordination of Intervention Steps

For stepped care to be effective, interventions of varying intensities need to be available, and patients need to be able to move between steps. However, in practice, coordination between levels of care can be challenging. In larger health-care systems, where a range of services of varying intensities may be available, services are often spread across departments such that teams are siloed with limited communication. Without communication across clinics and/or departments, it is unlikely that patients can be stepped up or down appropriately, in a timely manner, and without disruptions to care. Even when there is multidisciplinary communication, referring providers may not be able to make appropriate decisions on how to initiate treatment in stepped care, or it may not be clear what services different departments or teams offer. Further, providers may have different perspectives on what is most appropriate for the patient's clinical needs. When patients are referred to an inappropriate level of care, it can lead to increased wait times and delayed care. As a result, the flow between steps or services may not be consistent with the stepped-care model.

In smaller private practices or small clinics, providers may not be able to offer a range of services themselves, and they therefore need to depend on connections with other resources in their area. Depending on the community where the patient is receiving services, the availability and integration of levels of care can vary. For instance, patients in rural areas may only have access to primary care providers. While a large proportion of screening and initial treatment may be conducted in the primary care setting, a portion of these families would likely need a higher level of care than can be provided in this setting.

For providers interested in engaging in stepped care, but who work in settings where the range and integration of services are limited, they can still engage in stepped-care practices. Providers should consider how to integrate or recommend self-guided services (e.g., bibliotherapy, workbooks, and digital interventions) to their patients with mild symptoms. Providers can meet with these patients on a less frequent basis, such as monthly, while they are engaging in self-guided methods to monitor progress. If patients do not appear to benefit, then they may consider more frequent individual therapy with the provider. Providers can also offer or refer to group therapy as a lower-intensity treatment. Importantly, assessing progress and symptom change is imperative to stepped care and should be incorporated at all levels of integration and intervention intensities.

Reimbursement

Another practical consideration for stepped care is the lack of reimbursement for lower-intensity methods (Hermens et al., 2014) and higher-intensity treatment. Additionally, diagnosis is often needed for insurance coverage of treatment (Batstra et al., 2014), which creates the contingency in which evaluating and implementing lower levels of care is not financially sustainable. In a review of factors negatively impacting the implementation of low-intensity interventions, organizational demands such as time, cost, space, and capacity were among the most common barriers reported across studies (Roach et al., 2023).

Provider-Level Factors

The lack of familiarity with stepped care is another issue that is faced in translating stepped-care models into practice. Because clinical training programs tend to emphasize individual therapy, training in both lower- and higher-intensity interventions may be limited and variable. Without exposure to these methods, providers may be more likely to default to traditional 50-minute individual therapy without exploring the potential benefits of lower- or higher-step interventions. As discussed above, while increasing the number of providers trained in effective treatment is needed to address the unmet need for services, this is unlikely to be sufficient. Instead, training providers in both traditional services and effective, lower-intensity methods may be more effective in increasing the number of patients who receive needed services.

While some providers report openness to stepped care, others have cited concerns regarding the model. Some question whether it is feasible or realistic for a single provider/clinic to offer such an array of services of varying intensities. For example, can a single provider/clinic offer individual therapy, group therapy, computer-assisted therapy, and bibliotherapy and have the capacity and/or resources to switch methods appropriately? Care teams need to have specialists, nonspecialists, and options for a continuum of treatment intensities and/or collaborations with other institutions or systems that do. Others have felt that the model prioritizes cost-effectiveness and economic gain over clinical effectiveness and patient outcomes. While this belief likely stems from limited understanding and knowledge of the empirical support for lower-intensity methods, it reflects providers' concerns related to the risk of undertreatment described above.

Supervision Within Stepped-Care Models

A competency-based approach to clinical supervision, such as those outlined in Fouad et al. (2009), is an approach that defines core competencies within the professional field of psychology generally. Competencies of particular relevance to providing clinical care within stepped-care models include evaluation methods, case conceptualization and treatment planning, knowledge of interventions, intervention planning, intervention implementation, and progress evaluation.

Evaluation and Case Conceptualization

Diagnosis alone does not necessarily provide information about the severity or functional impairment to inform the level of care. As such, an individualized assessment is required to determine the level of care most appropriate for an individual patient and to monitor their progress throughout treatment. In addition to knowledge of and skills in screening, initial assessment, and progress monitoring as described above, case conceptualization is a foundational competency for learners who hope to engage in stepped care in pediatric settings.

Case conceptualization describes how key patient, family, and contextual variables interact to impact a patient's presenting concerns and maintain functional impairment to inform the treatment approach and level of care. Learners should be able to connect their case conceptualizations to specific active ingredients and behavior-change strategies. This is particularly important for brief interventions, as the goal is to provide the lowest dose of intervention rather than full treatment protocols for psychological disorders.

The components of the case conceptualization should inform the level of care. For example, more chronic and stable the symptoms and/or more functional impairment may require a higher level of care compared to patients with more localized symptoms or impairment. Patients with fewer maintaining factors may benefit from lower levels of care. Contextual factors that maintain symptoms (e.g., parental accommodation and coercive processes) may also play a role in deciding which level of care is most appropriate.

Intervention Knowledge, Planning, and Implementation

Consistent with Guideline 2 of the APA Guidelines on Evidence-Based Psychological Practice in Health Care, psychologists should "maintain and enhance their knowledge of the research and scholarly literature applicable to their practice" (American Psychological Association, 2021b, p. 8–10). There are a number of key areas that providers engaging in stepped care in pediatric settings should be knowledgeable of. In terms of treatment, process-based therapy (Hofmann & Hayes, 2019) and transdiagnostic approaches to psychological treatment (Barlow et al., 2010) are highly amenable to stepped care, as they focus on addressing multiple forms of psychopathology by targeting mechanisms of change and core dysfunctions rather than disorders. This approach can help learners flexibly implement components of treatments that will lead to the most change for that individual across all levels of care. Further, stepped care in the pediatric setting requires knowledge of both developmental psychology and population health. On the individual level, knowledge of developmental psychology is important for distinguishing between typical and atypical behavior across childhood and adolescence and the course of disorders across development. The application of developmental psychology in stepped care informs the development of social, emotional, and cognitive functioning of children and adolescents by tailoring treatment for the developmental stage. On the systems level, population health perspectives can support the scaling of interventions, developing technologies, and structuring systems to meet population-level needs (Singla et al., 2023).

Ideally, learners gain experience across the continuum of care for exposure to all symptom severities and most levels of care. Given that face-to-face hours are reported to internships and it is a commonly accepted metric of experience, learners may not gain experience in developing or delivering lower-intensity treatments that do not require or involve as much face-to-face time. Thus, choosing academic projects, such as thesis or dissertation projects, that allow learners to get more experience with these methods may increase their knowledge and comfort in prescribing and utilizing alternative interventions. Additionally, low-intensity intervention is a broad term that can encompass a variety of creative ways to deliver high-quality care in a format that is less resource intensive than weekly therapy and highly effective. For example, research has shown that caregivers of children with autism are less likely to engage in treatment that takes more than 6 months in duration (Call et al., 2015). Therefore, a higher-intensity, shorter-duration intervention may be more efficient and effective for some pediatric patients and families. This has also been referred to as highly concentrated psychological treatment and applied to children and adolescents with symptoms of anxiety (Öst & Ollendick, 2017).

Recommendations

There are significant gaps in the research on stepped care. Research is needed to build empirical support for the model related to its effectiveness in improving clinical outcomes and the cost and time effectiveness compared to treatment as usual, particularly in pediatric populations. In the absence of such support, the system may continue to default to traditional individual therapy and disregard the possible benefits of stepped care. In addition, future research is needed to inform guidelines for successful, effective implementation of stepped care in clinical practice. Ambiguity on how to engage in stepped care is a limiting factor for its use, as it is not clear what counts as stepped care, where to start, what lower-intensity methods to use, and how to make decisions regarding where to start treatment and when to step up or down.

To address limited training in and familiarity with stepped care, it is recommended that training programs increase exposure to and opportunities to engage in stepped-care models. This may include offering a variety of lower-intensity interventions in university department clinics, having supervisors trained in lower-intensity methods available to supervise students, and discussing the model in required coursework.

Finally, while some research has demonstrated patient openness to stepped care (Haugh et al., 2019), more is needed to clarify patient perceptions of lower-intensity methods more specifically. In practice, providers should discuss the approach with their patients and their families to gauge their interest, assess what types of interventions they may be interested in, and consider how they may collaboratively make decisions to change to a higher or lower step.

Select Bibliography

Bower, P., & Gilbody, S. (2005). Stepped care in psychological therapies: access, effectiveness, and efficiency: Narrative literature review. *The British Journal of Psychiatry, 186*(1), 11–17. https://doi.org/10.1192/bjp.186.1.11. A foundational article on stepped care describing the assumptions of stepped care and potential clinical and economic benefits of stepped care.

Davison, G. C. (2000). Stepped care: Doing more with less? *Journal of Consulting and Clinical Psychology, 68*(4), 580–585. https://doi.org/10.1037/0022-006X.68.4.580. An early description of important components of stepped care.

LaLonde, L., Good, J., Orkopoulou, E., Vriesman, M., & Maragakis, A. (2022). Tracing the missteps of stepped care: Improving implementation of stepped care through contextual behavioral science. *Journal of Contextual Behavioral Science, 23*, 109–116. https://doi.org/10.1016/j.jcbs.2022.01.001. A review of how a contextual behavioral approach to clinical treatment can enhance and improve stepped-care models.

Maragakis, A., & O'Donohue, W. T. (Eds.). (2018). *Principle-based stepped care and brief psychotherapy for integrated care settings*. Springer. https://doi.org/10.1007/978-3-319-70539-2. Descriptions of stepped care and brief treatment for a variety of concerns that are relevant to the pediatric population including ADHD, elimination disorders, oppositional defiant disorder, pediatric sleep resistance, and school refusal behavior.

Richards, D. A., Bower, P., Pagel, C., Weaver, A., Utley, M., Cape, J., Pilling, S., Lovell, K., Gilbody, S., Leibowitz, J., Owens, L., Paxton, R., Hennessy, S., Simpson, A., Gallivan, S., Tomson, D., & Vasilakis, C. (2012). Delivering stepped care: An analysis of implementation in routine practice. *Implementation Science, 7*(1), 1–11. https://doi.org/10.1186/1748-5908-7-3. An article reviewing the implementation of a stepped-care model in the United Kingdom.

Singla, D. R., Schleider, J. L., & Ptel, V. (2023). Democratizing access to psychological therapies and the role of psychologists. *Journal of Consulting and Clinical Psychology, 91*(11), 623–625. https://doi.org/10.1037/ccp0000850. A viewpoint paper on how to restructure mental health care to increase access and address unmet behavioral health-care needs.

References

Abrams, Z. (2022, October 26). *Why the benefits of annual anxiety and depression screenings for kids and teens outweigh the risks*. American Psychological Association. https://www.apa.org/monitor/2023/01/screening-kids-anxiety-depression

American Psychological Association. (2019, October 31). *Psychotherapy: Understanding group therapy*. https://www.apa.org/topics/psychotherapy/group-therapy

American Psychological Association. (2021a). *Decision-making within evidence-based practice in psychology: Using the APA clinical practice guideline for the treatment of depression in children and adolescents*. https://www.aafp.org/family-physician/patient-care/prevention-wellness/emotional-wellbeing/adhd-toolkit/assessment-and-diagnosis.html

American Psychological Association. (2021b). *Professional practice guidelines for evidence-based psychological practice in health care*. https://doi.org/10.1037/e503832021-001

Aspvall, K., Andersson, E., Melin, K., Norlin, L., Eriksson, V., Vigerland, S., Jolstedt, M., Silverberg-Mörse, M., Wallin, L., Sampaio, F., Feldman, I., Bottai, M., Lenhard, F., Mataix-Cols, D., & Serlachius, E. (2021). Effect of an internet-delivered stepped-care program vs in-person cognitive behavioral therapy on obsessive-compulsive disorder symptoms in children and adolescents: A randomized clinical trial. *JAMA, 325*(18), 1863. https://doi.org/10.1001/jama.2021.3839

Barlow, D. H., Ellard, K. K., & Fairholme, C. P. (2010). *Unified protocol for transdiagnostic treatment of emotional disorders: Workbook*. Oxford University Press.

Batstra, L., Nieweg, E. H., Pijl, S., Van Tol, D. G., & Hadders-Algra, M. (2014). Childhood ADHD: A stepped diagnosis approach. *Journal of Psychiatric Practice, 20*(3), 169–177. https://doi.org/10.1097/01.pra.0000450316.68494.20

Bower, P., & Gilbody, S. (2005). Stepped care in psychological therapies: Access, effectiveness and efficiency: Narrative literature review. *British Journal of Psychiatry, 186*(1), 11–17. https://doi.org/10.1192/bjp.186.1.11

Brewer, S., & Sarvet, B. (2011). Management of anxiety disorders in the pediatric primary care setting. *Pediatric Annals, 40*(11), 541–547. https://doi.org/10.3928/00904481-20111007-04

Broten, L. A., Naugle, A. E., Kalata, A. H., & Gaynor, S. T. (2011). Depression and a stepped care model. In C. Draper & W. T. O'Donohue (Eds.), *Stepped care and e-health* (pp. 17–43). Springer. https://doi.org/10.1007/978-1-4419-6510-3_2

Brown, J. D., & Wissow, L. S. (2010). Screening to identify mental health problems in pediatric primary care: Considerations for practice. *The International Journal of Psychiatry in Medicine, 40*(1), 1–19. https://doi.org/10.2190/PM.40.1.a

Bryan, C. J., Corso, M. L., Corso, K. A., Morrow, C. E., Kanzler, K. E., & Ray-Sannerud, B. (2012). Severity of mental health impairment and trajectories of improvement in an integrated primary care clinic. *Journal of Consulting and Clinical Psychology, 80*(3), 396–403. https://doi.org/10.1037/a0027726

Burkhart, K., Asogwa, K., Muzaffar, N., & Gabriel, M. (2020). Pediatric integrated care models: A systematic review. *Clinical Pediatrics, 59*(2), 148–153. https://doi.org/10.1177/0009922819890004

Burrows, K., & Rysdyk, S. (2022). Fast tracking assessment of attention deficit hyperactivity disorder in tertiary care. *Journal of Patient Experience, 9*, 237437352211122. https://doi.org/10.1177/23743735221112210

Call, N. A., Reavis, A. R., McCracken, C. E., Gillespie, S. E., & Scheithauer, M. C. (2015). The impact of delays on parents' perceptions of treatments for problem behavior. *Journal of Autism and Developmental Disorders, 45*(4), 1013–1025. https://doi.org/10.1007/s10803-014-2257-9

Centers for Disease Control and Prevention. (2023, November 4). *Ambulatory care use and physician office visits*. https://www.cdc.gov/nchs/fastats/physician-visits.htm

Cheung, A. H., Zuckerbrot, R. A., Jensen, P. S., Ghalib, K., Laraque, D., Stein, R. E. K., & The GLAD-PC Steering Group. (2007). Guidelines for adolescent depression in primary care (GLAD-PC): II. Treatment and ongoing management. *Pediatrics, 120*(5), e1313–e1326. https://doi.org/10.1542/peds.2006-1395

Cuijpers, P., Donker, T., van Straten, A., Li, J., & Andersson, G. (2010). Is guided self-help as effective as face-to-face psychotherapy for depression and anxiety disorders? A systematic review and meta-analysis of comparative outcome studies. *Psychological Medicine, 40*(12), 1943–1957. https://doi.org/10.1017/S0033291710000772

Cunningham, P. J. (2009). Beyond parity: Primary care physicians' perspectives on access to mental health care. *Health Affairs, 28*(Supplement 1), w490–w501. https://doi.org/10.1377/hlthaff.28.3.w490

Davison, G. C. (2000). Stepped care: Doing more with less? *Journal of Consulting and Clinical Psychology, 68*(4), 580–585. https://doi.org/10.1037/0022-006X.68.4.580

Fernandez, M. A., & Eyberg, S. M. (2005). Keeping families in once they've come through the door: Attrition in parent-child interaction therapy. *Journal of Early and Intensive Behavior Intervention, 2*(3), 207–212. https://doi.org/10.1037/h0100314

Fernandez, M. A., & Eyberg, S. M. (2009). Predicting treatment and follow-up attrition in parent–child interaction therapy. *Journal of Abnormal Child Psychology, 37*(3), 431–441. https://doi.org/10.1007/s10802-008-9281-1

Fouad, N. A., Grus, C. L., Hatcher, R. L., Kaslow, N. J., Hutchings, P. S., Madson, M. B., Collins, F. L., & Crossman, R. E. (2009). Competency benchmarks: A model for understanding and measuring competence in professional psychology across training levels. *Training and Education in Professional Psychology, 3*(4, Suppl), S5–S26. https://doi.org/10.1037/a0015832

Haaga, D. A. F. (2000). Introduction to the special section on stepped care models in psychotherapy. *Journal of Consulting and Clinical Psychology, 68*(4), 547–548. https://doi.org/10.1037/0022-006X.68.4.547

Hatfield, D., McCullough, L., Frantz, S. H. B., & Krieger, K. (2010). Do we know when our clients get worse? An investigation of therapists' ability to detect negative client change. *Clinical Psychology & Psychotherapy, 17*, 25. https://doi.org/10.1002/cpp.656

Haugh, J. A., Herbert, K., Choi, S., Petrides, J., Vermeulen, M. W., & D'Onofrio, J. (2019). Acceptability of the stepped care model of depression in primary care patients and providers. *Journal of Clinical Psychology in Medical Settings, 26*(4), 402–410. https://doi.org/10.1007/s10880-019-09599-2

Hermens, M. L., Muntingh, A., Franx, G., van Splunteren, P. T., & Nuyen, J. (2014). Stepped care for depression is easy to recommend, but harder to implement: Results of an explorative study within primary care in The Netherlands. *BMC Family Practice, 15*(1), 5. https://doi.org/10.1186/1471-2296-15-5

Hine, J. F., Allin, J., Allman, A., Black, M., Browning, B., Ramsey, B., Swanson, A., Warren, Z. E., Zawoyski, A., & Allen, W. (2020). Increasing access to autism spectrum disorder diagnostic consultation in rural and underserved communities: Streamlined evaluation within primary care. *Journal of Developmental & Behavioral Pediatrics, 41*(1), 16–22. https://doi.org/10.1097/DBP.0000000000000727

Hofmann, S. G., & Hayes, S. C. (2019). The future of intervention science: Process-based therapy. *Clinical Psychological Science, 7*(1), 37–50. https://doi.org/10.1177/2167702618772296

Horowitz, L. M., Snyder, D., Ludi, E., Rosenstein, D. L., Kohn-Godbout, J., Lee, L., Cartledge, T., Farrar, A., & Pao, M. (2013). Ask suicide-screening questions to everyone in medical settings: The asQ'em quality improvement project. *Psychosomatics, 54*(3), 239–247. https://doi.org/10.1016/j.psym.2013.01.002

Hyman, S. L., Levy, S. E., Myers, S. M., Council on Children With Disabilities, Section on Developmental and Behavioral Pediatrics, Kuo, D. Z., Apkon, S., Davidson, L. F., Ellerbeck, K. A., Foster, J. E. A., Noritz, G. H., Leppert, M. O., Saunders, B. S., Stille, C., Yin, L., Weitzman, C. C., Childers, D. O., Levine, J. M., Peralta-Carcelen, A. M., Poon, J. K., et al. (2020). Identification, evaluation, and management of children with autism spectrum disorder. *Pediatrics, 145*(1), e20193447. https://doi.org/10.1542/peds.2019-3447

Jolstedt, M., Vigerland, S., Mataix-Cols, D., Ljótsson, B., Wahlund, T., Nord, M., Högström, J., Öst, L.-G., & Serlachius, E. (2021). Long-term outcomes of internet-delivered cognitive behaviour therapy for paediatric anxiety disorders: Towards a stepped care model of health care delivery. *European Child & Adolescent Psychiatry, 30*(11), 1723–1732. https://doi.org/10.1007/s00787-020-01645-x

Kazdin, A. E. (2019). Annual research review: Expanding mental health services through novel models of intervention delivery. *Journal of Child Psychology and Psychiatry, 60*(4), 455–472. https://doi.org/10.1111/jcpp.12937

Kazdin, A. E., & Blase, S. L. (2011). Rebooting psychotherapy research and practice to reduce the burden of mental illness. *Perspectives on Psychological Science, 6*(1), 21–37. https://doi.org/10.1177/1745691610393527

Kendall, P. C., Makover, H., Swan, A., Carper, M. M., Mercado, R., Kagan, E., & Crawford, E. (2016). What steps to take? How to approach concerning anxiety in youth. *Clinical Psychology: Science and Practice, 23*(3), 211–229. https://doi.org/10.1111/cpsp.12156

Kennedy, S. M., Lanier, H., Salloum, A., Ehrenreich-May, J., & Storch, E. A. (2021). Development and implementation of a transdiagnostic, stepped-care approach to treating emotional disorders in children via telehealth. *Cognitive and Behavioral Practice, 28*(3), 350–363. https://doi.org/10.1016/j.cbpra.2020.06.001

LaLonde, L., Good, J., Orkopoulou, E., Vriesman, M., & Maragakis, A. (2022). Tracing the missteps of stepped care: Improving the implementation of stepped care through contextual behavioral science. *Journal of Contextual Behavioral Science, 23*, 109–116. https://doi.org/10.1016/j.jcbs.2022.01.001

Lauri, K. O., Andersson, E., Mataix-Cols, D., Norlin, L., Eriksson, V., Melin, K., Lenhard, F., Serlachius, E., & Aspvall, K. (2023). Long-term effect of stepped-care vs in-person cognitive behavioral therapy for pediatric obsessive-compulsive disorder. *Internet Interventions, 32*, 100613. https://doi.org/10.1016/j.invent.2023.100613

Lord, C., Luyster, R., Guthrie, W., & Pickles, A. (2012). Patterns of developmental trajectories in toddlers with autism spectrum disorder. *Journal of Consulting and Clinical Psychology, 80*(3), 477–489. https://doi.org/10.1037/a0027214

MacKenzie, K. T., Mazefsky, C. A., & Eack, S. M. (2023). Obtaining a first diagnosis of autism spectrum disorder: Descriptions of the diagnostic process and correlates of parent satisfaction from a national sample. *Journal of Autism and Developmental Disorders, 53*(10), 3799–3812. https://doi.org/10.1007/s10803-022-05673-1

Merikangas, K. R., He, J., Burstein, M., Swendsen, J., Avenevoli, S., Case, B., Georgiades, K., Heaton, L., Swanson, S., & Olfson, M. (2011). Service utilization for lifetime mental disorders in U.S. adolescents: Results of the National Comorbidity Survey–Adolescent Supplement (NCS-A). *Journal of the American Academy of Child & Adolescent Psychiatry, 50*(1), 32–45. https://doi.org/10.1016/j.jaac.2010.10.006

Mian, N. D., Carpenter, A. L., Pincus, D. B., Bair-Merritt, M., & Perrin, E. C. (2019). Training residents in the early identification of anxiety disorders: Feasibility and preliminary efficacy of a novel training program. *Clinical Pediatrics, 58*(7), 761–769. https://doi.org/10.1177/0009922819837356

Mojtabai, R., Olfson, M., & Han, B. (2016). National trends in the prevalence and treatment of depression in adolescents and young adults. *Pediatrics, 138*(6), e20161878. https://doi.org/10.1542/peds.2016-1878

Morris, L., Lovell, K., McEvoy, P., Emsley, R., Carter, L.-A., Edge, D., Bates, R., Wallwork, T., & Mansell, W. (2023). A brief transdiagnostic group (the take control course) compared to individual low-intensity CBT for depression and anxiety: A randomized non-inferiority trial. *Cognitive Behaviour Therapy, 52*(3), 176–197. https://doi.org/10.1080/16506073.2022.2143418

Mufson, L., Rynn, M., Yanes-Lukin, P., Choo, T. H., Soren, K., Stewart, E., & Wall, M. (2018). Stepped care interpersonal psychotherapy treatment for depressed adolescents: A pilot study in pediatric clinics. *Administration and Policy in Mental Health and Mental Health Services Research, 45*(3), 417–431. https://doi.org/10.1007/s10488-017-0836-8

O'Dell, S. M., Hosterman, S. J., Hostutler, C. A., Nordberg, C., Klinger, T., Petts, R. A., Whitehead, M. R., Faulkner, S. A., Garcia, A., & Massura, C. A. (2020). Retrospective cohort study of a novel acceptance and commitment therapy group intervention for adolescents implemented in integrated primary care. *Journal of Contextual Behavioral Science, 16*, 109–118. https://doi.org/10.1016/j.jcbs.2020.03.009

Öst, L.-G., & Ollendick, T. H. (2017). Brief, intensive and concentrated cognitive behavioral treatments for anxiety disorders in children: A systematic review and meta-analysis. *Behaviour Research and Therapy, 97*, 134–145. https://doi.org/10.1016/j.brat.2017.07.008

Pettit, J. W., Rey, Y., Bechor, M., Melendez, R., Vaclavik, D., Buitron, V., Bar-Haim, Y., Pine, D. S., & Silverman, W. K. (2017). Can less be more? Open trial of a stepped care approach for child and adolescent anxiety disorders. *Journal of Anxiety Disorders, 51*, 7–13. https://doi.org/10.1016/j.janxdis.2017.08.004

Plemmons, G., Hall, M., Doupnik, S., Gay, J., Brown, C., Browning, W., Casey, R., Freundlich, K., Johnson, D. P., Lind, C., Rehm, K., Thomas, S., & Williams, D. (2018). Hospitalization for suicide ideation or attempt: 2008–2015. *Pediatrics, 141*(6), e20172426. https://doi.org/10.1542/peds.2017-2426

Rapee, R. M., Lyneham, H. J., Wuthrich, V., Chatterton, M. L., Hudson, J. L., Kangas, M., & Mihalopoulos, C. (2017). Comparison of stepped care delivery against a single, empirically validated cognitive-behavioral therapy program for youth with anxiety: A randomized clinical trial. *Journal of the American Academy of Child & Adolescent Psychiatry, 56*(10), 841–848. https://doi.org/10.1016/j.jaac.2017.08.001

Richards, D. A., Bower, P., Pagel, C., Weaver, A., Utley, M., Cape, J., Pilling, S., Lovell, K., Gilbody, S., Leibowitz, J., Owens, L., Paxton, R., Hennessy, S., Simpson, A., Gallivan, S., Tomson, D., & Vasilakis, C. (2012). Delivering stepped care: An analysis of implementation in routine practice. *Implementation Science, 7*(1), 3. https://doi.org/10.1186/1748-5908-7-3

Roach, A., Cullinan, S., Shafran, R., Heyman, I., & Bennett, S. (2023). Implementing brief and low-intensity psychological interventions for children and young people with internalizing disorders: A rapid realist review. *British Medical Bulletin, 145*(1), 120–131. https://doi.org/10.1093/bmb/ldad001

Robins, D. L., Fein, D., & Barton, M. L. (1999). *Follow-up interview for the modified checklist for autism in toddlers (M-CHAT FUI)*. Self-published.

Robins, D. L., Fein, D., & Barton, M. L. (2009). *The modified checklist for autism in toddlers, revised with follow-up*. M-CHAT. https://mchatscreen.com/

Robins, D. L., Casagrande, K., Barton, M., Chen, C.-M. A., Dumont-Mathieu, T., & Fein, D. (2014). Validation of the modified checklist for autism in toddlers, revised with follow-up (M-CHAT-R/F). *Pediatrics, 133*(1), 37–45. https://doi.org/10.1542/peds.2013-1813

Salloum, A., Robst, J., Scheeringa, M. S., Cohen, J. A., Wang, W., Murphy, T. K., Tolin, D. F., & Storch, E. A. (2014). Step one within stepped care trauma-focused cognitive behavioral therapy for young children: A pilot study. *Child Psychiatry & Human Development, 45*(1), 65–77. https://doi.org/10.1007/s10578-013-0378-6

Salloum, A., Wang, W., Robst, J., Murphy, T. K., Scheeringa, M. S., Cohen, J. A., & Storch, E. A. (2016). Stepped care versus standard trauma-focused cognitive behavioral therapy for young children. *Journal of Child Psychology and Psychiatry, 57*(5), 614–622. https://doi.org/10.1111/jcpp.12471

Salloum, A., Lu, Y., Chen, H., Salomon, K., Scheeringa, M. S., Cohen, J. A., Swaidan, V., & Storch, E. A. (2022). Child and parent secondary outcomes in stepped care versus standard care treatment for childhood trauma. *Journal of Affective Disorders, 307*, 87–96. https://doi.org/10.1016/j.jad.2022.03.049

Schleider, J. L., & Weisz, J. R. (2017). Little treatments, promising effects? Meta-analysis of single-session interventions for youth psychiatric problems. *Journal of the American Academy of Child & Adolescent Psychiatry, 56*(2), 107–115. https://doi.org/10.1016/j.jaac.2016.11.007

Shafran, R., Myles-Hooton, P., Bennett, S., & Öst, L.-G. (2021). The concept and definition of low intensity cognitive behaviour therapy. *Behaviour Research and Therapy, 138*, 103803. https://doi.org/10.1016/j.brat.2021.103803

Silverman, W. K., & Albano, A. M. (1996). *Anxiety disorders interview schedule for DSM-IV: Child version*. Oxford University Press.

Singla, D. R., Schleider, J. L., & Patel, V. (2023). Democratizing access to psychological therapies: Innovations and the role of psychologists. *Journal of Consulting and Clinical Psychology, 91*(11), 623–625. https://doi.org/10.1037/ccp0000850

Snider, T. C., & Lyren, R. (2023, October 20). Thinking outside the box: Completing autism evaluations in the pediatric primary care setting. Collaborate family healthcare association annual conference, Phoenix, AZ.

Snipes, C., Maragakis, A., & O'Donohue, W. (2015). Team-based stepped care in integrated delivery settings. *Family Medicine and Community Health, 3*(1), 39–46. https://doi.org/10.15212/FMCH.2015.0101

Society for Adolescent Health and Medicine. (2020). Improving integration of behavioral health into primary care for adolescents and young adults. *Journal of Adolescent Health, 67*(2), 302–306. https://doi.org/10.1016/j.jadohealth.2020.05.016

Strosahl, K. (2005). Training behavioral health and primary care providers for integrated care: A core competencies approach. In W. T. O'Donohue, M. R. Byrd, N. A. Cummings, & D. A. Henderson (Eds.), *Behavioral integrative care: Treatments that work in the primary care setting* (pp. 15–52). Brunner-Routledge.

Stutzman, D. L. (2021). Long-term use of antidepressants, mood stabilizers, and antipsychotics in pediatric patients with a focus on appropriate deprescribing. *Mental Health Clinician, 11*(6), 320–333. https://doi.org/10.9740/mhc.2021.11.320

Thomas, C. R., & Holzer, C. E. (2006). The continuing shortage of child and adolescent psychiatrists. *Journal of the American Academy of Child & Adolescent Psychiatry, 45*(9), 1023–1031. https://doi.org/10.1097/01.chi.0000225353.16831.5d

US Preventive Services Task Force, Mangione, C. M., Barry, M. J., Nicholson, W. K., Cabana, M., Coker, T. R., Davidson, K. W., Davis, E. M., Donahue, K. E., Jaén, C. R., Kubik, M., Li, L., Ogedegbe, G., Pbert, L., Ruiz, J. M., Silverstein, M., Stevermer, J., & Wong, J. B. (2022). Screening for anxiety in children and adolescents: US preventive services task force recommendation statement. *JAMA, 328*(14), 1438. https://doi.org/10.1001/jama.2022.16936

van Straten, A., Hill, J., Richards, D. A., & Cuijpers, P. (2015). Stepped care treatment delivery for depression: A systematic review and meta-analysis. *Psychological Medicine, 45*(2), 231–246. https://doi.org/10.1017/S0033291714000701

Vassilopoulos, A., Mohammad, S., Dure, L., Kozlowska, K., & Fobian, A. D. (2022). Treatment approaches for functional neurological disorders in children. *Current Treatment Options in Neurology, 24*(2), 77–97. https://doi.org/10.1007/s11940-022-00708-5

Visser, S. N., Zablotsky, B., Holbrook, J. R., Danielson, M. L., & Bitsko, R. H. (2015). Diagnostic experiences of children with attention-deficit/hyperactivity disorder. *National Health Statistics Reports, 81*, 1–7.

Walter, H. J., Bukstein, O. G., Abright, A. R., Keable, H., Ramtekkar, U., Ripperger-Suhler, J., & Rockhill, C. (2020). Clinical practice guideline for the assessment and treatment of children and adolescents with anxiety disorders. *Journal of the American Academy of Child & Adolescent Psychiatry, 59*(10), 1107–1124. https://doi.org/10.1016/j.jaac.2020.05.005

Weisz, J. R., Chorpita, B. F., Frye, A., Ng, M. Y., Lau, N., Bearman, S. K., Ugueto, A. M., Langer, D. A., Hoagwood, K. E., & The Research Network on Youth Mental Health. (2011). Youth top problems: Using idiographic, consumer-guided assessment to identify treatment needs and to track change during psychotherapy. *Journal of Consulting and Clinical Psychology, 79*(3), 369–380. https://doi.org/10.1037/a0023307

Wilson, G. T., Vitousek, K. M., & Loeb, K. L. (2000). Stepped care treatment for eating disorders. *Journal of Consulting and Clinical Psychology, 68*(4), 564–572. https://doi.org/10.1037/0022-006X.68.4.564

Wolraich, M. L., Hagan, J. F., Allan, C., Chan, E., Davison, D., Earls, M., Evans, S. W., Flinn, S. K., Froehlich, T., Frost, J., Holbrook, J. R., Lehmann, C. U., Lessin, H. R., Okechukwu, K., Pierce, K. L., Winner, J. D., Zurhellen, W., & Subcommittee on Children and Adolescents with Attention-Deficit/Hyperactive Disorder. (2019). Clinical practice guideline for the diagnosis, evaluation, and treatment of attention-deficit/hyperactivity disorder in children and adolescents. *Pediatrics, 144*(4), e20192528. https://doi.org/10.1542/peds.2019-2528

Yeguez, C. E., Page, T. F., Rey, Y., Silverman, W. K., & Pettit, J. W. (2020). A cost analysis of a stepped care treatment approach for anxiety disorders in youth. *Journal of Clinical Child & Adolescent Psychology, 49*(4), 549–555. https://doi.org/10.1080/15374416.2018.1539913

Yonek, J., Lee, C.-M., Harrison, A., Mangurian, C., & Tolou-Shams, M. (2020). Key components of effective pediatric integrated mental health care models: A systematic review. *JAMA Pediatrics, 174*(5), 487. https://doi.org/10.1001/jamapediatrics.2020.0023

Part II
Specialty Settings

Chapter 5
Outpatient Behavioral Health Settings

Michelle Byrd

Addressing the Mental Health Crisis Among Children and Adolescents in America

In 2022, the American Academy of Pediatrics, the American Academy of Child and Adolescent Psychiatry, and the Children's Hospital Association jointly declared a national emergency in child and adolescent mental health, citing the significant upward trend in mental health problems among our nation's youth, further exacerbated by the COVID-19 pandemic (American Academy of Pediatrics, 2022). In any given year, it is estimated that 20% of America's children and adolescents will meet the diagnostic criterion for a mental health disorder, while 40% will at some point by age 18, which is approximately 5% higher than global rates (National Healthcare Quality and Disparities Report (NHQDP), 2022; Shim et al., 2022). From 2009 to 2019, suicidal behavior among adolescents increased by an estimated 40%, with psychological disorders being the leading cause of death and disability among adolescents (NHQDP, 2022). At the other end of the age spectrum, Bitsko et al. report that even very young children are showing a significant uptick in mental health problems and that minoritized children and those living in poverty are differentially at risk (2022). In a recent study which took into consideration geographical, racial/ethnic, and socioeconomic differences in prevalence rates of common psychological problems among school-aged (K-12) children across the United States, a range of 15–33% of children across sites met the criterion for an externalizing or internalizing disorder, in that order (Danielson et al., 2021). More specifically, oppositional

M. Byrd (✉)
Department of Psychiatry, UniversityVA Ann Arbor Healthcare System, Howell, MI, USA
e-mail: Michelle.Byrd@va.gov

A. Maragakis, M. Janikian (eds.), *Evidence-Based Behavioral Health Practices in Pediatric Specialty Settings*, https://doi.org/10.1007/978-3-031-71004-9_5

defiant disorder (ODD) and attention deficit hyperactivity disorder (ADHD) accounted for the highest proportion of diagnosable cases, followed by anxiety disorders and depressive disorders, respectively. Furthermore, Danielson et al. (2021) found significant demographic differences in children at risk for developing mental health problems (e.g., symptom elevation as rated by teachers), with males, non-Hispanic black children, and those receiving free or reduced lunch being differentially at risk at sites across the country. As clinicians working in outpatient behavioral health, knowledge and awareness of these epidemiological trends are critical and must be updated constantly in order to focus our attention on vulnerability factors among the children we serve.

Perhaps even more alarming than the prevalence rates of mental health problems among our nation's children is the gap between known cases and the receipt of appropriate care. According to the National Center for Health Statistics (2021), approximately 15% of children aged 5–17 received some form of mental health care, including both medication (8%) and/or psychotherapy (11.5%), with adolescents (aged 12–17) being much more likely to receive care than younger children. In the same dataset (2021), significant racial/ethnic differences also emerged, with Asian children being the least likely and white-non-Hispanic children being the most likely to receive mental health treatment. Perhaps somewhat surprisingly, children living in more rural areas were equally likely to have been engaged with psychotherapy as children in more urban/metropolitan areas, though rates of psychotropic medication use were positively correlated with rurality. Given these data, it is critical that therapists understand the demographics of the populations served in their clinical setting.

While some of these findings suggest a small gap between estimated prevalence and treatment rates, in a 2023 meta-analysis (pooled n = 310, 584; Wang et al., 2023), a more alarming trend emerged with only 38% of children/adolescents who met criteria for any disorder receiving any form of mental health treatment. Treatment rates vary significantly by disorder, with children presenting with ADHD being treated at the highest rate (58%), followed by behavior disorders (i.e., oppositional defiant disorder and conduct disorder; 49%), depressive disorders (36%), and, lastly, anxiety disorders (31%). It is perhaps not surprising that treatment rates map to the degree to which children's symptoms are likely to disrupt a family or school system's functioning; to put it bluntly, the more a child's mental health diagnosis becomes uncomfortable or inconvenient for the adults in their life, the more likely they are to receive care, which speaks to underlying cultural norms and values driving help-seeking behavior. Among children and adolescents seeking care, slightly more children are receiving psychotherapy (13%) vs. medication (9%; National Center for Health Statistics, 2022) and 85% of referring pediatricians found it difficult to obtain access to evidence-based practices (EBP) for children and families (Chien et al., 2022), which speaks to the need for more clinicians overall and, specifically more clinicians trained in child EBPs. This volume of work supports the critical need to train more clinicians in the workforce in EBPs.

Broad and General Foundations of Therapy for Clinicians in Child Outpatient Behavioral Care

The Society of Clinical and Child and Adolescent Psychology (American Psychological Association [APA], Division 53) has created a dynamic list of evidence-based approaches (created in 2017, last updated in 2024), which is a point of reference that all child clinicians should be familiar with. The following therapies are considered to be evidenced-based for children according to the APA: Applied Behavior Analysis (ABA), behavior therapy, cognitive behavior therapy (CBT), cognitive therapy, family therapy, dialectical behavior therapy (DBT), interpersonal therapy, motivational interviewing, and organizational skills training (2024). To take this list a step deeper and more specific, Friedberg (2018) offers some expert guidance on areas of competence, which he suggests are critical for CBT-based care. Specifically, Friedberg cites the following skills/abilities as being foundational to be successful in outpatient therapy (upon which more advanced and specific skills should be built): Socratic questioning, self-instruction, problem-solving, imagery, behavioral activation, social skills training, the ability to conduct behavioral experiments, and exposure/response prevention.

While a very helpful guidepost, a limitation of these suggested lists is that they fail to connect techniques and protocols to diagnostic groups. Moreover, it would be virtually impossible for a clinician to achieve true mastery in all of these approaches. Indeed, given the extensive depth of training required for some of these schools of therapy (e.g., ABA and CBT), most clinicians likely make the choice of what therapies they will practice de facto when they choose their graduate training program. Becoming an expert in one or two of these broader areas of competence is common and would generally be sufficient as a scope of practice. However, some of the approaches listed by APA nest into each other, such as specializing in DBT Motivational Interviewing in the context of learning CBT.

Mental health providers need to be mindful of the correspondence between the skill sets they are learning in their foundational training and the base rates of problems they are most likely to see in their likely later employment settings, taking into account the prevalence and socioeconomic correlates of the population they intend to serve. For example, if a clinician anticipates providing care as part of a treatment team, it may be helpful to specialize in modalities that are more likely to be utilized with a specific age range, such as elementary age or adolescence, assuming that other colleagues will have complementary expertise. It would also be common in outpatient settings for clinicians to develop specialization in particular kinds of clinical presentations/disorders, such as having expertise in ABA, if they intend to treat children with autism spectrum disorder (ASD) or developmental and/or cognitive disabilities. However, if a clinician is interested in working as a solo practitioner, particularly in a more rural area, it may be more beneficial for them to have more generalized training, across developmental stages, clinical presentations, and broad theoretical approaches, and be prepared to refer to other clinicians outside of their agency or geographic area for age ranges, clinical presentations, and particular therapeutic modalities cases that are outside of their scope of practice. Furthermore, outpatient clinicians must have the flexibility to apply techniques across treatment protocols competently and efficiently, mindful that comorbidities are common.

In the current practice landscape, clinicians must also expand their repertoire to provide some treatments digitally, as the demand for online care has been sustained at some level since the pandemic, though obviously lower than what was common during the height of the public health crisis. Unfortunately, the research literature has not yet caught up to current practices as to which treatments can be delivered with fidelity in an online format for children of particular developmental levels and clinical concerns. Without the guidance of empirical data, we rely solely on clinical judgment, which is the general state of the science regarding online care for children and adolescents.

As a general rule, younger children seem to benefit less from online care than their older counterparts. Children who are experiencing social isolation and are withdrawn also seem to be much more difficult to engage online. In addition, children with attentional problems are generally too distractable to have an optimal experience with online care. Therapists must be mindful to assess a particular client's learning history as it relates to online schooling before attempting online psychological services. Obviously, children had a range of experiences with mandatory online learning and some might even have had a highly aversive or even traumatic experience, which could jeopardize any therapeutic efforts. Other children may have had a very positive experience thus far with online engagement and might prefer to receive therapy in this format, as well. Given the diversity of techniques and the plethora of possible clinical presentations, achieving competence is a tall order and requires time, effort, and skill to build clinical expertise. Initial and ongoing clinical supervision is the key to achieving expertise.

Supervision

Competent and adequate clinical supervision is considered absolutely critical in building mastery of commonly deployed evidence-based practices in all settings; however, this is a pedagogical area of study that has not been as well-researched or documented as would be ideal. The assumption seems to be that clinicians will become trained in best practices during their graduate/early training, including internship and postdoctoral years; however, this assumption has not been critically examined, and state licensing laws necessitate at least preliminary supervision even when basic licensure requirements have been met. It is both alarming and shocking to note that according to numerous studies (e.g., Creed et al., 2016; Wiltsey Stirman et al., 2013), the *majority* of clinicians who self-identify as being competent in CBT are not able to meet criterion to be considered such, and half of "CBT" clinicians self-report that they routinely significantly utilize non-EBP approaches. It is beyond the scope of this chapter to speculate why or how such a derivation in practice occurs, but it is certain that these findings should necessitate a higher standard of ongoing supervision. Notably, research has suggested that the most basic and traditional didactic training methods are ineffective (Beidas et al., 2012; Edmunds et al., 2017). The extant supervision literature supports the need for ongoing consultation once basic didactic training has occurred (Beidas et al., 2012).

Building a Resource Library

Numerous high-quality texts have been written in support of outpatient pediatric behavioral care, particularly as this is the most common and most traditional treatment modality, and this author commends anyone reading this chapter to obtain and reference these works extensively. Notably, the majority of edited texts on pediatric outpatient care feature behavioral and/or CBT. Weisz and Kazdin's Evidence-Based Psychotherapies for Children and Adolescents (2017) is considered a classic and is an excellent resource. A second seminal text is Christopherson and VanScoyoc's Treatments That Work with Children: Empirically Supported Strategies for Managing Childhood Problems (2013), which provides a greater scope of problem areas and more concrete suggestions than many similar texts. For clinicians specializing in older children and adolescents, DBT Skills Manual for Adolescents (Rathus & Miller, 2015) is a seminal text, with chapters on classic DBT skills such as mindfulness, distress tolerance, emotion regulation, and interpersonal skills along with a special section for teens entitled "walking the middle path." This section is designed to help adolescents and their caregivers navigate the unique developmental tasks associated with seeking and achieving autonomy. It is the hope of this author that this volume will also be considered a critical text to practitioners.

The purpose of this chapter is to increase the probability of mental health providers delivering evidence-based modalities to children and adolescents presenting for care in outpatient behavioral health settings. An assumption of this chapter is that clinicians will already have a strong foundational repertoire, such as described earlier and consistent with APA's suggested competencies for evidence-based practice. As such, the author will focus this chapter on providing a "how to" reference for the highest base rate problems treated in outpatient pediatric mental health clinics: attentional disorders, disruptive behavior disorders, depression, and anxiety. However, before these problems can be addressed in outpatient settings, two significant metabolic comorbidities must be assessed for and treated if present, as failing to do so would decrease the efficacy of or thwart evidence-based approaches: sleep and feeding disorders.

Necessity of Pretreating Sleep and Feeding Disorders in Pediatric Outpatient Settings

The extant literature shows strong bidirectional relationships between sleep and feeding problems and behavioral health problems in children (Alfano & Gamble, 2009). Indeed, changes in sleep and appetite/eating patterns are included in the diagnostic criterion for some psychological disorders but can also be independent phenomena and, important to the context of this chapter, can significantly impact treatment efficacy for the most commonly seen behavioral concerns in outpatient clinics. For example, young children who are not getting enough sleep are 20% more likely to present with hyperactivity (Touchette et al., 2009). The old adage that "if you're not sleeping and you're not eating, nothing much else good can happen" really seems to hold true and provides foundational guidance for the pediatric clinician. Moreover, the interaction between sleep and nutrition has been established empirically (Diethelm et al., 2011) and when children are struggling with both, this negative synergy might be particularly detrimental to their psychological health. Therefore, it is critical to address these concerns prior to initiating evidence-based protocols for internalizing and externalizing disorders.

Sleep

Several recent global meta-analyses (Cai et al., 2024; Jahrami et al., 2022) indicated that 34–45% of healthy children/teens had significant sleep problems during the COVID-19 pandemic. Outside of the acute stress associated with that period of time, rates of sleep problems in this population are still significant, but data are not yet available to document what long-term impact COVID-19 had on sleep among people at various stages of development. Lewien et al. (2021) found that greater than 20% of healthy children and teens are getting less sleep and poorer quality sleep than recommended. Among school-aged children, the most common sleep problems are sleep resistance, difficulty falling or staying asleep, sleep-related anxiety, and parasomnia. Among adolescents, delayed sleep onset and daytime sleepiness are the most common problems. Hysing et al. (2013) reported that adolescents in their sample were getting about 2 h less sleep per night than recommended for optimal development, with 18% experiencing clinical levels of insomnia. All of these sleep-related concerns are more severe among children experiencing depression, anxiety, and ADHD (Ramtekkar & Ivanenko, 2015).

A recent meta-analysis found that two interventions have been shown to be especially effective in improving sleep among young children, settling methods and establishing a bedtime routine (Park et al., 2022). The goal of settling methods is to support children in learning to self-soothe to get themselves to sleep/back to sleep in a developmentally appropriate manner. These techniques can be deployed from the time of infancy when parents learn to put their baby down well-fed, dry, and drowsy, rather than fully asleep. As children grow, parents practice direct teaching and graduated extinction to support children to fall asleep by themselves and in their own beds, with minimal environmental intervention on the part of the parent such that the child can replicate the settling methods on their own when they wake up during the night. This training is seen as a form of self-regulation. Related, bedtime routines have been demonstrated to improve sleep in children of all ages. Interventions related to bedtime routines involve first setting a bedtime and then establishing a limited set of behaviors which routinely occur before bed to classically condition drowsiness. It is important that the elements of the bedtime routine are nourishing, soothing, and promoting good sleep hygiene. As such, the bedtime routine should ideally not involve any technology so as to prevent blue light exposure, which suppresses melatonin production and has been found to be twice as significant in children than in adults (Higuchi et al., 2016; Lee et al., 2019). For more suggestions and specific guidance on sleep training, Dr. Patrick Friman's self-help book, *Good Night, Sweet Dreams, I Love You, Now Get into Bed and Go to Sleep!* (2005) is an excellent resource for parents as well as clinicians and can be used for children through middle-school age.

Feeding

Malnutrition occurs in various forms including undernutrition and vitamin/mineral deficiencies. Depending on the type of malnutrition, children may present as either under or overweight and are more at risk if they are living in poverty. Furthermore, it is estimated that approximately 14% of young children are considered overweight/obese, with that number rising to about 20% of teenagers (Hales et al., 2017); these percentages translate to 37 million children globally World Health Organization (WHO, 2024) and children who are overweight/obese are also at risk of malnutrition. Roughly 150 million children worldwide have stunted growth, and 45 million are considered too thin for their height (WHO, 2024). Childhood malnutrition has been linked with numerous psychological concerns, including behavioral problems, and, therefore, must be a target of intervention in order to succeed in addressing more typical outpatient psychotherapy targets. Feeding disorders are characterized by either insufficient dietary intake or eating in ways that are developmentally or nutritionally inappropriate.

Feeding and nutrition problems are typically addressed by a multidisciplinary team, including medical doctors and registered dieticians, and may necessitate more involved assessment to determine etiology and treatment planning beyond the scope of this chapter. However, from a behavioral perspective, picky eating, mealtime tantrums, and food refusal are common clinical phenomena for which there are several techniques that clinicians should be familiar with. Clinical outcomes may be measured by the amount, caloric value, or weight of food eaten, broader food acceptance, or a change in appropriate feeding behavior (e.g., use of utensils and manners).

Treating feeding disorders involves using behavioral techniques and often begins with appetite manipulation. Simply stated, the goal of appetite manipulation is to withhold calorically dense but nutritionally void drinks and snacks so that children are motivated to eat healthier foods at mealtimes to reduce their hunger. A second key feeding technique is changing the valance of the feeding environment from mealtime being a battleground to a more naturally reinforcing situation. This often involves making mealtime more developmentally appropriate and engaging for children, including enhancing the interactions by engaging the child's senses and providing appropriately sized furniture and utensils, along with verbally engaging the child as long as they are behaving appropriately (contingent social attention). We encourage parents to actively ignore inappropriate behavior to create a contrast effect and differentially reinforce the desired behavior. When children are distracted at the table, we encourage parents to use consistent verbal and physical prompts to cue eating behavior (i.e., "Please take a bite."). Parents may also provide positive and negative tangible consequences contingent upon their child's feeding behavior such as offering bites of preferred foods as a reward for eating bites of nonpreferred food and even using tokens or toys as rewards for compliant behavior. Taken together, techniques used for improving feeding are highly effective (Kelly et al., 2014) and the course of treatment is typically around 1 month.

With sleep and feeding problems addressed, behavioral health providers can turn their clinical attention to externalizing and internalizing concerns.

Externalizing Disorders

Attentional Disorders

Research has indicated that the majority of children who present with ADHD are also diagnosed with a comorbid disorder, most commonly another externalizing disorder (66%; Reale & Bartoli, 2017). When clinicians are faced with children who have comorbid attentional and other either externalizing or internalizing disorders, clinical wisdom dictates that they treat attentional disorders first for two reasons. First, because attentional problems are most commonly addressed utilizing medication, often these concerns improve quite rapidly (Peterson et al., 2024), and ancillary concerns might spontaneously improve, as well. If ADHD symptoms are ameliorated at least to some extent, clinicians will have greater success in treating comorbidities and clients will be more able to effectively engage in behavioral treatment modalities. Thus, when faced with a child presenting with ADHD and other clinical concerns, the clinician should likely prioritize treating the attentional problems and, in doing so, should offer a medication referral as a first-line treatment.

There are also several well-established behavioral treatments for attentional problems including behavioral parent training (please see section below), CBT, organizational training, and school-based interventions, including peer interventions. Behavioral techniques include employing stimulus control in both the home and school environment as well as reinforcement of on-task behavior delivered by the parent, teacher, or child themselves. Token economies have been shown to be particularly effective in rewarding task completion. Cognitive strategies have also been well-studied and found to be useful for children of older elementary age and above. These include using positive self-statements and self-instructions to stay focused and intentional, as well as teaching self-evaluation of both process and outcome variables to help gain mastery over controlling one's attention. Mindfulness interventions are also emerging in practice as well as in the extant literature though they have not yet been as widely studied as more standard techniques. Some children with ADHD may also benefit from social skills training if their deficits have resulted in failing to learn age-appropriate interpersonal skills of engagement. By adolescence, the clinical goal is that individuals are able to taper interventions delivered by trusted adults for the most part because they have learned self-directed compensatory strategies such that their ADHD is well-managed.

Disruptive Behavior Disorders

Disruptive behavior disorders are highly comorbid with attentional disorders and are characterized by patterns of defiance, interpersonal argumentativeness, and persistent expressions of anger/aggression. There are two primary disruptive behavior disorders, Oppositional Defiant Disorder (ODD) and Conduct Disorder (CD), and clinicians will find themselves treating the former with much greater frequency than the latter in typical outpatient practice. In fact, ODD is considered to be the most prevalent clinical presentation in young children, with a peak diagnosis of age 6–7 (Loeber et al., 2000). Children with ODD often exhibit a willingness to break minor rules and are described as being strong-willed by their caregivers, often resorting to temper tantrums or "meltdowns" when they don't get their way. Notably, most cases of ODD do not result in a diagnosis of CD (only about 10%), which is considered to be a much more severe pattern of disruptive behavior and is characterized by a high level of aggressiveness and/or violation of norms, predictive of later criminal behavior.

Treatment of disruptive behavior is a family affair, as parenting style is considered the most robust predictor of disruptive behavior (Eddy et al., 2001). Specifically, counterpliant behavior is highly correlated with inconsistent and/or harsh disciplinary practices and caregivers having a negative attitude about their child, including portraying parental depression and demonstrating a lack of concern for the child's whereabouts, friends, interests, etc. Hence, a key element of treating disruptive behavior is teaching parent management techniques that reduce or eliminate ineffective disciplinary practices and help to heal the parent–child relationship(s). These goals are primarily achieved by employing parent management training (PMT).

Parent Management Training

There are numerous forms of parent management training (PMT; e.g., Barkley, 2007; Funderburk & Eyberg, 2011; McMahon & Forehand, 2003; Patterson, 1979; Webster-Stratton & Reid, 2003), all of which have considerable empirical support (Mingebach et al., 2018) and may include elements, which make them more appropriate for a particular family. In broad strokes, PMT is a behavioral treatment based on principles of operant learning within the context of developmental and, specifically, attachment theories, and, as such, is transtheoretical. Parent management protocols typically involve two phases of treatment: one to restore patterns of reinforcement and effective engagement between parent and child and one to create appropriate boundaries and predictable consequences for violations of those boundaries.

Within the first phase of PMT, designed to establish or improve the existing parent–child relationship, parents are encouraged to engage with their child in developmentally appropriate activities driven by the child's interests in an attempt to create a context for the parent being able to express interest in and appreciation for the child's behavior. During this phase, parents are taught (or retaught) how to convey appreciation for their child both verbally and nonverbally and nonjudgmentally/critically. This time together should be mutually reinforcing and create a contrast effect for when the parent needs to establish or enforce a boundary with the child.

Within the second phase of treatment, the parent is taught behavior management techniques to put limits on their child's behavior. The first step in this process is to establish clear and appropriate rules for their child and to communicate these rules. Parents are taught to build instructional control in their children by giving clear commands to establish expectations. They are then taught to apply consistent consequences for their child's behavior by rewarding compliance through social, token, and tangible means and punishing noncompliance through the removal of reinforcers such as attention, privileges, and material goods. This may take the form of time-out in younger children or grounding in older children.

When ODD escalates to CD, additional treatment strategies must be deployed, as a child meeting these diagnostic criteria is unlikely to return to baseline without clinical intervention. Psychopharmacology has not been shown to have direct empirical efficacy in treating CD outside of its utility in addressing a comorbid diagnosis; however, it may be considered an adjunctive treatment if the child is exhibiting violent behaviors. Parent management training, as described above, is also recommended for CD but should be undertaken along with child CBT.

CBT for CD involves numerous components, many of which focus on social dynamics both within and outside of the family and may incorporate DBT Interpersonal Effectiveness Skills (Rathus & Miller, 2015). Specifically, CBT for CD emphasizes teaching problem-solving skills to reduce conflict in interpersonal interactions, including teaching appropriate assertiveness. In doing so, behavioral techniques of role-playing, modeling, practice, and feedback are consistently used by the clinician, who may choose to strategically involve caregivers or other significant others in the child's life to help change patterns of interpersonal interactions. In addition, important adults in the child's life are taught and encouraged to use social and token reinforcement techniques to increase the probability of prosocial behavior. When parents and their children have been embroiled in a cycle of coercion (Patterson, 1982) where each has negatively reinforced the other's antisocial behavior, it is an important albeit Herculean feat to restore the ability for family members to actually deliver positive reinforcement to one another for desirable behavior.

Internalizing Disorders

Anxiety

As with adults, anxiety occurs on a continuum for children and adolescents from normative and even productive stress to that which is debilitating and ineffective (Beesdo et al., 2009). That said, there are numerous special considerations in treating anxiety in children and adolescents. In young children, there are some unique forms of anxiety that are unlikely to require treatment but which may later develop into treatable forms of anxiety, such as stranger anxiety and separation anxiety. While these types of anxiety are normative, parental response to their child's fears predicts either the development of resilience and a growth experience or the development of maladaptive avoidance behavior that gives rise to later diagnosable anxiety (Capriola-Hall et al., 2020). Because many of these fears are transient, typically anxiety is only the focus of clinical attention in children over age 8 or in those for whom their anxiety has led to impairments in important areas of functioning such as social isolation or school refusal. However, because parental anxiety is a major predictor of child anxiety, treatment may also be preventative in nature.

As clinicians, we must also be mindful to assess for traumatic experiences in a child's life which may precipitate anxiety, cognizant that trauma in the life of a child is entirely subjective and mediated by the parental ability to assist their child in coping with life's inevitable challenges and changes (Vanderzee et al., 2018). Guidance on providing trauma-informed care exceeds the scope of this chapter, but Acute Stress and Post-Traumatic Stress Disorder (PTSD) should be ruled out as foci of treatment before proceeding to treat more common presentations of anxiety.

There are several high base rate anxiety disorders in children and adolescents, and, indeed, anxiety is the most commonly diagnosed problem among children and adolescents (though it is not the most likely to be treated). As with other clinical presentations discussed in this chapter, anxiety disorders are also highly comorbid with other conditions including other forms of anxiety as well as externalizing disorders and medical conditions which also increase the probability of a child experiencing excessive worry. Moreover, anxiety in childhood predicts anxiety in adulthood (Copeland et al., 2014). The most common forms of anxiety are Generalized Anxiety Disorder (GAD), Social Phobia, and Simple Phobia.

As mentioned above, there are several predictors that have been linked with the development of clinical anxiety in children and these etiological factors guide the treatment. Many children with anxiety are reported to have longstanding patterns of behavior from toddlerhood characterized by behavioral inhibition in the form of excessive shyness or timidity. While some have theorized that these patterns are biological in nature, that is, impossible to determine given the hypothesized loading of parental factors on the development of diagnosable anxiety. To that end, parental anxiety is highly correlated with child anxiety, which may be genetic, transmitted through modeling of fearful responses, or, in all likelihood, a combination therein (Sweeney & Wilson, 2023). Other environmental factors such as having a highly chaotic and unpredictable home environment or a family that is highly critical also predict the development of an anxiety disorder in the next generation.

As there has not been an established biological mechanism of transmission of anxiety, it is not surprising that pharmacological treatments for anxiety are not recommended for children and adolescents except in the most severe cases and, even then, only long enough for other techniques to be used. CBT is by far the most well-established treatment for anxiety, though parent management training (as described above) has also been clinically indicated, especially in families that are characterized by being highly unpredictable or when children have comorbid externalizing disorders (James et al., 2020).

The first phase of CBT for anxiety in children focuses on examining patterns in the family that have inadvertently created anxious patterns of behavior. CBT for children starts with examining how anxious responses may be modeled and reinforced in the family environment. It is not uncommon for parents to require treatment for their own anxiety before their child can benefit from anxiety treatment, in fact. If an anxious child has an anxious parent, the therapist has to assume that that dyad does not have effective coping skills to deal with stressors and must begin by taking a psychoeducational approach to coping skills training for both child and caregiver. Prior to doing any exposure and response prevention, children (and their caregivers) often need to be taught self-regulation skills. Even young children can be taught diaphragmatic breathing techniques and progressive muscle relaxation, with numerous developmentally appropriate scripts available on the Internet. Depending on the age of the child, DBT Emotion Regulation skills may also be a useful skill set and may be taught to parent–child dyads simultaneously to encourage use. Clinicians should take care to ensure that there is some dyadic mastery of coping/regulation skills prior to undertaking any behavioral challenges to the existing anxious patterns of responding.

Much like adult CBT for anxiety, delivering CBT to children with anxiety involves conducting behavioral experiments and gradually exposing children to their feared stimuli while encouraging self-management, blocking avoidance, and reinforcing engagement (Seligman & Ollendick, 2011). Unlike adult CBT, conducting this work with children invites the opportunity for creativity in exposure and many clinicians will utilize nonverbal mechanisms for engaging feared stimuli, such as drawing or other artistic representations. In the treatment of social anxiety or specific phobia, modeling and role-playing may be critical components. It is important that clinicians ensure that parents actively reinforce their child staying in contact with the anxiety-producing stimuli, as the therapeutic efforts will be futile if not supported by the child's most salient source of social support. Specifically, parents should be coached to act as a co-therapist and be able to both model and reinforce non-anxious responses. In addition, parents should be encouraged to do planned ignoring of anxious behavior, including modifying their own parenting style to not accommodate (and thereby inadvertently reinforce) their child's anxiety. Depending on the specific nature of a child's anxiety, CBT in a group format may also be efficacious, particularly so for children who are struggling with social anxiety (Wolgensinger, 2015). CBT groups for anxiety have better outcomes when paired with parent training as an adjunctive therapy.

Depression

Unlike anxiety, depression in children often appears markedly differently than in adults (Bernaras et al., 2019). Specifically, children are more likely to manifest depression in symptoms such as irritability and defiance rather than having low mood being the most prominent feature. Children may also be more likely to report somatizing symptoms of depression such as stomach and head aches, which should be relatively uncommon in otherwise healthy youngsters. Other symptoms such as anhedonia, changes in metabolic functions, and impairment in functioning in important life domains are similar to adults (American Psychiatric Association, 2013). Because children's academic performance is continually graded and reported, academic changes may be one of the first signs of depression whereas job performance in adults may take much longer to become evident. As stated at the beginning of this chapter, depression is often comorbid with anxiety and clinicians should take care to rule out both trauma and ODD in addition. Rates of depression are positively correlated with developmental age; hence, therapists should be more concerned with etiology if they observe depression in younger children. Related to the age trajectory of incidence, therapists should also take care to assess for substance involvement in older children and adolescents who present with depression, as addiction would significantly change the treatment plan and increase the risk of impulsive and potentially dangerous behavior. A full discussion of suicide assessment and preventive care is beyond the scope of this chapter and worthy of its own dedicated conversation; however, it goes almost without saying that providers should be well-versed in assessing for and addressing any suicidal ideation, plan, or intent and be prepared to break confidentiality to protect their client's well-being (Weber et al., 2017).

As with anxiety disorders, the home or family context must be closely considered to determine if there are specific depressogenic factors that may be causing or contributing to an individual's mood disorder (Marroquín & Nolen-Hoeksema,

2015). For example, it would be critical to know if the family system is experiencing a significant life change or stressor, such as moving or a parent losing a job. It would also be important to record if any parents or caregivers are themselves struggling with depression, as this is a known risk factor for child/adolescent depression, as well. Because of the myriad changes associated with and expected during rapid periods of development in adolescence, it is important to distinguish between what may be an acute stress response or Adjustment Disorder vs. Major Depression, though treatment may still be warranted in the former cases as a form of secondary prevention.

The use of psychotropic medications in the treatment of childhood depression is a highly controversial topic in pediatric psychiatry, and a nuanced conversation about the risk/benefit profile cannot occur at the population level and must be considered on an individual basis (Zhou et al., 2020). While the role of the pediatric mental health provider would never be to prescribe medication, it may well be to monitor response to medication and possible side effects.

As with other internalizing disorders, CBT is the EBP of choice in addressing depression (Oud et al., 2019). Cognitive techniques are used to address symptoms of guilt, hopelessness, and worthlessness, in addition to challenging negative thoughts or beliefs. More specifically, cognitive techniques are used to discover automatic thoughts, underlying assumptions, and core beliefs. Clinicians then facilitate their clients in identifying cognitive distortions and questioning or refuting self-defeating thoughts. Because of the inherently social nature of childhood, many clinicians find that the "best friend" technique is particularly useful with children. This technique involves asking the child to imagine what they would say if their best friend espoused negative beliefs they have reported, which almost universally allows the person to defuse from and refute the thought in a more compassionate light. Training children in effective problem-solving is another cognitive technique used when treating depression that can contribute to a child's psychological repertoire of skills. When done in a sensitive and developmentally mindful manner, cognitive techniques can be employed across the age spectrum with good effect. There are also a number of commercially available therapeutic board games that further facilitate identifying cognitive aspects of depression treatment by providing the use of vignettes and giving children the opportunity to challenge cognitive distortions.

Behavioral techniques to combat depression involve bringing the child back in contact with available reinforcers in their environments such as home and school. Sometimes, increasing pleasant events requires parental involvement and support, as becoming more actively engaged often means a greater expenditure of resources such as time and money on the part of the parent, so clinicians must take care to ensure that parents are supportive of related treatment goals or these interventions could be counterproductive if the child has hope of being able to attend a potentially fun event only to be let down when a parent determines that doing so is not feasible. It is important that home environment also become more reinforcing if at all possible, which might involve either parent coaching or family therapy, depending on the specific dynamics at play.

Intertwined with increasing access to reinforcers is the third aspect of CBT for depression in children, improving social functioning with the goal of improving a sense of connectedness and social belonging. Children experiencing depression who have social skills deficits may benefit from direct social skills training in assertiveness or interpersonal effectiveness, both of which are directly addressed in DBT skills training. As with other aspects of CBT, in teaching social skills, using modeling, practice, role plays, and providing feedback are essential for skill acquisition. Parents and therapists should work together to help children reduce their social isolation by giving differential attention for prosocial behavior in the family context and encouraging active involvement in group settings outside of the family, such as joining teams, clubs, youth groups, and scout troops. If there are structural or logistic impediments to children being able to increase their social engagement, the clinician should engage in collaborative problem-solving with the caregiver to facilitate the child's ability to expand their social network and potential for reinforcement in a social context. As an expansion of this notion, for teenagers experiencing depression, interpersonal therapy (IPT) has also been shown to be an efficacious treatment modality, which also emphasizes improving the quality of relationships by targeting interpersonal communication, decision-making, and problem-solving (Jacobson et al., 2018).

Summary and Conclusions

Having the opportunity to work in an outpatient clinic for children and adolescents is a tremendous gift because it allows clinicians to treat a wide range of clinical presentations before these problems have had the time to behaviorally calcify, and while there is typically still the chance not only to intervene with the individual but to change their environment as well, to support improved mental health. These distinct advantages allow us to be more effective therapeutically in less time than would be required for adult care. We also often have the chance to introduce young people to the practice of psychotherapy and, if done well, can instill in them the experience of using EBPs with good outcomes and devoid of stigma. Doing so could literally create habits of mind and behavior that could change the course of a person's life; what a profound gift it is to have the potential to make that level of impact on a young life and a gift we should not squander. Rather, we as therapists must seize the opportunity to provide care to children and adolescents in the absolutely best manner possible, mindful of excellence in both process and outcomes. This chapter was written with these ideals in mind and with the hope of passing on traditions of exemplar psychotherapy to the next generation of clinicians who will be caring for our own children and grandchildren.

References

Alfano, C. A., & Gamble, A. L. (2009). The role of sleep in childhood psychiatric disorders. *Child & Youth Care Forum, 38*(6), 327–340. https://doi.org/10.1007/s10566-009-9081-y

American Academy of Pediatrics. (2022). *AAP-AACAP-CHA declaration of a national emergency in child and adolescent mental health.* Available at: https://www.aap.org/en/advocacy/child-and-adolescent-healthy-mental-development/aap-aacap-cha-declaration-of-a-national-emergency-in-child-and-adolescent-mental-health/. Accessed 31 Jan 2022.

American Psychiatric Association. (2013). *Diagnostic and statistical manual of mental disorders: DSM-5™* (DSM-5 task force) (5th ed.). American Psychiatric Publishing. https://doi.org/10.1176/appi.books.9780890425596

Barkley, R. A. (2007). School interventions for attention deficit hyperactivity disorder: Where to from here? *School Psychology Review, 36*(2), 279–286.

Beesdo, K., Knappe, S., & Pine, D. S. (2009). Anxiety and anxiety disorders in children and adolescents: Developmental issues and implications for DSM-V. *The Psychiatric Clinics of North America, 32*(3), 483–524. https://doi.org/10.1016/j.psc.2009.06.002

Beidas, R. S., Edmunds, J. M., Marcus, S. C., & Kendall, P. C. (2012). Training and consultation to promote implementation of an empirically supported treatment: A randomized trial. *Psychiatric Services, 63*(7), 660–665. https://doi.org/10.1176/appi.ps.201100401

Bernaras, E., Jaureguizar, J., & Garaigordobil, M. (2019). Child and adolescent depression: A review of theories, evaluation instruments, prevention programs, and treatments. *Frontiers in Psychology, 10*, 543. https://doi.org/10.3389/fpsyg.2019.00543

Bitsko, R. H., Claussen, A. H., Lichstein, J., Black, L. I., Jones, S. E., Danielson, M. L., Hoenig, J. M., Davis Jack, S. P., Brody, D. J., Gyawali, S., Maenner, M. J., Warner, M., Holland, K. M., Perou, R., Crosby, A. E., Blumberg, S. J., Avenevoli, S., Kaminski, J. W., Ghandour, R. M., & Contributor. (2022). Mental health surveillance among children – United States, 2013–2019. *MMWR Supplements, 71*(2), 1–42. https://doi.org/10.15585/mmwr.su7102a1

Cai, H., Chen, P., Jin, Y., Zhang, Q., Cheung, T., Ng, C. H., Xiang, Y. T., & Feng, Y. (2024). Prevalence of sleep disturbances in children and adolescents during COVID-19 pandemic: A meta-analysis and systematic review of epidemiological surveys. *Translational Psychiatry, 14*(1), 12. https://doi.org/10.1038/s41398-023-02654-5

Capriola-Hall, N. N., Booker, J. A., & Ollendick, T. H. (2020). Parent- and child-factors in specific phobias: The interplay of overprotection and negative affectivity. *Journal of Abnormal Child Psychology, 48*(10), 1291–1302. https://doi.org/10.1007/s10802-020-00662-3

Chien, A. T., Leyenaar, J., Tomaino, M., Woloshin, S., Leininger, L., Barnett, E. R., McLaren, J. L., & Meara, E. (2022). Difficulty obtaining behavioral health services for children: A national survey of multiphysician practices. *Annals of Family Medicine, 20*(1), 42–50. https://doi.org/10.1370/afm.2759

Christophersen, E. R., & VanScoyoc, S. M. (2013). *Treatments that work with children: Empirically supported strategies for managing childhood problems* (2nd ed.). American Psychological Association. https://doi.org/10.1037/14137-000

Copeland, W. E., Angold, A., Shanahan, L., & Costello, E. J. (2014). Longitudinal patterns of anxiety from childhood to adulthood: The Great Smoky Mountains Study. *Journal of the American Academy of Child and Adolescent Psychiatry, 53*(1), 21–33. https://doi.org/10.1016/j.jaac.2013.09.017

Creed, T. A., Frankel, S. A., German, R. E., Green, K. L., Jager-Hyman, S., Taylor, K. P., Adler, A. D., Wolk, C. B., Stirman, S. W., Waltman, S. H., Williston, M. A., Sherrill, R., Evans, A. C., & Beck, A. T. (2016). Implementation of transdiagnostic cognitive therapy in community behavioral health: The Beck Community Initiative. *Journal of Consulting and Clinical Psychology, 84*(12), 1116–1126. https://doi.org/10.1037/ccp0000105

Danielson, M. L., Bitsko, R. H., Holbrook, J. R., Charania, S. N., Claussen, A. H., McKeown, R. E., Cuffe, S. P., Owens, J. S., Evans, S. W., Kubicek, L., & Flory, K. (2021). Community-based prevalence of externalizing and internalizing disorders among school-aged children and adolescents in four geographically dispersed school districts in the United States. *Child Psychiatry and Human Development, 52*(3), 500–514. https://doi.org/10.1007/s10578-020-01027-z

Diethelm, K., Bolzenius, K., Cheng, G., Remer, T., & Buyken, A. E. (2011). Longitudinal associations between reported sleep duration in early childhood and the development of body mass index, fat mass index and fat free mass index until age 7. *International Journal of Pediatric Obesity, 6*(2), e114–e123. https://doi.org/10.3109/17477166.2011.566338

Eddy, J. M., Leve, L. D., & Fagot, B. I. (2001). Coercive family processes: A replication and extension of Patterson's Coercion Model. *Aggressive Behavior, 27*(1), 14–25. https://doi.org/10.1002/1098-2337(20010101/31)27:1<14::AID-AB2>3.0.CO;2-2

Edmunds, J. M., Brodman, D. M., Ringle, V. A., Read, K. L., Kendall, P. C., & Beidas, R. S. (2017). Examining adherence to components of cognitive-behavioral therapy for youth anxiety after training and consultation. *Professional Psychology, Research and Practice, 48*(1), 54–61. https://doi.org/10.1037/pro0000100

Friedberg, R. D. (2018). Best practices in supervising cognitive behavioral therapy with youth. *World Journal of Clinical Pediatrics, 7*(1), 1–8. https://doi.org/10.5409/wjcp.v7.i1.1

Friman, P. C. (2005). *Good night, sweet dreams, I love you: Now get into bed and go to sleep!* Boys Town Press.

Funderburk, B. W., & Eyberg, S. (2011). Parent–child interaction therapy. In J. C. Norcross, G. R. VandenBos, & D. K. Freedheim (Eds.), *History of psychotherapy: Continuity and change* (2nd ed., pp. 415–420). American Psychological Association. https://doi.org/10.1037/12353-021

Hales, C. M., Carroll, M. D., Fryar, C. D., & Ogden, C. L. (2017). Prevalence of obesity among adults and youth: United States, 2015-2016. *NCHS Data Brief, 288*, 1–8.

Higuchi, S., Lee, S. I., Kozaki, T., Harada, T., & Tanaka, I. (2016). Late circadian phase in adults and children is correlated with use of high color temperature light at home at night. *Chronobiology International, 33*(4), 448–452. https://doi.org/10.3109/07420528.2016.1152978

Hysing, M., Pallesen, S., Stormark, K. M., Lundervold, A. J., & Sivertsen, B. (2013). Sleep patterns and insomnia among adolescents: A population-based study. *Journal of Sleep Research, 22*(5), 549–556. https://doi.org/10.1111/jsr.12055

Jacobson, C. M., Mufson, L. H., & Young, J. F. (2018). Treating adolescent depression using interpersonal psychotherapy. In J. R. Weisz & A. E. Kazdin (Eds.), *Evidence-based psychotherapies for children and adolescents* (3rd ed., pp. 66–82). The Guilford Press.

Jahrami, H. A., Alhaj, O. A., Humood, A. M., Alenezi, A. F., Fekih-Romdhane, F., AlRasheed, M. M., Saif, Z. Q., Bragazzi, N. L., Pandi-Perumal, S. R., BaHammam, A. S., & Vitiello, M. V. (2022). Sleep disturbances during the COVID-19 pandemic: A systematic review, meta-analysis, and meta-regression. *Sleep Medicine Reviews, 62*, 101591. https://doi.org/10.1016/j.smrv.2022.101591

James, A. C., Reardon, T., Soler, A., James, G., & Creswell, C. (2020). Cognitive behavioural therapy for anxiety disorders in children and adolescents. *The Cochrane Database of Systematic Reviews, 11*(11), CD013162. https://doi.org/10.1002/14651858.CD013162.pub2

Kelly, N. R., Shank, L. M., Bakalar, J. L., & Tanofsky-Kraff, M. (2014). Pediatric feeding and eating disorders: Current state of diagnosis and treatment. *Current Psychiatry Reports, 16*(5), 446. https://doi.org/10.1007/s11920-014-0446-z

Lee, S., Hale, L., Chang, A. M., Nahmod, N. G., Master, L., Berger, L. M., & Buxton, O. M. (2019). Longitudinal associations of childhood bedtime and sleep routines with adolescent body mass index. *Sleep, 42*(1), zsy202. https://doi.org/10.1093/sleep/zsy202

Lewien, C., Genuneit, J., Meigen, C., Kiess, W., & Poulain, T. (2021). Sleep-related difficulties in healthy children and adolescents. *BMC Pediatrics, 21*(1), 82. https://doi.org/10.1186/s12887-021-02529-y

Loeber, R., Burke, J. D., Lahey, B. B., Winters, A., & Zera, M. (2000). Oppositional defiant and conduct disorder: A review of the past 10 years, part I. *Journal of the American Academy of Child and Adolescent Psychiatry, 39*(12), 1468–1484. https://doi.org/10.1097/00004583-200012000-00007

Marroquín, B., & Nolen-Hoeksema, S. (2015). Emotion regulation and depressive symptoms: Close relationships as social context and influence. *Journal of Personality and Social Psychology, 109*(5), 836–855. https://doi.org/10.1037/pspi0000034

McMahon, R. J., & Forehand, R. L. (2003). *Helping the noncompliant child: Family-based treatment for oppositional behavior* (2nd ed.). Guilford Press.

Mingebach, T., Kamp-Becker, I., Christiansen, H., & Weber, L. (2018). Meta-meta-analysis on the effectiveness of parent-based interventions for the treatment of child externalizing behavior problems. *PLoS One, 13*(9), e0202855. https://doi.org/10.1371/journal.pone.0202855

National Center for Health Statistics. (2021) National Health Interview Survey.

National Center for Health Statistics. (2022). National Health Interview Survey, 2019 and 2022.

National Healthcare Quality and Disparities Report. (2022, October). *Agency for healthcare research and quality*. AHRQ Pub. No. 22(23)-0030.

Oud, M., de Winter, L., Vermeulen-Smit, E., Bodden, D., Nauta, M., Stone, L., et al. (2019). Effectiveness of CBT for children and adolescents with depression: A systematic review and meta-regression analysis. *European Psychiatry, 57*, 33–45. https://doi.org/10.1016/j.eurpsy.2018.12.008

Park, J., Kim, S. Y., & Lee, K. (2022). Effectiveness of behavioral sleep interventions on children's and mothers' sleep quality and maternal depression: A systematic review and meta-analysis. *Scientific Reports, 12*(1), 4172. https://doi.org/10.1038/s41598-022-07762-8

Patterson, G. R. (1979). A performance theory for coercive family interaction. In R. B. Cairns (Ed.), *The analysis of social interactions: Methods, issues, and illustrations* (pp. 119–162). Lawrence Erlbaum.

Patterson, G. R. (1982). *Coercive family process*. Castalia.

Peterson, B. S., Trampush, J., Maglione, M., Bolshakova, M., Brown, M., Rozelle, M., Motala, A., Yagyu, S., Miles, J., Pakdaman, S., Gastelum, M., Nguyen, B. T., Tokutomi, E., Lee, E., Belay, J. Z., Schaefer, C., Coughlin, B., Celosse, K., Molakalapalli, S., Shaw, B., et al. (2024). ADHD diagnosis and treatment in children and adolescents. In *Comparative effectiveness review* (Number 267). Agency for Healthcare Research and Quality.

Ramtekkar, U., & Ivanenko, A. (2015). Sleep in children with psychiatric disorders. *Seminars in Pediatric Neurology, 22*(2), 148–155. https://doi.org/10.1016/j.spen.2015.04.004

Rathus, J. H., & Miller, A. L. (2015). *DBT®skills manual for adolescents*. Guilford Press.

Reale, L., Bartoli, B., Cartabia, M., Zanetti, M., Costantino, M. A., Canevini, M. P., Termine, C., Bonati, M., & Lombardy ADHD Group. (2017). Comorbidity prevalence and treatment outcome in children and adolescents with ADHD. *European Child & Adolescent Psychiatry, 26*(12), 1443–1457. https://doi.org/10.1007/s00787-017-1005-z

Seligman, L. D., & Ollendick, T. H. (2011). Cognitive-behavioral therapy for anxiety disorders in youth. *Child and Adolescent Psychiatric Clinics of North America, 20*(2), 217–238. https://doi.org/10.1016/j.chc.2011.01.003

Shim, R., Szilagyi, M., & Perrin, J. M. (2022). Epidemic rates of child and adolescent mental health disorders require an urgent response. *Pediatrics, 149*(5), e2022056611. https://doi.org/10.1542/peds.2022-056611

Sweeney, S., & Wilson, C. (2023). Parental anxiety and offspring development: A systematic review. *Journal of Affective Disorders, 327*, 64–78. https://doi.org/10.1016/j.jad.2023.01.128

Touchette, E., Côté, S. M., Petit, D., Liu, X., Boivin, M., Falissard, B., Tremblay, R. E., & Montplaisir, J. Y. (2009). Short nighttime sleep-duration and hyperactivity trajectories in early childhood. *Pediatrics, 124*(5), e985–e993. https://doi.org/10.1542/peds.2008-2005

Vanderzee, K. L., Sigel, B. A., Pemberton, J. R., & John, S. G. (2018). Treatments for early childhood trauma: Decision considerations for clinicians. *Journal of Child & Adolescent Trauma, 12*(4), 515–528. https://doi.org/10.1007/s40653-018-0244-6

Wang, S., Li, Q., Lu, J., Ran, H., Che, Y., Fang, D., Liang, X., Sun, H., Chen, L., Peng, J., Shi, Y., & Xiao, Y. (2023). Treatment rates for mental disorders among children and adolescents: A systematic review and meta-analysis. *JAMA Network Open, 6*(10), e2338174. https://doi.org/10.1001/jamanetworkopen.2023.38174

Weber, A. N., Michail, M., Thompson, A., & Fiedorowicz, J. G. (2017). Psychiatric emergencies: Assessing and managing suicidal ideation. *The Medical Clinics of North America, 101*(3), 553–571. https://doi.org/10.1016/j.mcna.2016.12.006

Webster-Stratton, C., & Reid, M. J. (2003). The incredible years parents, teachers and children training series: A multifaceted treatment approach for young children with conduct problems. In A. E. Kazdin & J. R. Weisz (Eds.), *Evidence-based psychotherapies for children and adolescents* (pp. 224–240). The Guilford Press.

Weisz, J. R., & Kazdin, A. E. (Eds.). (2017). *Evidence-based psychotherapies for children and adolescents* (3rd ed.). The Guilford Press.

Wiltsey Stirman, S., Calloway, A., Toder, K., Miller, C. J., Devito, A. K., Meisel, S. N., Xhezo, R., Evans, A. C., Beck, A. T., & Crits-Christoph, P. (2013). Community mental health provider modifications to cognitive therapy: Implications for sustainability. *Psychiatric Services, 64*(10), 1056–1059. https://doi.org/10.1176/appi.ps.201200456

Wolgensinger, L. (2015). Cognitive behavioral group therapy for anxiety: Recent developments. *Dialogues in Clinical Neuroscience, 17*(3), 347–351. https://doi.org/10.31887/DCNS.2015.17.3/lwolgensinger

World Health Statistics (2024): monitoring health for the SDGs, Sustainable Development Goals. Geneva: World Health Organization; 2024. Licence: CC BY-NC-SA 3.0 IGO.

Zhou, X., Teng, T., Zhang, Y., Del Giovane, C., Furukawa, T. A., Weisz, J. R., Li, X., Cuijpers, P., Coghill, D., Xiang, Y., Hetrick, S. E., Leucht, S., Qin, M., Barth, J., Ravindran, A. V., Yang, L., Curry, J., Fan, L., Silva, S. G., Cipriani, A., et al. (2020). Comparative efficacy and acceptability of antidepressants, psychotherapies, and their combination for acute treatment of children and adolescents with depressive disorder: A systematic review and network meta-analysis. *The Lancet Psychiatry, 7*(7), 581–601. https://doi.org/10.1016/S2215-0366(20)30137-1

Chapter 6
Evidence-Based Behavioral Health Practices in School Settings

Jamie M. Lawler, Maitri Jain, Sarah E. Freeman, Elizabeth Freehling, Aidan P. Schmitt, and Celeste Hawkins

Most children and adolescents spend more of their waking hours at school than anywhere else. Thus, schools play an integral role in supporting children's development and well-being. Over the last decade, increasing rates of depression, suicidality, aggression, anxiety, and behavior problems have resulted in a "state of emergency" in child and adolescent mental health (Children's Hospital Association, 2021). These behavioral health problems interact bidirectionally with academic and social functioning (Weeks et al., 2016). Therefore, schools are motivated and uniquely situated to provide access to behavioral health services for children in need. Through tiered systems of support, schools can provide universal socio-emotional learning, targeted group interventions, and intensive one-on-one services depending on student needs. Increased acknowledgment of the scope and consequences of youth mental health problems over the last several decades has led to an increase in funding for behavioral health in schools, and in fact, the majority of youth behavioral health services now take place in schools (Duong et al., 2021). However, a sufficient workforce with expertise in evidence-based interventions for school settings is still lacking and funding continues to be an issue (Ogburn et al., 2023).

This chapter will provide a review of evidence-based behavioral health practices in school settings. First, we will provide an overview of the unique characteristics

---

J. M. Lawler (✉) · M. Jain · S. E. Freeman · E. Freehling
Department of Psychology, Eastern Michigan University, Ypsilanti, MI, USA
e-mail: jlawler1@emich.edu

A. P. Schmitt
Department of Child & Adolescent Outpatient Psychiatry,
Weill Cornell New York-Presbyterian Hospital, New York, NY, USA

C. Hawkins
School of Social Work, Eastern Michigan University, Ypsilanti, MI, USA

A. Maragakis, M. Janikian (eds.), *Evidence-Based Behavioral Health Practices in Pediatric Specialty Settings*, https://doi.org/10.1007/978-3-031-71004-9_6

of the school setting. Next, we will review the recent literature on evidence-based clinician-delivered behavioral health interventions for the school setting. The third section will review the implementation of behavioral health services in this setting, including strengths, challenges, and supervision considerations, and will briefly describe a pilot program integrating behavioral health services into early learning settings. Finally, we will end with recommendations and conclusions for future directions in school-based behavioral health.

Overview of School-Based Settings

School-based settings offer unique opportunities and challenges related to evidence-based behavioral health. Much (but not all) of school-based behavioral health interventions in the United States operate using tiered systems, typically within the Multi-tiered Systems of Support (MTSS) model. MTSS is based on the public health model of primary, secondary, and tertiary prevention wherein the most resource-intensive interventions are reserved for the highest-need individuals (Bruns et al., 2016). This model is used across different areas of learning and development in schools (e.g., reading and behavior). In MTSS, the first level of interventions (Tier 1) includes universal programs that reach all students in all settings. In the case of behavioral health, these are primary prevention programs focused on socio-emotional learning, as well as setting school-wide expectations for students. Tier 2 interventions include targeted approaches that are provided to students who demonstrate emerging mental health concerns or some risk factors for socio-emotional or academic difficulties (approximately 15% of all students; Wayne RESA, n.d.). Lastly, Tier 3 interventions include intensive, individualized interventions that are provided based on need and lack of response to Tier 2 interventions. They are delivered only to the students with the highest needs, with about 1% of students falling into this category.

In schools without fully implemented MTSSs, often only the most intensive interventions are implemented, which leads to a "wait-to-fail" scenario. In this case, students are failing to achieve their appropriate milestones and often do not become eligible to receive services until they have already experienced significant negative consequences. MTSS, on the other hand, advocates school-wide interventions for preventing, reducing, and managing behavior (Bruns et al., 2016). MTSS models also suggest implementing universal screening of all students to assess mental strengths and weaknesses. However, this has only been implemented in a small fraction of schools (Wood & McDaniel, 2020). Tier 1 and 2 interventions are typically delivered by nonclinical school personnel, most often teachers, and thus are outside the scope of this review. Other Tier 2 and nearly all Tier 3 interventions are delivered by mental health professionals, either through an "in-house" model (i.e., delivered by school-employed social workers, counselors, behavior specialists, and school psychologists) or "outside-in" model, where partnerships with community providers or behavioral health consultants implement the interventions (Doll et al.,

2017). These "outside-in" models are also sometimes called "Expanded School Mental Health" programs, though the field lacks universal definitions for many of these constructs, and independently developed models share many common features (Raviv et al., 2022). Specific models of MTSS and specific clinician-delivered interventions within those models will be discussed below.

It is important to note that structures may be different in specialized schools (e.g., schools in which all students receive special education services) and in non-public (i.e., charter and private) schools. Additionally, certain aspects of implementation are unique to certain countries. In the United States, special education is provided following guidelines stipulated by federal law, and it is funded by the federal government. Thus, other countries may have different systems for implementing behavioral health interventions in schools. Some school-based mental health services in the United States are billed to insurance (specifically Medicaid), while others are funded by special education funds. Each leads to unique challenges, which we will return to in the implementation section.

Implementation of school-based mental health care is unique compared to other pediatric settings for multiple reasons. First, interventions are closely tied to and operate within the academic setting. In the United States, many interventions are provided within the context of special education services and/or in response to difficulties functioning in the classroom. Therefore, treatment goals typically relate to academic functioning and classroom behavior, at least at a high level. This may lead to a focus on externalizing problems (that may be getting in the way of academic work and disturbing the classroom) and a lack of attention to internalizing problems. Another main difference between school-based behavioral health and other settings is the level of parent involvement, which is a challenge in school settings and may be drastically less than one would find in outpatient mental health settings (Santiago et al., 2013). Further, constraints on sessions (e.g., taking place during school hours and within the confines of school campuses, being taken out of class by providers, time constraints on sessions and treatment) are different from other settings and affect school-based mental health services. Lastly, mental health professionals within school settings may have various levels of training, and schools may lack qualified staff to address the needs of their populations, especially depending on federal and state funding (Ogburn et al., 2023).

Schools in higher socioeconomic status areas may have more resources available for "outside-in" services such as contracting with community agencies to provide services in addition to standard Individualized Education Program (IEP) interventions. However, even when there are ample resources available, students may not receive care when they are experiencing mental health difficulties because decisions about eligibility for services are often made by a whole school team, including parents, teachers, administrators, and service providers. For example, if a parent requests therapy but a majority of the school team deems therapy unnecessary, the child may be found ineligible due to their academic functioning being average and school resources being limited.

Another aspect of school-based behavioral health care that differs from other pediatric settings is the composition of its multidisciplinary teams. School teams can be composed of parents, teachers, administrators, and health care providers including psychologists, social workers, occupational therapists, speech-language pathologists, and physical therapists. Decisions about eligibility and treatment are often made in collaboration with both the student's parents and multiple professionals weighing in about the child's functioning and needs, both at home and in school. Naturally, this necessity for collaboration can also cause conflict that may not occur within outpatient and other pediatric settings. Nonetheless, it is typically highly advantageous for psychologists and other mental health providers within the school to have the opportunity to consult with teachers on a daily basis and to work collaboratively to implement interventions, especially those that have been specifically designed with school-based settings in mind (e.g., daily report cards for students with behavioral difficulties). Further strengths and limitations of implementing behavioral health services in schools will be discussed in the implementation section. The next section turns to specific evidence-based models and interventions in school settings.

Evidence-Based Interventions for Elementary, Middle, and High Schools

Models of School-Based Behavioral Health

Positive Behavior Intervention and Supports (PBIS) is an MTSS school-based intervention that has been relatively widely implemented and strives for an improved school climate in the school-based setting that can prevent and treat externalizing behavior problems (Michigan Department of Education, 2023; Santiago-Rosario et al., 2023). PBIS aims to promote positive academic and social outcomes for students, and it has been implemented in all 50 US states in over 25,000 schools (Kern et al., 2022). Initially, PBIS was a system created for schools in urban settings to address the unique needs of these populations (Goodman-Scott et al., 2017). As an MTSS model, PBIS aims to prevent and intervene against disruptive behaviors by utilizing a three-tier system. PBIS has been primarily studied in the context of Tier 1 (i.e., universal) interventions or as a full model across all three tiers (Lee & Gage, 2020), though specific interventions implemented at Tier 2 and 3 have also been evaluated (see below). PBIS interventions at each tier largely seek to decrease disruptive behavior and increase prosocial behavior. At the highest tier (i.e., Tier 3), individualized services (typically functional behavior analysis) are used to identify ways to support students with the most acute needs. Overall, PBIS has been shown to decrease problematic behaviors and emotional disturbance among students (Lee & Gage, 2020). The PBIS system is used across the developmental range, though it is less prevalent in high schools, with factors such as school sizes and increased levels of student independence making these settings more difficult for implementation (Freeman et al., 2016). Like in younger age ranges, PBIS has been shown to improve behavioral outcomes in this age group when implemented with fidelity, including increases in attendance and fewer office discipline referrals (Freeman et al., 2016), and has even been shown to effectively reduce diagnostic severity levels for students deemed "at risk" (Eiraldi et al., 2019). While a limitation of most PBIS systems is its exclusive focus on externalizing behavior, the Interconnected Systems Framework (ISF) has been developed to integrate traditional PBIS models with School Mental Health Systems with the goal of creating a broader, more well-rounded program that can meet students' needs across externalizing, internalizing, and other symptoms (Eber et al., 2019).

Similar to the Interconnected Systems Framework, the Behavioral Health Team (BHT) model can be used to supplement traditional PBIS models to be more inclusive of the full array of student needs (Raviv et al., 2022). Unlike the ISF, however, the BHT model can also be implemented as a stand-alone model for schools that have not already implemented PBIS. The BHT also explicitly relies on an outside-in model where community agencies and clinicians are engaged in providing services within the school setting, in line with the "Expanded School Mental Health" model. Developed in collaboration with a large urban school district and a smaller suburban school district, the BHT model is designed to be flexible to meet the needs of individual schools while also adhering to best practices (Raviv et al., 2022). BHT includes mechanisms for selecting and implementing evidence-based interventions at each level, along with built-in collaboration and communication between stakeholders.

Community and School-Based Behavioral Health Service (CSBBH) is another school-based behavioral health service model that has been used in Pennsylvania. In CSBBH, schools contract with community mental health providers to deliver school-based behavioral health services to Medicaid-eligible students with emotional, behavioral, and developmental disabilities (Herschell et al., 2021). CSBBH's clinical model does not operate under a specific modality, but rather emphasizes four key clinical priorities: the therapeutic alliance; developing a clinical formulation; offering, but not requiring, treatment options; and an iterative feedback process to inform implementation. Clinicians may use a variety of evidence-based practices under this model. CSBBH is associated with improvements in overall student functioning, family functioning, behavioral difficulties, hyperactivity, and conduct problems in both rural and urban settings (Herschell et al., 2021). Of note, results are more mixed for emotion problems and prosocial behavior (Herschell et al., 2021).

On the island of Kauai in Hawaii, the Mokihana program, which is a collaboration between the Department of Health and the Department of Education, provides culturally responsive, full-service behavioral health care to children with behavioral or emotional symptoms in every public school, including the integration of family-based services (Klontz et al., 2015). This program offers a wide variety of interventions, including case management, psychological evaluations, medication monitoring, and individual, group, and family therapies such as multisystemic therapy and functional family therapy. Children receiving Mokihana services showed significant improvements in hyperactivity, aggression, conduct problems, depression, somatization, attention problems, learning problems, and withdrawal, whereas anxiety symptoms only showed a nonsignificant decrease (Klontz et al., 2015).

Tier 1 Interventions

While the focus of this chapter is on clinician-delivered interventions, there are a few noteworthy Tier 1 (universal) interventions that are worth mentioning. Promoting Alternative Thinking Strategies for aggression (PATHS; Crean & Johnson, 2013) is a well-established curriculum that aims to promote emotional and social competencies and to reduce aggression and behavior problems in elementary school children. At middle and high school levels, other evidence-based approaches to prevention and intervention have been demonstrated in the school setting, such as preventing substance use/misuse (e.g., Life Skills Training; Botvin & Griffin, 2014), reducing bullying (Gaffney et al., 2021), and promoting socio-emotional development (Cipriano et al., 2023). While mental health clinicians do not typically deliver these interventions, they are often involved in designing such interventions and training nonclinical staff (e.g., teachers) to deliver the content.

Tier 2 Interventions

Tier 2 interventions are targeted small-group interventions for students assessed to be at-risk. While some Tier 2 interventions are teacher- or paraprofessional-delivered (such as Check, Connect, and Expect [Cheney et al., 2009] and other "check-in/checkout" style interventions), there are a number of evidence-based Tier 2 interventions that are clinician-delivered as well.

Cognitive Behavioral Intervention for Trauma in Schools (CBITS) is a Tier 2 evidence-based intervention for students from fourth grade through high school who have been exposed to trauma (Jaycox et al., 2012). CBITS was originally developed for students who were recent immigrants but has since been adapted to meet the needs of many different trauma-exposed groups. CBITS incorporates group, individual, and teacher and parent education sessions and uses CBT skills and concepts to prepare students for developing a trauma narrative (Jaycox et al., 2012). This intervention has been found to result in improvements in PTSD symptoms, depressive symptoms, and psychosocial functioning (Stein et al., 2003).

Another Tier 2 intervention called Project SOARing (Mazzone et al., 2023) integrates Dialectical Behavior Therapy (DBT) within a school-based framework. Project SOARing utilizes DBT skills social-emotional curriculum including mindfulness, distress tolerance, emotion regulation, and interpersonal effectiveness, in high school settings in order to address and help students navigate symptoms of various mental health disorders, emotional dysregulation, and stressful life events

(Mazza et al., 2016). While preliminary studies of feasibility and acceptability show promise for Project SOARing (Mazzone et al., 2023), further research is necessary to determine if DBT skills delivered in schools show similar efficacy and effectiveness as outpatient-based DBT interventions.

Tier 3 Interventions

The most common Tier 3 intervention within the PBIS system is functional behavioral analysis (FBA). Numerous single-subject design studies have demonstrated the efficacy of FBA techniques for improving behaviors (see review by Little & Akin-Little, 2019). A randomized controlled trial demonstrated that the Prevent–Teach–Reinforce model (also aligned with the principles of applied behavior analysis) resulted in lower problem behavior and improved socio-emotional skills (Iovannone et al., 2009).

As in other pediatric settings, Cognitive Behavioral Therapy (CBT) has also been shown to be effective in treating anxiety (Neil & Christensen, 2009) and depression (Calear & Christensen, 2010) symptoms when offered in schools. While various CBT-based programs have been offered across universal, selected, and indicated levels, Tier 3 programs, which are more akin to CBT in outpatient settings, show greater effects, especially for depression. Still, the evidence base for CBT in schools lags behind evidence of effectiveness in other settings, with mixed findings and small effects (Caldwell et al., 2019). When studies do show positive results, most show efficacy but not yet effectiveness in real-world school settings (Bruns et al., 2016). Another evidence-based intervention for adolescents, Interpersonal Psychotherapy for Adolescents (IPT-A), has also faced challenges when being implemented in the school setting (Mufson et al., 2004). Despite these challenges, IPT-A performs better than "services as usual" in school settings (Tang et al., 2009). We return to this discussion of implementation below.

Developmental Considerations

The behavioral health needs of students vary across developmental levels. Behavioral/conduct problems are the most frequently diagnosed in the elementary age group (i.e., 9.1% of students of ages 6–11), though anxiety is also quite common (6.6% of students), followed by depression (1.9%; Ghandour et al., 2019). During adolescence, internalizing symptoms increase dramatically, especially for females (Lewis et al., 2020). For older children and adolescents, there are many studied benefits of school-based mental health services, including access to care, preventative services, and the ability to access services in a more natural setting (Iachini et al., 2015). This age group differs from earlier childhood in that the framework and demands for interventions vary greatly. Many interventions in this age range focus on topics such as behavioral interventions, anxiety, and depression, as well as substance use and misuse. It is important to prioritize interventions during this developmental period because emerging signs and symptoms of various mental health disorders develop during adolescence (Merikangas et al., 2010). This makes high school an opportune time to implement interventions with the goal of providing support for students in their academic, emotional, behavioral, and social endeavors before they meet diagnostic criteria for a mental health disorder (Iachini et al., 2015).

Evidence-Based Interventions for Preschoolers in School Settings

Before children are even old enough to enter the public school system (i.e., 5-years old), many of them are already facing behavioral health challenges. It has been difficult to estimate the prevalence of behavioral health conditions among preschool-aged children, in part due to a historical emphasis on diagnostic classification systems that are not as sensitive to the unique developmental needs of this population (Egger & Emde, 2011). Nevertheless, reviews have estimated that psychiatric disorders occur in preschoolers at rates similar to those reported among older children (i.e., 14–26.4%), with any type of anxiety disorder occurring with the greatest prevalence, followed by oppositional defiant disorder and attention-deficit/hyperactivity disorder (ADHD) (Egger & Angold, 2006). Many interventions delivered to preschool-aged children have historically been delivered in the home or have taken more of a universal/preventive approach in schools (Kay-Lambkin et al., 2007). Interventions also tend to focus more on behavioral problems than emotional problems (Bayer et al., 2009), perhaps because behavioral problems may be considered more disruptive in the school setting than problems that are more internalizing in nature. Evidence suggests that efforts to treat and prevent behavioral health disorders early in childhood can reduce the prevalence and burden of these disorders later in life (Forbes et al., 2019), underscoring the importance of early intervention.

One of the most well-studied and widely adopted school-based intervention models addressing the mental and behavioral health problems of preschoolers is Early Childhood Mental Health Consultation (ECMHC; Cohen & Kaufmann, 2000). In contrast to traditional one-on-one therapeutic services, mental health experts in ECMHC programs seek to address the mental and behavioral health needs of individuals and schools indirectly by consulting with children's various caregivers (i.e., teachers, parents, and school administrators) to build their capacity to provide appropriate supports and environmental changes (Duran et al., 2009). A research synthesis summarizing the impact of ECMHC on children's problems and prosocial behaviors across 14 studies found that ECMHC was consistently associated with reductions in externalizing behaviors and improvements in terms of prosocial behaviors (e.g., social skills; Perry et al., 2010). In contrast, only about half of the included studies that reported on internalizing behaviors found evidence of improvement within this domain, a noteworthy finding given the relatively higher prevalence of internalizing disorders (Egger & Angold, 2006). More recent evaluations have reported similar findings (i.e., improvements in externalizing but not internalizing problems), additionally noting no significant changes in terms of overall classroom mental health climate or likelihood of student expulsion (Gilliam et al., 2016; Reyes & Gilliam, 2021). However, improvements in terms of child initiative and interpersonal relationships have been described (Reyes & Gilliam, 2021).

Many other school-based behavioral health interventions for preschoolers are play-based, such as Child-Centered Play Therapy (CCPT; Bratton et al., 2013) and Relationships for Growth and Learning (RfGL; Bekar et al., 2017). While these interventions have shown promise for reducing symptomatology and improving social competence (Bekar et al., 2017; Bratton et al., 2013), additional research is necessary to designate these programs as evidence-based. A pilot "outside-in" model for delivering evidence-based behavioral health services in preschool settings will be discussed in the implementation section.

Implementation

There are certain similarities between school-based behavioral health interventions and behavioral interventions in other settings. For instance, the principles of confidentiality, the importance of rapport building, and the basic structures of individual and group therapy share similarities across settings. However, there are aspects of implementing behavioral interventions that are unique to the school setting. One particular benefit of school-based interventions is that they can be made accessible to a large number of students at the same time. Furthermore, school-based services can reduce and alleviate many of the common barriers to treatment in the community, such as cost, transportation, and time. Normalizing mental health services by colocating them in schools can help reduce stigmatization. Another strength is the ease of consulting with teachers to implement interventions. However, there are multiple challenges in implementing interventions within schools. These challenges, as well as their potential solutions, are discussed below under four categories: the outer settings, the inner settings, the individuals, and the interventions.

Outer Settings

The outer setting refers to the broader context in which the school is placed (e.g., district, resources available, demographic composition, and government policies). Schools often have differing policies with regard to the role of the school psychologist, time given to nonacademic tasks, and the involvement of parents, which can affect the delivery of behavioral health care. The biggest challenge associated with the outer setting is often a lack of funding and resources to effectively implement school-based interventions. Resources may not be available to have a full-time psychologist at a school (Fazel et al., 2014). Even when school psychologists are in place, billing for mental health services in school settings poses a challenge, though guidelines were recently developed to aid in the Medicaid billing process in an attempt to make school-based mental health services more accessible (Ogburn et al., 2023). Recent U.S. legislation signed into law in 2022 included numerous provisions to support school-based behavioral health services, signaling a broader policy shift toward increased support for these areas. Thus, a new challenge is now assisting schools in taking advantage of the available funding in such a way as to make the most of these resources within their unique setting.

In addition to funding, other logistical procedures need to be put in place to support the effective implementation of interventions in school settings. For example, it is essential to deliver interventions in such a way that they fit into the school schedule and academic needs of the students rather than adding to their burden on top of their academic work. Furthermore, targeted dissemination of information about interventions among parents, legislators, and school staff members should be prioritized so that all stakeholders are in agreement and interventions can be integrated with community activities (Fazel et al., 2014).

Inner Settings

The inner setting refers to the environment of the school and how conducive it is to the implementation of behavioral health interventions. Splett et al. (2022) highlighted that the readiness of schools is determined by both the motivation of the school to implement the intervention and the capacity of the organization in two areas: the capacity to implement any innovation and the innovation-specific capacity to implement the particular intervention at hand. There are multiple challenges associated with the inner setting. Unlike interventions implemented in outpatient settings, students receiving interventions within their schools often have to return to class after their therapy sessions. This necessity makes it difficult to treat complex issues such as trauma in a school setting. Moreover, therapy sessions in schools can be as short as 30 minutes, which limits intervention time compared to 45 to 60 minutes in other behavioral health settings (Gee et al., 2021). Based on requirements for special education services, interventions funded in this way must be tied to the academic outcomes of the students, meaning that not all behavioral and emotional problems can always be addressed. Other logistics related to conducting therapy in schools also pose challenges. There may be limited space at the school, and, thus, there may be difficulty in coordinating room availability. Moreover, rooms being used for multiple purposes may also reduce student confidentiality (Perales et al., 2023).

To tackle these problems, the school leadership should prioritize behavioral health interventions so that appropriate resources and time can be allocated for these services. This may also require making time in students' schedules for therapy and adjusting their assignment deadlines (March et al., 2022). The importance of interventions should also be communicated to other school staff to create a culture of support for behavioral health interventions (Bruns et al., 2016). Psychoeducation about behavioral health for teachers, administrators, and parents can help gain buy-in and avoid stigmatization for youth taking part in services. When partnerships with community agencies are involved, communication and collaboration between school staff and the community partners is essential.

Individuals

"Individuals" refer to those who will be involved in administering and receiving the intervention (March et al., 2022). Students receiving behavioral health interventions at school may face stigma from their peers and teachers and thus may be vulnerable to bullying and victimization (Bowers et al., 2013; Fazel et al., 2014; Foulkes & Stapley, 2022). Additionally, students may not fully understand why they are being treated differently from other students and may experience self-stigma. Students who have difficulties with tardiness and absences from school (due to various factors like physical or mental health, lack of motivation, family, or community factors) will have lapses in care. Students might also actively refuse to attend therapy sessions for a number of reasons, including low motivation or not wanting to miss classes. Personnel implementing interventions (e.g., social worker, school psychologist, and clinical psychologist) may also have received different levels of training in school behavioral health interventions, which may affect the way in which the interventions are delivered. Concerns about whose "turf" the school setting is can lead to challenges in navigating interprofessional collaboration and can lead to conflict (Mellin et al., 2010), especially when "outside-in" systems bring in community practitioners. These community practitioners contracted by the school may have varying levels of familiarity with school systems which can also lead to challenges.

Implementation of behavioral health interventions in schools also differs from other settings due to fewer opportunities for parent involvement, in part because sessions take place in school, and potentially lower engagement by parents even when encouraged. This is in part because treatment in school settings is sometimes recommended by school staff rather than being requested by parents as in outpatient settings, although parents typically must still provide consent for school-based services. While parent consultations are sometimes required for younger students, it still may not be at the level expected in other settings. For example, Parent Management Training (PMT) interventions, a first-line evidence-based treatment for disruptive behavior, are difficult to deliver in the school setting due to PMTs heavy emphasis on parenting. Further, given lower levels of parent involvement and available resources, traditional intake sessions do not necessarily precede treatment, and access to previous health records is limited, so background history and information about the child's functioning outside of school, including their relationship with their parents, may not be known. Given that these factors often play an essential role in psychopathology and thus conceptualization and treatment, this lack of information can affect care. Disagreement between teachers and parents about the child's behavior may also make it difficult to agree upon and implement interventions (Fazel et al., 2014). Collaboration between teachers, school administrators, caregivers, and mental health professionals (whether internal or community partners) is essential to the successful delivery of services.

Since mental health professionals may come from a variety of educational backgrounds, it is essential that all professionals be trained in the specific intervention prior to delivery to maintain the quality of implementation. Further, ongoing support and consultation are often essential to delivering evidence-based interventions rather than "train and hope" style implementation (Lyon & Bruns, 2019). Implementers also need to develop adequate self-efficacy to administer the intervention. In addition, parents and students need to believe in their capacity to change their behavior and thus have the motivation to engage in the intervention (March et al., 2022).

Interventions

The intervention itself needs to be adapted to the specific school in which it is being implemented so that the resources available in the school can be appropriately used to be most helpful to its students (Crooks et al., 2022). Schools in high vs. low socioeconomic (SES) environments, and different types of schools (e.g., private vs. public vs. charter schools) have different demographics of students and provide different services. Schools in high SES neighborhoods usually have more resources, more training for their staff, and families with higher levels of education, a higher overall quality of life, and better general health care access from which children can receive therapy services outside of school (APA, 2017). However, children from lower SES families may have Medicaid insurance, and school may be one of the few places where students are able to obtain therapeutic services. Moreover, while special education services, including behavioral supports, are integrated within public schools and thus available to students in at least some capacity, the same is not always true of private schools (CHADD, 2017). However, private and high SES schools are often more able to pilot new interventions for behavioral health and may have more resources available for Tier 1-level universal preventive interventions.

Ten "best practice principles" have been established for school-based mental health, which include accessibility of services, use of evidence-based interventions, progress monitoring across a continuum of care, ethical and culturally informed implementation and distribution, strength-based, and interdisciplinary collaboration across stakeholders within the school and the community (Weist et al., 2017). For schools to effectively implement interventions, it is essential to make the core components of the intervention clearly identifiable to the professional team. Clear definitions and protocols should be made for all practices. At the same time, there should be enough flexibility in the program to adapt it to the needs of each school and student. One part of making interventions flexible is minimizing the number of classes students have to miss to participate in the behavioral health intervention. Missing classes can negatively impact students' academic performance, reduce opportunities for socialization, and put students in a conflict of prioritizing their studies or their mental health. Therefore, interventions would have a better uptake if they are scheduled around students' classes whenever possible.

Examples of Implementation Models in Schools

Tier 2 Implementation: Comprehensive Implementation Training and Support Model

An example of a Tier 2 implementation model is the Comprehensive Implementation Training and Support Model (CITAS; Goldenthal et al., 2021), which focuses on developing school-based provider's knowledge, skills, and resources to successfully implement quality Tier-2 interventions in schools. CITAS incorporates the training principles of expert coaching through direct feedback, pretraining assessment, interpersonal education opportunities, peer collaboration, problem-based learning, and reminders for weekly session content (Lyon et al., 2011). These principles are accomplished through monthly site visits by clinicians, monthly calls with the implementation staff at school, and weekly emails for reminders of session content, tips, resources, and supporting documents. It also involves two 90-minute in-person meetings of the professional learning committee (PLC). PLC meetings focus on the review of behavior management strategies, content-based discussions and role plays for anger management sessions, peer-to-peer learning, troubleshooting problems, and celebrating successes. The CITAS model also provides additional resources and videos shared online to help with the implementation of the intervention. CITAS emphasizes intensity of training and coaching as well as contextual factors at the organizational level for the effective implementation of interventions. In a test of the CITAS model with the Anger Coping Groups intervention, it was found to have adequate acceptability, feasibility, fidelity, and satisfaction with school psychologists.

Tier 3 Implementation: Cognitive Behavioral Therapy (CBT)

As mentioned above, evidence for CBT in the school setting is growing but is not as robust as the evidence base in outpatient settings. One reason for this might be that CBT is not always being implemented with fidelity in school-based settings, and school-based clinicians may not have adequate training in effective delivery of CBT (Forman et al., 2012). Kilbourne et al. (2018) developed a model for improving the standardized implementation of CBT-based interventions in schools. The model contains three components for effective implementation of interventions: replicating effective programs, coaching, and facilitation. Replicating effective programs means that the interventions should be based on established evidence-based treatments but customized to the local needs. The second component is coaching, which includes live supervision of delivery and feedback to improve implementation. The last component is facilitation, in which consultation is provided by organizational experts in strategic thinking for implementation providers. It also highlights increasing organizational and leadership support for the implementation of the intervention. The staff is taught the background of CBT, facilitators and barriers are reviewed, and measurable goals are set with the lead psychologist monitoring the intervention and the school staff. The implementers are mentored on overcoming barriers and aligning the intervention to the available resources through weekly meetings. The facilitator reaches out to school administrators to understand their priorities for the school and the community and helps the clinician align the use and goals of CBT with these existing priorities. The facilitator also helps the school-based mental health provider explain the added value of the CBT program to the school employees. Throughout this process, there is ongoing marketing of the program in which the facilitator, school-based mental health provider, and school leadership summarize the progress of the program to other school staff and develop sustainability plans for the intervention.

Pilot Program Example: Eastern Michigan University Early Childhood Satellite Clinic

Given the cascading effects of early intervention, it is important to consider and evaluate novel approaches to providing behavioral health services in preschool settings. Eastern Michigan University (EMU) in Ypsilanti, Michigan U.S., is currently operating an integrated school-based behavioral health program that could be replicated in other settings. As a part of this program, EMU clinical psychology graduate students provide behavioral health services at "satellite" clinics located in local preschools primarily serving low-income families of children of ages two to six. Trainees are supervised by a licensed psychologist with expertise in early childhood behavioral health and are able to provide services directly in schools. The advantages of this model include both convenience and practicality. As families do not have to travel to an additional location to receive services and may be referred directly by teachers, this significantly reduces access-related barriers. Additionally, clinicians are able to provide in vivo practice of social skills in the classroom or can easily incorporate parent-focused work by scheduling sessions directly before or after school. The program also supports direct consultation with teachers to improve classroom management with individual children's needs in mind. Since the direct services are provided by student clinicians and logistic support (e.g., electronic medical record system, scheduling, and clinical resources) is supplied by a training clinic, the program is quite cost-effective and families can receive services free of charge, removing another barrier to care. Finally, this work can be supported by training grants aiming to bolster the early childhood behavioral health workforce and provide highly specialized training at the graduate level.

Embedding Equity into School-Based Behavioral Health Interventions

Students from marginalized backgrounds face disadvantages in their academic outcomes as well as in the receipt of mental health services (Lipari et al., 2016), which are associated with systemic racism rather than individual, family, or community deficits (Exner-Corterns et al., 2022). Specifically, research has found that while culturally and linguistically diverse (CLD) populations often present with a higher prevalence of mental health needs, they are less likely to seek and have access to services (Castro-Olivo, 2017). Therefore, it is important to ensure that mental health interventions are equitable and responsive to issues of diversity, equity, and inclusion (DEI). Exner-Cortens et al. (2022) developed a model for defining equity in school mental health research, which included both horizontal equity (equal services for equal needs) as well as vertical equity (enhanced services for greater need). It also included concepts of social justice, including fair distribution of resources (distributive justice), fair procedure for this distribution (procedural justice), non-discriminatory practice, and advocacy. The authors assert that these processes and outcomes are situated within the broader context of children's rights, use of collaborative anti-oppressive practices and policies, and understanding and challenging one's own privilege, power, and biases, as well as harmful social norms (Exner-Cortens et al., 2022).

Lazarus et al. (2022) also developed a culturally responsible dual-factor model for transforming school mental health services. The model mentions that school mental health interventions should focus both on the presence of well-being and the absence of pathology (thus "dual factor"). In other words, the services provided should be strength-focused rather than weakness-focused. The term "culturally responsible" implies that behavioral health intervention should act in ways that are ethical, effective, and dismantle existing discrepancies in students' mental health. The term "responsible" is used to signify that the intervention does not need to be just "responsive" to cultural factors, but it is the duty (or responsibility) of those implementing the intervention to foster the well-being of students from marginalized backgrounds. The model takes a cultural-ecological-psychological approach to changing the environments and systems that affect all students' success. Finally, the model emphasizes access by promoting the need to convene family, community providers, and other supports to meet both the mental health and prevention needs of all students (Lazarus et al., 2022).

Supervision in School Settings

School settings provide unique challenges related to clinical supervision. First, given the multidisciplinary nature of behavioral health services in schools, multiple different types of clinical supervision may be needed. For example, if a clinical psychology trainee is interested in gaining experience providing intervention in a school setting, the trainee may have trouble finding supervision from a licensed psychologist since schools primarily employ social workers and school psychologists (who are typically certified but not licensed). A trainee in such a position would likely need to seek out a school system with an "outside-in" model that contracts with clinical psychologists who would be able to provide appropriate supervision. Relatedly, as many schools are still building their mental health workforce, school psychologists may find themselves to be the only behavioral health workers in an entire school or district. That position may make it difficult to obtain supervision and consultation, which are recommended throughout school psychologists' careers (National Association of School Psychologists, 2018). Supervisors in school settings need to be not only well versed in evidence-based practice but also familiar with the school setting and the implications of delivering clinical services in this context. To work toward solving these issues, increased communication and collaboration between disciplines is essential, and further workforce development is crucial.

Limitations of the Literature

The majority of literature on school-based behavioral health interventions is composed of nonrandomized studies, and, thus, the results may not be fully reliable or valid. Many of the outcome measures used are also not well validated or do not measure outcomes that are relevant for the student's academic success in the school (Fazel et al., 2014). There have also been many interventions piloted with a small sample of schools, which have not yet been manualized and/or disseminated, and, thus, it is difficult to generalize these interventions to other school contexts. There is limited research evaluating and replicating these interventions in multiple school contexts. Even though there is some research on the efficacy of these interventions, research on fidelity is scarce, and the cost-benefit analyses of most interventions have not been conducted. Therefore, the literature on school mental health should be interpreted with caution, and more research needs to be conducted to better understand school behavioral health interventions.

Recommendations and Conclusion

There is great promise in school-based behavioral health services to reach many children and adolescents who would not otherwise receive services. As August and colleagues (2018) state, "schools have significant potential and opportunity to lead prevention efforts for children and youth, and never has there been a better time to harness that opportunity within the current climate of MTSS and emphasis on school mental health" (p. 95). Much work is still left to be done in terms of evaluating the effectiveness of behavioral health interventions in the school setting and disseminating well-established practices. Policymakers should continue increasing funds for school-based behavioral health services and implement policies to make sure services are being distributed equitably. Additionally, more attention should be paid to early intervention services, such as providing school-based services in early childhood settings. School administrators should devote time and money to establishing MTSS programs in their schools and implementing evidence-based interventions at each tier. School stakeholders, including administrators, teachers, and staff should work to make behavioral health a priority in the school culture. Clinicians across disciplines should work together to implement behavioral health services and should strive to offer supervision to clinicians-in-training whenever possible to grow the workforce with expertise in this setting. Student clinicians should seek out specific training in school-based behavioral health interventions to be prepared to deliver services in this setting. Depending on the student's background and training area, additional preparation may be needed in clinical interventions (for those with school psychology backgrounds) or in education and socio-emotional learning (for those with clinical psychology or clinical social work backgrounds) in order to be most effective in the school setting. Student clinicians may want to seek out training through specific centers (e.g., UW SMART center, UMD National Center for School Mental Health, UCLA Center for Mental Health in Schools) and/or programs (e.g., Behavioral Health Team model, CSBBH, and Mokihana program) that are known for their commitment to evidence-based intervention.

In conclusion, behavioral health interventions in the school setting can be highly beneficial for the mental health of youth. Developing trust and building relationships with students, families, and teachers/staff is a critical component of support and access as it will impact how students and families learn about, engage with, and directly connect to mental health services, resources, and interventions to alleviate barriers and reduce stigma that is associated with receiving mental health treatment. Taken together, the extant literature suggests that many different levels of school-based intervention (i.e., universal or targeted) may be successful at improving behavioral health outcomes among children and adolescents. The evidence base for school-based interventions that are associated with reductions in externalizing symptoms appears somewhat stronger, especially among younger children, and more work is needed to identify and rigorously evaluate programs that target internalizing symptoms, such as depression and anxiety. There are still many challenges to the effective and equitable implementation of evidence-based behavioral health interventions that remain to be solved in order to make school-based behavioral health services as impactful as possible.

Statements and Declarations The authors have no relevant financial or nonfinancial interests to disclose.

Select Bibliography

Bruns, E. J., Duong, M. T., Lyon, A. R., Pullmann, M. D., Cook, C. R., Cheney, D., & McCauley, E. (2016). Fostering SMART partnerships to develop an effective continuum of behavioral health services and supports in schools. *American Journal of Orthopsychiatry, 86*(2), 156–170. https://doi.org/10.1037/ort0000083. Describes the multi-tiered systems of support (MTSS) model to increase access and effectiveness of school behavioral health interventions. It also discusses a case of community-academic partnership in Washington state that exemplifies the use of the MTSS framework.

Doll, B., Nastasi, B. K., Cornell, L., & Song, S. Y. (2017). School-based mental health services: Definitions and models of effective practice. *Journal of Applied School Psychology, 33*(3), 179–194. Describes the use of the Participatory Culture-Specific Intervention Model in enhancing community-school partnerships to provide culturally and ecologically relevant school-based behavioral health services. It emphasizes the use of community resources and strengths, data-driven decision making, and improving the capacity of schools to implement behavioral health interventions.

Fazel, M., Hoagwood, K., Stephan, S., & Ford, T. (2014). Mental health interventions in schools 1. *The Lancet Psychiatry, 1*(5), 377–387. https://doi.org/10.1016/S2215-0366(14)70312-8. This paper discusses strategies for effective implementation of school-based behavioral health interventions. It highlights the need for integrated interventions that combine student-level, classroom-level, and system-level supports to make interventions accessible to more number of students.

Kilbourne, A. M., Smith, S. N., Choi, S. Y., Koschmann, E., Liebrecht, C., Rusch, A., et al. (2018). Adaptive School-based Implementation of CBT (ASIC): Clustered-SMART for building an optimized adaptive implementation intervention to improve uptake of mental health interventions in schools. *Implementation Science, 13*(1), 1–15. This is a protocol paper that describes an implementation model aimed at improving the adoption of, and effective integration of CBT interventions, in school settings. The model highlights three strategies: replicating effective intervention, coaching, and facilitation.

Lazarus, P. J., Doll, B., Song, S. Y., & Radliff, K. (2022). Transforming school mental health services based on a culturally responsible dual-factor model. *School Psychology Review, 51*(6), 755–770. https://doi.org/10.1080/2372966X.2021.1968282. This paper describes the culturally responsible dual-factor model for school behavioral health interventions. It is a proactive and prevention oriented model that focuses on equity and population-based interventions.

References

Bayer, J., Hiscock, H., Scalzo, K., Mathers, M., McDonald, M., Morris, A., Birdseye, J., & Wake, M. (2009). Systematic review of preventive interventions for children's mental health: What would work in Australian contexts? *Australian and New Zealand Journal of Psychiatry, 43*(8), 695–710.

Bekar, Ö., Shahmoon-Shanok, R., Steele, M., Levy, J., Defressine, L., Giuseppone, K., & Steele, H. (2017). Effectiveness of school-based mental health playgroups for diagnosable and at-risk preschool children. *American Journal of Orthopsychiatry, 87*(3), 304.

Botvin, G. J., & Griffin, K. W. (2014). Life skills training: Preventing substance misuse by enhancing individual and social competence. *New Directions for Youth Development*, 57–65. https://doi.org/10.1002/yd.20086

Bowers, H., Manion, I., Papadopoulos, D., & Gauvreau, E. (2013). Stigma in school-based mental health: Perceptions of young people and service providers. *Child and Adolescent Mental Health, 18*(3), 165–170. https://doi.org/10.1111/j.1475-3588.2012.00673.x

Bratton, S. C., Ceballos, P. L., Sheely-Moore, A., Meany-Walen, K., Pronchenko, Y., & Jones, L. D. (2013). Head start early mental health intervention: Effects of child-centered play therapy on disruptive behaviors. *International Journal of Play Therapy, 22*(1), 28–42. https://doi.org/10.1037/a0030318

Bruns, E. J., Duong, M. T., Lyon, A. R., Pullmann, M. D., Cook, C. R., Cheney, D., & McCauley, E. (2016). Fostering SMART partnerships to develop an effective continuum of behavioral health services and supports in schools. *American Journal of Orthopsychiatry, 86*(2), 156–170. https://doi.org/10.1037/ort0000083

Caldwell, D. M., Davies, S. R., Hetrick, S. E., Palmer, J. C., Caro, P., López-López, J. A., Gunnell, D., Kidger, J., Thomas, J., French, C., Stockings, E., Campbell, R., & Welton, N. J. (2019). School-based interventions to prevent anxiety and depression in children and young people: A systematic review and network meta-analysis. *The Lancet Psychiatry, 6*(12), 1011–1020.

Calear, A. L., & Christensen, H. (2010). Systematic review of school-based prevention and early intervention programs for depression. *Journal of Adolescence, 33*(3), 429–438.

Castro-Olivo, S. M. (2017). Introduction to special issue: Culturally responsive school-base mental health interventions. *Contemporary School Psychology, 21*, 177–180.

Cheney, D. A., Stage, S. A., Hawken, L. S., Lynass, L., Mielenz, C., & Waugh, M. (2009). A 2-year outcome study of the check, connect, and expect intervention for students at risk for severe behavior problems. *Journal of Emotional and Behavioral Disorders, 17*(4), 226–243.

Children's Hospital Association. (2021, October 19). *Children's hospitals declare national emergency in children's mental health.* https://www.childrenshospitals.org/news/newsroom/2021/11/childrens-hospitals-declare-national-emergency-in-childrens-mental-health

Cipriano, C., Strambler, M. J., Naples, L. H., Ha, C., Kirk, M., Wood, M., Sehgal, K., Zieher, A. K., Eveleigh, A., McCarthy, M., Funaro, M., Ponnock, A., Chow, J. C., & Durlak, J. (2023). The state of evidence for social and emotional learning: A contemporary meta-analysis of universal school-based SEL interventions. *Child Development, 94*(5), 1181–1204.

Cohen, E., & Kaufmann, R. K. (2000). *Early childhood mental health consultation* (Vol. Report No. No. CMHS-SVP0151). Center for Mental Health Services.

Crean, H. F., & Johnson, D. B. (2013). Promoting Alternative Thinking Strategies (PATHS) and elementary school aged children's aggression: Results from a cluster randomized trial. *American Journal of Community Psychology, 52*, 56–72.

Crooks, C. V., Dunlop, C., & Short, K. (2022). A structured conceptualization of implementation-sensitive interventions for school mental health. *Canadian Journal of Community Mental Health, 41*(3), 1–17. https://doi.org/10.7870/cjcmh-2022-018

Doll, B., Nastasi, B. K., Cornell, L., & Song, S. Y. (2017). School-based mental health services: Definitions and models of effective practice. *Journal of Applied School Psychology, 33*(3), 179–194.

Duong, M. T., Bruns, E. J., Lee, K., Cox, S., Coifman, J., Mayworm, A., & Lyon, A. R. (2021). Rates of mental health service utilization by children and adolescents in schools and other common service settings: A systematic review and meta-analysis. *Administration and Policy in Mental Health and Mental Health Services Research, 48*, 420–439.

Duran, F., Hepburn, K., Irvine, M., Kaufmann, K., Anthony, B., Horen, N., & Perry, D. (2009). *What works? A study of effective early childhood mental health consultation programs.* Georgetown University Center for Child and Human Development.

Eber, L., Barrett, S., Perales, K., Jeffrey-Pearsall, J., Pohlman, K., Putnam, R, Splett, J., & Weist, M. D.(2019). *Advancing education effectiveness: Interconnecting school mental health and school-wide PBIS, Volume 2: An implementation fuide.* Center for Positive Behavior Interventions and Supports (funded by the Office of Special Education Programs, U.S. Department of Education). University of Oregon Press.

Education and socioeconomic status factsheet. (2017). American Psychological Association. https://www.apa.org/pi/ses/resources/publications/education

Educational rights: Public vs. private schools. (2017). CHADD. Retrieved November 19, 2023, from https://chadd.org/adhd-weekly/educational-rights-public-vs-private-schools/

Egger, H. L., & Angold, A. (2006). Common emotional and behavioral disorders in preschool children: Presentation, nosology, and epidemiology. *Journal of Child Psychology and Psychiatry, 47*(3–4), 313–337.

Egger, H. L., & Emde, R. N. (2011). Developmentally sensitive diagnostic criteria for mental health disorders in early childhood: The diagnostic and statistical manual of mental disorders—IV, the research diagnostic criteria—preschool age, and the Diagnostic Classification of Mental Health and Developmental Disorders of Infancy and Early Childhood—Revised. *American Psychologist, 66*(2), 95.

Eiraldi, R., McCurdy, B., Schwartz, B., et al. (2019). Pilot study for the fidelity, acceptability, and effectiveness of a PBIS program plus mental health supports in under-resourced urban schools. *Psychology in the Schools, 56*, 1230–1245. https://doi-org.ezproxy.emich.edu/10.1002/pits.22272

Exner-Cortens, D., Gaias, L., Splett, J. W., Jones, J., & Walker, W. (2022). Embedding equity into school mental health theory, research, and practice: An introduction to the special issue series. *Psychology in the Schools, 59*(10), 1941–1947. https://doi.org/10.1002/pits.22679

Fazel, M., Hoagwood, K., Stephan, S., & Ford, T. (2014). Mental health interventions in schools 1. *The Lancet Psychiatry, 1*(5), 377–387. https://doi.org/10.1016/S2215-0366(14)70312-8

Forbes, M. K., Rapee, R. M., & Krueger, R. F. (2019). Opportunities for the prevention of mental disorders by reducing general psychopathology in early childhood. *Behaviour Research and Therapy, 119*, 103411.

Forman, S. G., Fagley, N. S., Chu, B. C., & Walkup, J. T. (2012). Factors influencing school psychologists'"willingness to implement" evidence-based interventions. *School Mental Health, 4*, 207–218.

Foulkes, L., & Stapley, E. (2022). Want to improve school mental health interventions? Ask young people what they actually think. *Journal of Philosophy of Education, 56*(1), 41–50. https://doi.org/10.1111/1467-9752.12649

Freeman, J., Simonsen, B., McCoach, D. B., Sugai, G., Lombardi, A., & Horner, R. (2016). Relationship between school-wide Positive Behavior Interventions and Supports and academic, attendance, and behavior outcomes in high schools. *Journal of Positive Behavior Interventions, 18*(1), 41–51. https://doi.org/10.1177/1098300715580992

Gaffney, H., Ttofi, M. M., & Farrington, D. P. (2021). What works in anti-bullying programs? Analysis of effective intervention components. *Journal of School Psychology, 85*, 37–56.

Gee, B., Wilson, J., Clarke, T., Farthing, S., Carroll, B., Jackson, C., King, K., Murdoch, J., Fonagy, P., & Notley, C. (2021). Review: Delivering mental health support within schools and colleges – a thematic synthesis of barriers and facilitators to implementation of indicated psychological interventions for adolescents. *Child and Adolescent Mental Health, 26*(1), 34–46. https://doi.org/10.1111/camh.12381

Ghandour, R. M., Sherman, L. J., Vladutiu, C. J., Ali, M. M., Lynch, S. E., Bitsko, R. H., & Blumberg, S. J. (2019). Prevalence and treatment of depression, anxiety, and conduct problems in US children. *The Journal of Pediatrics, 206*, 256–267.e3. https://doi.org/10.1016/j.jpeds.2018.09.021

Gilliam, W. S., Maupin, A. N., & Reyes, C. R. (2016). Early childhood mental health consultation: Results of a statewide random-controlled evaluation. *Journal of the American Academy of Child & Adolescent Psychiatry, 55*(9), 754–761.

Goldenthal, H. J., Raviv, T., Baker, S., Holley, C., Summersett Williams, F., & Gouze, K. R. (2021). Development of a training and implementation model for school-based behavioral health interventions. *Psychology in the Schools, 58*(7), 1299–1319. https://doi.org/10.1002/pits.22504

Goodman-Scott, E., Hays, D., & Cholewa, B. (2017). "It Takes a Village": A case study of positive behavioral interventions and supports implementation in an exemplary urban middle school. Retrieved September 7, 2023, from https://link.springer.com/article/10.1007/s11256-017-0431-z

Herschell, A. D., Schake, P. L., Hutchison, S. L., Karpov, I. O., Gavin, J. G., Crisan, T. B., & Wasilchak, D. S. (2021). Evaluating the effectiveness of a statewide school-based behavioral health program for rural and urban elementary-aged students. *School Mental Health*, 1–13.

Iachini, A. L., Brown, E. L., Ball, A., Gibson, J. E., & Lize, S. E. (2015). School mental health early interventions and academic outcomes for at-risk high school students: A meta-analysis. *Advances in School Mental Health Promotion, 8*(3), 156–175. https://doi.org/10.1080/1754730X.2015.1044252

Iovannone, R., Greenbaum, P. E., Wang, W., Kincaid, D., Dunlap, G., & Strain, P. (2009). Randomized controlled trial of the Prevent—Teach—Reinforce (PTR) tertiary intervention for students with problem behaviors: Preliminary outcomes. *Journal of Emotional and Behavioral Disorders, 17*(4), 213–225.

Jaycox, L. H., Kataoka, S. H., Stein, B. D., Langley, A. K., & Wong, M. (2012). Cognitive Behavioral Intervention for Trauma in Schools. *Journal of Applied School Psychology, 28*(3), 239–255.

Kay-Lambkin, F., Kemp, E., Stafford, K., & Hazell, T. (2007). Mental health promotion and early intervention in early childhood and primary school settings: A review. *The Journal of Student Wellbeing, 1*(1), 31–56.

Kern, L., George, H. P., Fintel, N., & Baton, E. (2022). *PBIS implementation in rural schools in the U.S. Center on PBIS*. https://assets-global.website-files.com/5d3725188825e071f1670246/62fd270c0fee5e799a38a6ab_PBIS%20Implementation%20in%20Rural%20Schools%20in%20the%20US.pdf

Kilbourne, A. M., Smith, S. N., Choi, S. Y., Koschmann, E., Liebrecht, C., Rusch, A., Abelson, J. L., Eisenberg, D., Himle, J. A., Fitzgerald, K., & Almirall, D. (2018). Adaptive School-based Implementation of CBT (ASIC): Clustered-SMART for building an optimized adaptive implementation intervention to improve uptake of mental health interventions in schools. *Implementation Science, 13*(1), 1–15.

Klontz, B. T., Bivens, A., Michels, S., DeLeon, P. H., & Tom, L. (2015). The Mokihana Program: The effectiveness of an integrated department of education and department of health school-based behavioral health approach. *Psychological Services, 12*(2), 101–111. https://doi.org/10.1037/a0038015

Lazarus, P. J., Doll, B., Song, S. Y., & Radliff, K. (2022). Transforming school mental health services based on a culturally responsible dual-factor model. *School Psychology Review, 51*(6), 755–770. https://doi.org/10.1080/2372966X.2021.1968282

Lee, A., & Gage, N. A. (2020). Updating and expanding systematic reviews and meta-analyses on the effects of school-wide positive behavior interventions and supports. *Psychology in the Schools, 57*(5), 783–804.

Lewis, A. J., Sae-Koew, J. H., Toumbourou, J. W., & Rowland, B. (2020). Gender differences in trajectories of depressive symptoms across childhood and adolescence: A multi-group growth mixture model. *Journal of Affective Disorders, 260*, 463–472.

Lipari, R. N., Hedden, S., Blau, G., & Rubenstein, L. (2016). Adolescent mental health service use and reasons for using services in speciality, educational, and general medical settings. The CBHSQ Report. https://www.ncbi.nlm.nih.gov/books/NBK362074/#SR-222_RB-1973.s.s3

Little, S. G., & Akin-Little, A. E. (2019). *Behavioral interventions in schools: Evidence-based positive strategies* (pp. xix–422). American Psychological Association.

Lyon, A. R., & Bruns, E. J. (2019). From evidence to impact: Joining our best school mental health practices with our best implementation strategies. *School Mental Health, 11*, 106–114.

Lyon, A. R., Stirman, S. W., Kerns, S. E. U., & Bruns, E. J. (2011). Developing the mental health workforce: Review and application of training approaches from multiple disciplines. *Administration and Policy in Mental Health and Mental Health Services Research, 38*(4), 238–253. https://doi.org/10.1007/s10488-010-0331-y

March, A., Stapley, E., Hayes, D., Town, R., & Deighton, J. (2022). Barriers and facilitators to sustaining school-based mental health and wellbeing interventions: A systematic review. *International Journal of Environmental Research and Public Health, 19*(6), 6. https://doi.org/10.3390/ijerph19063587

Mazza, J. J., Dexter-Mazza, E. T., Miller, A. L., Rathus, J. H., & Murphy, H. E. (2016). *DBT Skills in Schools: Skills training for emotional problem solving for adolescents (DBT steps-A).*

Mazzone, E., Antommarchi, C., Bagnara, G., Jutzy, H., Alido, A., & Boustani, M. (2023). School-based opportunities for adolescent recovery (SOARing): The acceptability and feasibility of a DBT skills training program for at-risk adolescents. *Evidence-Based Practice in Child and Adolescent Mental Health*. https://doi.org/10.1080/23794925.2023.2224113

Mellin, E. A., Bronstein, L., Anderson-Butcher, D., Amorose, A. J., Ball, A., & Green, J. (2010). Measuring interprofessional team collaboration in expanded school mental health: Model refinement and scale development. *Journal of Interprofessional Care, 24*(5), 514–523.

Merikangas, K. R., He, J., Burstein, M., Swanson, S. A., Avenevoli, S., Cui, L., Benjet, C., Georgiades, K., & Swendsen, J. (2010). Lifetime prevalence of mental disorders in U.S. adolescents: Results from the National Comorbidity Survey Replication–Adolescent Supplement (NCS-A). *Journal of the American Academy of Child & Adolescent Psychiatry, 49*(10), 980–989. https://doi.org/10.1016/j.jaac.2010.05.017

Michigan Department of Education. (2023). *Positive Behavioral Interventions and Supports (PBIS) at School.* Family Matters. https://www.michigan.gov/-/media/Project/Websites/mde/specialeducation/familymatters/FM1/PBIS_School_FactSheet.pdf?rev=568afa854d054a79ac1f23a1736ef215

Mufson, L., Dorta, K. P., Wickramaratne, P., Nomura, Y., Olfson, M., & Weissman, M. M. (2004). A randomized effectiveness trial of interpersonal psychotherapy for depressed adolescents. *Archives of General Psychiatry, 61*(6), 577–584.

National Association of School Psychologists. (2018). *Supervision in School Psychology [Position statement].* Author.

Neil, A. L., & Christensen, H. (2009). Efficacy and effectiveness of school-based prevention and early intervention programs for anxiety. *Clinical Psychology Review, 29*(3), 208–215.

Ogburn, J. Gratale, D., Hogan, L., Peoples, K., & Comizio, C. (2023). *Fostering school-based behavioral health services: Key recommendations for policymakers and philanthropy.* Nemours Children's Health. https://www.nemours.org/content/dam/nemours/shared/collateral/policy-briefs/policy-brief-fostering-school-based-behavioral-health-services-key-recommendations.pdf

Perales, K., Eber, L., Raulerson, C., & Kern, L. (2023). *Addressing confidentiality while supporting the social-emotional- behavioral needs of students within schools.* Center on Positive Behavioral Interventions and Supports.

Perry, D. F., Allen, M. D., Brennan, E. M., & Bradley, J. R. (2010). The evidence base for mental health consultation in early childhood settings: A research synthesis addressing children's behavioral outcomes. *Early Education and Development, 21*(6), 795–824.

Raviv, T., Smith, M., Hurwitz, L., Gill, T. L., Baker, S., Torres, S. A., et al. (2022). Supporting school-community collaboration for the implementation of a multi-tiered school mental health program: The Behavioral Health Team model. *Psychology in the Schools, 59*(6), 1239–1258.

Reyes, C. R., & Gilliam, W. S. (2021). Addressing challenging behaviors in challenging environments: Findings from Ohio's early childhood mental health consultation system. *Development and Psychopathology, 33*(2), 634–646.

Santiago, C. D., Pears, G., Baweja, S., Vona, P., Tang, J., & Kataoka, S. H. (2013). Engaging parents in evidence-based treatments in schools: Community perspectives from implementing CBITS. *School Mental Health, 5*, 209–220.

Santiago-Rosario, M. R., McIntosh, K., Izzard, S., Lissman, D. C., & Calhoun, E. (2023). *Is Positive Behavioral Interventions and Supports (PBIS) an evidence-based practice?* Center on PBIS. https://assets-global.website-files.com/5d3725188825e071f1670246/651c4d7d5174726ec6da1d86_Is%20Positive%20Behavioral%20Interventions%20and%20Supports%20(PBIS)%20an%20Evidence-Based%20Practice.pdf

Splett, J. W., Perales, K., Miller, E., Hartley, S. N., Wandersman, A., Halliday, C. A., & Weist, M. D. (2022). Using readiness to understand implementation challenges in school mental health research. *Journal of Community Psychology, 50*(7), 3101–3121. https://doi.org/10.1002/jcop.22818

Stein, B. D., Jaycox, L. H., Kataoka, S. H., Wong, M., Tu, W., Elliott, M. N., & Fink, A. (2003). A mental health intervention for schoolchildren exposed to violence: A randomized controlled trial. *Journal of the American Medical Association, 290*(5), 603–611.

Tang, T. C., Jou, S. H., Ko, C. H., Huang, S. Y., & Yen, C. F. (2009). Randomized study of school-based intensive interpersonal psychotherapy for depressed adolescents with suicidal risk and parasuicide behaviors. *Psychiatry and Clinical Neurosciences, 63*(4), 463–470.

Wayne RESA. (n.d.). *Quick guide for multi-tiered system of supports: The classroom.* Wayne RESA. https://resources.finalsite.net/images/v1568836530/resanet/drbszjpnchsgxle0u5cq/QuickguideforMTSSTheDistrictLevel.pdf

Weeks, M., Ploubidis, G. B., Cairney, J., Wild, T. C., Naicker, K., & Colman, I. (2016). Developmental pathways linking childhood and adolescent internalizing, externalizing, academic competence, and adolescent depression. *Journal of Adolescence, 51*, 30–40. https://doi.org/10.1016/j.adolescence.2016.05.009

Weist, M. D., Bruns, E. J., Whitaker, K., Wei, Y., Kutcher, S., Larsen, T., Holsen, I., Cooper, J. L., Geroski, A., & Short, K. H. (2017). School mental health promotion and intervention: Experiences from four nations. *School Psychology International, 38*(4), 343–362. https://doi.org/10.1177/0143034317695379

Wood, B. J., & McDaniel, T. (2020). A preliminary investigation of universal mental health screening practices in schools. *Children and Youth Services Review, 112*, 104943.

Chapter 7
Early Childhood Intensive Intervention Settings

Lillian V. Pelios and Konstantina Psiachou

Outline of Chapter

This chapter provides a review of early intervention (EI) services for children with autism spectrum disorders (ASDs) and the relevant settings in which they take place. Emphasis is placed on describing interventions that meet best practices in EI, in particular, early intensive behavioral interventions (EIBI) and naturalistic developmental behavioral interventions (NDBI). First, we provide a definition of EI as well as eligibility criteria for services, including information pertinent to assessment, diagnosis, and referral for EI services; overall benefits to learners; and a brief discussion on autism diagnosis and prevalence. Next, we introduce the three models of EI, i.e., biological, psychodynamic, and educational, and provide a thorough discussion on educational models, with an emphasis on EIBI and NDBIs. This discussion addresses a variety of parameters contributing to the effectiveness of these models, including parent training. Finally, the cardinal matters of ongoing training and supervision are addressed as experimental evidence acknowledges both of these components as determinants of evidence-based and thus effective interventions.

L. V. Pelios (✉)
School of Graduate and Professional Education, The American College of Greece, Athens, Greece
e-mail: lpelios@acg.edu

K. Psiachou
Department of Psychology, Panteion University, Athens, Greece
e-mail: k.psiachou@panteion.gr

A. Maragakis, M. Janikian (eds.), *Evidence-Based Behavioral Health Practices in Pediatric Specialty Settings*, https://doi.org/10.1007/978-3-031-71004-9_7

Early Intervention and Who It Concerns

Early intervention (EI) provides services to babies, toddlers, and young children with developmental delays or disabilities. It aims at facilitating the development of skills that typically occur during the first 3 years of life. These skills may be physical, communication, social/emotional, cognitive, and self-care. Early intervention services are available to all children and their families throughout the United States; these services are supported by the Individuals with Disabilities Education Act (IDEA, 2004), which was enacted by Congress in 1975 to ensure that children with disabilities are given the same opportunities to a free and appropriate public education like all other children. Services to infants and toddlers with delays and disabilities, from birth through 2 years of age, are ensured by IDEA Part C, while IDEA Part B ensures special education and related services to children and youth aged 3–21 years.

Assessment, Diagnosis, Referral, and Services for EI Eligibility for EI is determined by local or territory service providers to children with or without a diagnosis. The referring agent may be the child's pediatrician, day care provider, preschool teacher, or the parents themselves. A doctor's referral is not required for assessment; however, assessment requires parental permission regardless of who the referring agent is. Many children who receive EI services do not necessarily have a formal diagnosis. If the assessment reveals mild or serious delays, the child may be considered at risk for developing a disability, and publicly funded services will be prescribed, depending on the severity of the delays, for free or at a reduced cost. Services may include speech, occupational, and physical therapy or other specialized therapies, such as applied behavior analytic (ABA) interventions, depending on the child's needs. For more details about eligibility, visit the CDC's online site at: https://www.cdc.gov/ncbddd/autism/treatment.html.

Overall Benefits of EI to Learners A multitude of research supports the effectiveness of general EI services to children with a wide range of difficulties, such as sensory impairments (i.e., hearing, vision), low birth weight, neglect and abuse, poverty, as well as intellectual and behavioral difficulties (Burger, 2010; Campbell & Ramey, 1994; Guralnick, 1997, 2017; Reynolds, 1994). Three important questions that parents and mental health professionals continue to ask despite the extensive research available on the effectiveness of EI, involve (a) the timeliness of the services, i.e., how early they are provided; (b) the type of services offered; and (c) the intensity that is required for the children eligible for these services to achieve their potential. All these questions are strongly associated with outcomes with respect to a number of disabilities, especially ASDs, which will be the focus of this chapter.

When it comes to providing services to children with an ASD diagnosis in particular, researchers and practitioners both have expressed concern with the fact that benefits can be seriously compromised unless timeliness, intensity, and the most effective types of services are made available to children with ASD (Richardson et al., 2019; Richardson et al., 2020; Woods & Wetherby, 2003).

Autism Spectrum Disorder: Diagnosis ASD is a neurodevelopmental disorder characterized by impairments in social communication and social interactions across many contexts and by restrictive and repetitive behavior, interests, or activities, including stereotyped speech or echolalia. Symptoms are present in early childhood, although they may not become apparent until later in life as the social demands become increasingly more challenging. The symptoms of ASD fall on a spectrum from mild to severe (APA, 2013). It is noteworthy that 1/3 of the children who are 8 years old with an ASD diagnosis have IQs in the range of intellectual disability, i.e., 70 or below (CDC, 2018).

Autism Spectrum Disorder: Prevalence Li et al. (2022), based on the Global Burden of Disease database, provide epidemiological estimates of the prevalence of ASD at 28.3 million cases worldwide. The aforementioned authors report that increases in ASD have been noted in prevalent cases from 1990 to 2019. More recently, according to the Centers for Disease Control and Prevention (CDC, 2023a, b), in the United States, it is estimated that 1 in 36 children, up from 1 in 44 (CDC, 2023a), has autism, with 4 in 100 being boys, while 1 in 100 girls. Autism prevalence in the United States is now lower among white children (2.4%) than other racial and ethnic groups, with black children being diagnosed at 2.9%, Hispanic children at 3.2%, and Asian or Pacific Islander children at 3.3% (CDC, 2023b). The increasing trend of ASD diagnosis amongst children of minority communities is viewed in a positive light as it is believed to reflect improvements in outreach, screening, and reduced stigma of the autism diagnosis (CDC, 2023b). Research on prevalence indicates a consistent upward trend of ASD cases in the United States (Fombonne, 2009; Newschaffer et al., 2005) and around the world (United Kingdom; Baird et al., 2003; Iceland, Saemundsen et al., 2013; South Korea, Kim et al., 2011) for several decades now. Notably, in their study of the epidemiology of ASD, Li et al. (2022) found that from 1990 to 2019, the prevalent cases of ASD globally increased by almost 40%.

The Need for Effective Services The increasing prevalence trends and the pervasive characteristics of the disorder across all levels of functioning, combined with the fact that it is a lifelong condition for most persons diagnosed, underscores the necessity for identifying effective interventions (Zeidan et al., 2022).

Models of Early Intervention

The interventions classified under EI are divided into biological, psychodynamic, and educational (Roberts & Prior, 2006). *Biologically based interventions* for autism include medical interventions and complementary and alternative medicine

(CAM). According to Gringras (2000), the "tailor-made" medical approach to addressing autism that targets specific symptoms, i.e., behavioral excesses like hyperactivity, aggression, and obsessive-compulsive behaviors, seems to be more accepted by the medical and scientific community. More information about the psychopharmacology of ASD and the specific treatments most commonly used (i.e., atypical antipsychotics, tricyclic antidepressants, SSRIs, etc.), as well as their efficacy and side effects, can be found in Persico et al.'s (2021) thorough review.

Complementary and alternative medication (CAM) interventions are hypothesized to cause physiological alterations that target the biological causes of autism rather than address specific symptoms. Levy and Hyman (2005) identify the following four alternative medical interventions for autism: treatments for the immune system, treatments for the gastrointestinal system, supplements affecting neurotransmitter regulation, and nonbiological treatments. The most used CAM treatments are customized diets (i.e., gluten-free and casein-free diets and the use of vitamin supplements), administration of the hormone secretin, and craniosacral manipulation. However, research supporting the above treatments is very limited (Christon et al., 2010) and methodologically compromised due to small sample sizes, multicomponent interventions, and undefined placebo conditions (Brondino et al., 2015). Furthermore, interpretations of findings hinge on correlation rather than causation (Levy & Hyman, 2005).

Psychodynamic models endorse the premise that autism is an emotional disorder caused by distant mothers who reject their children, causing them to become seriously traumatized (Bruno, 1967; Kanner, 1943). Subsequently, this trauma is reflected by the children turning inward and refusing to interact with the world. Nowadays, research is clear that autism is a neurodevelopmental rather than an emotional disorder, thus rendering psychoanalytic interventions, such as holding therapy (Crawford, 1986) and pheraplay (mentioned in Mesibov, 1992), as rare choices for early intervention for autism (Roberts & Prior, 2006).

The most effective treatments for early intervention, supported by empirical evidence, are broadly classified as educational treatments. *Educational interventions* are treatments that focus on teaching new skills and reducing unwanted behavior through environmental manipulation, i.e., through operant teaching strategies and the use of extensive and specialized curriculum sequences that are implemented by well-trained staff. Educational interventions may be eclectic, focused, or comprehensive in nature. The eclectic approach draws upon different theoretical orientations and borrows from their techniques and interventions, sometimes incorporating evidence-based approaches as well. The focused approach involves the use of evidence-based interventions that target specific areas of development.

Eclectic and focused approaches refer to interventions that tend to isolate one skill area, for instance, emotional responding or language. Eclectic approaches use combined methods and a trial-and-error decision-making process for selecting interventions. A primary characteristic of eclectic approaches is to combine treatment as usual (TAU) or sensory integration training with the occasional use of one or more of the following evidence-based approaches: discrete trial training/instruction (DTI), token economies, pivotal response training (PRT), to name a few, to address the skill deficit (Howard et al., 2005; Smith & Iadarola, 2015); or in the case of focused teaching evidence-based interventions such as prompting, reinforcement, DTI, may be introduced to TAU for a brief period of time as part of an unspecified special education curriculum with the intent to teach new behavior and then discontinue the intervention (Odom et al., 2010), or they may use speech therapy, occupational therapy, physiotherapies, again either individually or combined with each other and/or with TAU (Roberts & Prior, 2006).

Stated differently, eclectic or focused treatment models are supplementary to standard educational services or TAU approaches and are commonly offered as EI services in public special education schools. As mentioned earlier, they are based on combining different methods and tactics at will, thus making it easier to use varying materials, places, and professionals. However, at the same time, the consistency of services delivered through eclectic models is limited. To illustrate what delivery of services from this perspective would look like, consider that throughout an ordinary school day, the learner would be exposed to TAU that is most likely not developed or taught from a behavior analytic perspective, with the exception of specific protocols such as a toilet training protocol, or the use of picture exchange communication (PECS, Bondy & Frost, 1998) for teaching manding, or the use of visual cues for structuring the environment and the learner's responses. These protocols, although behavior analytic in nature, are not part of a comprehensive and coherent intervention, which would take into consideration every aspect of the learner's development and circumstances, including systematic attempts to understand the function of behavior. Furthermore, these protocols may be introduced and used by staff who are not trained in behavior analysis or supervised by qualified behavior analysts. Thus, the treatment fidelity of these protocols is highly questionable. Furthermore, these protocols tend to treat best practice approaches as a group of interventions from which practitioners pick and choose without consideration for the need for theoretical coherence, which should be the guiding force behind problem-solving.

During the past few years, the emergence of early intervention models combining behavioral and developmental approaches has been noted. Some of these methods have been used for years, while others are relatively new both in the literature and in clinical practice. The term naturalistic developmental behavioral interventions (NDBIs) introduced by Schreibman et al. (2015) highlights the fact that they combine principles from both the behavioral and developmental sciences. NDBIs can be focused or comprehensive in their scope of intervention, such as EIBI. Some examples of focused NDBIs are Incidental Teaching (Hart & Risley, 1975), Enhanced Milieu Training (Kaiser & Hester, 1994), Social Communication/Emotional Regulation/Transactional Support (Prizant et al., 2003), Pivotal Response Training (Koegel & Koegel, 2006), Joint Attention Symbolic Play Engagement and Regulation (Kasari et al., 2006), Reciprocal Imitation Training (Ingersoll, 2010), Early Achievements (Landa et al., 2011), and Project ImPACT (Ingersoll & Wainer, 2013). Thus far, to the author's knowledge, only one example of a comprehensive NDBI is identified in the literature and that is the early start Denver model (Smith & Iadarola, 2015).

Comprehensive early intervention refers to "treatments that simultaneously address core deficits in autism, including language, socialization, cognition, and play" (Rogers & Vismara, 2008, p. 10). In other words, they are designed to address a broader number of areas of functioning, if not all the areas of functioning, where skill deficits may be present, thereby achieving a greater impact on learning. Rogers and Vismara (2008) describe these programs as "branded" interventions as they use a specific name to identify themselves, such as the Denver Model, Lovaas Model, or Walden Model. Furthermore, these models maintain intervention over an extended period (often up to 2 years or more), they are intensive (20–25 h per week or more), and they employ multiple components.

When compared with eclectic or focused models, research shows that comprehensive behavioral models such as EIBI are more effective in improving cognitive, language, adaptive, and academic skills in children with autism, especially when the intervention is intensive and begins before the third year of life (Eikeseth et al., 2002; Howard et al., 2005, 2014). Overall, as far as both short- and long-term positive outcomes are concerned, there seems to be considerable evidence supporting comprehensive interventions such as EIBI. However, when it comes to long-term outcomes, there seems to be little to no evidence supporting treatment as usual approaches and an emerging body of evidence supporting certain eclectic or focused approaches grouped under the general term NDBIs, particularly ESDM and Pivotal Response Training (PRT) approaches combined with ABA (Howard et al., 2005, 2014; Smith & Iadarola, 2015).

Comprehensive models of intervention that are grounded in the philosophy of ABA, as described by the early work of B.F. Skinner (1965) and Baer et al. (1968) have gained considerable scientific recognition as evidence-based interventions and best practice approaches to autism intervention (Gould et al., 2011; Howard et al., 2005, 2014; Odom et al., 2010; Rogers & Vismara, 2008; Virués-Ortega, 2010; Warren et al., 2011). One of the most important characteristics of ABA that is indicative of best practice is that it provides a framework of decision-making at the center, which is the evaluation of the evidence available and relevant to the problem at hand (Smith, 2013). The strategy of compiling lists of effective treatments and labeling them as evidence-based practices, although useful, encompasses the danger of limiting the way practitioners use research findings. Contrary to relying on lists, relying on a framework of problem-solving allows the practitioner to combine the best available evidence, their clinical expertise, and client values and context to design effective interventions (Slocum et al., 2014). Unlike TAU and most eclectic approaches, EIBI, and more recently ESDM are examples of comprehensive models of intervention that rely on a framework of problem-solving and that meet best practice criteria.

EIBI Meets the Requirements for Evidence-Based Practice

There was a time when children with autism were believed to be untrainable, and parents were advised to institutionalize them (Cattell-Gordon & Cattell-Gordon, 1998). However, treatment options for persons with ASD and their families changed considerably and in a positive direction after Lovaas' 1987 publication. In this seminal publication, Lovaas provided evidence that with early, intensive, behavior-analytic interventions all the children, who participated in his study, learned new skills while some children demonstrated considerable improvements in both IQ and adaptive functioning. Furthermore, the children who demonstrated marked improvements in a follow-up study that took place several years later were not distinguishable from their peers (McEachin et al., 1993). Since then, numerous publications have followed confirming the findings from Lovaas' original study (Cohen et al., 2006; Eikeseth et al., 2007; Eldevik et al., 2009; Howard et al., 2005; Makrygianni et al., 2018; McEachin et al., 1993; Remington et al., 2007; Rogers & Vismara, 2008; Smith et al., 2000; Smith & Iadarola, 2015; Virués-Ortega, 2010). Moreover, research findings have corroborated Lovaas' prediction for optimal outcomes, i.e., some people with a clearly documented history of autism no longer meeting criteria for the diagnosis according to the autism diagnostic observation schedule (ADOS, Lord et al., 2000) and clinical judgment (Fein et al., 2013). Thus, the intervention that Lovaas designed and tested in his 1987 study, which has come to be known as EIBI despite the various disputes regarding the methodological problems of this study, is now considered an empirically validated intervention for addressing autism.

EIBI services constitute a comprehensive intervention offered to young children diagnosed at risk for autism or with autism, from 2 to 8 years old, who usually experience marked delays and/or disabilities across multiple areas of functioning. Thus, EIBI can be viewed as a service provided by early intervention services. EIBI programs consist of several characteristics that must be present to achieve optimal outcomes. These characteristics include: (a) commencing prior to the age of four, preferably as close to two as possible (Klintwall et al., 2015; Klintwall & Eikeseth, 2014); (b) intensive 1:1 therapy, 25–40 h weekly (Eldevik et al., 2006; Lovaas, 1987; Love et al., 2009); (c) therapy provision by extensively trained therapists in the principles of ABA, preferably by behavior technicians, i.e., therapists who are certified and supervised by master's or Ph.D. level behavior analysts (Gould et al., 2011); and (d) use of a comprehensive and individualized curriculum addressing all areas of functioning (Gould et al., 2011).

ESDM Meets the Requirements for Possibly Efficacious

Smith and Iadarola (2015) classify NDBIs as a combination of ABA and developmental social-pragmatic (DSP) methods, explaining that these methods use a combination of structured teacher-led activities and play time child-led interventions.

NDBIs, when used as separate interventions and not as part of a comprehensive curriculum, would be considered focused or eclectic approaches (Smith & Iadarola, 2015) since they target specific areas of learning and development. An exception to the above is ESDM, which adopts a comprehensive approach to intervention, targeting all aspects of behavior (Dawson et al., 2010). ESDM incorporates operant learning principles, utilizes shared control between the child and the therapist, relies heavily on natural contingencies of reinforcement, and emphasizes the use of developmentally appropriate target behaviors, which are identified by paying special attention to prerequisite skills. Finally, from the beginning, intervention is implemented in naturalistic settings (Schreibman et al., 2015).

In the past decade, attempts have been made to evaluate various NDBIs. The findings thus far are optimistic, particularly for ESDM, which is of particular interest as it is considered a comprehensive intervention for young children with ASD. Smith and Iadarola (2015) state that in a well-designed randomized controlled study assessing the effectiveness of ESDM, conducted by Dawson et al. (2010), medium to large effects were documented on measures of developmental quotient and adaptive behavior for 48 toddlers with ASD aged 18–30 months at the beginning of the intervention, as compared to TAU. This finding allows for the classification of ESDM as possibly efficacious, a term introduced by Rogers and Vismara in their 2008 Journal of Clinical Child and Adolescent Psychology (JCCAP) review, which is used again by Smith and Iadarola in their 2015 article, published in the same journal. Smith and Iadarola (2015) based their classification of various models of intervention as best practice or as emerging as potentially efficacious on interventions that had been manualized and evaluated through experimental or quasi-experimental group studies or through systematic reviews of single-subject studies. Furthermore, the effectiveness of ESDM is augmented by the fact that it incorporates elements that enhance learning. Examples of such elements are the fact that children tend to learn better in contexts that are meaningful to them (Kuhl et al., 2003), when they are actively engaged (Kuhl et al., 2003), and finally, children learn better when they participate in developmentally appropriate activities (Charlesworth, 1998; Filler & Xu, 2006). In addition to the above practices, ESDM utilizes operant principles in the natural environment, taking into account the child's developmental stage. Schreibman et al. (2015) argue that NDBIs enhance the teaching and learning process while building more easily towards generalization, increasing spontaneity, and decreasing escape/avoidant behaviors and prompt dependency. Finally, ESDM and other NDBIs' share the same goals as EIBI of targeting socially significant behaviors, while objectively recording behaviors before, during, and after intervention (Schreibman et al., 2015; Vivanti & Stahmer, 2021).

EIBI and ESDM Components/Moderators and Mediators

When it comes to EIBI, the following moderators/mediators have been identified: timeliness and intensity of intervention, relying on the principles of applied behavior analysis, the development of a comprehensive curriculum, the necessity of ongoing assessment, the setting of the intervention, parent training, and supervision by certified behavior analysts. Each of these variables is discussed below with the intention of drawing connections between these moderators and mediators and outcome.

Regarding NDBIs and particularly ESDM, since it is a developing field, the research has focused mostly on evaluating outcomes rather than identifying the main common components that are prerequisite for successful intervention (Crank et al., 2021). In the section that follows, EIBI quality indicators are discussed thoroughly as a multitude of studies have attempted to identify moderators and mediators of effective outcomes. Similarly, a discussion of ESDM quality indicators is attempted. As with EIBI, core elements that affect the intervention outcomes of ESDM, addressed in the section that follows, are chronological and developmental age, intensity of intervention, curriculum, parent involvement, and supervision. However, it should be noted that research on these quality indicators is still very limited, and results are often inconsistent.

Timeliness and Intensity The timeliness of an intervention refers to how early services are prescribed for a child with special needs. Contrary to EI, EIBI services are generally made available after the diagnosis of ASD. The diagnosis tends to be provided to children of 2–5 years of age, with the average age at diagnosis being closer to 5 years (Baird et al., 2003) despite the fact that the symptoms may be present for several years prior to the diagnosis (Neimy & Pelaez, 2021) and despite the fact that assessment and diagnosis are possible starting at 18 months (Baird et al., 2003; Neimy & Pelaez, 2021; Ozonoff et al., 2010; Rogers & Pennington, 1991). As a matter of fact, Neimy and Pelaez (2021) argue that when taking into consideration the certainty that the younger the age at the onset of intervention, the better the outcome (Granpeesheh et al., 2009a; Harris & Handleman, 2000; Lovaas et al., 1973), combined with the evidence supporting the effectiveness of behavioral interventions, EIBI for infants at risk for an autism diagnosis and other developmental delays or disorders, should be made available as a preventative intervention.

When it comes to the ESDM, chronological age seems to affect intervention outcome, suggesting that the earlier the onset of intervention, the better and faster the results (Dawson et al., 2010). At the same time, developmental age (cognitive abilities before the start of intervention) seem to contribute to better results; however, research is still inconclusive (Contaldo et al., 2020).

Intensity refers to both the weekly hours of intervention as well as the number of years for which children with an ASD diagnosis receive services. One of the four variables that have been identified by researchers as a key characteristic of best practice is the delivery of 1:1 sessions for 30 h per week, or more, for a minimum of at least 2 years (Eldevik et al., 2006, 2010; Klintwall & Eikeseth, 2014; Lovaas, 1987). Intensity is an important moderator even when comparing low- and high-intensity EIBI programs. Specifically, Eldevik et al. (2006) found that low-intensity ABA programs lead to fewer gains compared to high-intensity EIBI, but still both low- and high-intensity EIBIs lead to greater gains than those achieved by children receiving other services. Similarly, Dawson et al. (2010) state that the delivery of ESDM services should be intensive, i.e., a minimum of 20 h per week for up to 2 years. Furthermore, Dawson et al. found that high-intensity ESDM produced better outcomes than low-intensity ESDM. For other NDBIs, the literature is still lacking support for the intensity of intervention.

Behavior Analytic Services delivered by EIBI programs involve the use of evidence-based procedures originating from ABA to help young children acquire adaptive, functional, and cognitive skills (Klintwall & Eikeseth, 2014; Lovaas & Smith, 1989). Of course, behavior analysts can and should operate as part of a multidisciplinary team whenever necessary. For instance, speech and occupational therapy services may be particularly useful in situations where the learner is experiencing difficulties with articulation and/or musculoskeletal or neurological difficulties, respectively. In such situations, the respective discipline can work with the child directly and provide consultative services to the team under the supervision of the certified behavior analyst. Examples of adaptive, functional, and cognitive skills include imitation, communication, language, social, play, preacademic and academic skills, and self-care skills. To achieve the targeted changes in behavior, the aforementioned areas of functioning must be targeted simultaneously. Learning takes place through the use of reinforcement, shaping, extinction, prompting and prompt fading, behavioral chains, visual support systems, incidental teaching, natural environment training, and generalization training (see Cooper et al., 2020 for detailed descriptions of these procedures).

Important components of EIBI programs are pre-teaching assessment, ongoing data collection, data-based evaluation of programs and interventions, and criterion-based learning. Thus, initial assessments to determine children's level of functioning as well as ongoing evaluation to assess the efficacy of programming are integral to effective interventions. In addition to teaching children's new skills, interventions target the reduction of problem behavior, such as tantrums and stereotypic and self-injurious behavior, through the use of functional analysis (FA) or functional behavioral assessments (FBA) (Witt et al., 2000). Functional analysis allows the practitioner to identify both the triggers of problem behavior as well as the consequences that maintain it (Dunlap et al., 1993; Hanley, 2012; Hanley et al., 2003; Kern et al., 1994). Once this information is acquired, alternative responses that serve the same purpose can be identified and taught; in this manner, problem behavior can be replaced with socially acceptable and functional responses without the use of punishment (Pelios et al., 1999).

In addition to the aforementioned, important behavior analytic components of EIBI are programming for maintenance and generalization of mastered skills as well as planning for the assessment of treatment integrity and social validity. In a survey of clinical practices conducted by Love et al. (2009), 98% of the respondents indicated that their program design included measures of maintenance and generalization of learning. Furthermore, as evidenced by many of the studies that evaluated either branded or nonbranded EIBI programs, manualized interventions that allow for the assessment of treatment integrity, i.e., whether procedures are implemented reliably, was an important characteristic of best practice programs (Odom et al., 2010; Smith et al., 2007). Finally, social validity, i.e., as defined by Wolf (1978), consists of three components: the social significance of treatment goals, social appropriateness of procedures, and social importance of treatment effects, and is a core principle of any ABA-related activities, including EIBI (Huntington et al., 2023; Park & Blair, 2019). As mentioned previously, NDBIs also share the same behavior analytic core elements as EIBI, are based on operant learning principles and teaching methods, and follow objective data collection throughout the intervention, while focusing on socially significant behaviors (Schreibman et al., 2015).

A Comprehensive Curriculum A comprehensive curriculum has been identified as one of the four key components that effective and empirically validated EIBI programs offer (Hancock et al., 2000; Lovaas, 2003; Love et al., 2009). Hundreds of different skills must be taught to children with ASD in the course of receiving EIBI services, and dozens of curricular programs exist to teach these skills (Love et al., 2009). Nevertheless, to the best knowledge of the authors, nowhere in the literature can one find what constitutes a comprehensive curriculum nor how one can be designed (Gould et al., 2011).

Goal setting, or more broadly speaking, curriculum analysis and development, is one of the more complex and demanding aspects of EIBI, which in practice is often misunderstood as being equivalent to compiling lists of skills or target behaviors. Although target behaviors and curriculum are informed by developmental sequences, the relationship between them is not always linear and hierarchical. On the one hand, as indicated by the concept of cumulative and hierarchical learning, one repertoire provides the basis for learning another repertoire. On the other hand, it is also necessary for several repertoires to be combined to facilitate the development of a new, more complex repertoire. Thus, the concept of a basic behavioral repertoire (BBR) emerges, serving an important practical function of explaining how the hundreds of individual skills that are taught early on during the intervention combine in systematic ways to yield coherent behavioral repertoires and generative responses (Lund, 2001; Staats, 1981, 1994). Learning how to develop a comprehensive and individualized curriculum takes place during clinical practice under the mentorship of highly experienced behavior analysts.

Curriculum Models The very first EIBI curriculum that was ever published was developed by O. Ivar Lovaas and his colleagues at the University of Los Angeles (UCLA) (Lovaas, 1981; Lovaas & Buch, 1997). All subsequent curricula are to a greater or lesser degree influenced by these two original publications (Petursdottir & Carr, 2011). Presently, published curricula fall under one of two models: the UCLA model or the verbal behavior (VB) model. Although both models follow the same logic of breaking down complex language skills into simpler skills and teaching them through the use of positive reinforcement, discrimination training, and transfer of stimulus control procedures (Petursdottir & Carr, 2011) they are different with respect to which skills they prioritize. The UCLA model follows the logic of teaching receptive language prior to expressive language, as well as targeting social, self-help, and pre-academic skills simultaneously, whereas the verbal behavior model prioritizes mand training and generally follows Skinner's logic as described by his theory of the analysis of verbal behavior (1957). Examples of UCLA curricula are those published by Leaf and McEachin (1999), Lovaas (1981, 2003), Lovaas and Buch (1997), and Maurice et al. (2001).

An unusual but necessary characteristic of curricula for children with ASD is that for some of these learners, linguistic terms must be taught out of context and experientially. As many children with ASD do not attend to environmental stimuli pertinent to language development, nor do they naturally imitate, their learning is seriously hindered. Thus, to teach the concept of actions/verbs, for instance, the interventionist will deliver the instruction "jump" and physically prompt the learner to jump. It may be necessary to run dozens or hundreds of such trials before the learner can respond to "jump" with jumping without assistance from the interventionist. Next, the learner will be taught to "clap"; once this instruction can be performed independently, clap and jump will be taught in discrimination against each other. Gradually, a complete, age-appropriate list of verbs/actions will be taught in this manner, following careful assessment of each learner so that at some point, the learner will have acquired the expected number of actions for their age. Similarly, the multitude of constructs necessary to build language will be taught in this manner. Examples of such constructs are colors, shapes, numbers, letters, and all the basic categories, such as vegetables, fruits, animals, vehicles, furniture, clothing, and anything and everything that can be placed in a specific category. Additionally, more abstract linguistic concepts such as prepositions and pronouns will be taught along with hundreds of other programs addressing the multitude of constructs necessary to build language comprehension and subsequently speech.

Developing a curriculum for a specific child depends on several variables. Age and level of functioning play a protagonist role in this process. With respect to age, one must consider that the younger the child, most likely, the fewer the skills. Regardless, with young children, the curriculum tends to have an ascending direction. In other words, the interventionist will assess the child's skills and knowledge and will develop an individualized curriculum that begins a little below the child's current level of functioning and builds upward. Another important component of curriculum development is that it can be conceptualized as linear when building a specific skill. For instance, in order for a child to develop an understanding of object-to-number correspondence, the curriculum sequence may begin with matching identical numbers, receptively identifying numbers, labeling numbers, and then matching a number to a manipulative, for instance, matching the number 3 to 3 pencils, then for generalization purposes, the child may be asked to first match numerals with the dots on a dice, and then identify and label the dots on the dice, then they may be asked to give someone a certain number of items or place a certain number of objects in various places, finally, they may be asked to perform a number of actions. These programs will persist until the learner builds fluency with the object-to-number correspondence both receptively and expressively up to a certain number, depending on their age, for instance, all possible combinations of correspondence up to the number 10. Linear curriculum development must be carefully planned so that several skills are selected and taught simultaneously in a manner that allows their recombination leading to a higher level and more complex knowledge. The type of curriculum development required for teaching higher-order skills, although dependent on linear curriculum development, is considerably different. A metaphor that captures the essence of this type of curriculum development is the spider web. Initially, numerous different skills must be taught simultaneously and in a specific order so that subsequently, these skills can recombine into more complex curricula or programs. For instance, a curriculum sequence that illustrates the spider web is the following: teach colors and shapes following a similar sequence as above, i.e., teach first matching identical colors and shapes, then matching nonidentical colors and shapes, then receptive and expressive identification of colors and shapes, then introduce a conditional discrimination program where the learner has to use both color and shape by responding to the instruction "point to red circle" when in their field there is also a red square and a blue circle. Conditional discrimination will also include alternating combinations of instructions like point to house (name), point to square (shape), and point to yellow (color), given the picture of a house. Now, the learner is developing an understanding that the object, house, has other dimensions, such as a shape and a color. Furthermore, the learner may be asked to point to different parts of the house or identify the material of which the house is made (brick, glass, iron, wood, etc.), extending their understanding of the constructed house even further.

When working with older learners, curriculum development may follow an ascending direction as described above, but it will also require the interventionist to work backward, otherwise described as deconstructing. In other words, with older learners, it is very difficult to conduct thorough curriculum assessments to determine in detail what they know and what is missing from their repertoire. Thus, when a skill deficit is documented, for instance, a learner cannot describe an object, the interventionist deconstructs the skill of "describing an object" by identifying all the skills a learner must have to describe an object, working their way backward. For instance, to the instruction "describe this" when displaying an apple, the learner is expected to say: "This is an apple. The apple is a fruit. It is round and red. It is crunchy, and it tastes sweet. It grows on trees." Thus, the programs that must be taught for a learner to successfully describe an object would include in the case of the apple: labeling an object, object category, shape, color, textures, and flavors of foods. Having deconstructed the higher-order program into the prerequisite skills necessary, the interventionist can begin to assess the learner and determine what needs to be taught next. It should be noted that the individual programs necessary to be taught will vary depending on the object category, for instance, with animals, the learner would have to know the parts of the animal, perhaps what the animal produces, for instance, a cow gives us milk, or where they live, a cow lives on the farm, etc.

NDBIs have their own curriculum, each focusing on specific areas of functioning and behavior. ESDM, which is considered the most comprehensive NDBI model so far, utilizes a 446-item curriculum checklist (Rogers & Dawson, 2020), which constitutes the basis for the target behaviors to be taught throughout the intervention. This curriculum covers areas of receptive language, expressive language, fine motor, joint attention, imitation, and social and play skills. The ESDM curriculum checklist is administered to children receiving intervention every 12 weeks, as a means of assessing progress and setting future goals.

Setting goals or designing a child's curriculum is a highly individualized process that is based on observing the child, interviewing parents and teachers, and conducting a systematic and comprehensive assessment of skills across all areas of functioning (Hancock et al., 2000). This underscores the importance of assessment in curriculum analysis and development.

Assessment Designing an individualized comprehensive curriculum, a challenging task, requires extensive assessment across all areas of functioning to identify skills and deficits, which will serve as the basis upon which new target behaviors will be built. As it is not possible to teach every single skill a child needs to function independently, in addition to having access to a hierarchically organized and developmentally sequenced curriculum (Love et al., 2009), priority should be given to teaching behavioral cusps (Rosales-Ruiz & Baer, 1997) and pivotal responses (Koegel et al., 1987), which will facilitate learning across settings and over time.

According to Gould et al. (2011), assessments should (a) be compressive, i.e., addressing all areas of functioning (social, motor, language, daily living, play, executive functions, social cognition, and academic skills), (b) target early childhood development, (c) prioritize the function of responding over its topography, (d) link directly to behavioral targets, and (e) allow tracking progress over time. In addition to the absence of comprehensive curricula in the literature, there are no comprehensive assessments specifically developed for EIBI. For this purpose, clinicians and behavior analysts use any instrument that produces information regarding the aforementioned areas of functioning. There appears to be considerable agreement between both national education and treatment centers as well as ABA-based providers on the use of these assessment instruments for a variety of purposes including curriculum design (Luiseli et al., 2001). In their review of assessments for determining the content of EIBI, Gould et al. (2011) provide a detailed description of existing assessments. They begin their discussion by indicating that although the golden standard for assessment within ABA programs is direct observation, rather than relying on third-party reports, a more feasible choice is indirect assessment, such as rating scales. As they point out, direct observation comes with insurmountable practical constrains, such as the need for many hours of observation, possibly ranging from dozens to hundreds, to obtain a complete picture of levels of functioning, whereas indirect assessments combined with some direct observation data, when necessary, are more practical.

In their paper, Gould et al. (2011) list several rating scales addressing all domains of functioning such as developmental/educational, social skills, motor function, speech and language/communication, daily living skills, play skills, and intelligence. Under each of these domains, they list a number of well-known rating scales summarizing important information such as areas of functioning addressed by each assessment scale, the age range they cover, how long it takes to administer the assessment, and whether the assessment can be administered by a qualified professional or by someone who has been trained to do so. Here are some examples of assessments across the domain areas identified previously: for D*evelopmental/ Educational Purposes* the (a) Batele Developmental Inventory-Second edition (BDI-2, Newborg, 2005, (b) Bayley Scales of Infant and Toddler Development-Third Edition (Bayley-III, Bayley, 2006), (c) Brigance Diagnostic Inventory of Early Development-II (Brigance IED-II. Brigance, 2004), (d) Denver II (Frankenburg et al., 1990, (e) Psychoeducational Profile-Revised (PEP-R; Schopler et al., 1990); for *Social skills* the (a) Social Responsiveness scale (SRS; Constantino & Gruber, 2005), (b) Social Skills Rating Scale (SSRS; Gresham & Elliot, 1990); for *Motor Function* the (a) Beery–Buktenica Developmental Test of Visual-Motor integration-Fifth Edition (Beery VMI; Beery & Beery, 2004), (b) Peabody Developmental Motor Scales-Second edition (PMDS-2; Folio & Fewell, 2000); for *Speech and Language/communication* the (a) Assessment of Basic Language and Learning Skills-Revised (ABLLS-R; Partington, 2008), (b) Clinical Evaluation of Language Fundamentals-Fourth Edition (CELF-4; Semel et al., 2003), (c) Peabody Picture Vocabulary Test-Fourth Edition (PPVT-4; Dunn & Dunn, 2007), (d) Pragmatics Profile of Everyday Communication Skills in Children Revised Edition (Dewart & Summers, 1995); for *Daily Living Skills* (a) The Scales of Independent Behavior-Revised (SIB-R; Bruininks et al., 1996) and (b) Vineland Adaptive Behavior Scales-Second Edition (VABS-II; Sparow et al., 2005); for *Play Skills* (a) Symbolic Play Test-Second Edition (SPT; Lowe & Costello, 1988); for *Academic Achievement* (a) Brigance Diagnostic Comprehensive Inventory of Basic Skills-Revised (CIBS-R; Brigance, 1999), (b) Peabody Individual Achievement Test-Revised (PIAT-Revised; Markwardt, 1997), (c) Woodcock-Johnson III Normative Update (WJ III NU; McGrew et al., 2007); for *Intelligence* (a) Wechsler Intelligence Scale for Children-Fourth Edition Integrated (WISC-IV Integrated; Wechsler et al., 2004) and (b) Wechsler Preschool and Primary Scale of Intelligence-Third Edition (WPPSI-III; Wechsler, 2002).

Despite the usefulness of these commonly used screening and assessment tools, as mentioned above, Gould et al. (2011) caution practitioners with regard to the fact that they do not measure changes in behavior rather they measure changes in functioning. This is not very useful to the practitioner as poor performance in a specific domain is most often indicative of the need for further assessment to identify the respective skills that need to be taught. As these assessments do not provide this information, clinicians are left to their own means to identify which skills will lead to improved functioning in the respective areas.

Intervention Settings In designing what turned out to be their seminal study in EIBI, Lovaas et al. (1973) were influenced by Colligan and Belamy's 1968 study, which was the first published study demonstrating the effectiveness of behavioral interventions for children with ASD (Klintwall & Eikeseth, 2014). Lovaas et al. among other things noticed that the outcomes, in the Colligan and Belamy study, were situation specific, so in their seminal study, they moved the intervention to the home rather than conducting it in centers. Since then, a plethora of research documents the effectiveness of EIBI across settings in addition to the home, i.e., centers and schools (Love et al., 2009; Smith et al., 2000). Thus, EIBI services currently are delivered through home- or center-based programs, and taught skills are gradually generalized to classroom and community settings, with the intention to fade supports so that the child eventually functions independently, thereby becoming integrated into the respective settings (Howard et al., 2005; Sallows & Graupner, 2005).

Similarly to EIBI, NDBIs can be accessed as part of EI services or combined with EIBI. NDBIs rely on the principles of operant learning, and the core features of their intervention can be incorporated into various disciplines involved in neurodevelopmental conditions such as ASD (Ingersoll et al., 2023). This makes NDBIs suitable for intervention across multiple settings, such as home-based therapy, school contexts, and centers, similar to EIBI services. However, it is still unclear in the literature if intervention setting affects treatment outcomes for NDBIs.

Parent Training Lovaas et al. (1973) noticed that clinical gains for the children receiving EIBI interventions could be maximized when their parents became skilled teachers. As a result of parent training, child outcomes were accelerated and maintained over time (Klintwall & Eikeseth, 2014). Since then, numerous studies underscore the importance and necessity of parent training as a supplement to EIBI services (Solish & Perry, 2008; Strauss et al., 2012, 2013), while at the same time, it is suggested that parents be a part of supervisory meetings along with the interventionists (Lovaas & Buch, 1997; National Research Council, 2001). Similarly, some NDBIs include parent training as a key aspect of intervention (Frost et al., 2020), as in the case of ESDM, which aims to promote and foster generalized skills in everyday settings. While other NDBI models, like project imPACT (Ingersoll & Wainer, 2013), are based solely on parental involvement and training. It is advised that providers to children with ASD, whether they are delivering services from an EIBI or NDBI-based framework, should plan for parent training and parental involvement from the beginning of the intervention, making it a prerequisite and an integral part of the intervention and utilizing individualized strategies to meet each child's and family's unique needs. Behavioral Skills Training, video modeling, and one-on-one or group training sessions have been demonstrated to be highly effective methods in both staff and parent training (Booth et al., 2018).

The Importance of Supervision In addition to the characteristics of EIBI, i.e., timeliness and intensity, use of behavior analytic interventions, and an extensive and comprehensive curriculum, regular supervision by well-trained and experienced behavior analysts has been added to the list. This was not an arbitrary decision rather it has arisen from need, and it is maintained by empirical evidence (Eikeseth, 2009; Gould et al., 2011).

The need for well-trained and experienced supervisors, in part, stems from the fact that EIBI services are provided by a team. Depending on the hours of intervention, four to five different therapists may be involved in delivering 1:1 services to a child. In other words, the same programs may be implemented by four or five different therapists on any given day. This creates the need for a division of responsibilities, which in turn creates the need for a hierarchy to ensure proper design and implementation of services. Furthermore, comprehensive early intervention curricula, as discussed previously, are complex, need constant adjustments, and are highly individualized. Thus, there is a great need for professional guidance, ongoing training, and continuous supervision of the therapists both in the use of behavior analytic principles and tactics as well as in selecting goals and designing interventions to teach them.

NDBI, a fairly new approach to EI services, similarly to EIBI, can be utilized by a multidisciplinary team (including a psychologist, speech therapist, occupational therapist, special educator, etc.), making the need for thorough training necessary to increase competency and thereby prevent the ineffective implementation of NDBI techniques (Hampton & Sandbank, 2022). Consequently, the need for ongoing training, in theory and in practice, and supervision of NDBI providers is crucial (Ingersoll et al., 2023). The frequency of supervision for NDBI practices depends on the provider's level of experience, and it should be constantly adjusted and individualized to meet the learners' and supervisees' needs (Longenecker, 2010).

Experimental evidence involving EIBI's effectiveness clearly demonstrates that appropriate and adequate supervision has a great impact on the quality of services provided, the proper design of learning objectives, and the effectiveness of treatment and the overall outcome of the intervention (Dixon et al., 2016; Gibson et al., 2009; LeBlanc & Luiselli, 2016). As a result, supervision is now considered an integral part of ABA in general and EIBI in particular (Valentino, 2021). Similar research for NDBIs is yet to be conducted.

Supervision Requirements When developing and practicing EIBI and NDBI, finding suitable personnel and supervisors is of utmost importance. Psychologists, speech and language therapists, occupational therapists, special education teachers, or other practitioners involved in EIBI and ASD treatments, who wish to implement best practice interventions, should seek certified behavior analysts as supervisors to ensure the best possible outcomes for their learners. To the authors' knowledge, there are no formal routes to certification when it comes to NDBI services. It appears that the scientist who introduced the model tends to also be the training and certification agent in the respective NDBI. Consequently, all the above specialties, who are interested in utilizing NDBIs, should seek training and supervision from a practitioner trained in the NDBI of interest.

The Behavior Analyst Certification Board (BACB), established in 1998 with the goal of meeting professional certification needs both for behavior analysts and consumers, has established specific criteria for supervisors. Supervisors must be board-certified behavior analysts (BCBAs), i.e., they must have a masters (BCBA) or doctorate level (BCBA-D) degree in a behavior analysis-related field, at least 1500 h of supervised practical experience, and they must pass the BCBA exams. Since 2012, specifically for BCBAs and BCBA-Ds providing supervision services to others pursuing certification or professionals practicing behavior analysis, an 8-h supervision training is required, along with the continuing education units that must be completed within the 2-year recertification period. The BACB registry can be used for finding behavior analysts that meet the aforementioned requirements (www.bacb.com). Also, additional guidelines for identifying qualified EIBI supervisors can be found at the Autism Special Interest Group (1998), as mentioned in Davis et al. (2002).

When it comes to NDBIs, for the time being, the authors' knowledge, training, certification, and supervision are available through the founders of each model and the institutions they have established. Many BCBA supervisors and practitioners seem to have limited knowledge and training in behavioral developmental approaches like NDBIs (Hampton & Sandbank, 2022); thus, it is recommended to seek professionals who are trained in ABA with additional training in naturalistic developmental models. In any case, supervisors should maintain their knowledge and practices up to date, taking into account the most recent developments in the field.

Best Practice in Supervision Regarding best supervision practices, although the literature is still developing, three major components of supervision have been identified: (a) intensity of supervision, (b) expertise of supervisors and consultants, and (c) use of behavior skills training (written and oral instructions, modeling, rehearsal, and feedback).

It is clearly documented in the literature that the intensity of supervision (Eikeseth et al., 2009) can positively affect the intervention outcomes. In view of the existing data, it is suggested that supervision be provided weekly (Brown et al., 2023; Fraidlin et al., 2023; Lovaas, 1987) or at least biweekly (Bibby et al., 2002). For those seeking certification, the BACB requires at least 5% of practicum hours to be supervised in a mixed group and in a one-to-one format.

The expertise of supervisors and consultants (Davis et al., 2002; Eikeseth, 2010) is also of utmost importance for the quality and effectiveness of services provided. As mentioned earlier, EIBI supervisors should be BCBAs or BCBA-Ds and must have completed an 8-h training in supervision, along with ongoing education regarding supervision. In terms of demonstrating proficiency, supervisors should be able to successfully complete the following tasks described in Davis et al. (2002): selecting new instructional programs for the child's curriculum, working with the child to start these programs, providing parents and aides with information about autism and EIBI, evaluating aides' implementation of instructional procedures, and interacting with parents and aides. It is also necessary that supervisors remain up-to-date regarding the latest research and developments in the literature, as well as demonstrate professionalism and soft skills (Cowan & Kodak, 2024). Finally, behavioral skills training (Miltenberger, 2016), and consistent assessments both for staff and for treatment goals (Eikeseth, 2010), play a major role in the efficacy of the intervention.

As with everything else in EIBI, specific supervision practices, frequency, and format, i.e., group, individual, or peer supervision, should be individualized and subject to changes following systematic assessment of the data. Given that EIBI and NDBI address all areas of functioning, endorsing interdisciplinary collaboration when designing treatment plans and goals is advisable. However, it is important to maintain a coherent behavior analytic intervention to meet best practice criteria. This can be achieved by adopting a consultative model of collaboration where psychologists, special educators, speech and occupational therapists, and physical therapists consult with the supervising BCBA in their specific area. It is then the responsibility of the BCBA to incorporate the recommendations from the different disciplines into the child's curriculum. This process will more likely result in a comprehensive intervention that is holistic yet remains behavior analytic.

Useful Information Regarding ABA Certification and Certified Supervisors Information regarding pursuing certification, finding a supervisor, and providing ethical and efficient behavior analytic services for people in the United States, United Kingdom, and Australia can be found on the BACB website (www.bacb.com). The Association for Behavior Analysis International (ABAI) (www.abainternational.org), established in 1974, also provides information regarding the philosophy, science, application, and teaching of behavior analysis. The Association of European Behavior Analysis (EuroBA) (www.euroba.org) was recently founded, aiming to ensure recognition of behavior analysts in Europe. The Qualified Applied Behavior Analysis Credentialing Board (QABA) (www.qababoard.com), as well as the International Behavior Analysis Organization (IBAO) (www.theibao.com), both offer internationally recognized certifications for those practicing ABA-related services.

Conclusion

This chapter offered an overview of EI, with a focus on EIBI and NDBIs. Our science continues to have a long way to go in terms of providing definitive and precise answers with respect to a number of important questions that scientists, practitioners, and parents ask such as who will benefit from what type of services, when is it best to begin intervention, how intensive should intervention be, what should be taught, which is the best setting for teaching, and which approach produces best short- and long-term outcomes. Inevitably, more precise answers to these questions will emerge through systematic research evaluating the specific moderators and mediators involved in EIBI and NDBI. For instance, in addition to age at onset of intervention and intensity of services, pretreatment IQ, family characteristics, and biological variables may also play important roles in moderating outcomes.

Then again, it is clear from the evidence that has been reviewed in this chapter that long-term comprehensive EIBI results in positive improvements in intellectual functioning, language development, acquisition of daily living skills, and social functioning for many children with ASD (Eikeseth, 2009; Makrygianni et al., 2018; Rogers & Vismara, 2008; Virués-Ortega, 2010). Furthermore, EIBI produces better learning outcomes across the aforementioned domains of functioning compared to treatment as usual (Eckes et al., 2023), as well as compared to eclectic and focused approaches (Howard et al., 2005; Odom et al., 2010; Rogers & Vismara, 2008), with outcomes being significantly higher for children between 2 and 7 years of age (Granpeesheh et al., 2009b). It is important to underscore the fact that recent empirical evidence supports integrated approaches, such as NDBI plus ABA, as emerging best practices (Smith & Iadarola, 2015). Specifically, ESDM has been found to produce good outcomes, particularly in the areas of social communication, cognitive development, adaptive functioning, and school readiness (Dawson et al., 2010).

In conclusion, for those interested in providing effective, ethical, and comprehensive EIBI or NDBI services, this chapter has provided a discussion of all the parameters one has to consider when designing and implementing a comprehensive intervention, including a holistic curriculum, addressing selecting personnel and supervisors, and whether to involve parents or guardians and teachers and other professionals in the intervention. Finally, it is important to remember that EIBIs and NDBIs involve dynamic processes that require individualizing services, ongoing review, and data informed adjustments to maximize the outcome and benefits of the intervention.

Selected Bibliography

Bruinsma, Y. E., Minjarez, M. B., Schreibman, L., & Stahmer, A. C. (2020). *Naturalistic developmental behavioral interventions for autism spectrum disorder*. Brookes Publishing Company. This handbook is a summary of the latest information on NDBIs, including the models' framework, best strategies to be used, parent training and inclusion.

Cooper, J. O., Heron, T. E., & Heward, W. L. (2020). *Applied behavior analysis*. Pearson UK. This book offers a detailed description of the field of applied behavior analysis, while analyzing the principles and procedures for changing socially important behavior.

Lovaas, I. O. (1981). *Teaching developmentally disabled children: The "me" book*. Pro-Ed Australia.

Lovaas, O. I. (2003). *Teaching individuals with developmental delays: Basic intervention techniques*. Pro-ed. These two books by Lovaas offer detailed descriptions of current programs and a complete and comprehensive curriculum, based on proven practices, to teach individuals with developmental delays and disabilities.

Leaf, R. B., & McEachin, J. (Eds.). (1999). *A work in progress: Behavior management strategies and a curriculum for intensive behavioral treatment of autism*. DRL Books. The authors offer a complete guide and curriculum to teach individuals on the spectrum, using the principles of applied behavior analysis.

Rogers, S. J., & Dawson, G. (2020). *Early Start Denver Model for young children with autism: Promoting language, learning, and engagement*. Guilford Publications. This book covers all basic aspects of the most comprehensive NDBI model so far, offering the theoretical framework, teaching methods and the curriculum checklist to be used for assessment and intervention.

Sundberg, M. L., & Partington, J. W. (1998). *Teaching language to children with autism or other developmental disabilities*. Behavior Analysts, Incorporated. These authors offer an introduction to Skinner's verbal behavior and a brief language assessment system, as well as the basic teaching procedures for developing early and intermediate verbal skills.

Sundberg, M. L. (2014). *The verbal behavior milestones assessment and placement program: The VB-MAPP* (2nd ed.). AVB Press. Based on Skinner's analysis of verbal behavior, contains a language assessment, the assessment instructions, and the specific scoring criteria for each of the 170 milestones, the 24 language and learning barriers, the transition to less restrictive settings assessment, and the over 900 tasks contained in the task analysis of each skill area.

Tarbox, J., Dixon, D., Sturmey, P., & Matson, J. L. (2014). *Handbook of early intervention for autism spectrum disorders*. Springer. The authors reference evidence-based research on early intervention for autism while analyzing the autism spectrum disorder, and discussing methods of assessment and treatment.

Vargas, J. S. (2013). *Behavior analysis for effective teaching*. Routledge. *For those interested to progress from early intervention and integrate the principles of behavior analysis* into classrooms and other school settings.

References

American Psychiatric Association, D. S. M. T. F., & American Psychiatric Association, D. S. (2013). *Diagnostic and statistical manual of mental disorders: DSM-5* (Vol. 5, No. 5). Washington, DC: American psychiatric association.

Baer, D. M., Wolf, M. M., & Risley, T. R. (1968). Some current dimensions of applied behavior analysis. *Journal of Applied Behavior Analysis, 1*(1), 91–97. https://doi.org/10.1901/jaba.1968.1-91

Baird, G., Cass, H., & Slonims, V. (2003). Diagnosis of autism. *BMJ, 327*(7413), 488–493. https://doi.org/10.1136/bmj.327.7413.488

Bayley, N. (2006). *Bayley Scales of Infant and Toddler Development (3rd ed) administration manual*. PsychCorp.

Beery, K. E., & Beery, N. A. (2004). *The Beery-Buktenica developmental test of visual-motor integration* (5th ed.). Pearson.

Bibby, P., Eikeseth, S., Martin, N. T., Mudford, O. C., & Reeves, D. (2002). Progress and outcomes for children with autism receiving parent-managed intensive interventions. *Research in Developmental Disabilities, 23*(1), 81–104. https://doi.org/10.1016/S0891-4222(02)00095-1

Bondy, A. S., & Frost, L. A. (1998). The picture exchange communication system. *Seminars in Speech and Language, 19*(4), 373–388; quiz 389; 424. https://doi.org/10.1055/s-2008-1064055

Booth, N., Gallagher, S., & Mickey, K. (2018). Autism, Interventions and Parent Training. *Psichologija, 57*, 74–94. https://doi.org/10.15388/Psichol.2018.0.11904

Brigance, A. H. (1999). *Brigance diagnostic comprehensive inventory of basic skills revised*. Curriculum Associates.

Brigance, A. H. (2004). *Brigance diagnostic inventory of early development-II*. Curriculum Associates.

Brondino, N., Fusar-Poli, L., Rocchetti, M., Provenzani, U., Barale, F., & Politi, P. (2015). Complementary and alternative therapies for autism spectrum disorder. *Evidence-based Complementary and Alternative Medicine, 2015*. https://doi.org/10.1155/2015/258589

Brown, K. R., Randall, K. R., Smith, S. W., & Rios, D. (2023). Supervision for early career BCBAs. *Behavior Analysis in Practice*, 1–12. https://doi.org/10.1007/s40617-023-00786-0

Bruininks, R. H., Woodcock, R. W., Weatherman, R. F., & Hill, B. K. (1996). *Scales of independent behavior-revised comprehensive manual*. Riverside Publishing.

Bruno, B. (1967). *The empty fortress: Infantile autism and the birth of the self*. The Free Press.

Burger, K. (2010). How does early childhood care and education affect cognitive development? An international review of the effects of early interventions for children from different social backgrounds. *Early Childhood Research Quarterly, 25*(2), 140–165. https://doi.org/10.1016/j.ecresq.2009.11.001

Campbell, F. A., & Ramey, C. T. (1994). Effects of early intervention on intellectual and academic achievement: A follow-up study of children from low-income families. *Child Development, 65*(2), 684–698. https://doi.org/10.2307/1131410

Cattell-Gordon, D., & Cattell-Gordon, D. (1998). The development of an effective applied behavioral analysis program for a young child with autism: A parent's perspective. *Infants & Young Children, 10*(3), 79–85.

Centers for Disease Control and Prevention. (2018). *Spotlight on a new pattern in racial and ethnic differences emerge in Autism Spectrum Disorder (ASD). Identification among 8-year-old children*. https://www.cdc.gov/ncbddd/autism/addm-community-report/spotlight-on-racial-ethnic-differences.html

Centers for Disease Control and Prevention. (2023a). Prevalence and characteristics of autism spectrum disorder among children aged 8 years — Autism and developmental disabilities monitoring network, 11 sites, United States. *Surveillance Summaries, 72*(2), 1–14. https://www.cdc.gov/mmwr/volumes/72/ss/ss7202a1.html

Centers for Disease Control and Prevention. (2023b). *Data and statistics on autism spectrum disorder*. https://www.cdc.gov/ncbddd/autism/data.html

Charlesworth, R. (1998). Developmentally appropriate practice is for everyone. *Childhood Education, 74*(5), 274–282. https://doi.org/10.1080/00094056.1998.10521951

Christon, L. M., Mackintosh, V. H., & Myers, B. J. (2010). Use of complementary and alternative medicine (CAM) treatments by parents of children with autism spectrum disorders. *Research in Autism Spectrum Disorders, 4*(2), 249–259. https://doi.org/10.1016/j.rasd.2009.09.013

Cohen, H., Amerine-Dickens, M., & Smith, T. (2006). Early intensive behavioral treatment: Replication of the UCLA model in a community setting. *Journal of Developmental & Behavioral Pediatrics, 27*(2), S145–S155.

Colligan, R. C., & Bellamy, C. M. (1968). Effects of a two-year treatment program for a young autistic child. *Psychotherapy: Theory, Research & Practice, 5*(4), 214–219. https://doi.org/10.1037/h0088692

Constantino, J. N., & Gruber, C. P. (2005). *Social Responsiveness Scale (SRS) manual*. Western Psychological Services.

Contaldo, A., Colombi, C., Pierotti, C., Masoni, P., & Muratori, F. (2020). Outcomes and moderators of Early Start Denver Model intervention in young children with autism spectrum disorder delivered in a mixed individual and group setting. *Autism, 24*(3), 718–729. https://doi.org/10.1177/1362361319888344

Cooper, J. O., Heron, T. E., & Heward, W. L. (2020). *Applied behavior analysis*. Pearson UK.

Cowan, L. S., & Kodak, T. (2024). Professional skills for behavior analysts: A survey on the proficiency and importance of hard and soft skills. *Behav Analysis Practice, 17*, 199–211. https://doi.org/10.1007/s40617-023-00823-y

Crank, J. E., Sandbank, M., Dunham, K., Crowley, S., Bottema-Beutel, K., Feldman, J., & Woynaroski, T. G. (2021). Understanding the effects of naturalistic developmental behavioral interventions: A project AIM meta-analysis. *Autism Research, 14*(4), 817–834. https://doi.org/10.1002/aur.2471

Crawford, S. (1986). Holding therapy. *Adoption & Fostering, 10*(4), 43–46. https://doi.org/10.1177/030857598601000411

Davis, B. J., Smith, T., & Donahoe, P. (2002). Evaluating supervisors in the UCLA treatment model for children with autism: Validation of an assessment procedure. *Behavior Therapy, 33*(4), 601–614. https://doi.org/10.1016/S0005-7894(02)80019-3

Dawson, G., Rogers, S., Munson, J., Smith, M., Winter, J., Greenson, J., et al. (2010). Randomized, controlled trial of an intervention for toddlers with autism: the Early Start Denver Model. *Pediatrics, 125*(1), e17–e23. https://doi.org/10.1542/peds.2009-0958

Dewart, H., & Summers, S. (1995). *The pragmatics profile of everyday communication skills in children (Rev. ed.) manual*. NFER-Nelson.

Dixon, D. R., Linstead, E., Granpeesheh, D., Novack, M. N., French, R., Stevens, E., et al. (2016). An evaluation of the impact of supervision intensity, supervisor qualifications, and caseload on outcomes in the treatment of autism spectrum disorder. *Behavior Analysis in Practice, 9*, 339–348. https://doi.org/10.1007/s40617-016-0132-1

Dunlap, G., Kern, L., Deperczel, M., Clarke, S., Wilson, D., Childs, K. E., et al. (1993). Functional analysis of classroom variables for students with emotional and behavioral disorders. *Behavioral Disorders, 18*(4), 275–291. https://doi.org/10.1177/019874299301800403

Dunn, I. M., & Dunn, D. M. (2007). *Peabody picture vocabulary test* (4th ed.). Pearson.

Eckes, T., Buhlmann, U., Holling, H. D., & Möllmann, A. (2023). Comprehensive ABA-based interventions in the treatment of children with autism spectrum disorder–a meta-analysis. *BMC Psychiatry, 23*(1), 1–19. https://doi.org/10.1186/s12888-022-04412-1

Eikeseth, S. (2009). Outcome of comprehensive psycho-educational interventions for young children with autism. *Research in Developmental Disabilities, 30*(1), 158–178. https://doi.org/10.1016/j.ridd.2008.02.003

Eikeseth, S. (2010). Examination of qualifications required of an EIBI professional. *European Journal of Behavior Analysis, 11*(2), 239–246. https://doi.org/10.1080/15021149.2010.11434348

Eikeseth, S., Smith, T., Jahr, E., & Eldevik, S. (2002). Intensive behavioral treatment at school for 4-to 7-year-old children with autism: A 1-year comparison-controlled study. *Behavior Modification, 26*(1), 49–68. https://doi.org/10.1177/0145445502026001004

Eikeseth, S., Smith, T., Jahr, E., & Eldevik, S. (2007). Outcome for children with autism who began intensive behavioral treatment between ages 4 and 7: A comparison-controlled study. *Behavior Modification, 31*(3), 264–278. https://doi.org/10.1177/0145445506291396

Eikeseth, S., Hayward, D., Gale, C., Gitlesen, J. P., & Eldevik, S. (2009). Intensity of supervision and outcome for preschool aged children receiving early and intensive behavioral interventions: A preliminary study. *Research in Autism Spectrum Disorders, 3*(1), 67–73. https://doi.org/10.1016/j.rasd.2008.04.003

Eldevik, S., Eikeseth, S., Jahr, E., & Smith, T. (2006). Effects of low-intensity behavioral treatment for children with autism and mental retardation. *Journal of Autism and Developmental Disorders, 36*, 211–224. https://doi.org/10.1007/s10803-005-0058-x

Eldevik, S., Hastings, R. P., Hughes, J. C., Jahr, E., Eikeseth, S., & Cross, S. (2009). Meta-analysis of early intensive behavioral intervention for children with autism. *Journal of Clinical Child & Adolescent Psychology, 38*(3), 439–450. https://doi.org/10.1080/15374410902851739

Eldevik, S., Hastings, J., Hughes, C., Jahr, E., Eikeseth, S., & Cross, S. (2010). Using participant data to extend the evidence base for intensive behavioral intervention for children with autism. *American Journal on Intellectual and Developmental Disorders, 115*(5), 381–405. https://doi.org/10.1352/1944-7558-115.5.381

Fein, D., Barton, M., Eigsti, I. M., Kelley, E., Naigles, L., Schultz, R. T., Stevens, M., Helt, M., Orinstein, A., Rosenthal, M., Troyb, E., & Tyson, K. (2013). Optimal outcome in individuals with a history of autism. *The Journal of Child Psychology and Psychiatry, 54*(2), 195–205. https://doi.org/10.1111/jcpp.12037

Filler, J., & Xu, Y. (2006). Including children with disabilities in early childhood education programs: Individualizing developmentally appropriate practices. *Childhood Education, 83*(2), 92–98. https://doi.org/10.1080/00094056.2007.10522887

Folio, M. R., & Fewell, R. R. (2000). *Peabody development motor scales* (2nd ed.). Pro-Ed.

Fombonne, E. (2009). Epidemiology of pervasive developmental disorders. *Pediatric Research, 65*(6), 591–598. https://doi.org/10.1203/PDR.0b013e31819e7203

Fraidlin, A., McElroy, A., Moses, K. A., Jenssen, K., & Van Stratton, J. E. (2023). Designing a successful supervision journey: Recommendations and resources for new BCBA supervisors. *Behavior Analysis in Practice, 16*(2), 374–387. https://doi.org/10.1007/s40617-022-00728-2

Frankenburg, W. K., Dodds, J., Archer, P., Bresnick, B., Maschka, P., Edelman, N., et al. (1990). *Denver II*. Denver Developmental Materials.

Frost, K. M., Brian, J., Gengoux, G. W., Hardan, A., Rieth, S. R., Stahmer, A., & Ingersoll, B. (2020). Identifying and measuring the common elements of naturalistic developmental behavioral interventions for autism spectrum disorder: Development of the NDBI-Fi. *Autism, 24*(8), 2285–2297. https://doi.org/10.1177/1362361320944011

Gibson, J. A., Grey, I. M., & Hastings, R. P. (2009). Supervisor support as a predictor of burnout and therapeutic self-efficacy in therapists working in ABA schools. *Journal of Autism and Developmental Disorders, 39*, 1024–1030. https://doi.org/10.1007/s10803-009-0709-4

Gould, E., Dixon, D. S., Najdowski, A. C., Smith, M. N., & Tarbox, J. (2011). A review of assessments for determining the content of early intensive behavioral intervention programs for autism spectrum disorders. *Research in Autism Spectrum Disorders, 5*, 990–1002. https://doi.org/10.1016/j.rasd.2011.01.012

Granpeesheh, D., Dixon, D. R., Tarbox, J., Kaplan, A. M., & Wilke, A. E. (2009a). The effects of age and treatment intensity on behavioral intervention outcomes for children with autism spectrum disorders. *Research in Autism Spectrum Disorders, 3*(4), 1014–1022. https://doi.org/10.1016/j.rasd.2009.06.007

Granpeesheh, D., Tarbox, J., & Dixon, D. R. (2009b). Applied behavior analytic interventions for children with autism: A description and review of treatment research. *Annals of Clinical Psychiatry, 21*(3), 162–173.

Gresham, F. M., & Elliot, S. N. (1990). *Social skills rating system manual*. American Guidance Service.

Gringras, P. (2000). Practical paediatric psychopharmacological prescribing in autism: The potential and the pitfalls. *Autism, 4*(3), 229–247. https://doi.org/10.1177/1362361300004003002

Guralnick, M. J. (1997). Effectiveness of early intervention for vulnerable children: A developmental perspective. *American Journal on Mental Retardation, 102*(4), 319–345. https://doi.org/10.1352/0895-8017(1998)102<0319:eoeifv>2.0.co;2

Guralnick, M. J. (2017). Early intervention for children with intellectual disabilities: An update. *Journal of Applied Research in Intellectual Disabilities, 30*(2), 211–229. https://doi.org/10.1111/jar.12233

Hampton, L. H., & Sandbank, M. P. (2022). Keeping up with the evidence base: Survey of behavior professionals about naturalistic developmental behavioral interventions. *Autism, 26*, 875–888. https://doi.org/10.1177/13623613211035233

Hancock, M. A., Cautilli, J. D., Rosenwasser, B., & Clark, K. (2000). Four tactics for improving behavior analytic services. *The Behavior Analyst Today, 1*(2), 35. https://doi.org/10.1037/h0099873

Hanley, G. P. (2012). Functional assessment of problem behavior: Dispelling myths, overcoming implementation obstacles and developing new lore. *Behavior Analysis in Practice, 5*, 54–72. https://doi.org/10.1007/BF03391818

Hanley, G. P., Iwata, B. A., & McCord, B. E. (2003). Functional analysis of problem behavior: A review. *Journal of Applied Behavior Analysis, 36*(2), 147–185. https://doi.org/10.1901/jaba.2003.36-147

Harris, S. L., & Handleman, J. S. (2000). Age and IQ at intake as predictors of placement for young children with autism: A four-to six-year follow-up. *Journal of Autism and Developmental Disorders, 30*, 137–142. https://doi.org/10.1023/A:1005459606120

Hart, B., & Risley, T. R. (1975). Incidental teaching of language in the preschool 1. *Journal of Applied Behavior Analysis, 8*(4), 411–420. https://doi.org/10.1901/jaba.1975.8-411

Howard, J. S., Sparkman, C. R., Cohen, H. G., Green, G., & Stanislaw, H. (2005). A comparison of intensive behavior analytic and eclectic treatments for young children with autism. *Research in Developmental Disabilities, 26*(4), 359–383. https://doi.org/10.1016/j.ridd.2004.09.005

Howard, J. S., Stanislaw, H., Green, G., Sparkman, C. R., & Cohen, H. G. (2014). Comparison of behavior analytic and eclectic early interventions for young children with autism after three years. *Research in Developmental Disabilities, 35*(12), 3326–3344. https://doi.org/10.1016/j.ridd.2014.08.021

Huntington, R. N., Badgett, N. M., Rosenberg, N. E., Greeny, K., Bravo, A., Bristol, R. M., et al. (2023). Social validity in behavioral research: A selective review. *Perspectives on Dehavior Science, 46*(1), 201–215. https://doi.org/10.1007/s40614-022-00364-9

IDEA. (2004). *Individuals with disabilities education improvement act of 2004*. ProQuest.

Ingersoll, B. (2010). Brief report: Pilot randomized controlled trial of reciprocal imitation training for teaching elicited and spontaneous imitation to children with autism. *Journal of Autism and Developmental Disorders, 40*, 1154–1160. https://doi.org/10.1007/s10803-010-0966-2

Ingersoll, B., & Wainer, A. (2013). Initial efficacy of Project ImPACT: A parent-mediated social communication intervention for young children with ASD. *Journal of Autism and Developmental Disorders, 43*, 2943–2952. https://doi.org/10.1007/s10803-013-1840-9

Ingersoll, B., Douglas, S. N., Brodhead, M. T., Barber, A., & Kaczmarek, L. A. (2023). Interdisciplinary competencies for implementing NDBIs with young children with autism and other social communication challenges. *Journal of Early Intervention*. https://doi.org/10.1177/10538151231218928

Kaiser, A. P., & Hester, P. P. (1994). Generalized effects of enhanced milieu teaching. *Journal of Speech, Language, and Hearing Research, 37*(6), 1320–1340. https://doi.org/10.1044/jshr.3706.1320

Kanner, L. (1943). Autistic disturbances of affective contact. *Nervous Child, 2*(3), 217–250.

Kasari, C., Freeman, S., & Paparella, T. (2006). Joint attention and symbolic play in young children with autism: A randomized controlled intervention study. *Journal of Child Psychology and Psychiatry, 47*(6), 611–620. https://doi.org/10.1111/j.1469-7610.2005.01567.x

Kern, L., Dunlap, G., Clarke, S., & Childs, K. E. (1994). Student-assisted functional assessment interview. *Diagnostique, 19*(2–3), 29–39. https://doi.org/10.1177/073724779401900203

Kim, Y. S., Leventhal, B. L., Koh, Y. J., Fombone, E., Laska, E., Lim, E. C., Cheon, K. A., Kim, S. J., Kim, Y. K., Lee, H., Song, D. H., & Richard Grinker, R. (2011). Prevalence of autism spectrum disorder in a total population sample. *American Journal of Psychiatry, 168*, 904–912. https://doi.org/10.1176/appi.ajp.2011.10101532

Klintwall, L., & Eikeseth, S. (2014). Early and intensive behavioral intervention (EIBI) in autism. In V. B. Patel et al. (Eds.), *Comprehensive guide to autism*. Springer Science+Business Media.

Klintwall, L., Eldevik, S., & Eikeseth, S. (2015). Narrowing the gap: Effects of intervention on developmental trajectories in autism. *Autism, 19*(1), 53–63. https://doi.org/10.1177/1362361313510067

Koegel, R. L., & Koegel, L. K. (2006). *Pivotal response treatments for autism: Communication, social, & academic development*. Paul H. Brookes Publishing Co.

Koegel, R. L., O'dell, M. C., & Koegel, L. K. (1987). A natural language teaching paradigm for nonverbal autistic children. *Journal of Autism and Developmental Disorders, 17*(2), 187–200. https://doi.org/10.1007/BF01495055

Kuhl, P. K., Tsao, F. M., & Liu, H. M. (2003). Foreign-language experience in infancy: Effects of short-term exposure and social interaction on phonetic learning. *Proceedings of the National Academy of Sciences, 100*(15), 9096–9101. https://doi.org/10.1073/pnas.1532872100

Landa, R. J., Holman, K. C., O'Neill, A. H., & Stuart, E. A. (2011). Intervention targeting development of socially synchronous engagement in toddlers with autism spectrum disorder: A randomized controlled trial. *Journal of Child Psychology and Psychiatry, 52*(1), 13–21. https://doi.org/10.1111/j.1469-7610.2010.02288.x

Leaf, R. B., & McEachin, J. (1999). *A work in progress: Behavior management strategies and a curriculum for intensive behavioral treatment of autism*. Springer.

LeBlanc, L. A., & Luiselli, J. K. (2016). Refining supervisory practices in the field of behavior analysis: Introduction to the special section on supervision. *Behavior Analysis in Practice, 9*(4), 271–273. https://doi.org/10.1007/s40617-016-0156-6

Levy, S. E., & Hyman, S. L. (2005). Novel treatments for autistic spectrum disorders. *Mental Retardation and Developmental Disabilities Research Reviews, 11*(2), 131–142. https://doi.org/10.1002/mrdd.20062

Li, Q., Li, Y., Liu, B., Chen, Q., Xing, X., Xu, G., & Yang, W. (2022). Prevalence of autism spectrum disorder among children and adolescents in the United States from 2019 to 2020. *JAMA Pediatrics, 176*(9), 943–945. https://doi.org/10.1001/jamapediatrics.2022.1846

Longenecker, C. O. (2010). Coaching for better results: Key practices of high-performance leaders. *Industrial and Commercial Training, 42*(1), 32–40.

Lord, C., Risi, S., Lambrecht, L., Cook, E. H., Leventhal, B. L., DiLavore, P. C., et al. (2000). The Autism Diagnostic Observation Schedule—Generic: A standard measure of social and communication deficits associated with the spectrum of autism. *Journal of Autism and Developmental Disorders, 30*, 205–223. https://doi.org/10.1023/A:1005592401947

Lovaas, I. O. (1981). *Teaching developmentally disabled children: The "me" book.* Pro-Ed Australia.

Lovaas, O. I. (1987). Behavioral treatment and normal educational and intellectual functioning in young autistic children. *Journal of Consulting and Clinical Psychology, 55*(1), 3. https://doi.org/10.1037/0022-006X.55.1.3

Lovaas, O. I. (2003). *Teaching individuals with developmental delays: Basic intervention techniques.* Pro-ed.

Lovaas, O. I., & Buch, G. (1997). Intensive behavioral intervention with young children with autism. In N. N. Singh (Ed.), *Prevention and treatment of severe behavior problems: Models and methods in developmental disabilities* (pp. 61–86). Thomson Brooks/Cole Publishing Co.

Lovaas, O. I., & Smith, T. (1989). A comprehensive behavioral theory of autistic children: Paradigm for research and treatment. *Journal of Behavior Therapy and Experimental Psychiatry, 20*(1), 17–29. https://doi.org/10.1016/0005-7916(89)90004-9

Lovaas, O. I., Koegel, R., Simmons, J. Q., & Long, J. S. (1973). Some generalization and follow-up measures on autistic children in behavior therapy. *Journal of Applied Behavior Analysis, 6*(1), 131–165. https://doi.org/10.1901/jaba.1973.6-131

Love, J. R., Carr, J. E., Almason, S. M., & Petursdottir, A. I. (2009). Early and intensive behavioral intervention for autism: A survey of clinical practices. *Research in Autism Spectrum Disorders, 3*(2), 421–428. https://doi.org/10.1016/j.rasd.2008.08.008

Lowe, M., & Costello, A. J. (1988). *Symbolic play test* (2nd ed.). Windsor, NFER-Nelso.

Luiseli, J. K., Campbell, B., DiPietro, E., Ellis, J. T., Taras, M., et al. (2001). Assessment instruments used in the education and treatment of persons with autism: Brief report of a survey of national service center. *Research in Developmental Disabilities, 22*, 389–398. https://doi.org/10.1016/S0891-4222(01)00079-8

Lund, S. K. (2001). Content and contingencies: Considerations regarding curriculum development for young children with autism. *The Behavior Analyst Today, 2*(3), 187. https://psycnet.apa.org/fulltext/2014-44007-006.pdf

Makrygianni, M. K., Gena, A., Katoudi, S., & Galanis, P. (2018). The effectiveness of applied behavior analytic interventions for children with autism spectrum disorder: A meta-analytic study. *Research in Autism Spectrum Disorders, 51*, 18–31. https://doi.org/10.1016/j.rasd.2018.03.006

Markwardt, Jr, F. C. (1997). Peabody Individual Achievement Test--Revised--Normative Update.

Maurice, C., Green, G., & Foxx, R. M. (Eds.). (2001). *Making a difference: Behavioral intervention for autism.* PRO-ED.

McEachin, J. J., Smith, T., & Lovaas, O. I. (1993). Long-term outcome for children with autism who received early intensive behavioral treatment. *American Journal of Mental Retardation, 97*(4), 359–372.

McGrew, K. S., Dailey, D. E. H., & Schrank, F. A. (2007). *Woodcock-Johnson III/Woodcock-Johnson III Normative update score differences: What the user can expect and why* (Woodcock-Johnson III Assessment Service Bulletin No. 9). Rolling Meadows, IL: Riverside Publishing.

Mesibov, G. B. (1992). Treatment issues with high-functioning adolescents and adults with autism. In E. Schopler & G. B. Mesiibov (Eds.), *High-functioning individuals with autism* (pp. 143–155). Springer US.

Miltenberger, R. G. (2016). *Behavior modification: Principles and procedures*. Cengage Learning.

National Research Council. (2001). *Educating children with autism*. National Academies Press.

Neimy, H., & Pelaez, M. (2021). Early intervention for infants at risk for autism spectrum disorders. In A. Maragakis, C. Droseel, & T. Waltz (Eds.), *Application of behavior analysis in healthcare and beyond*. Springer.

Newborg, J. (2005). *Battelle Devlopmental Inventoy* (2nd ed.). Riverside Publishing.

Newschaffer, C. J., Falb, M. D., & Gurney, J. G. (2005). National autism prevalence trends from United States special education data. *Pediatrics, 115*(3), 277–282. https://doi.org/10.1542/peds.2004-1958

Odom, S. L., Boyd, B. A., Hall, L. J., & Hume, K. (2010). Evaluation of comprehensive treatment models for individuals with autism spectrum disorders. *Journal of Autism and Developmental Disorders, 40*, 425–436. https://doi.org/10.1007/s10803-009-0825-1

Ozonoff, S., Iosif, A. M., Baguio, F., Cook, I. C., Hill, M. M., Hutman, T., et al. (2010). A prospective study of the emergence of early behavioral signs of autism. *Journal of the American Academy of Child & Adolescent Psychiatry, 49*(3), 256–266. https://doi.org/10.1016/j.jaac.2009.11.009

Park, E. Y., & Blair, K. S. C. (2019). Social validity assessment in behavior interventions for young children: A systematic review. *Topics in Early Childhood Special Education, 39*(3), 156–169. https://doi.org/10.1177/0271121419860195

Partington, J. W. (2008). *The assessment of basics language and learning skills-revised: Scoring instruction and IEP development guide*. Behavior Analysts.

Pelios, L., Morren, J., Tesch, D., & Axelrod, S. (1999). The impact of functional analysis methodology on treatment choice for self-injurious and aggressive behavior. *Journal of Applied Behavior Analysis, 32*(2), 185–195. https://doi.org/10.1901/jaba.1999.32-185

Persico, A. M., Ricciardello, A., Lamberti, M., Turriziani, L., Cucinotta, F., Brogna, C., et al. (2021). The pediatric psychopharmacology of autism spectrum disorder: A systematic review-Part I: The past and the present. *Progress in Neuro-Psychopharmacology and Biological Psychiatry, 110*, 110326. https://doi.org/10.1016/j.pnpbp.2021.110326

Petursdottir, A. I., & Carr, J. E. (2011). A review of recommendations for sequencing receptive and expressive language instruction. *Journal of Applied Behavior Analysis, 44*(4), 859–876. https://doi.org/10.1901/jaba.2011.44-859

Prizant, B. M., Wetherby, A. M., Rubin, E., & Laurent, A. C. (2003). The SCERTS model: A transactional, family-centered approach to enhancing communication and socioemotional abilities of children with autism spectrum disorder. *Infants & Young Children, 16*(4), 296–316.

Remington, B., Hastings, R. P., Kovshoff, H., degli Espinosa, F., Jahr, E., Brown, T., et al. (2007). Early intensive behavioral intervention: Outcomes for children with autism and their parents after two years. *American Journal on Mental Retardation, 112*(6), 418–438. https://doi.org/10.1352/0895-8017(2007)112[418:EIBIOF]2.0.CO;2

Reynolds, A. J. (1994). Effects of a preschool plus follow-on intervention for children at risk. *Developmental Psychology, 30*(6), 787–804. https://doi.org/10.1037/0012-1649.30.6.787

Richardson, Z. S., Khetani, M. A., Scully, E., Dooling-Litfin, J., Murphy, N. J., & McManus, B. M. (2019). Social and functional characteristics of receipt and service use intensity of core early intervention services. *Academic Pediatrics, 19*(7), 722–732. https://doi.org/10.1016/j.acap.2019.02.004

Richardson, Z. S., Scully, E. A., Dooling-Litfin, J. K., Murphy, N. J., Rigau, B., Khetani, M. A., & McManus, B. M. (2020). Early intervention service intensity and change in children's functional capabilities. *Archives of Physical Medicine and Rehabilitation, 101*(5), 815–821. https://doi.org/10.1016/j.apmr.2019.10.188

Roberts, J. M., & Prior, M. (2006). *A review of the research to identify the most effective models of practice in early intervention for children with autism spectrum disorders*. Australian Government Department of Health and Ageing, Australia.

Rogers, S. J., & Dawson, G. (2020). *Early start Denver model for young children with autism: Promoting language, learning, and engagement*. Guilford Publications.

Rogers, S. J., & Pennington, B. F. (1991). A theoretical approach to the deficits in infantile autism. *Development and Psychopathology, 3*(2), 137–162. https://doi.org/10.1017/S0954579400000043

Rogers, S. J., & Vismara, L. A. (2008). Evidence-based comprehensive treatments for early autism. *Journal of Clinical Child & Adolescent Psychology, 37*(1), 8–38. https://doi.org/10.1080/15374410701817808

Rosales-Ruiz, J., & Baer, D. M. (1997). Behavioral cusps: A developmental and pragmatic concept for behavior analysis. *Journal of Applied Behavior Analysis, 30*(3), 533–544. https://doi.org/10.1901/jaba.1997.30-533

Saemundsen, E., Magnússon, P., Georgsdóttir, I., Egilsson, E., & Rafnsson, V. (2013). Prevalence of autism spectrum disorders in an Icelandic birth cohort. *BMJ Open, 3*(6), e002748. https://doi.org/10.1136/bmjopen-2013-002748

Sallows, G. O., & Graupner, T. D. (2005). Intensive behavioral treatment for children with autism: Four-year outcome and predictors. *American Journal on Mental Retardation, 110*(6), 417–438. https://doi.org/10.1352/0895-8017(2005)110[417:IBTFCW]2.0.CO;2

Schopler, E., Reichler, R. J., Bashford, A., Lansing, M. D., & Marcus, L. M. (1990). *Individualized assessment and treatment for autistic and developmentally disabled children: Vol. 1. Psychoeducational Profile-Revised (PEP-R)*. Pro-Ed.

Schreibman, L., Dawson, G., Stahmer, A. C., Landa, R., Rogers, S. J., McGee, G. G., et al. (2015). Naturalistic developmental behavioral interventions: Empirically validated treatments for autism spectrum disorder. *Journal of Autism and Developmental Disorders, 45*, 2411–2428. https://doi.org/10.1007/s10803-015-2407-8

Semel, E., Wiig, E. H., & Secord, W. A. (2003). *Clinical evaluation of language fundamentals* (4th ed.). PsychCorp.

Skinner, B. F. (1957). *Verbal behavior*. Skinner Foundation Reprint.

Skinner, B. F. (1965). *Science and human behavior*. Simon and Schuster.

Slocum, T. A., Detrich, R., Wilczynski, S. M., Spencer, T. D., Lewis, T., & Wolfe, K. (2014). The evidence-based practice of applied behavior analysis. *The Behavior Analyst, 37*, 41–56. https://doi.org/10.1007/s40614-014-0005-2

Smith, T. (2013). What is evidence-based behavior analysis? *The Behavior Analyst, 36*(1), 7–33. https://doi.org/10.1007/BF03392290

Smith, T., & Iadarola, S. (2015). Evidence base update for autism spectrum disorder. *Journal of Clinical Child and Adolescent Psychology, 44*(6), 897–922. https://doi.org/10.1080/15374416.2015.1077448

Smith, T., Groen, A. D., & Wynn, J. W. (2000). Randomized trial of intensive early intervention for children with pervasive developmental disorder. *American Journal on Mental Retardation, 105*(4), 269–285. https://doi.org/10.1352/0895-8017(2000)105<0269:RTOIEI>2.0.CO;2

Smith, T., Scahill, L., Dawson, G., Guthrie, D., Lord, C., Odom, S., et al. (2007). Designing research studies on psychosocial interventions in autism. *Journal of Autism and Developmental Disorders, 37*, 354–366. https://doi.org/10.1007/s10803-006-0173-3

Solish, A., & Perry, A. (2008). Parents' involvement in their children's behavioral intervention programs: Parent and therapist perspectives. *Research in Autism Spectrum Disorders, 2*(4), 728–738. https://doi.org/10.1016/j.rasd.2008.03.001

Sparow, S. S., Cicchetti, D. V., & Balla, D. A. (2005). *Vineland adaptive behavior scales: Survey forms manual* (2nd ed.). American Guidance Service.

Staats, A. W. (1981). Paradigmatic behaviorism, unified theory, unified theory construction methods, and the Zeitgeist of separatism. *American Psychologist, 36*(3), 239. https://doi.org/10.1037/0003-066X.36.3.239

Staats, A. W. (1994). Psychological behaviorism and behaviorizing psychology. *The Behavior Analyst, 17*, 93–114. https://doi.org/10.1007/BF03392655

Strauss, K., Vicari, S., Valeri, G., D'Elia, L., Arima, S., & Fava, L. (2012). Parent inclusion in early intensive behavioral intervention: The influence of parental stress, parent treatment fidelity and parent-mediated generalization of behavior targets on child outcomes. *Research in Developmental Disabilities, 33*(2), 688–703. https://doi.org/10.1016/j.ridd.2011.11.008

Strauss, K., Mancini, F., Fava, L., & SPC Group. (2013). Parent inclusion in early intensive behavior interventions for young children with ASD: A synthesis of meta-analyses from 2009 to 2011. *Research in Developmental Disabilities, 34*(9), 2967–2985. https://doi.org/10.1016/j.ridd.2013.06.007

Valentino, A. L. (2021). Supervision and mentoring. In J. K. Luiselli, R. M. Gardner, F. L. Bird, & H. Maguire (Eds.), *Organizational behavior management (OBM) approaches for intellectual and developmental disabilities*. Routledge.

Virués-Ortega, J. (2010). Applied behavior analytic intervention for autism in early childhood: Meta-analysis, meta-regression and dose–response meta-analysis of multiple outcomes. *Clinical Psychology Review, 30*(4), 387–399. https://doi.org/10.1016/j.cpr.2010.01.008

Vivanti, G., & Stahmer, A. C. (2021). Can the early start Denver model be considered ABA practice? *Behavior Analysis in Practice, 14*(1), 230–239. https://doi.org/10.1007/s40617-020-00474-3

Warren, Z., McPheeters, M. L., Sathe, N., Foss-Feig, J. H., Glasser, A., & Veenstra-Vanderweele, J. (2011). A systematic review of early intensive intervention for autism spectrum disorders. *Pediatrics, 127*(5), 1303–1311. https://doi.org/10.1542/peds.2011-0426. Epub 2011 Apr 4.

Wechsler, D. (2002). *WPPSI-III administration and scoring manual*. PsychCorp.

Wechsler, D., Kaplan, E., Fein, D., Krmer, J., Morris, R., Delis, D., et al. (2004). *Wechsler intelligence scale for children—Integrated administration and scoring manual* (4th ed.). PsychCorp.

Witt, J. C., Daly, E. M., & Noell, G. (2000). *Functional assessments: A step-by-step guide to solving academic and behavior problems*. Sopris West.

Wolf, M. M. (1978). Social validity: The case for subjective measurement or how applied behavior analysis is finding its heart. *Journal of Applied Behavior Analysis, 11*(2), 203–214. https://doi.org/10.1901/jaba.1978.11-203

Woods, J. J., & Wetherby, A. M. (2003). Early identification of and intervention for infants and toddlers who are at risk for autism spectrum disorder. *Language, Speech & Hearing Services in Schools, 34*(3), 180. https://doi.org/10.1044/0161-1461(2003/015)

Zeidan, J., Fombonne, E., Scorah, J., Ibrahim, A., Durkin, M. S., Saxena, S., et al. (2022). Global prevalence of autism: A systematic review update. *Autism Research, 15*(5), 778–790.

Chapter 8
Evidence-Based Practice in Integrated Pediatric Primary Care

Blake Lancaster, Dana Gadaire, Phoebe Jordan, Brigid Behrens, Amanda Hicks, and Marie Rodriguez

Pediatric behavioral health (BH) represents a global health priority due to increasing needs and insufficient resources to meet demands (The United States Government, 2022; Forsman et al., 2015; Keynejad et al., 2021). Recent estimates suggest that 40% of teenagers and 25% of children experience clinically elevated symptoms of anxiety and/or depression that adversely impact their daily functioning (Jones et al., 2022; Racine et al., 2021). Prior to the COVID-19 pandemic, nearly 20% of all children were expected to meet the diagnostic criteria for a BH issue before they reach the age of majority (Ghandour et al., 2019). In recent years, that number has nearly doubled (Harrison et al., 2022; Racine et al., 2021) along with emergency department use for pediatric BH crises (Burstein et al., 2019; Hoffmann & Foster, 2020; Manuel et al., 2023). Despite receiving regular medical care from a pediatrician, approximately 80% of these children never receive BH treatment (Chien et al., 2022) with pronounced disparities for disadvantaged patient populations, including racial/ethnic minorities, non-English speakers, rural, and low-income families (Aratani & Cooper, 2012; Chang & Slopen, 2024; Rodgers et al., 2022). Untreated and undertreated BH problems often increase in severity along with the risk for later substance use disorders, suicide, and chronic needs for ongoing (physical and behavioral) health-care services (Das et al., 2016). For these reasons, the United States has initiated strategic plans to strengthen system capacity, connecting more Americans to care and creating a continuum of support (The United States Government, 2022). The World Health Organization (WHO, 2008a, b) is also working with low- and middle-income countries to increase the capacity of primary care providers to meet BH needs. This chapter provides the rationale behind the emergence of integrated primary care models in pediatrics and reviews

B. Lancaster (✉) · D. Gadaire · P. Jordan · B. Behrens · A. Hicks · M. Rodriguez
Department of Pediatric Psychology, University of Michigan Health, Ann Arbor, MI, USA
e-mail: blancast@med.umich.edu

A. Maragakis, M. Janikian (eds.), *Evidence-Based Behavioral Health Practices in Pediatric Specialty Settings*, https://doi.org/10.1007/978-3-031-71004-9_8

some barriers to the model's expansion. The unique treatment considerations that apply to integrated primary care models are reviewed, and the features of quality training experiences, including competencies, are provided.

Primary care providers represent a key pathway for facilitating broader access to timely evidence-based BH treatment because they functionally reduce many of the most cited barriers to care (e.g., stigma, concerns about confidentiality/trust, and logistical barriers; Radez et al., 2021). Over 90% of children receive regular care from a primary care physician, with the vast majority reporting a positive, trusting relationship with their provider (Cooper et al., 2006; Kelleher & Stevens, 2009). Indeed, pediatricians are estimated to provide 80% of the mental health care for children in the United States and are often described as "de facto" behavioral health providers. The most common reasons for off-cycle (i.e., "sick") primary care visits are primarily related to BH rather than physiological concerns, with physicians estimating they spend approximately 37% of their time addressing behavioral and school issues (Lancaster et al., 2019). Even if BH is not the primary reason for a visit, most visits involve consultation regarding at least one BH concern (Lancaster et al., 2019). Unfortunately, increasing pediatric BH needs and emergency department usage leaves pediatricians reeling to assess and triage patients' safety risks, treatment needs, and access to needed services post-discharge. What is more, pediatricians often report that they lack the specialized training, time, and structural support (e.g., financial reimbursement) to address their patients' complex BH issues effectively and efficiently (McMillan et al., 2020).

Integrated Primary Care

Integrating behavioral health specialists into the primary care environment serves multiple purposes including: (1) increased access to services for children and families with behavioral health concerns, (2) timely intervention to prevent/mitigate emerging problems, and (3) reduced frustration and inefficiency in the provision of medical care (Stancin & Perrin, 2014).

First, external referrals for BH services often result in daunting wait lists and poor follow-through on the part of patients and caregivers. Indeed, studies suggest that rates of attendance for external BH referrals are often as low as 10% (American Psychological Association [APA], 2014; Collins et al., 2010). Embedding behavioral health providers (BHPs with specialized training in short-term, targeted care in the medical home significantly enhances the likelihood that patients will access these much-needed specialty services. When BH services are available in primary care clinics, approximately 80% of referrals attend their initial appointments (APA, 2014; Auxier et al., 2012; Valleley et al., 2020; Vickers et al., 2013). It is noteworthy that Valleley et al. (2020) found these high levels of follow-through generalized across diverse demographics and geographical locations.

Second, early identification and intervention for commonly occurring BH issues can substantially improve downstream outcomes in academic performance and professional achievement while reducing the risks of delinquency and substance abuse. To this end, integrated behavioral health (IBH) teams not only increase access but research comparing IBH within primary care against conventional primary care practices has also yielded compelling results. Asarnow et al. (2015) conducted a meta-analysis demonstrating treatment effects for depression, anxiety, and disruptive behavior concerns in IBH settings. Additionally, Weersing et al. (2017) compared the efficacy and functional impact of a brief behavioral therapy intervention for anxiety and depression across integrated and community behavioral health providers. Relative to community BH providers, they found IBH in primary care to be associated with differential improvement in symptoms, functional impact, and retention. Thus, brief behavioral interventions in primary care settings are not only effective, but they potentially outperform traditional therapeutic approaches. The advantages are particularly pronounced considering the large number of cases involving the co-occurring (and often interrelated) medical and BH conditions (Vogel et al., 2017).

Finally, the integration of BHPs facilitates a "physician leveraging effect" (Cummings et al., 2009), allowing physicians to dedicate more time to medical care, increase patient volume, and deliver care within their established training and comfort zone. Leveraging the skills of medical providers will continue to be a critical factor given global shortages in workforce capacity, not to mention increasing concerns about physician stress and burnout that have been exponentially exacerbated by the COVID-19 pandemic (Committee on Pediatric Workforce et al., 2013; Nigri et al., 2023). Fortunately, research consistently suggests that medical providers value the expertise and practical support BHPs offer, reporting that it enhances their abilities to provide excellent patient care and increases practice productivity (Gouge et al., 2016; Hine et al., 2017; Lancaster et al., 2019; Miller-Matero et al., 2016).

There is some level of variability with respect to the integration of BHPs into primary care. In colocated practices, BHPs are physically located at the same site and use a shared records system, but they maintain a separate schedule and do not see patients when they are present for medical visits (Heath et al., 2013); in colocated models, visits can be kept to a consistent length and patients are seen after referral by a physician. In a fully integrated system, BHPs are consulted in real time when patients present for medical visits and may conduct joint visits with medical providers; this can lead to brief visits and more patients seen for a one-time consultation or through warm handoffs than in a colocated practice (e.g., Herbst et al., 2020), but can lead to more population-level reach as one more barrier to care is reduced (referral/scheduling).

Barriers to Implementation

Despite the many well-established benefits of integrating BH in pediatric primary care (Asarnow et al., 2015), many obstacles limit the widespread dissemination of effective care models. Nationwide, behavioral health specialists are formally embedded in a minority of primary clinics, with practices citing obstacles related to administrative constraints and professional/training needs (Katzelnick & Williams, 2015). First, initial setup and coordination with new providers can be cumbersome and often garners resistance from administrative personnel anticipating workload expansion. Complex and incongruous systems of billing and insurance paneling can exacerbate these tensions. BHPs must become credentialed with relevant insurance carriers, bearing in mind the involvement of third-party "carve out" policies that may require additional documentation. Second, coding represents challenges given that medical and BH rely on unique classification systems (i.e., International Classification of Disease [ICD] and Diagnostic and Statistical Manual of Mental Disorders [DSM]). Fortunately, substantial overlap and online crosswalks can help BHPs adapt their billing processes rather seamlessly.

A more significant barrier involves the differential coverage of BH care and the requirement of many insurance carriers for pre-authorization of BH services. Administrative teams are encouraged to verify insurance benefits specific to BH and communicate them to patients before scheduling. Communicating frustrating information (e.g., when services are not covered at the same rates as medical services) may represent a significant burden on administrative staff. For example, many states and insurance providers have adopted prohibitive regulations regarding autism diagnosis and documentation required to access early intervention services (Grace et al., 2014; Stuart, 2011). Providers (both medical and BH) should familiarize themselves with local regulations and policies so they can qualify patient expectations regarding services that are likely to be denied and support patients in navigating complex service delivery systems.

The sheer volume of patients served in primary care presents a challenge in effectively screening all patients for BH concerns. Under-resourced primary care providers may lack adequate metrics for identifying at-risk patients in the absence of imminent BH concerns and may require openness to new workflows and temporary inefficiencies. For example, BHPs can be instrumental in navigating crisis situations and assessing patient safety risk. However, overreliance on crisis mitigation reduces their capacity to provide primary and secondary prevention of BH problems (i.e., catch them before they become crises).

Finally, providers trained in traditional psychotherapy models must adapt to large caseloads and rapid pace of primary care settings. Simply moving traditional behavioral health into primary care practices will lead to full schedules, referral backlogs, and long waiting lists. In other words, the same barriers to mental health service will be present despite the colocation of behavioral health providers. What is more, a considerable portion of patients attend only a single visit (Valleley et al., 2020), underscoring the imperative nature of providing single-session interventions.

For maximum effect, BHPs need evidence-based, solution-focused approaches that efficiently target well-defined problems. In many cases, rigorous assessment can be replaced with functional transdiagnostic interventions (e.g., common elements; Cuijpers et al., 2019) that target specific symptoms or concerns. Much like the screening process, valid measurement of target outcomes is critical to determining whether treatment goals are being met or if more intensive approaches are needed. Thus, BHPs must efficiently prioritize and operationally define target behaviors (e.g., frequency and/or intensity of emotional outbursts). They require training in brief functional assessment of behavior to identify contingencies that may be maintaining or exacerbating the problem (e.g., parental attention and escape from tasks). Understanding basic mechanisms underlying learning and behavior also promotes environmental engineering to support more desired behaviors. Finally, because parents will be critical to intervention delivery, BHPs must be fluent in communicating and adapting interventions to meet the unique characteristics and preferences of families.

Brief Evidence-Based Intervention

Brief, transdiagnostic, evidence-based interventions facilitate efficient population-based treatment for a variety of BH concerns (Campo et al., 2015; Dalgleish et al., 2020; Evans et al., 2020). Despite their comprehensive approach to addressing complex pediatric BH issues, these therapies are strategically streamlined for efficient delivery within the time-constrained environment of integrated primary care settings, ensuring both effectiveness and practicality in the provision of care to young patients. Research increasingly suggests several common factors (e.g., structured intervention, therapeutic relationship; Cuijpers et al., 2019; Norcross & Wampold, 2011) to be integral in obtaining meaningful therapeutic outcomes. Additionally, research into drivers of pediatric health outcomes consistently identifies several interventions that can be applied universally (Barlow et al., 2010) or targeted to specific functional problems (Chorpita et al., 2005; Evans et al., 2020). Such interventions cut across diagnostic boundaries, thus reducing resource-intensive assessment processes and obviating the need for stigmatizing labels. Rather, clinicians enjoy the flexibility to focus on mutable behavioral processes across a spectrum of pediatric behavioral health concerns, often reducing the need for more intrusive diagnostic assessment and/or psychiatric intervention (Walter et al., 2019). Collaborative problem definition and treatment development involve patients and families in prioritizing targeted clinical goals, fostering a sense of ownership and engagement in the therapeutic process. BHPs then draw their specialized training in behavioral science to (1) identify variables precipitating and/or maintaining BH concerns; (2) adapt empirically supported interventions to meet the unique needs of patients and families; and (3) facilitate progress monitoring to refine and align treatment with evolving patient needs (Bruhn & Hirsch, 2017).

Parent Consultation Model

Brief BH therapies emphasize the pivotal role of family dynamics in pediatric care, integrating family therapy and parent consultation as integral components of the treatment paradigm. This inclusion acknowledges the profound impact of familial relationships on a child's behavioral well-being. Patient factors such as the type/severity of the presenting problem, age of the child, and family configuration can influence the degree to which parents are involved in treatment protocols (Manassis, 2005).

Billing practices, insurance authorization procedures, and in-person vs. telehealth services delivery options can impact the modality through which protocols are delivered. With all these factors considered, the most employed treatment delivery models in primary care typically include parent consultation. Even when addressing prevalent concerns such as anxiety and depression in older children, these therapies consistently incorporate parental involvement, recognizing their influential role in supporting treatment efficacy (Dowell & Ogles, 2010).

Parent consultation models are particularly well-suited for the pediatric integrated primary care setting, largely because of the age and developmental level of the patients (Bond et al., 2013; Richards et al., 2008). When providing BH services in the pediatric primary care setting, BHPs can expect to receive referrals for children before their first birthday through young adulthood (Hostutler et al., 2023). Considering that most referrals will be made before a child becomes a teen (average age ~ 9 y/o), BHPs should expect to involve parents in the process of assessment, psychoeducation, and the delivery of specific recommendations. The rationale for the inclusion of parents largely revolves around the developmental level and the degree of insight that a typical preteen brings to the clinical situation.

In situations where the presenting issues concern externalizing behavior problems, the patient oftentimes is unaware that their behavior is a problem. In cases of this nature, the patient may not be motivated to attend the session, engage in therapy, or possess the necessary cognitive reasoning skills to understand psychoeducation or specific components of treatment. In other situations where the presenting problem is related to anxiety or depression, parent involvement in the implementation of treatment recommendations can be critical. The rationale for exposure-based therapies (for anxiety) and behavior activation (for depression) can be counterintuitive for patients and families (Abramowitz et al., 2019). Once an understanding is achieved, motivation issues are best addressed by having all members of the family in agreement regarding the rational and specific components of treatment. The best chances at successfully addressing BH issues in pediatric primary care hinge on members of the family unit working together.

Presenting Problems and Targeted Interventions

Consistent with national prevalence data on child mental health (Bitsko et al., 2022), the most common referrals to integrated primary care involve behavioral concerns (20–30%), anxiety (11–36%), and Attention Deficit/Hyperactivity Disorder (ADHD) (12–18%). However, various other concerns are also treated including mood disorders (12–17%), medically related concerns such as sleep and toileting (13.1%), school-related issues (5%), adjustment disorders (4–6%), and developmental delays (4–5%; Okoroji et al., 2023; Vechiu et al., 2023). To address these challenges, BHPs often develop treatment plans involving components from existing evidence-based practices based on cognitive behavioral, behavioral-analytic, and social learning theory (Allen et al., 1993; Baer et al., 1968; Bandura, 1977; Beck, 1970; Davis et al., 2019; Hoffman et al., 2012; Wodarski, 2014). We describe several of these interventions below.

Etiology of Evidence-Based Protocols in Integrated Primary Care

The etiology of brief, evidence-based behavioral health protocols delivered within the primary care pediatric setting is a topic of increasing interest and importance in the field. Despite the growing recognition of the need for such interventions, there remains a scarcity of literature offering direct guidance on the development and implementation of these protocols (Briggs et al., 2016). As a result, behavioral health providers (BHPs) often draw from the broader literature on behavioral principles to construct individualized protocols tailored to the unique needs of their patients and settings.

Much of the existing literature addressing commonly presenting behavioral health concerns in pediatric primary care originates from the behavioral and behavioral analytic traditions. This body of research, often informed by data collected by behavior analysts, serves as a foundational resource for developing evidence-based practices in this context. Consequently, many brief behavioral health protocols utilized in integrated primary care settings are grounded in well-established behavioral principles such as reinforcement, extinction, and differential attention. These principles provide a systematic framework for understanding and addressing a range of behavioral issues, including externalizing behaviors, attention-deficit/hyperactivity disorder (ADHD), anxiety, depression, toileting difficulties, feeding problems, and sleep disturbances.

Central to the development and implementation of these protocols is the use of transdiagnostic methods, including functional assessment techniques, during the initial intake process (Vogel et al., 2017). By identifying specific target behaviors and elucidating the antecedents and consequences maintaining these behaviors, BHPs can construct a comprehensive case conceptualization that informs the selection and refinement of evidence-based interventions. This behavioral framework not only facilitates the design of treatment protocols aimed at disrupting maladaptive response–reinforcer relations but also serves as a valuable tool for monitoring progress and troubleshooting when clinical goals are not being met. Furthermore, by involving parents in a collaborative and solution-oriented manner through parent consultation and problem-solving approaches, BHPs empower families to play an active role in their child's behavioral health care (Dowell & Ogles, 2010).

Table 8.1 presents examples of commonly used brief protocols in integrated pediatric primary care, highlighting the behavioral principles underpinning each intervention. Through the systematic application of these principles and the integration of parent consultation methods, BHPs can deliver effective, evidence-based interventions that promote positive outcomes for children and families within the primary care setting.

Table 8.1 Examples of commonly used brief protocols in integrated pediatric primary care

Protocol	ABA principles	Commonly used for…
Time in with differential attention Attend to the child when not engaging in problem behavior and ignore previously R^+ attention-maintained behavior	DRO, EXT, R^+	Behavior problems, anxiety
Child-directed interaction (CDI)—"child's game" Labeled and descriptive praise (R^+) with no questions or demands	NCR, MO	Behavior problems, anxiety
Time-out/job card grounding Removes all access to R^+ in response to problem behavior for a defined period of time	Punishment	Behavior problems, school issues
Daily report card (DRC) Objective daily behavior ratings for school target Bx (e.g., completing work, staying in seat, and hands to self) with R^+ at home	R+, shaping	School issues, ADHD
Exposure with response prevention Use systematic desensitization to gradually expose the patient to increasingly aversive stimuli while preventing escape	Escape EXT and respondent EXT	Anxiety, obsessive-compulsive disorder
Behavioral activation Increased access to R^+ through pleasurable event scheduling	R^+	Depression

Bruni et al. (2016, May 27–31)

DRO differential reinforcement of other behavior, EXT extinction, R+ reinforcement, NCR non-contingent reinforcement, MO motivational operation

Secondary Gain

A critical consideration in parent-consultation models involves the iterative nature of behavioral concerns in dynamic family contexts. In other words, behaviors or concerns may initially arise under a specific context (e.g., fear or stress), but social reactions to the behaviors result in additional benefits to the child in terms of attention, avoidance of aversive tasks, etc. Initially described by Freud as "interpersonal or social advantage attained because of illness (1959), secondary gain is not necessarily intentional manipulation, but a byproduct of adventitious social reinforcement. Thus, behavioral health consultants often work with parents to develop responsive practices that do not inadvertently reinforce problematic behavior patterns. Common interventions include *Acknowledge and Defer* and/or scheduling *Worry Time* (Nash, 2014) that is temporally distinct from the immediate stressor. These approaches facilitate communication of parental care and concern without allowing the symptom or concern to dominate interpersonal interactions and routines. When they are not effective in attaining secondary social reinforcers, *deferred* stressors may resolve on their own for many pediatric patients. If not, scheduling time to discuss and problem-solve when parties are calm and focused (i.e., not experiencing the physiological sequelae of stress) can often facilitate and can then be addressed more effectively during a specified time when all parties are more focused on effective problem-solving.

Diagnostic Assessment

Despite the benefits of transdiagnostic, functional interventions in integrated behavioral health, many patients undoubtedly benefit from diagnostic clarification afforded by the integration of behavioral health specialists in primary care. Differential diagnosis may help patients and caregivers achieve greater insight into their behavior patterns and tailor behavior management strategies to their unique needs (Antony & Barlow, 2020). Pediatric patients with Attention Deficit/Hyperactivity Disorder (ADHD) may also benefit from pharmacological intervention as part of a comprehensive treatment plan (Boland et al., 2020) However, the controlled nature and potential for these medications to be overused underscore the need for efficient and valid assessment to determine if a child's symptoms are consistent with this diagnosis (American Academy of Pediatrics, 2000). Rapid screeners (e.g., Vanderbilt ADHD Assessments; Wolraich et al., 2003) that are commonly used in pediatric practices have high false-positive rates and tend to be sensitive measures in terms of detecting gross deviations from expected developmental norms, but they lack sensitivity in identifying the precise nature of these difficulties (Becker et al., 2012; Langberg et al., 2010). Integrated behavioral health providers can offer more comprehensive assessment to clarify the nature and extent to which symptoms are consistent with ADHD and whether medication may be warranted and/or useful in treating presenting problems. Specifically, the American Academy of Pediatrics (Wolraich et al., 2019) recommends ADHD assessment, which includes measures of core symptoms and impairment in multiple settings (e.g., school and home). Broadband measures of social-emotional functioning are also recommended to rule out alternative causes and screen for comorbid conditions (e.g., anxiety, depression and autism). A review of educational records and/or academic performance is also indicated to identify potential developmental or learning considerations.

Case Example

As discussed previously, ADHD and externalizing behavior concerns are one of the most common reasons for referral to IBH (Okoroji et al., 2023). To illustrate what IBH would look like for the presenting concern, we can consider a case example. AB is an 8-year-old male referred to IBH for school concerns and emotional lability in addition to adjustment concerns (recent parental separation). His primary care provider is hoping to differentiate whether his presentation is more consistent with ADHD or an adjustment disorder. At the family's first session, the BHP initiates a comprehensive assessment including an interview to assess frequency, intensity and duration of presenting concerns, developmental and family history, and broadband and narrowband rating scales completed by multiple raters to evaluate internalizing and externalizing symptoms. At the first session, the BHP also provides initial behavior management recommendations, including introducing the functions of behavior (attention, escape, access to tangible items, and automatic reinforcement) and encouraging both antecedent and consequence management strategies to promote desired behavior.

At the following visit, the BHP reviews the results of all measures completed and determines that symptoms related to both hyperactivity and inattention predated parental separation symptoms are present across settings and are functionally impairing, and they are not better explained by an underlying learning disability, internalizing disorder, or another medical condition. The BHP reviews the diagnosis of ADHD with the child and parent, being sure to provide comprehensive psychoeducation related to the diagnosis (such as how developmental expectations for independence in activities of daily living and academic tasks may need to be adjusted to account for deficits in executive functioning) and best practices for treatment, including a combination of medication management, behavioral parent training, and behavioral classroom management strategies (Wolraich et al., 2019). The BHP supports the parent in requesting a school-based evaluation to determine eligibility for any additional support through an Individualized Education Plan (IEP) or 504 plan and in scheduling an appointment with their primary care provider to discuss medication management. The BHP also evaluates AB's response to behavioral strategies implemented by the family by measuring the frequency of the target behavior and the degree of fidelity with which the interventions were delivered.

Finally, throughout intervention delivery, the BHP sets the expectation with the family that treatment progress will determine next steps—if IBH is not sufficient, care can be "stepped" up as needed by the addition of medication management and/or referral to more intensive community referrals.

If behaviors have not yet met target levels following initial two to three sessions, additional sessions could be scheduled to advance progress toward behavioral goals. However, if behavioral goals have been met, the family may also elect to return to IBH as needed if new presenting concerns arise or their progress is not maintained. This is consistent with a stepped model of care advocated for by Maragakis et al. (2018) that is determined by progress monitoring and by a shared understanding between the provider and the family. Any future sessions would use a modular and behavioral approach to deliver protocols tailored to the presenting concern. For example, if AB's mother reports that he is having difficulty completing his morning routine on time, treatment could use both antecedent strategies (increasing visual supports, practicing routine, using timers, etc.) as well as consequence strategies (daily reinforcement available contingent on completion of morning routine) to increase compliance with this activity.

Training

Integrating BHPs into the primary care environment significantly increases access to care for children. Addressing BH issues early in primary care allows clinicians to address behavior health issues early, when severity is at the mild to moderate level, significantly increasing the likelihood that the issues can be effectively addressed (Colizzi et al., 2020; McGorry & Mei, 2018). Many BH issues, if identified early, can be relatively easily addressed by brief, evidence-based care in the primary care setting. Catching these issues early and delivering evidence-based protocols efficiently can alleviate many downstream issues such as chronic and severe behavior problems, severe anxiety and depression, adult delinquency, and substance abuse, and can improve social, academic, and professional functioning. All of these downstream issues present significant barriers for individuals later in life, and many of these effects can be mitigated or eliminated altogether in some cases if the individual receives access to quality, evidence-based care, early in life. There is perhaps no better environment than the primary care setting for children to receive such care.

BHPs who seek to engage in this work need specific training in brief assessment and brief evidence-based approaches to treating the most common BH issues (e.g., anxiety and depression) (Hunter et al., 2009). Traditional outpatient BH training models will likely not prepare trainees for practice in the primary care environment. The most effective protocols in integrated care settings are short-term, evidence-based, parent continuation-driven protocols with established, concrete, and objective treatment goals. There is no handbook for these types of protocols, as they must be adapted from the current evidence-based literature and modified for efficient delivery in the primary care environment. Students need specific training to effectively perform this type of protocol development, as it is not part of the curriculum in standard clinical, school, or counseling psychology programs.

Indeed, training in IBH should focus on four unique, core competencies. These include knowledge-based competencies (e.g., common physical and behavioral pediatric health concerns), clinical competencies in assessment, intervention and consultation, interprofessional competencies (e.g., ways to establish and maintain productive partnerships with medical providers), and practice management (e.g., time constraints, authorization, billing and sharing of medical records, level of integration, and an understanding of the administrative structure) competencies (see Ham et al., 2022 for a full review). While it is important for a BHP to be proficient in all areas summarized above, education in administrative issues is critical in establishing and maintaining an IBH program. Financial viability (i.e., billing) is essential in establishing a sustainable program. There are also idiosyncratic aspects of brief consultation with pediatricians and disseminating general treatment approaches within a particular clinic that are critical to learning before trying to enter this field (Ham et al., 2022).

Most importantly, interested trainees should look for a program that offers opportunities to deliver these evidence-based protocols to patients in general pediatrics clinics to gain experience working with interdisciplinary medical teams. Most providers who are trained as generalists lack specialized training in evidence-based care and lack experience working with physicians (Horevitz & Manoleas, 2013). Finding a program that embodies these elements is essential if one wants to enter the field of IBH in the primary care setting. Though programs that offer this type of training are sparse, the American Psychological Association provides a Primary Care Psychology Training Program Directory on their website (APA, 2023). Interested trainees should be prepared to do their "homework" and assess each program for the aforementioned curriculum recommendations before entering the field.

Conclusion

The integration of BHPs into pediatric primary care clinics is a growing field, and the model has garnered support from all stakeholders. This model of care is not necessarily a new idea but there is significant momentum regarding this model for mental health-care delivery. All relevant parties: families, providers (pediatricians and BHPs), and payers are all in favor of the Integrated Primary Care model, but the lack of a well-trained workforce remains the most significant barrier. Further expansion of training programs, improved payer rules, and significant demand make it a ripe field for BHPs to enter.

Acknowledgments We have no conflicts of interest to disclose.

Selected Bibliography

Campo, J. V., Geist, R., & Kolko, D. J. (2018). Integration of pediatric behavioral health services in primary care: improving access and outcomes with collaborative care. *The Canadian Journal of Psychiatry, 63*(7), 432–438 This paper explores collaborative care interventions integrating pediatric mental health services into primary care, aiming to overcome barriers to service delivery and improve access and outcomes. It reviews literature on structural and attitudinal barriers, emphasizing the potential of integration to address shortages and stigma. Findings suggest that collaborative interventions enhance access, outcomes, and patient satisfaction, but implementing them in practice may require changes to reimbursement models and healthcare culture.

Lines, M. M. (2022). Pediatric integrated primary care: A population health approach to meeting the behavioral health needs of children and families. *Delaware Journal of Public Health, 8*(2), 6–9. https://doi.org/10.32481/djph.2022.05.002 PMID: 35692992; PMCID: PMC9162403. This paper emphasizes the significance of integrating behavioral health into primary care to address access barriers. A model program has been developed in Delaware for children and youth, underscoring the necessity of policy support to sustain and expand similar initiatives.

Njoroge, W. F., Hostutler, C. A., Schwartz, B. S., & Mautone, J. A. (2016). Integrated behavioral health in pediatric primary care. *Current Psychiatry Reports, 18*(12), 106. https://doi.org/10.1007/s11920-016-0745-7 PMID: 27766533. This paper discusses barriers to accessing quality behavioral health care for youth and proposes integrated services within pediatric primary care as a solution. It highlights a joint position paper by the American Academy of Pediatrics (AAP) and the Academy of Child and Adolescent Psychiatry (AACAP) advocating for initial screening and treatment in the medical home. The paper aims to review literature on integrated behavioral health services in pediatrics, focusing on innovative models.

Oppenheim, J., Stewart, W., Zoubak, E., Donato, I., Huang, L., & Hudock, W. (2016). Launching forward: The integration of behavioral health in primary care as a key strategy for promoting young child wellness. *American Journal of Orthopsychiatry, 86*(2), 124 This paper explores insights from Project LAUNCH, a SAMHSA initiative launched in 2008 to enhance behavioral health and developmental outcomes for young children. It focuses on one of the core strategies—integrating behavioral health into primary care—and analyzes experiences from Project LAUNCH grantees. The paper outlines ten common elements of integration approaches and discusses challenges in promoting health and preventing social, emotional, and behavioral issues at a population level.

Stancin, T., & Perrin, E. C. (2014). Psychologists and pediatricians: Opportunities for collaboration in primary care. *American Psychologist, 69*(4), 332. This paper underscores the potential for greater impact of psychologists in pediatric primary care, highlighting their unique contributions alongside pediatricians. It emphasizes the importance of preventive health behavior change for children and adolescents, advocating for psychologists' leadership in clinical, research, and advocacy roles to leverage family relationships for promoting children's health..

References

Abramowitz, J. S., Deacon, B. J., & H., W. S. P. (2019). *Exposure therapy for anxiety: Principles and practice*. The Guilford Press.

Allen, K. D., Barone, V. J., & Kuhn, B. R. (1993). A behavioral prescription for promoting applied behavior analysis within pediatrics. *Journal of Applied Behavior Analysis, 26*(4), 493–502. https://doi.org/10.1901/jaba.1993.26-493

American Academy of Pediatrics. (2000). Clinical practice guideline: Diagnosis and evaluation of the child with attention-deficit/hyperactivity disorder. *Pediatrics, 105*(5), 1158–1170. https://doi.org/10.1542/peds.105.5.1158

American Psychological Association. (2014). *Briefing series on the role of psychology in healthcare*. Retrieved from https://www.apa.org/health/briefs/primary-care.pdf
American Psychological Association. (2023). *Primary care psychology training programs directory*. Retrieved from https://www.apa.org/ed/graduate/primary-care-program-directory
Antony, M. M., & Barlow, D. H. (2020). *Handbook of assessment and treatment planning for psychological disorders*. Guilford Publications.
Aratani, Y., & Cooper, J. L. (2012). Racial and ethnic disparities in the continuation of community-based children's mental health services. *The Journal of Behavioral Health Services & Research, 39*, 116–129. https://doi.org/10.1007/s11414-011-9261-z
Asarnow, J. R., Rozenman, M., Wiblin, J., & Zeltzer, L. (2015). Integrated medical-behavioral care compared with usual primary care for child and adolescent behavioral health: A meta-analysis. *JAMA Pediatrics, 169*(10), 929–937. https://doi.org/10.1001/jamapediatrics.2015.1141
Auxier, A., Runyan, C., Mullin, D., Mendenhall, T., Young, J., & Kessler, R. (2012). Behavioral health referrals and treatment initiation rates in integrated primary care: A collaborative care research network study. *Translational Behavioral Medicine, 2*(3), 337–344. https://doi.org/10.1007/s13142-012-0141-8
Baer, D. M., Wolf, M. M., & Risley, T. R. (1968). Some current dimensions of applied behavior analysis. *Journal of Applied Behavior Analysis, 1*(1), 91–97. https://doi.org/10.1901/jaba.1968.1-91
Bandura, A. (1977). *Social learning theory*. Prentice-Hall, Inc.
Barlow, D. H., Ellard, K. K., Fairholme, C. P., Farchione, T. J., Boisseau, C. L., Allan, L. B., & Ehrenreich-May, J. T. (2010). *Unified protocol for transdiagnostic treatment of emotional disorders: Workbook* (1st ed.). Oxford University Press.
Beck, A. T. (1970). Cognitive therapy: Nature and relation to behavior therapy. *Behavior Therapy, 1*(2), 184–200. https://doi.org/10.1016/S0005-7894(70)80030-2
Becker, S. P., Langberg, J. M., Vaughn, A. J., & Epstein, J. N. (2012). Clinical utility of the Vanderbilt ADHD diagnostic parent rating scale comorbidity screening scales. *Journal of Developmental & Behavioral Pediatrics, 33*(3), 221–228.
Bitsko, R. H., Claussen, A. H., Lichstein, J., Black, L. I., Everett Jones, S., Danielson, M. L., Hoenig, J. M., Davis Jack, S. P., Brody, D. J., Gyawali, S., Maenner, M. J., Warner, M., Holland, K. M., Perou, R., Crosby, A. E., Blumberg, S. J., Avenevoli, S., Kaminski, J. W., & Ghandour, R. M. (2022). *Mental health surveillance among children—United States, 2013–2019*. Centers for Disease Control and Prevention. https://www.cdc.gov/mmwr/volumes/71/su/su7102a1.htm
Boland, H., DiSalvo, M., Fried, R., Woodworth, K. Y., Wilens, T., Faraone, S. V., & Biederman, J. (2020). A literature review and meta-analysis on the effects of ADHD medications on functional outcomes. *Journal of Psychiatric Research, 123*, 21–30.
Bond, C., Woods, K., Humphrey, N., Symes, W., & Green, L. (2013). Practitioner review: The effectiveness of solution focused brief therapy with children and families: A systematic and critical evaluation of the literature from 1990–2010. *Journal of Child Psychology and Psychiatry and Allied Disciplines, 54*(7), 707–723. https://doi.org/10.1111/jcpp.12058
Briggs, R. D., German, M., Schrag Hershberg, R., Cirilli, C., Crawford, D. E., & Racine, A. D. (2016). Integrated pediatric behavioral health: Implications for training and intervention models. *Professional Psychology: Research and Practice, 47*(4), 312.
Bruhn, A. L., & Hirsch, S. E. (2017). From good intentions to great implementation. *Report on Emotional and Behavioral Disorders in Youth, 173*(3), 64–70.
Bruni, T. P., Lancaster, B. M., & Cook, A. (2016). *The application of applied behavior analysis within integrated behavioral health care [poster presentation]*. Presented at the Annual Meeting of the Association for Applied Behavior Analysis, 27–31 May 2016.
Burstein, B., Agostino, H., & Greenfield, B. (2019). Suicidal attempts and ideation among children and adolescents in US emergency departments, 2007–2015. *JAMA Pediatrics, 173*(6), 598–600. https://doi.org/10.1001/jamapediatrics.2019.0464
Campo, J. V., Bridge, J. A., & Fontanella, C. A. (2015). Access to mental health services. *JAMA Pediatrics, 169*(4), 299. https://doi.org/10.1001/jamapediatrics.2014.3558
Chang, A. R., & Slopen, N. (2024). Racial and ethnic disparities for unmet needs by mental health condition: 2016 to 2021. *Pediatrics, 153*(1). https://doi.org/10.1542/peds.2023-062286

Chien, A. T., Leyenaar, J., Tomaino, M., Woloshin, S., Leininger, L., Barnett, E. R., McLaren, J. L., & Meara, E. (2022). Difficulty obtaining behavioral health services for children: A national survey of multiphysician practices. *The Annals of Family Medicine, 20*(1), 42–50. https://doi.org/10.1370/afm.2759

Chorpita, B. F., Daleiden, E. L., & Weistaz, J. R. (2005). Modularity in the design and application of therapeutic interventions. *Applied & Preventive Psychology, 11*, 141–156. https://doi.org/10.1016/j.appsy.2005.05.002

Colizzi, M., Lasalvia, A., & Ruggeri, M. (2020). Prevention and early intervention in youth mental health: Is it time for a multidisciplinary and trans-diagnostic model for care? *International Journal of Mental Health Systems, 14*, 1–14. https://doi.org/10.1186/s13033-020-00356-9

Collins, C., Hewson, D. L., Munger, R., & Wade, T. (2010). *Evolving models of behavioral health integration in primary care*. Milbank Memorial Fund.

Committee on Pediatric Workforce, Basco, W. T., Rimsza, M. E., Rimsza, M. E., Hotaling, A. J., Sigrest, T. D., & Simon, F. A. (2013). Pediatrician workforce policy statement. *Pediatrics, 132*(2), 390–397. https://doi.org/10.1542/peds.2013-1517

Cooper, S., Valleley, R. J., Polaha, J., Begeny, J., & Evans, J. H. (2006). Running out of time: Physician management of behavioral health concerns in rural pediatric primary care. *Pediatrics, 118*, e132–e138. https://doi.org/10.1542/peds.2005-2612

Cuijpers, P., Reijnders, M., & Huibers, M. J. H. (2019). The role of common factors in psychotherapy outcomes. *Annual Review of Clinical Psychology, 15*, 207–231. https://doi.org/10.1146/annurev-clinpsy050718-095424

Cummings, N. A., O'Donohue, W. T., & Cummings, J. L. (2009). The financial dimension of integrated behavioral/primary care. *Journal of Clinical Psychology in Medical Settings., 16*(1), 31–39. https://doi.org/10.1007/s10880-008-9139-2

Dalgleish, T., Black, M., Johnston, D., & Bevan, A. (2020). Transdiagnostic approaches to mental health problems: Current status and future directions. *Journal of Consulting and Clinical Psychology, 88*(3), 179–195. https://doi.org/10.1037/ccp0000482

Das, J. K., Salam, R. A., Lassi, Z. S., Kahn, M. N., Mahmood, W., Patel, V., & Bhutta, Z. A. (2016). Interventions for adolescent mental health: An overview of systematic reviews. *Journal of Adolescent Health, 59*(4S), S49–S60. https://doi.org/10.1016/j.jadohealth.2016.06.020

Davis, J. P., Palitz, S. A., Knepley, M., & Kendall, P. C. (2019). Cognitive-behavioral therapy with youth. In K. S. Dobson & D. J. A. Dozois (Eds.), *Handbook of cognitive-behavioral therapies* (4th ed., pp. 349–382). Guilford Press.

Dowell, K. A., & Ogles, B. M. (2010). The effects of parent participation on child psychotherapy outcome: A meta-analytic review. *Journal of Clinical Child & Adolescent Psychology, 39*(2), 151–162. https://doi.org/10.1080/15374410903532585

Evans, S., Weisz, J. R., Carvalho, A. C., Garibaldi, P. M., Bearman, S. K., & Chorpita, B. F., & The Research Network for Youth Mental Health. (2020). Effects of standard and modular psychotherapies in the treatment of youth with severe irritability. *Journal of Consulting and Clinical Psychology, 88*, 255–268. https://doi.org/10.1037/ccp0000456

Forsman, A. K., Wahlbeck, K., Aarø, L. E., Alonso, J., Barry, M. M., Brunn, M., Cardoso, G., Cattan, M., de Girolamo, G., Eberhard-Gran, M., Evans-Lacko, S., Fiorillo, A., Hansson, L., Haro, J. M., Hazo, J.-B., Hegerl, U., Katschnig, H., Knappe, S., Luciano, M., Miret, M., Nordentoft, N., Obradors-Tarragó, C., Pilgrim, D., Ruud, T., Salize, H. J., Stewart-Brown, S. L., Tómasson, K., van der Feltz-Cornelis, C. M., Ventus, D. B. J., Vuori, J., Värnik, A., & ROAMER Consortium. (2015). Research priorities for public mental health in Europe: Recommendations of the ROAMER project. *European Journal of Public Health, 25*(2), 249–254. https://doi.org/10.1093/eurpub/cku232

Ghandour, R. M., Sherman, L. J., Vladutiu, C. J., Ali, M. M., Lynch, S. E., Bitsko, R. H., & Blumberg, S. J. (2019). Prevalence and treatment of depression, anxiety, and conduct problems in US children. *The Journal of Pediatrics, 206*, 256–267. https://doi.org/10.1016/j.jpeds.2018.09.021

Gouge, N., Polaha, J., Rogers, R., & Harden, A. (2016). Integrating behavioral health into pediatric primary care: Implications for provider time and cost. *Southern Medical Journal, 109*(12), 774–778. https://doi.org/10.14423/SMJ.0000000000000564

Grace, A. M., Noonan, K. G., Cheng, T. L., Miller, D., Verga, B., Rubin, D., & Rosenbaum, S. (2014). The ACA's pediatric essential health benefit has resulted in a state-by-state patchwork of coverage with exclusions. *Health Affairs, 33*(12), 2136–2143. https://doi.org/10.1377/hlthaff.2014.0743

Ham, H. L., Lancaster, B., Maragakis, A., Harris, S. D., Turnier, L., & Bruni, T. (2022). Competencies for the practice of pediatric integrated primary care: A continuum of training. *Training and Education in Professional Psychology, 16*(2), 190.

Harrison, L., Carducci, B., Klein, J. D., & Bhutta, Z. A. (2022). Indirect effects of COVID-19 on child and adolescent mental health: An overview of systematic reviews. *BMJ Global Health, 7*(12), e010713. https://doi.org/10.1136/bmjgh-2022-010713

Heath, B. W. R. P., Wise Romero, P., & Reynolds, K. (2013). *A standard framework for levels of integrated healthcare*. SAMHSA-HRSA Center for Integrated Health Solutions.

Herbst, R. B., McClure, J. M., Ammerman, R. T., Stark, L. J., Kahn, R. S., Mansour, M. E., & Burkhardt, M. C. (2020). Four innovations: A robust integrated behavioral health program in pediatric primary care. *Families, Systems & Health, 38*(4), 450–463. https://doi.org/10.1037/fsh0000537

Hine, J. F., Grennan, A. Q., Menousek, K. M., Robertson, G., Valleley, R. J., & Evans, J. H. (2017). Physician satisfaction with integrated behavioral health in pediatric primary care: Consistency across rural and urban settings. *Journal of Primary Care & Community Health, 8*(2), 89–93. https://doi.org/10.1177/2150131916668115

Hoffmann, J. A., & Foster, A. A. (2020). Ready or not, here I come: Emergency department readiness for pediatric mental health visits. *Pediatrics, 145*(6). https://doi.org/10.1542/peds.2019-3542

Hoffmann, S. G., Asnaani, A., Vonk, I. J., Sawyer, A. T., & Fang, A. (2012). The efficacy of cognitive behavioral therapy: A review of meta-analyses. *Cognitive Therapy and Research, 36*(5), 427–440. https://doi.org/10.1007/s10608-012-9476-1

Horevitz, E., & Manoleas, P. (2013). Professional competencies and training needs of professional social workers in integrated behavioral health in primary care. *Social Work in Health Care, 52*(8), 752–787. https://doi.org/10.1080/00981389.2013.791362

Hostutler, C., Wolf, N., Snider, T., Butz, C., Kemper, A. R., & Butter, E. (2023). Increasing access to and utilization of behavioral health care through integrated primary care. *Pediatrics, 152*(6). https://doi.org/10.1542/peds.2023-062514

Hunter, C. L., Goodie, J. L., Oordt, M. S., & Dobmeyer, A. C. (2009). *Integrated behavioral health in primary care: Step-by-step guidance for assessment and intervention*. American Psychological Association.

Jones, S. E., Ethier, K. A., Hertz, M., Den Gue, S., Le, V. D., Thornton, J., Lim, C., Dittus, P. J., & Geda, S. (2022). Mental health, suicidality, and connectedness among high school students during the COVID-19 pandemic—Adolescent behaviors and experiences survey, United States, January–June 2021. *MWR Supplements*. https://pubmed.ncbi.nlm.nih.gov/35358165/

Katzelnick, D. J., & Williams, M. D. (2015). Large-scale dissemination of collaborative care and implications for psychiatry. *Psychiatry Services, 66*(9), 904–906. https://doi.org/10.1176/appi.ps.201400529

Kelleher, K. J., & Stevens, J. (2009). Evolution of child mental health services in primary care. *Academic Pediatrics, 9*, 7–14. https://doi.org/10.1016/j.acap.2008.11.008

Keynejad, R., Spagnolo, J., & Thornicroft, G. (2021). WHO mental health gap action programme (mhGAP) intervention guide: Updated systematic review on evidence and impact. *BMJ Mental Health, 24*(3), 124–130. https://doi.org/10.1136/ebmental-2021-300254

Lancaster, B., Cook, A., Bruni, T., Sturza, J., Sevecke, J., Ham, H., Knight, R., Hoffses, K., Wickham, C. A., & Orringer, K. A. (2019). Comparing primary care pediatricians' perceptions of clinics with and without integrated behavioral health. *Primary Health Care Research & Development, 20*, e63. https://doi.org/10.1017/S1463423618000579

Langberg, J. M., Vaughn, A. J., Brinkman, W. B., Froehlich, T., & Epstein, J. N. (2010). Clinical utility of the Vanderbilt ADHD rating scale for ruling out comorbid learning disorders. *Pediatrics, 126*(5), e1033–e1038.

Manassis, K. (2005). Family involvement in psychotherapy: What's the evidence? In J. Hudson & R. Rapee (Eds.), *Psychopathology and the family* (pp. 283–300). Elsevier.

Manuel, M. M., Yen, K., Feng, S. Y., & Patel, F. (2023). The burden of mental and behavioral health visits to the pediatric ED: A 3-year tertiary care center experience. *Child and Adolescent Mental Health, 28*(4), 488–496. https://doi.org/10.1111/camh.12638

Maragakis, A., Lindeman, S., & Nolan, J. (2018). Evidence based and intensity specific services in the integrated care setting: Ethical considerations for a developing field. *Behavior Analysis: Research and Practice, 18*(4), 425–435. https://doi.org/10.1037/bar0000127

McGorry, P. D., & Mei, C. (2018). Early intervention in youth mental health: Progress and future directions. *BMJ Mental Health, 21*(4), 182–184. https://doi.org/10.1136/ebmental-2018-300060

McMillan, J. A., Land, M., Tucker, A. E., & Leslie, L. K. (2020). Preparing future pediatricians to meet the behavioral and mental health needs of children. *Pediatrics, 145*(1). https://doi.org/10.1542/peds.2018-3796

Miller-Matero, L. R., Dykuis, K. E., Albujoq, K., Martens, K., Fuller, B. S., Robinson, V., & Willens, D. E. (2016). Benefits of integrated behavioral health services: The physician perspective. *Families, Systems & Health: The Journal of Collaborative Family Healthcare, 34*(1), 51. https://doi.org/10.1037/fsh0000182

Nigri, L., Carrasco-Sanz, A., Pop, T. L., Giardino, I., Vural, M., Ferrara, P., Indrio, F., & Pettoello-Mantovani, M. (2023). Burnout in primary care pediatrics and the additional burden from the COVID-19 pandemic. *The Journal of Pediatrics, 260*. https://doi.org/10.1016/j.jpeds.2023.113447

Norcross, J. C., & Wampold, B. E. (2011). Evidence-based therapy relationships: Research conclusions and clinical practices. *Psychotherapy, 48*(1), 98–102. https://doi.org/10.1037/a0022161

Okoroji, C., Mack Kolsky, R., Williamson, A. A., & Mautone, J. A. (2023). Integrated behavioral health in pediatric primary care: Rates of consultation requests and treatment duration. *Child & Youth Care Forum*. https://doi.org/10.1007/s10566-023-09769-2

Racine, N., McArthur, B. A., Cooke, J. E., Eirich, R., Zhu, J., & Madigan, S. (2021). Global prevalence of depressive and anxiety symptoms in children and adolescents during COVID-19: A meta-analysis. *JAMA Pediatrics, 175*(11), 1142–1150. https://doi.org/10.1001/jamapediatrics.2021.2482

Radez, J., Reardon, T., Creswell, C., Lawrence, P. J., Evdoka-Burton, G., & Waite, P. (2021). Why do children and adolescents (not) seek and access professional help for their mental health problems? A systematic review of quantitative and qualitative studies. *European Child & Adolescent Psychiatry, 30*(2), 183–121. https://doi.org/10.1007/s00787-019-01469-4

Richards, M. M., Bowers, M. J., Lazicki, T., Krall, D., & Jacobs, A. K. (2008). Caregiver involvement in the intensive mental health program: Influence on changes in child functioning. *Journal of Child and Family Studies, 17*(2), 241–252. https://doi.org/10.1007/s10826-007-9163-0

Rodgers, C. R. R., Flores, M. W., Bassey, O., Augenblick, J. M., & Cook, B. L. (2022). Racial/ethnic disparity trends in children's mental health care access and expenditures from 2010-2017: Disparities remain despite sweeping policy reform. *Journal of the American Academy of Child & Adolescent Psychiatry, 61*(7), 915–925. https://doi.org/10.1016/j.jaac.2021.09.420

Stancin, T., & Perrin, E. C. (2014). Psychologists and pediatricians: Opportunities for collaboration in primary care. *American Psychologist, 69*(4), 332. https://doi.org/10.1037/a0036046

Stuart, M. (2011). Autism insurance reform: A comparison of state initiatives. *Indiana Health Law Review, 8*(2). https://doi.org/10.18060/2021

The United States Government. (2022, May 31). Fact sheet: Biden-Harris administration highlights strategy to address the national mental health crisis. *The White House*. https://www.whitehouse.gov/briefing-room/statements-releases/2022/05/31/fact-sheet-biden-harris-administration-highlights-strategy-to-address-the-national-mental-health-crisis/

Valleley, R. J., Meadows, T. J., Burt, J., Menousek, K., Hembree, K., Evans, J., Gathje, R., Kupzyk, K., Sevecke, J. R., & Lancaster, B. (2020). Demonstrating the impact of colocated behavioral health in pediatric primary care. *Clinical Practice in Pediatric Psychology, 8*, 13–24. https://doi.org/10.1037/cpp0000284

Vechiu, C., Zimmermann, M., Zepeda, M., O'Donohue, W. T., & Broten, L. (2023). Referral patterns and sociodemographic predictors of adult and pediatric behavioral health referrals in a federally qualified health center. *The Journal of Behavioral Health Services & Research, 51*(1), 101–113. https://doi.org/10.1007/s11414-023-09855-2

Vickers, K. S., Ridgeway, J. L., Hathaway, J. C., Egginton, J. S., Kaderlik, A. B., & Katzelnick, D. J. (2013). Integration of mental health resources in a primary care setting leads to increased provider satisfaction and patient access. *General Hospital Psychiatry, 35*(5), 461–467. https://doi.org/10.1016/j.genhosppsych.2013.06.011

Vogel, M. E., Kanzler, K. E., Aikens, J. E., & Goodie, A. L. (2017). Integration of behavioral health and primary care: Current knowledge and future directions. *Journal of Behavioral Medicine, 40*(1), 69–84. https://doi.org/10.1007/s10865-016-9798-7

Walter, H. J., Vernacchio, L., Trudell, E. K., Bromberg, J., Goodman, E., Barton, J., Young, G. J., DeMaso, D. R., & Focht, G. (2019). Five-year outcomes of behavioral health integration in pediatric primary care. *Pediatrics, 144*(1). https://doi.org/10.1542/peds.2018-3243

Weersing, V. R., Brent, D. A., Rozenman, M. S., Gonzalez, A., Jeffreys, M., Dickerson, J. F., Lynch, F. L., Porta, G., & Iyengar, S. (2017). Brief behavioral therapy for pediatric anxiety and depression in primary care: A randomized clinical trial. *JAMA Psychiatry, 74*(6), 571–578. https://doi.org/10.1001/jamapsychiatry.2017.0429

Wodarski, J. S. (2014). The integrated behavioral health service delivery system model. *Social Work in Public Health, 29*(4), 301–317. https://doi.org/10.1080/19371918.2011.622243

Wolraich, M. L., Lambert, W., Doffing, M. A., Bickman, L., Simmons, T., & Worley, K. (2003). Psychometric properties of the Vanderbilt ADHD diagnostic parent rating scale in a referred population. *Journal of Pediatric Psychology, 28*(8), 559–568. https://doi.org/10.1093/jpepsy/jsg046

Wolraich, M. L., Hagan, J. F., Allan, C., Chan, E., Davison, D., Earls, M., et al. (2019). Clinical practice guideline for the diagnosis, evaluation, and treatment of attention-deficit/hyperactivity disorder in children and adolescents. *Pediatrics, 144*(4), E20192528. https://doi.org/10.1542/peds.2019-2528

World Health Organization. (2008a). *The World Health Report 2008- primary health care: Now more than ever.* https://reliefweb.int/report/world/world-health-report-2008-primary-health-care-now-more-ever

World Health Organization. (2008b). *mhGAP mental health gap action programme: Scaling up care for mental, neurological and substance use disorders.* https://www.who.int/publications/i/item/9789241596206

Chapter 9
Specialty Medical Settings

Ulrick Vieux, Andrew Shaw Benotakeia, Jayasudha Gude, Ilona Fishkin, and Gary Small

Introduction

Specialty medical settings involve a team of medical specialists who diagnose and treat patients with diseases specific to their field. In many health care systems, a patient will receive a referral from a primary care provider to a specialist for management of their condition. Medical specialists can be independent providers or employees of tertiary centers, such as university hospitals. In this chapter, we will focus on the role and use of mental health providers within tertiary systems as this is where multiple medical specialists may initially interface with mental health professionals. We also will provide an example of a specialty medical setting, providing a template on how to establish a sports psychiatry clinic.

A patient who seeks specialist care typically needs management due to an acute, exacerbated, or uncontrolled medical condition. The goal of specialty-based care often involves evaluation and diagnosis of new symptoms or exacerbation of chronic disease. Treatment and care are crafted by multiple disciplines involving medical specialists, nurses, social workers, physical therapists, and others. In hospital-based settings, mental health has traditionally been integrated as a consultation-liaison service meant to help navigate teams through comorbid mental health issues that arise from patients in medical care.

Given the myriad of medical issues present in hospital and outpatient medical settings, mental health services have a special role in elucidating the mental and physical nuances behind diseases, both acute and chronic. Mental health providers should be aware of how medical conditions can change the dynamic of one's life. Inflammatory bowel disease, diabetes mellitus, chronic traumatic encephalopathy,

U. Vieux (✉) · A. S. Benotakeia · J. Gude · I. Fishkin · G. Small
Department of Psychiatry, Hackensack Meridian Health-Hackensack University Medical Center, Hackensack, NJ, USA

A. Maragakis, M. Janikian (eds.), *Evidence-Based Behavioral Health Practices in Pediatric Specialty Settings*, https://doi.org/10.1007/978-3-031-71004-9_9

and many other chronic illnesses are well-known to disrupt mental well-being. As patients navigate through their disease, input from mental health professionals is integral. An interdisciplinary approach requires effective communication among team members, and an understanding of the perspectives of multiple disciplines aiming to improve patient care.

In the United States, consultation-liaison psychiatry established itself with its pioneering research on the innate relationship between the body and mind. An estimated 26–45% of patients who are hospitalized in nonpsychiatric settings have a psychiatric disorder (Vincze et al., 2004). Numbers are similar in pediatric settings where anxiety disorders and disruptive behaviors are reported in 7–40% of children with serious and chronic physical illnesses (Pao et al., 2011). Although the use of structured mental health services can improve disease outcomes, it also can increase health care costs. For example, common presentations in medical settings are medically unexplained symptoms and illnesses. In Canada, the cost of care in pediatric populations with somatic symptom-related disorders was estimated to be greater than $300,000,000 over a two-year period (Saunders et al., 2020). Care for somatic symptoms and disorders often involves multiple medical specialties, such as neurology, gastroenterology, and specialties to rule out a medical etiology. Early intervention and involvement of mental health professionals for functional disorders are associated with lower hospital visits and overall health care cost.

Bio-Psycho-Social Model of Disease

George Engel was a psychiatrist who introduced the biopsychosocial model of disease. His 1977 article opposed the prevalent ideals of the biomedical model that derived from a Cartesian paradigm. This model proposed a separation of body and mind. Engel argued that biomedical markers could not account for the degree and severity of "illness" seen in many patients (Engel, 1977). Engel acknowledged that our understanding of disease and illness likely needed further exploration beyond this model. Further nuances to this model have been introduced, and the one that is relevant to emphasize within medical settings is an integration of the patient–physician relationship (Borrell-Carrió et al., 2004).

Of course, biological variables involve pathophysiological mechanisms that can be understood at a molecular level. For example, type 1 diabetes mellitus is an autoimmune disorder that attacks pancreatic beta cells, resulting in a lack of insulin that is necessary for glucose usage. The disease leads to high glucose in the bloodstream, which has a myriad of physical issues, such as kidney disease, cardiovascular disease, loss of eyesight, and diabetic ketoacidosis to name a few. The definitive treatment is lifestyle changes (i.e., dietary habits) and insulin. Insulin must sometimes be administered three times a day with a short-acting insulin and/or a long-acting insulin once a day. This is a drastic shift in many children's lives.

It is well known that adolescents have difficulty maintaining glycemic control due to multiple factors, such as poor eating behaviors and lack of familial support

(Borus & Laffel, 2010). Such social factors can play a major role in the course of many diseases. Furthermore, the psychological burden of the disease can play a role in patients with comorbid depression and anxiety. A way of connecting such psychological and social variables is that lower education, financial challenges, and discrimination among peers related to depressive episodes in those with diabetes (Van Duinkerken et al., 2020). There are a multitude of different dynamics that can alter this one disease of diabetes. Mental health clinicians can help improve illness outcomes by shedding light on these and other psychological and social determinants of health.

Consultation Steps

Psychiatrists and psychologists often act in dual capacity as medical consultants and as liaisons among patients, their families, and their medical teams (Carter et al., 2009). This consultation-liaison model is one of the most well-recognized behavioral health frameworks in nonpsychiatric clinical settings (Gautam et al., 2022). The model involves both direct psychiatric consultations to the medical team as well as fostering a collaborative relationship between the consultant and consultee (i.e., liaison).

Reasons for consults can range anywhere from questions about management of mood disorders to suicidal behavior (Kullgren et al., 2018). Consultants use clinical tools such as standardized rating scales, structured interviews, and the mental status examination to assess psychiatric illness and make recommendations on further management. Psychiatrists may also provide medication recommendations.

Within the consultation-liaison model, the liaison role is needed when there is miscommunication, confusion, or other rifts among medical teams, patients, and their families (Gautam et al., 2022). In these situations, mental health professionals work to identify patient or family concerns and facilitate communication among patients, families, and health care teams (Shaw & De Maso, 2020). It is important to understand that mental health consultation requires a sensitive approach involving collaboration among all health care professionals (Carter et al., 2009).

In the pediatric population, mental health consultations are sought for various reasons often related to the intersections of mental and physical health. Some of the common reasons and diagnoses for psychiatric consultation in a hospital setting for children and adolescents include:

- Behavioral and emotional disturbances: Disruptive and Impulse control disorders, such as oppositional defiant disorder (ODD) and conduct disorder; other developmental disorders such as autism spectrum disorder (ASD), intellectual disabilities that can result in behavioral issues (Creon et al., 2006; Thom et al., 2019).
- Mood disorders: Anxiety related disorders (Cobam et al., 2020); Unipolar depression or bipolar disorder (Kullgren et al., 2018; Malas et al., 2019).

- Suicidal ideation or self-harm: Safety risk assessment and appropriate intervention for children and adolescents experiencing thoughts of self-harm or suicide (Becker et al., 2020; Shaw et al., 2016).
- Psychiatric symptoms associated with medical conditions: Psychiatric evaluation of children with chronic medical conditions such as diabetes, epilepsy, or autoimmune disorders where mental health may impact treatment adherence and well-being.
- Adjustment disorders: Support for children and families with significant life changes, such as new diagnosis, hospitalization, or traumatic events (Compas et al., 2012; Van Schoors et al., 2016).
- Neurological conditions: Evaluations of psychiatric symptoms in children with neurological conditions such as epilepsy or neurogenetic disorders (Bujoreanu et al., 2011).
- Eating disorders: Assessment and intervention for children with eating disorders such as anorexia nervosa, bulimia nervosa, avoidant/restrictive food intake disorder (Eddy & Thomas, 2019).
- Trauma and Post-traumatic Stress Disorder (PTSD): Support for children who experienced trauma, abuse, or adverse childhood events.
- Psychosomatic disorders: Assessment of conditions where psychological factors contribute to physical symptoms such as somatic symptoms disorder, Psychogenic Non-Epileptic Seizures (PNES) (Kullgren et al., 2018; Sztein & Lane, 2016).
- Sleep disorders: Evaluations and management of sleep disturbances impacting a child's mental and physical health.
- Psychiatric medication management: Consultation for medication management when prescribed in conjunction to medical treatment (Becker et al., 2020).
- Anxiety and pain tolerance related to medical condition/procedure: Support for children experiencing significant anxiety related to hospitalization, procedures, treatment, or surgeries (Doupnik et al., 2016; Gatchel et al., 2014).

The purpose of the consultation typically determines the nature and extent of the evaluation. The evaluation is focused on addressing the question posed by the consultee, assessing the clinical presentation, and reviewing the treatment history. This process is highly beneficial for providing diagnostic clarification and treatment recommendations. A general overview to complete a consultation is discussed below.

Initial Communication

Consultations may be requested either in person, via telephone, pager, or electronic medical record, to the consultation service. The consultant needs to first establish direct communication with the consultee, either by telephone or in person, to clarify the reason for consultation and obtain relevant clinical information. During the initial communication, important topics to discuss include determining the urgency of the consult, assessing any immediate danger to the patient or staff, and identifying if the patient is suicidal or homicidal (Carter et al., 2009). At this stage, the consultant needs to find information about the medical history of the patient, reason for admission, medical status, any known psychiatric history, and ongoing treatment with the physician ordering the consult (Gautam et al., 2022). Another task is to obtain clarification regarding the patient and parent's understanding of why a consult was placed and further obtain assent from the child and consent from the parents (Becker et al., 2020). Orders such as one-to-one observation may be placed if necessary, especially in cases involving suicidal/homicidal ideation, agitation, disorganized behavior, or a high risk of elopement even before the actual evaluation. One-to-one observation may be discontinued after evaluation if deemed unnecessary (Carter et al., 2009).

Review

The consultant then needs to collect data through chart review prior to the consultation. Despite challenges in emergency cases, such as agitation or disruptive behaviors, it is imperative to review the following:

- Chief complaint and circumstances of admission
- History of present illness
- Past medical, surgical, and psychiatric histories
- Substance use history
- Notes from nursing staff and social worker
- Physical exam and laboratory test results
- Medication regimens and treatment history
- Review of hospital course until the point of psychiatric consultation

For patients with a significant past psychiatric history, a review of the previous medical record for diagnostic impressions and treatment regimens is important. Information needs to be further gathered from nursing staff, one-to-one observation sitters, and other staff members about the patient's behaviors and attitude toward the medical team. Compliance with medications, sleeping, and eating habits needs to be determined (Gautam et al., 2022).

Visit

The bedside evaluation of the patient begins with an assessment of the patient's comfort level, irritability, agitation, distress, depression, or other mental states. Establishing rapport, introducing oneself, explaining the reason for the visit, and making the patient feel as comfortable as possible are essential steps before proceeding with the interview (Becker et al., 2020). The interview should focus on the chief complaint, the patient's current mental status, and a psychiatric review of systems (Gautam et al., 2022). Additional information can be obtained as necessary, following the template of interviewing, which includes past psychiatric and medical history, substance use, and psychosocial history.

In the case of the pediatric population, collateral history is an essential part of the evaluation. Questions should be addressed to the caregivers of the child including their perspective on the child's behavior. Other background information from caregivers should include developmental, birth history, and family history. In addition, it is important that if one is obtaining information from parents who are divorced, any medical decisions are made by the custodial parent.

Recommendations

After the evaluation, recommendations should be directly communicated to the primary medical team. Urgent and acute conditions such as agitation or suicidal behavior should take priority. Suggestions for additional consultations, necessary lab tests for diagnostic clarity, and specific recommendations for psychiatric care while in the hospital can be made (Becker et al., 2020; Gautam et al., 2022). Contact information, such as a pager or phone number, should be provided for the mental health team to address any questions from the primary team. Clear instructions about follow-up care and the need for outpatient psychiatric care should also be given. This communication also provides an opportunity to educate the physician, which can be helpful in educating the patient and their family (Becker et al., 2020).

Follow-Up

If further mental health care is deemed necessary, the next step is follow-up. Appropriate handoff should be provided to the consultation and liaison psychiatry team for proper follow-up by on-call attendings outside of regular work hours, weekends, or holidays.

Common Psychiatric Manifestations of Medical Disorders

There is a large overlap of primary psychiatric disorders and chronic medical conditions. Listed are some common psychiatric manifestations associated with certain medical conditions in pediatric patients. This list is not meant to be exhaustive but to give an idea of the vast array of comorbid conditions that can occur.

Neurological disorders (Lyketsos et al., 2007; Welch et al., 2018)	Epilepsy Cerebral Palsy Multiple Sclerosis	Behavioral and emotional difficulties, anxiety, and mood disorders Increased risk of attention-deficit hyperactivity disorder (ADHD) and emotional disturbances
Endocrine disorders (Solomon & Conner, 2017)	Diabetes Thyroid disorder	Increased risk of mood and anxiety disorders Can impact mood and behavior, depression, and anxiety
Genetic disorders (Anderson et al., 2013; Demily & Sedel, 2014; Tsiouris & Brown, 2004)	Downs syndrome Fragile X syndrome Huntington disease Lysosomal storage disorders	Higher prevalence of attention deficits, mood disorders, and obsessive-compulsive behaviors Social anxiety, attention difficulties, and autistic-like behaviors
Metabolic disorders (Demily & Sedel, 2014)	Phenylketonuria Mitochondrial disorders	Cognitive and behavioral issues Developmental delays and behavioral issues
Autoimmune disorders (Weiss et al., 2005)	Lupus Rheumatoid arthritis Sarcoidosis Vasculitis	Neuropsychiatric symptoms, mood disorder, cognitive dysfunction, and psychosis Anxiety and depression
Cardiovascular disorders (Chen et al., 2022; Gonzalez et al., 2021)	Congenital heart defects Kawasaki disease	Anxiety and depression, ADHD Behavioral changes
Infectious Diseases (Munjal et al., 2017)	HIV/AIDS PANDAS Syphilis Lyme disease Prion disease	Neurocognitive and psychiatric symptoms Obsessive-compulsive disorder (OCD) symptoms
Gastrointestinal disorders (Hommel et al., 2010)	Inflammatory bowel disease Celiac disease	Anxiety and depression Mood disorders and ADHD-like symptoms
Renal Disorders (Dryjańska & Kills-Pstrusinska, 2023)	Chronic kidney disease	Depression and anxiety
Haematologic disorders (Bhatt-Poulouse et al., 2016; Myrvik et al., 2012)	Sickle cell disease	Anxiety and depression
Dermatologic disorders (Kern et al., 2021; Miniksar et al., 2021)	Acne Atopic dermatitis Hidradenitis Suppurativa	Depression, anxiety, and adjustment disorder

Evidence-Based Interventions

Mental health professionals are able to provide interventions that may benefit outcomes for medical diseases. Prior we listed common psychiatric diagnoses comorbid with medical issues. These issues should be addressed with evidence-based therapeutic modalities that the clinician is trained in. If medications are indicated, then a mental health or medical professional can prescribe them as well. If medications are prescribed, an appropriate medical follow-up should be arranged and side effects monitored.

Medical Adherence

Medical adherence refers to how a patient follows medical advice whether it be medication, appointments, or lifestyle changes. The literature consistently shows that about half of pediatric patients with chronic diseases report suboptimal adherence (Rapoff, 2010). This can drastically worsen the disease burden and is an area where mental health professionals can help alter disease courses. Examples of this may be a patient with asthma not taking their inhalers or a child with epilepsy who is not taking their anti-seizure medications. There are multiple reasons that can explain medication nonadherence. The World Health Organization (WHO) utilizes five main categories including socioeconomic barriers, provider-patient factors, factors regarding the medical therapy itself, factors related to the disease, and factors related to the patient (WHO, 2003).

Educational approaches focus on providing information to families. The medical professional can inform patients and their families verbally, provide documents with relevant information about their illness, or both. Due to the complexity of medical disease, it is a clinician's responsibility to try and make the information they specialize in understandable to their patients. Educational approaches can be limited by multiple factors, including time the physician has with the patient, the skill of the physician in conveying information, and the health literacy of the patient.

Providing information about a complicated disease and treatment may not be enough to actually promote disease-modifying behavior. The literature suggests that education is more effective when combined with behavioral interventions (Wu et al., 2013). Some individuals respond well to contracts for behavior modification, which may include listing expectations and consequences (Rapoff, 2010). In pediatrics, having family members actively involved with behavior changes is important as well. For behavior contracts of reinforcement to work, families must be consistent. Across different age ranges, practical environmental changes such as reminders on cell phones (this can be done for the child and the parent) is a simple behavioral modification, which can help with adherence. Finding basic motivations and focusing on the goals of the patient are key to creating lifelong behavioral changes.

When patients are nonadherent, we may often see frustration directed towards them. This can come from caregivers as well as medical team members. Reminders of the significant psychosocial burden associated with the particular disease can also be part of a mental health professional's role. Finding specific barriers for each patient may be crucial in tailoring changes. We recommend working with families and teams for problem-based approaches that each patient may bring up. It may be useful to implement techniques found in motivational interviewing (Berhe et al., 2020) or cognitive behavioral therapy in having patients solve some of these barriers as the consultant helps guide them through the process.

Distress and Coping

In the setting of a significant stressor such as a medical illness, both patients and caregivers usually experience emotional responses. The severity of such responses varies, and medical specialists may be able to identify signs of significant psychological distress. At times medical providers may consult a mental health professional viewing the distress as "anxiety" or "depression". Mental health professionals, after obtaining a comprehensive history, should examine areas where they can assist.

A new medical diagnosis or an exacerbation of an existing condition can have both direct and indirect consequences on mental health. Physical debilitation from a condition such as chronic pain can cause significant mental distress. A diagnosis of a condition such as cancer can have serious implications on the livelihood of the patient and family. We suggest organizing the specific stressors (both from the disease and not related to the disease) and seeing the severity of these stressors. This is especially important as past life stressors increase the risk of more severe mental health disease when a medical disease is diagnosed (Currier et al., 2009).

In the literature, there is a discussion on the loci of control and types of coping strategies that can alter disease course. This is especially studied in pain. There are several studies looking at differing types of coping. Accommodative coping refers to changing goals and perceptions given outside constraints. Passive coping refers to more maladaptive strategies such as avoidance. Active coping skills, such as emotional regulation and problem-solving, have been linked to less depression and pain compared to passive coping strategies (Brown, 1987). These indicate efforts to help with one's stress or emotions. Previous evidence indicates that the practice of active coping strategies has many benefits on different chronic diseases. In type 1 diabetes mellitus, active coping strategies are associated with better quality of life and metabolic control (Jaser et al., 2011). In comparison with passive coping, secondary coping skills are associated with fewer somatic complaints and anxiety in recurrent abdominal pain (Compas et al., 2012). Children with cancer with avoidance coping strategies have higher rates of reported depression (Frank et al., 1997).

Caregiver Considerations

In pediatrics, there likely is an importance for mental health providers to be aware of and assist with caregivers as well. Caregivers of chronically ill children can be twice as likely to have issues with mental and physical health compared to caregivers with healthy children (Brehaut et al., 2009). Coping strategies can also influence the quality of life of parents (Fairfax et al., 2019). There is limited data in the literature about techniques that improve parent well-being. Problem-solving therapy has been shown to improve the mental health symptoms of caregivers and perhaps indirectly help with outcomes for children (Law et al., 2014). An explanation of the benefits of problem-solving therapy could be in developing empowering skills for the many inevitable difficulties associated with caregiving.

Other techniques can involve family-focused interventions to improve the structure of the parent–child relationship (Parrish et al., 2020). Relations among family members can exacerbate any familial stressors, let alone one that is medical. Those skilled in family therapy may focus on underlying family structure dynamics that can impact disease courses. A child or adolescent who already does not feel comfortable speaking about issues to their parents may not go to them in times of medical necessity.

Situations of Special Considerations

Psychiatric Emergencies

It is not uncommon for psychiatric emergencies to arise in specialty settings. Psychiatric emergencies are often accompanied by alterations in sensorium or behavior, aggression, agitation, and/or self-injurious behaviors. Some psychiatric emergencies, such as alcohol withdrawal and catatonia, can be lethal if they are not managed promptly with the appropriate level of care.

In psychiatric emergencies, the standard of care is to involve family and start with the least invasive and restrictive approach. Verbal de-escalation and supportive language should be used initially. Staff should use nonconfrontational speech and body language toward a patient. Meeting basic needs for the patient, such as food or water, can also be helpful. Restrictive methods, such as physical restraints, can lead to escalating agitation and injury and are a last resort. Within a safe medical environment, where vital signs can be monitored and medical staff are available, medications may also mitigate disruptive behaviors (Georgadarellis et al., 2023).

Crisis teams can also be helpful when patients are agitated, altered, or express suicidal ideation and/or intent. Hospitalized patients who express suicidality require prompt assessment to determine suicide risk (Margret & Hilt, 2018). These patients may require 24-hour monitoring, environmental modifications, and visitor and/or item restrictions to ensure their safety.

Psychiatric emergencies may further be compounded by prior traumatic events and can become especially stressful for staff and families who do not have a background in mental health or medicine. Specialty settings may have multidisciplinary teams available to support families and staff through these emotionally charged situations.

Boarding

Psychiatric hospitals have a finite number of beds and are unable to readily admit every patient who requires inpatient-level psychiatric care. Furthermore, not all psychiatric hospitals are equipped to treat all diagnoses. If a patient has a behavioral problem such as aggression, a developmental delay such as ASD, or special health needs such as an eating disorder, they may need to wait even longer for an open bed in a unit that is specifically tailored to treat them (Hoffman et al., 2019; O'Donnell et al., 2020).

Boarding occurs when there are no psychiatric beds available for a patient who requires inpatient-level psychiatric care and therefore close, constant monitoring in a hospital takes place. A child who is boarding may spend days to weeks in a less-than-ideal setting, such as on a stretcher in a hallway in the emergency room or isolated in a room in a pediatric or medical ward. While pediatric wards and emergency settings are not optimized for psychiatric care, multidisciplinary teams made up of psychiatrists, psychologists, mental health technicians, and social workers can offer a range of therapeutic support. With appropriate engagement of behavioral health teams, patients who are boarding are less likely to clinically deteriorate (Gallagher, 2017; Wolff et al., 2023).

Gender Dysphoria

According to Diagnostic and Statistical Manual of Mental Disorders, Fifth Edition (DSM-5) (Nokoff, 2022) gender dysphoria is defined as a strong desire to eliminate one's primary and/or secondary sex characteristics due to a significant incongruence with one's experienced/expressed gender (or, in young adolescents, a desire to prevent the development of anticipated secondary sex characteristics). It was found that among individuals having gender dysphoria, 53% of them suffer from an Axis I mental disorder (Freitas et al., 2019). The most prevalent mental disorders in this population are mood and anxiety disorders, with frequencies of 42 and 27%, respectively. Substance use/abuse disorders rank third in frequency, affecting 15% of individuals (Freitas et al., 2019).

For young people diagnosed with gender dysphoria, the risk of suicidal thoughts is severe, with half of this population report having considered suicide at least once, and almost a third have made at least one attempt (Olson et al., 2015). Even in individuals with gender dysphoria who do not have another psychiatric disorder diagnosis, there is a high rate of suicidal behavior and self-mutilation (Hoshiai et al., 2010).

The management of gender dysphoria ideally involves a multidisciplinary health care team, including psychiatrists, endocrinologists, and surgeons. Treatment options may include psychotherapy, hormone therapy, and gender reassignment surgeries. The preferred approach typically begins with a psychological assessment, followed by medical interventions. Endocrine treatment includes two phases: the initiation of Gonadotropin releasing agonist to prevent pubertal development (a fully reversible intervention) and the addition of gender-affirming hormones, leading to irreversible changes. Surgical interventions are considered as necessary (de Vries & Cohen-Kettenis, 2012).

In a consultation-liaison setting, mental health professionals can focus on psychoeducation, such as explaining to parents that gender exploration is part of a developmental process for many children and does not necessarily result in persistent gender dysphoria in adolescence. Balancing watchful waiting with steps toward gender-affirming interventions is a crucial goal in children with gender dysphoria (de Vries & Cohen-Kettenis, 2012). Childhood interventions aim to decrease distress related to gender incongruence and support the child and parents in exploring and developing potential steps as endogenous puberty begins. Transgender mental and medical health care is a lengthy process involving counseling for the child/adolescent with gender incongruence and their parents in making choices about social, medical, and legal transitions.

For adolescents, medical interventions and psychological counseling guide and support the adolescent and parents throughout the process. After the initial diagnostic phase, discussions about medical treatments, including hormonal treatment, surgery, and fertility preservation, consider the expectations of both the adolescent and the parents. Continuous support during medical interventions, including seeking peer group support through support groups or self-help organizations, is advisable.

Substance Use

Substance use disorder has continued to be a worldwide issue, especially for adolescents. In the United States, there was a substantial increase in substance-related deaths in those aged 15–24 years in 2020 (Hedegaard et al., 2020). A 2012 study looking at American children's hospital admissions indicated that about 22% of them were related to substance usage (Egorova et al., 2018). Furthermore, screening in primary care settings is challenging as it is estimated that less than 30% of these settings offer screening for substance use (Klein et al., 1993).

From a pharmacological standpoint, substances such as alcohol can interact with multiple medications or exacerbate side effects. For example, a child with Juvenile Idiopathic Arthritis may need to take methotrexate, an immunomodulator that carries the side effect of hepatotoxicity. This risk is increased with alcohol usage

(Weathermon, 1999). Children with chronic diseases are also found to start substance use at a younger age compared to their peers. With adherence already being an issue in chronic disease, this can be worsened by substances such as alcohol use (Weitzman et al., 2015).

Encouraging facilities to offer screening for substance use, and encouraging patients to have open discussions about usage with medical professionals can be helpful in identifying patients. Having professionals communicate confidentiality and building rapport and trust is essential. An example where confidentiality may need to be broken is after an overdose. In such cases, the health care professional, may notify the patient's family, friends, or caregivers of the patient's condition.

Exploring reasons why patients may be using substances can be useful. For patients with good insight into their disease, exploring how substances can affect their own chronic illness may be a motivating factor to stop. From a systems standpoint, encouraging regular screening in pediatric subspecialties can prove useful. A prospective intervention study to increase risky behavior screening for pediatric rheumatologists showed a statistically significant increase in interventions for alcohol use and follow-up counseling (Britto et al., 2000).

Sports Psychiatry Clinic

This chapter has provided a general overview of what a specialty medical setting would look like. The rest of the chapter will introduce what a specific specialty medical setting clinic will entail. The example will be the formation of a sports psychiatry specialty medical clinic.

Sports psychiatry is a developing subspecialty in the field of psychiatry because of the need to treat the unique psychiatric conditions of athletes. Historically, sports psychologists have taken the lead in providing care for athletes starting in the 1920s with the work of psychologist Coleman Griffith. In 1967, psychiatrist Arnold Beisser published a landmark series of case studies of athletes in psychotherapy. Professional athletic leagues have begun to acknowledge the importance of addressing the mental health needs of their athletes. The formation of a sports psychiatry clinic follows many of the steps outlined in this chapter. When planning a clinic the following questions should be answered.

What is the need?	Athletes have unique psychological needs and a population exists that would utilize this resource
What is unique about the clinic?	Distinctive biopsychosocial needs of athletes can be addressed in a clinic that has a specific niche that provides a competitive edge in the community. This is also an opportunity for the recruitment of skilled providers
Who are the stakeholders?	CEO, Chair of Psychiatry Department, Board of Directors, Chief Medical Officer, Department Manager, Board of Directors, Marketing Department, Human Resources, IT, and Head of Psychology

What are the start-up costs?	Costs vary depending on whether this service will be strictly hospital-based, outpatient-based, virtual, or a hybrid
What is the cost of maintaining the clinic?	Costs include space, staff salaries (including benefits), marketing, continuing medical education for eligible staff, and IT needs
What will be the return on investment (ROI)	The ROI will depend on the service model (e.g., concierge, cash-based solely, insurance plans, etc.)
What will be the referral source?	Sports medicine, family medicine, orthopedic surgery psychiatry, addiction medicine, child and adolescent psychiatry, pediatrics, schools, coaches, sports agents, and other mental health practitioners (e.g., psychology, social work, marriage, and family therapy)
What will be the staffing needs?	Psychiatrists, psychologists, and ancillary administrative support

Sports is usually viewed as society's great equalizer. Numerous Horatio Alger stories exist of the impoverished athlete coming from meager means reaching the pinnacle of success in their chosen sport thus earning tremendous wealth; or, the wealthy high school student who uses sports as a way to connect with their peers and ultimately develop a healthy hobby. For many, the attraction of being an elite athlete provides a certain amount of comfort and a sense of accomplishment with the belief that only athletic merit is needed to excel in a world that many perceive as providing the same opportunities for all. The reality is that pursuing a career in sports is not always so straightforward, and thus a need exists to safeguard the mental health needs of athletes. What sports psychiatrists bring to this field is the in-depth expertise of psychopathology, an ability to practice in an integrative model with other medical specialties, such as orthopedic surgery, family medicine, pediatrics, internal medicine, cardiology, and many others. Psychiatrists are also able to prescribe medicine that reduces symptoms of psychiatric disorders. The sports psychiatrist aims to optimize health, improve the patient's athletic performance, and manage psychiatric symptoms or disorders (Glick et al., 2009).

Eating and substance use disorders are the most common conditions found in athletes. However, more research is needed on the following diagnoses as they relate to athletes: overtraining syndrome, bipolar disorder, suicidality, anxiety disorders, attention-deficit hyperactivity disorder (ADHD), concussions, body dysmorphic disorder, traumatic brain injuries (TBI), chronic traumatic encephalopathy (CTE), and psychosis. In addition, elite professional athletes who suffer from mental illness face a quagmire regarding the role of taking medications with unknown performance and safety effects (Reardon & Factor, 2010).

As a result, sports psychiatrists are more likely to prescribe nonsedating medications that are less likely to cause weight gain, cardiac side effects, and tremor. The sports psychiatrist should also take into consideration side effects that relate to performance and anti-doping policies. (Reardon & Creado, 2016) Top choices for athletes are bupropion for depression without anxiety and without bipolar spectrum disorder; escitalopram for generalized anxiety disorder; melatonin for insomnia;

atomoxetine for ADHD; lamotrigine for bipolar spectrum disorders; and aripiprazole for psychotic disorders.

Among women in sports that focus on sports like gymnastics or figure skating, eating and body dysmorphic disorder are more common. College athletes have been reported to use alcohol at higher levels than the general public despite its detriment to performance. Common substance use disorders are anabolic steroids. Despite outward appearances of strength and wellness athletes are not immune to mental illness. Depression and anxiety disorders may occur in athletes at least as commonly in the general population and disorders such as eating disorders, attention-deficit/hyperactivity disorder, and substance use disorders may occur even more frequently in athletes than in the general population (Reardon & Creado, 2016).

The aforementioned diagnoses are the most common in sports psychiatry and in addition to the suggested psychotropics that will not have as much of a deleterious effect on the physical ability of the athlete, it is imperative to understand the psychodynamic perspective that athletes face in order to help them reach their full potential.

Conclusion

Psychologists, clinical social workers, and other behavioral health providers are critical members of the interdisciplinary health team across pediatric specialty medical settings and should be familiar with common comorbid health conditions faced by children with mental health diagnoses who require specialty-level care. Excelling in specialty medical settings requires a fundamental understanding of commonly used models of care, such as consult-liaison psychiatry, as well as a knowledge of factors that may be encountered across a wide range scope of practice settings such as a sports psychiatry specialty clinic. Choosing to care for children across specialty medical settings nearly always means meeting children and their families where they are in treatment, ensuring that psychiatric and behavioral health treatment does not stop when a child is seen in a specialist clinic or requires admission to a tertiary hospital.

Selected Bibliography

Becker, J. E., Smith, J. R., & Hazen, E. P. (2020). Pediatric consultation-liaison psychiatry: An update and review. *Psychosomatics, 61*(5), 467–480. https://doi.org/10.1016/j.psym.2020.04.015 This is a paper that provides a good overview of frequently encountered clinical presentations, diagnostic entities and recent advances in the field of pediatric c-l settings. It also emphasizes evidence-based care and effective interventions in psychiatrically ill pediatric medical patients. It also discusses the psychiatric consultation models, updates in clinical management along with systems level of care and future directions.

Compas, B. E., Jaser, S. S., Dunn, M. J., & Rodriguez, E. M. (2012). Coping with chronic illness in childhood and adolescence. *Annual Review of Clinical Psychology, 8*(1), 455–480. https://doi.

org/10.1146/annurev-clinpsy-032511-143108 The authors review coping and stress on medical illness in children. This article is a great overview on literature and current thinking. There is well organized sections of certain diseases (i.e. diabetes) and literature behind each one.

Galanter, C. A., & Jensen, P. S. (Eds.). (2016). *DSM-5 (R) casebook and treatment guide for child mental health*. American Psychiatric Association Publishing Twenty nine cases in pediatric mental health presented as vignettes with cross-disciplinary commentary. Each case helps illustrate a DSM-5 diagnosis with input from child-adolescent psychiatrists, clinical psychologists and other mental health professionals. Some vignettes include complex patient presentations with co-morbid diagnoses which may help contextualize some of the concepts presented in this chapter.

Rapoff, M. A. (2010). *Adherence to pediatric medical regimens* (2nd ed., p. xii, 231). Springer Publishing Co.. https://doi.org/10.1007/978-1-4419-0570-3 A comprehensive book about medication adherence. Assessments, strategies and review of literature about adherence is addressed.

Shaw, R. J., & DeMaso, D. R. (2019). *Clinical manual of pediatric consultation-liaison psychiatry* (2nd ed.). American Psychiatric Association Publishing This manual provides a comprehensive and focused overview of psychiatric and psychological assessment of children and adolescents in a consult setting. It also addresses the clinical questions that commonly arise before a pediatric consultation. It discusses the assessment, clinical presentations, specific psychiatric conditions, treatment and assessment of physically ill children with mental health conditions.

References

Anderson, T., Buterbaugh, A., Love, K., & Visootsak, J. (2013). A male with co-occurrence of down syndrome and fragile x syndrome. *Case Reports in Genetics, 2013*, 504695. https://doi.org/10.1155/2013/504695

Becker, J. E., Smith, J. R., & Hazen, E. P. (2020). Pediatric consultation-liaison psychiatry: An update and review. *Psychosomatics, 61*(5), 467–480. https://doi.org/10.1016/j.psym.2020.04.015

Beisser, A. (1967). The madness in sports: Psychosocial observations in sports. *Appleton-Century-Crofts*.

Berhe, K. K., Gebru, H. B., & Kahsay, H. B. (2020). Effect of motivational interviewing intervention on HbA1C and depression in people with type 2 diabetes mellitus (systematic review and meta-analysis). *PLoS One, 15*(10), e0240839. https://doi.org/10.1371/journal.pone.0240839

Bhatt-Poulose, K., James, K., Reid, M., Harrison, A., & Asnani, M. (2016). Increased rates of body dissatisfaction, depressive symptoms, and suicide attempts in Jamaican teens with sickle cell disease. *Pediatric Blood & Cancer, 63*(12), 2159–2166. https://doi.org/10.1002/pbc.26091

Borrell-Carrió, F., Suchman, A. L., & Epstein, R. M. (2004). The biopsychosocial model 25 years later: Principles, practice, and scientific inquiry. *Annals of Family Medicine, 2*(6), 576–582. https://doi.org/10.1370/afm.245

Borus, J. S., & Laffel, L. (2010). Adherence challenges in the management of type 1 diabetes in adolescents: Prevention and intervention. *Current Opinion in Pediatrics, 22*(4), 405–411. https://doi.org/10.1097/MOP.0b013e32833a46a7

Brehaut, J. C., Kohen, D. E., Garner, R. E., Miller, A. R., Lach, L. M., Klassen, A. F., & Rosenbaum, P. L. (2009). Health among caregivers of children with health problems: Findings from a Canadian population-based study. *American Journal of Public Health, 99*(7), 1254–1262. https://doi.org/10.2105/AJPH.2007.129817

Britto, M. T., Rosenthal, S. L., Taylor, J., & Passo, M. H. (2000). Improving rheumatologists' screening for alcohol use and sexual activity. *Archives of Pediatrics & Adolescent Medicine, 154*(5), 478–483. https://doi.org/10.1001/archpedi.154.5.478

Brown, G. K., & Nicassio, P. M. (1987). Development of a questionnaire for the assessment of active and passive coping strategies in chronic pain patients. *Pain, 31*(1), 53–64. https://doi.org/10.1016/0304-3959(87)90006-6

Bujoreanu, I. S., Ibeziako, P., & Demaso, D. R. (2011). Psychiatric concerns in pediatric epilepsy. *Pediatric Clinics of North America, 58*(4), 973–988., xii. https://doi.org/10.1016/j.pcl.2011.06.001

Carter, B. D., Kronenberger, W. G., Scott, E., & Ernst, M. M. (2009). Inpatient pediatric consultation-liaison. *Handbook of pediatric. Psychology, 4*, 114–129. https://psycnet.apa.org/fulltext/2009-18462-008.pdf

Chen, D. T.-L., Chang, J. P.-C., Cheng, S.-W., Chang, H.-C., Hsu, J.-H., Chang, H.-H., Chiu, W.-C., & Su, K.-P. (2022). Kawasaki disease in childhood and psychiatric disorders: A population-based case-control prospective study in Taiwan. *Brain, Behavior, and Immunity, 100*, 105–111. https://doi.org/10.1016/j.bbi.2021.11.017

Cobham, V. E., Hickling, A., Kimball, H., Thomas, H. J., Scott, J. G., & Middeldorp, C. M. (2020). Systematic review: Anxiety in children and adolescents with chronic medical conditions. *Journal of the American Academy of Child and Adolescent Psychiatry, 59*(5), 595–618. https://doi.org/10.1016/j.jaac.2019.10.010

Compas, B. E., Jaser, S. S., Dunn, M. J., & Rodriguez, E. M. (2012). Coping with chronic illness in childhood and adolescence. *Annual Review of Clinical Psychology, 8*(1), 455–480. https://doi.org/10.1146/annurev-clinpsy-032511-143108

Creon, L. A., Najjar, D. V., Ray, G. T., Lotspeich, L., & Bernal, P. (2006). A comparison of health care utilization and costs of children with and without autism spectrum disorders in a large group-model health plan. *Pediatrics, 118*(4), e1203–e1211. https://doi.org/10.1542/peds.2006-0127

Currier, J. M., Jobe-Shields, L. E., & Phipps, S. (2009). Stressful life events and posttraumatic stress symptoms in children with cancer. *Journal of Traumatic Stress, 22*(1), 28–35. https://doi.org/10.1002/jts.20382

de Vries, A. L. C., & Cohen-Kettenis, P. T. (2012). Clinical management of gender dysphoria in children and adolescents: The Dutch approach. *Journal of Homosexuality, 59*(3), 301–320. https://doi.org/10.1080/00918369.2012.653300

Demily, C., & Sedel, F. (2014). Psychiatric manifestations of treatable hereditary metabolic disorders in adults. *Annals of General Psychiatry, 13*, 27. https://doi.org/10.1186/s12991-014-0027-x

Doupnik, S. K., Lawlor, J., Zima, B. T., Coker, T. R., Bardach, N. S., Hall, M., & Berry, J. G. (2016). Mental health conditions and medical and surgical hospital utilization. *Pediatrics, 138*(6). https://doi.org/10.1542/peds.2016-2416

Dryjańska, N., & Kiliś-Pstrusińska, K. (2023). Depression in children and adolescents with chronic kidney disease-review of available literature. *Journal of Clinical Medicine, 12*(10). https://doi.org/10.3390/jcm12103554

Eddy, K. T., & Thomas, J. J. (2019). Introduction to a special issue on child and adolescent feeding and eating disorders and avoidant/restrictive food intake disorder. *The International Journal of Eating Disorders, 52*(4), 327–330. https://doi.org/10.1002/eat.23052

Egorova, N. N., Pincus, H. A., Shemesh, E., & Kleinman, L. C. (2018). Behavioral health diagnoses among children and adolescents hospitalized in the United States: Observations and implications. *Psychiatric Services (Washington, D.C.), 69*(8), 910–918. https://doi.org/10.1176/appi.ps.201700389

Engel, G. L. (1977). The need for a new medical model: A challenge for biomedicine. *Science (New York, N.Y.), 196*(4286), 129–136. https://doi.org/10.1126/science.847460

Fairfax, A., Brehaut, J., Colman, I., Sikora, L., Kazakova, A., Chakraborty, P., & Potter, B. (2019). A systematic review of the association between coping strategies and quality of life among caregivers of children with chronic illness and/or disability. *BMC Pediatrics, 19*. https://doi.org/10.1186/s12887-019-1587-3

Frank, N. C., Blount, R. L., & Brown, R. T. (1997). Attributions, coping, and adjustment in children with cancer. *Journal of Pediatric Psychology, 22*(4), 563–576. https://doi.org/10.1093/jpepsy/22.4.563

Freitas, L. D., Léda-Rêgo, G., Bezerra-Filho, S., & Miranda-Scippa, Â. (2019). Psychiatric disorders in individuals diagnosed with gender dysphoria: A systematic review. *Psychiatry and Clinical Neurosciences, 74*(2), 99–104. https://doi.org/10.1111/pcn.12947

Gallagher, K. A. S., Bujoreanu, I. S., Cheung, P., Choi, C., Golden, S., Brodziak, K., Andrade, G., & Ibeziako, P. (2017). Psychiatric boarding in the pediatric inpatient medical setting: A retrospective analysis. *Hospital Pediatrics, 7*(8), 444–450. https://doi.org/10.1542/hpeds.2017-0005

Gatchel, R. J., McGeary, D. D., McGeary, C. A., & Lippe, B. (2014). Interdisciplinary chronic pain management: Past, present, and future. *The American Psychologist, 69*(2), 119–130. https://doi.org/10.1037/a0035514

Gautam, S., Gautam, M., Jain, A., & Yadav, K. (2022). Overview of practice of consultation-liaison psychiatry. *Indian Journal of Psychiatry, 64*(8), 201. https://doi.org/10.4103/indianjpsychiatry.indianjpsychiatry_1019_21

Georgadarellis, A. G., & Baum, C. R. (2023). De-escalation techniques for the agitated pediatric patient. *Pediatric Emergency Care, 39*(7), 535–539. https://doi.org/10.1097/PEC.0000000000002989

Glick, I. D., Kamm, R., & Morse, E. (2009). The evolution of sport psychiatry, circa 2009. *Sports Medicine (Auckland, N.Z.), 39*(8), 607–613. https://doi.org/10.2165/00007256-200939080-00001

Gonzalez, V. J., Kimbro, R. T., Cutitta, K. E., Shabosky, J. C., Bilal, M. F., Penny, D. J., & Lopez, K. N. (2021). Mental health disorders in children with congenital heart disease. *Pediatrics, 147*(2), e20201693. https://doi.org/10.1542/peds.2020-1693

Hedegaard, H., Miniño, A., & Warner, M. (2020). Drug overdose deaths in the United States, 1999–2018 Key findings data from the national vital statistics system, Mortality. https://www.cdc.gov/nchs/data/databriefs/db356-h.pdf

Hoffmann, J. A., Stack, A. M., Monuteaux, M. C., Levin, R., & Lee, L. K. (2019). Factors associated with boarding and length of stay for pediatric mental health emergency visits. *The American Journal of Emergency Medicine, 37*(10), 1829–1835. https://doi.org/10.1016/j.ajem.2018.12.041

Hommel, K. A., McGraw, K. L., Ammerman, R. T., Heubi, J. E., Hansen, M., Dunlap, E., & Beidel, D. C. (2010). Psychosocial functioning in children and adolescents with gastrointestinal complaints and disorders. *Journal of Clinical Psychology in Medical Settings, 17*(2), 159–166. https://doi.org/10.1007/s10880-010-9193-4

Hoshiai, M., Matsumoto, Y., Sato, T., Ohnishi, M., Okabe, N., Kishimoto, Y., Terada, S., & Kuroda, S. (2010). Psychiatric comorbidity among patients with gender identity disorder. *Psychiatry and Clinical Neurosciences, 64*(5), 514–519. https://doi.org/10.1111/j.1440-1819.2010.02118.x

Jaser, S. S., & White, L. E. (2011). Coping and resilience in adolescents with type 1 diabetes. *Child: Care, Health and Development, 37*(3), 335–342. https://doi.org/10.1111/j.1365-2214.2010.01184.x

Kern, C., Wan, J., LeWinn, K. Z., Ramirez, F. D., Lee, Y., McCulloch, C. E., Langan, S. M., & Abuabara, K. (2021). Association of atopic dermatitis and mental health outcomes across childhood: A longitudinal cohort study. *JAMA Dermatology (Chicago, Ill.), 157*(10), 1200–1208. https://doi.org/10.1001/jamadermatol.2021.2657

Klein, J. D., Slap, G. B., Elster, A. B., & Cohn, S. E. (1993). Adolescents and access to health care. *Bulletin of the New York Academy of Medicine, 70*(3), 219–235.

Kullgren, K. A., Sullivan, S. K., & Bravender, T. (2018). Understanding the unique needs of hospitalized adolescents and young adults referred for psychology consults. *Clinical Pediatrics, 57*(11), 1286–1293. https://doi.org/10.1177/0009922818774339

Law, E. F., Fisher, E., Fales, J., Noel, M., & Eccleston, C. (2014). Systematic review and meta-analysis of parent and family-based interventions for children and adolescents with chronic medical conditions. *Journal of Pediatric Psychology, 39*(8), 866–886. https://doi.org/10.1093/jpepsy/jsu032

Lyketsos, C. G., Kozauer, N., & Rabins, P. V. (2007). Psychiatric manifestations of neurologic disease: Where are we headed? *Dialogues in Clinical Neuroscience, 9*(2), 111–124. https://doi.org/10.31887/dcns.2007.9.2/clyketsos

Malas, N., Plioplys, S., & Pao, M. (2019). Depression in medically ill children and adolescents. *Child and Adolescent Psychiatric Clinics of North America, 28*(3), 421–445. https://doi.org/10.1016/j.chc.2019.02.005

Margret, C. P., & Hilt, R. (2018). Evaluation and management of psychiatric emergencies in children. *Pediatric Annals, 47*(8), e328–e333. https://doi.org/10.3928/19382359-20180709-01

Miniksar, D. Y., Özcan, Ö. Ö., Cenk, H., Kapıcıoğlu, Y., & Polat, A. (2021). Psychiatric morbidity in children and adolescents with dermatological disorders. *Scandinavian Journal of Child and Adolescent Psychiatry and Psychology, 9*(1), 1–8. https://doi.org/10.21307/sjcapp-2021-001

Munjal, S., Ferrando, S. J., & Freyberg, Z. (2017). Neuropsychiatric aspects of infectious diseases: An update. *Critical Care Clinics, 33*(3), 681–712. https://doi.org/10.1016/j.ccc.2017.03.007

Myrvik, M. P., Campbell, A. D., Davis, M. M., & Butcher, J. L. (2012). Impact of psychiatric diagnoses on hospital length of stay in children with sickle cell anemia. *Pediatric Blood & Cancer, 58*(2), 239–243. https://doi.org/10.1002/pbc.23117

Nokoff, N. J. (2022, January 19). Table 2. [DSM-5 criteria for gender dysphoria]. [Text]. MDText.com, Inc. https://www.ncbi.nlm.nih.gov/books/NBK577212/table/pediat_transgender.T.dsm5_criteria_for_g/

O'Donnell, E. P., Yanek, L., Reynolds, E., Manning Ryan, L., & Ngo, T. L. (2020). Characteristics of mental health patients boarding for longer than 24 hours in a pediatric emergency department. *JAMA Pediatrics, 174*(12), 1206–1208. https://doi.org/10.1001/jamapediatrics.2019.5991

Olson, J., Schrager, S. M., Belzer, M., Simons, L. K., & Clark, L. F. (2015). Baseline physiologic and psychosocial characteristics of transgender youth seeking care for gender dysphoria. *The Journal of Adolescent Health: Official Publication of the Society for Adolescent Medicine, 57*(4), 374–380. https://doi.org/10.1016/j.jadohealth.2015.04.027

Pao, M., & Bosk, A. (2011). Anxiety in medically ill children/adolescents. *Depression and Anxiety, 28*(1), 40–49. https://doi.org/10.1002/da.20727

Parrish, C., Bartolini, E., Song, Y., Hernandez, E., Green, K., & Ostrander, R. (2020). Pediatric medical psychology. *International Review of Psychiatry (Abingdon, England), 32*(3), 284–297. https://doi.org/10.1080/09540261.2019.1705258

Rapoff, M. A. (2010). *Adherence to pediatric medical regimens* (2nd ed., p. xii, 231). Springer Publishing. https://doi.org/10.1007/978-1-4419-0570-3

Reardon, C. L., & Creado, S. (2016). Psychiatric medication preferences of sports psychiatrists. *The Physician and Sportsmedicine, 44*(4), 397–402. https://doi.org/10.1080/00913847.2016.1216719

Reardon, C. L., & Factor, R. M. (2010). Sport psychiatry: A systematic review of diagnosis and medical treatment of mental illness in athletes. *Sports Medicine (Auckland, N.Z.), 40*(11), 961–980. https://doi.org/10.2165/11536580-000000000-00000

Saunders, N. R., Gandhi, S., Chen, S., Vigod, S., Fung, K., De Souza, C., Saab, H., & Kurdyak, P. (2020). Health care use and costs of children, adolescents, and young adults with somatic symptom and related disorders. *JAMA Network Open, 3*(7), e2011295. https://doi.org/10.1001/jamanetworkopen.2020.11295

Shaw, R. J., & DeMaso, D. R. (2020). *Clinical manual of pediatric consultation-liaison psychiatry* (2nd ed.). American Psychiatric Association Publishing.

Shaw, R. J., Pao, M., Holland, J. E., & DeMaso, D. R. (2016). Practice patterns revisited in pediatric psychosomatic medicine. *Psychosomatics, 57*(6), 576–585. https://doi.org/10.1016/j.psym.2016.05.006

Solomon, S., & Conner, S. H. (2017). Psychiatric manifestations of endocrine disorders. *Human Endocrinology, 2*(1), 1–7. https://doi.org/10.24966/he-9640/100007

Sztein, D. M., & Lane, W. G. (2016). Examination of the comorbidity of mental illness and somatic conditions in hospitalized children in the United States using the kids' inpatient database, 2009. *Hospital Pediatrics, 6*(3), 126–134. https://doi.org/10.1542/hpeds.2015-0117

Thom, R. P., McDougle, C. J., & Hazen, E. P. (2019). Challenges in the medical care of patients with autism spectrum disorder: The role of the consultation-liaison psychiatrist. *Psychosomatics, 60*(5), 435–443. https://doi.org/10.1016/j.psym.2019.04.003

Tsiouris, J. A., & Brown, W. T. (2004). Neuropsychiatric symptoms of fragile X syndrome: Pathophysiology and pharmacotherapy. *CNS Drugs, 18*(11), 687–703. https://doi.org/10.2165/00023210-200418110-00001

van Duinkerken, E., Moreno, A. B., Eto, F. N., Lotufo, P., Barreto, S. M., Giatti, L., Viana, M. C., Nunes, M. A., Chor, D., & Griep, R. H. (2020). Biopsychosocial factors associated with a current depressive episode in diabetes: The ELSA-Brasil study. *Diabetic Medicine, 37*(10), 1742–1751. https://doi.org/10.1111/dme.14349

Van Schoors, M., Caes, L., Knoble, N. B., Goubert, L., Verhofstadt, L. L., & Alderfer, M. A. (2016). Systematic review: Associations between family functioning and child adjustment after pediatric cancer diagnosis: A meta-analysis. *Journal of Pediatric Psychology, jsw070*. https://doi.org/10.1093/jpepsy/jsw070

Vincze, G., Túry, F., Murányi, I., & Kovács, J. (2004). Psychiatric symptoms in general medical hospital units—Assessment of the need for psychiatric consultation-liaison in Hungary. *Neuropsychopharmacologia Hungarica: A Magyar Pszichofarmakologiai Egyesulet Lapja = Official Journal of the Hungarian Association of Psychopharmacology, 6*(3), 127–132.

Weathermon, R., & Crabb, D. W. (1999). Alcohol and medication interactions. *Alcohol Research & Health, 23*(1), 40–54.

Weiss, D. B., Dyrud, J., House, R. M., & Beresford, T. P. (2005). Psychiatric manifestations of autoimmune disorders. *Current Treatment Options in Neurology, 7*(5), 413–417. https://doi.org/10.1007/s11940-005-0033-z

Weitzman, E. R., Ziemnik, R. E., Huang, Q., & Levy, S. (2015). Alcohol and marijuana use and treatment nonadherence among medically vulnerable youth. *Pediatrics, 136*(3), 450–457. https://doi.org/10.1542/peds.2015-0722

Welch, K. A., & Carson, A. J. (2018). When psychiatric symptoms reflect medical conditions. *Clinical Medicine, 18*(1), 80–87. https://doi.org/10.7861/clinmedicine.18-1-80

Wolff, J. C., Maron, M., Chou, T., Hood, E., Sodano, S., Cheek, S., Thompson, E., Donise, K., Katz, E., & Mannix, M. (2023). Experiences of child and adolescent psychiatric patients boarding in the emergency department from staff perspectives: Patient journey mapping. *Administration and Policy in Mental Health, 50*(3), 417–426. https://doi.org/10.1007/s10488-022-01249-4

World Health Organization (Ed.). (2003). *Adherence to long-therm therapies: Evidence for action*. WHO.

Wu, Y. P., Rohan, J. M., Martin, S., Hommel, K., Greenley, R. N., Loiselle, K., Ambrosino, J., & Fredericks, E. M. (2013). Pediatric psychologist use of adherence assessments and interventions. *Journal of Pediatric Psychology, 38*(6), 595–604. https://doi.org/10.1093/jpepsy/jst025

Chapter 10
Evidence-Based Practices in Substance Use Settings for Adolescent Populations

Remos Armaos

Substance Use: Current Trends and Treatment Modalities

Addiction treatment encompasses a wide range of services and approaches designed to help individuals struggling with addiction to drugs, alcohol, or other behavioral problems. Recently, the coronavirus disease 2019 (COVID-19) pandemic has impacted people with Substance Use Disorders (SUDs) worldwide, and healthcare systems had to adapt and reorganize their services accordingly. SUD causes substantial suffering both for the individual, like premature death, and for the significant others, like family dysfunctioning, and other systems' irregularities, it also bears huge societal costs. Clinical observations and treatment studies have shown that SUD tends to be a persistent condition, often with the need for long trajectories of treatment efforts to obtain drug-free and/or good quality lives (Lauritzen & Nordfjaern, 2018).

There is growing concern among public health authorities in many countries about overprescription of and self-medication with psychotropic medicines. Annual prevalence data on the global misuse of opioids indicate that such use has nearly doubled over the past decade and that illicit markets for such drugs are expanding. Also, there is a growing threat posed by a surge in illicit cocaine production and trafficking. Despite the COVID-19 pandemic, global seizures of cocaine have increased exponentially since 2015, reaching a record of high level of 1424 tons in 2020 (International Narcotics Control Board, 2023). Novel emerging dangerous substances like non-fentanyl opioids with no legitimate uses are increasingly scanned on various online platforms, such as discussion groups, social media,

R. Armaos (✉)
Department of Psychology, The American College of Greece, Athens, Greece
e-mail: rarmaos@acg.edu

A. Maragakis, M. Janikian (eds.), *Evidence-Based Behavioral Health Practices in Pediatric Specialty Settings*, https://doi.org/10.1007/978-3-031-71004-9_10

chemical research sites, and business-to-business e-commerce platforms (International Narcotics Control Board, 2023).

At the same time, the stigma surrounding addiction remains a significant barrier to treatment in many countries. Awareness campaigns and advocacy efforts to reduce stigma and promote access to treatment are ongoing.

A brief overview of the SUD treatment field on a global scale includes diverse approaches, in which SUD treatment programs and methods vary widely across the world, with a combination of medical, psychological, and social approaches. These may range from prevention interventions, detoxification, medication-assisted treatment (pharmacological), chemical dependency programs therapeutic communities, outpatient model, social model, 12-step, other counseling services, aftercare, and harm reduction strategies. Most of the above modalities satisfy certain social criteria for treatment success, that is, to diminish crime in community, tax consumptive behavior, illicit substance abuse, or increase tax productive behavior or personal well-being.

Meanwhile, advances in addiction medicine and therapy are ongoing, with research into new treatments and strategies for prevention and harm reduction. Although, there is growing evidence in the field supporting that treatment satisfaction is a critical component of client-centered care, yet little is known about the degree to which client satisfaction is linked to subsequent outcomes, especially in SUD treatments and self-help or 12-step groups.

Client pathways through drug treatment are often characterized by the use of different services, multiple entries, and varying lengths of stay. In Europe, the SUD treatment field has its own distinct characteristics and challenges. In the field of harm reduction, several European countries have adopted relevant policies that prioritize the minimization of the negative consequences of drug use, including needle exchange programs and supervised injection facilities. The provision of more than one opioid agonist treatment medication like methadone or buprenorphine is common in Europe to manage opioid addiction (European Monitoring Center for Drugs and Drug Addiction, 2023). In the field of health and drug policy management, goals and priorities, the structure and funding of addiction treatment services differ among European countries, as some have integrated health care systems while others rely on distinct addiction treatment services. The European Union has made efforts to address the issue of drug addiction on a transnational level, with initiatives and partnerships aimed at harmonizing policies and sharing best practices.

While Greece's addiction issues may not be at the same scale as larger European countries or the United States, addiction in Greece still causes serious issues, especially in large cities like Athens and Thessaloniki. As of 2019, the two primary drugs in Greece are still heroin (the drug of abuse for 55% of adults seeking help with their drug problems) followed by marijuana, the reason why 27% of adults seek help in treatment programs. Addiction treatment services are provided by both the public health care system and private organizations, either nongovernmental organizations (NGOs) or other entities. Public services are generally accessible and are offered for free albeit with limited resources in certain geographical areas. Since over two decades, Greece has established harm reduction programs, such as needle exchange units, in order to address the consequences of drug addiction. The stigma surrounding addiction remains a challenge, too, often leading to delayed or inadequate treatment for those with SUDs. In any case, there are ongoing efforts to improve the accessibility and quality of SUD treatment services in Greece, with a focus on harm reduction and public health interventions and full recovery.

In Greece, there is an extensive network of drug-free therapeutic communities as well as substitution services, operated mainly by two public treatment organizations, KETHEA and OKANA, respectively, coupled with local services operated by drug treatment units within the Psychiatric Hospitals in Athens (18 ANO), and in Thessaloniki (IANOS, ARGO) as well as other treatment units across the mainland and the islands.

Motivation and addiction behaviors are closely considered. Motivation is not viewed as a static quality that clients either have or have not; rather is more effectively perceived as the balance between the costs and benefits of undertaking any behavior, juxtaposed with the costs and benefits of ceasing the behavior (Saunders & Wilkinson, 1990). Motivational interventions bring a different perspective where practitioners are encouraged to adopt a different style from solving problems for people to encouraging them to solve them for themselves (Miller & Rollnick, 2023). Using a certain strategy, motivational interviewing (MI), practitioners of MI treat motivation as an interpersonal process as opposed to a personality trait, suggesting motivation is a mindset that can be taught and encouraged. It is intended to be used in addition to other forms of therapy, such as cognitive behavioral therapy (CBT), for MI does not focus on the underlying causes of an addiction such as traumas or mental illnesses. MI is meant to inspire change in clients who may otherwise feel reluctant to do so. It is, overall, a client-centered process that aims to help them identify their goals and begin to work toward them. Using Miller and Rolnick's (2013) four processes and a series of interpersonal communication skills, like OARS, practitioners are expected to facilitate clear conversations and a collaborative working relationship. OARS—asking Open questions, Affirming, Reflecting, and Summarizing—are counseling techniques that may be used primarily (but not exclusively) in the engaging process of MI.

When holistically working with addictions or changing behavior patterns, practitioners are necessary to also reflect on the aspect of relapse prevention (RP). This is a self-control program model aspiring to change excessive or addictive behavior patterns focusing more on the maintenance phase of the habit change process (Marlatt & George, 1984). As such, relapse is not viewed merely as an indicator of treatment failure. Instead, potential and actual episodes are key targets for both proactive and reactive intervention strategies. RP treatment includes specific intervention techniques designed to teach the individual to effectively anticipate and cope with potential relapse situations. RP is based on the principle of social learning theory (Bandura, 1977), viewed as a psychoeducational program that combines behavioral skill-training procedures with cognitive intervention techniques (Marlatt & Donovan, 2005). In addictions, relapse prevention procedures can be applied either in the form of a specific maintenance strategy to prevent relapse or as a more general program of lifestyle change (Parks et al., 2001).

It is important to note that the SUD treatment field is dynamic, and the landscape may evolve over time in response to changing drug trends, health care policies, and public attitudes. This brief overview offers just a glimpse of the SUD treatment field at a global, European, and Greek level, but specific details may vary depending on the region and the time frame.

Epidemiology of Use, Treatment, and Special Considerations in Adolescents

Evidence indicates that for the majority of individuals, substance use and abuse initiates and increases rapidly from early to late adolescence—a stage that encompasses elements of biological growth and major social role transitions (Sawyers et al., 2018)—peaks during the transition to young adulthood and declines as they get into adulthood (Bachman et al., 2014; Griffin & Botwin, 2010). Although patterns of substance use and abuse vary greatly from person to person, epidemiological patterns are quite consistent and predictable. The onset of substance use in early adolescence contributes to a higher level of use and abuse later in life and is also related to future negative health, social, and behavioral problems, such as physical and mental health, aggression, violence, and other adjustment problems in the workplace and family (Newcomb & Locke, 2005). The latest US survey data (National Center for Drug Abuse Statistics, 2024) demonstrate that among Americans aged 12 years and older,

- 37.31 million were current illegal drug users (used within the last 30 days) as of 2020.
- 13.5% of Americans aged 12 and above used drugs in the last month, a 3.8% increase year-over-year (YoY).
- 59.28 million or 21.4% have used illegal drugs or misused prescription drugs within the last year.
- 138.54 million or 50% of people have illicitly used drugs in their lifetime.
- usership is down 0.4% YoY.
- 138.52 million Americans aged 12 and over drink alcohol.
- 28.32 million or 20.4% of them have an alcohol use disorder.
- 57.28 million people use tobacco or nicotine products (vape).
- 25.4% of illegal drug users have a drug disorder.
- 24.7% of those with drug disorders have an opioid disorder; this includes prescription pain relievers or “pain killers” and heroin.

Various studies have shown that the factors influencing the trajectory of adolescent substance use can be categorized into three large groups, i.e., individual, family, and societal (Nawi, 2021). Family and peers however have been identified by many studies as the most important ones, since they have a strong and direct effect on the life of the adolescent, and because they can serve either as protective or as risk factors (EMCDDA, 2023; Trucco, 2020). In Europe, the latest report from the European School Survey Project on Alcohol and Other Drugs states that smoking and drinking among 15–16-year-old school students is showing signs of decline (ESPAD Group, 2021). According to the same source, however, there is growing concern over potentially risky cannabis use and the challenges posed by new patterns of addiction. Nonmedical opiates prevalence rates have increased in recent years too, whereas there is already strong evidence of the abuse of over-the-counter medications among adolescents (Abraham & Chmielinski, 2018; Johnston et al., 2009).

It is a commonplace for young people to engage in some level of experimentation with substances. It is the developmental stage that favors experimentation and it is an innate human nature to do so. This does not necessarily mean one will become addicted (Khurana et al., 2015). The onset of substance use and abuse occurs typically within a social context early in adolescence and involves substances readily available, such as alcohol, tobacco, and inhalants (Kandel, 2002). Substance use often coincides with significant developmental milestones and objectives. The extent of a teenager's involvement with substance use is typically influenced by both negative prodrug social pressures they encounter, which are central, powerful factors that promote experimentation or initiation of use, and their personal susceptibilities to these pressures (Griffin & Botwin, 2010). During adolescence, individuals often engage in a variety of behaviors and lifestyle choices as part of their developmental journey toward independence, maturity, and social acceptance, among other developmental objectives. This phase is marked by experimentation with different behaviors and patterns as adolescents seek to establish their identity and achieve various developmental goals (Nawi et al., 2021).

The most successful prevention strategies recognize that substance use behaviors can serve various developmental needs at the level of the individual, family, school, and community (Whitesell et al., 2013). Given the well-established pattern of onset and progression of substance use and abuse during adolescence, a series of prevention initiatives have been developed targeting primarily to children and adolescents. In the school context, current effective methods for preventing substance use in schools can be classified into three main categories: a) training in social resistance skills; b) education on norms; and c) training to enhance competence skills (Alarcó-Rosales et al., 2021; Griffin & Botwin, 2010; Botwin, 2000).

In the family context, effective family-based prevention strategies for adolescent substance abuse come in various forms. Some concentrate solely on equipping parents with the necessary skills to steer their children away from drugs (Kumpfer, 2014; Lochman & Steenhoven, 2002). Another type of family-based prevention involves teaching family skills with both parents and children present. These programs aim to enhance overall family functioning, communication abilities, and provide instruction on developing family policies regarding substance abuse (Nawi et al., 2021). However, a significant challenge in family-based prevention efforts is engaging parents, particularly those of teenagers who are at a higher risk of drug abuse (Griffin & Botwin, 2010). Recent findings indicate that psychosocial treatments like family-based therapy, CBT, and multicomponent approaches continue to be the most successful treatment methods. Nonetheless, enhancing these treatment strategies could involve innovative approaches such as digital interventions and interventions tailored to specific cultural contexts. Additionally, advancements in adjunctive treatments like pharmacotherapy, exercise, mindfulness, and recovery-oriented educational centers may offer some clinical benefits (Fadus et al., 2019).

In the community context, evidence-based prevention programs targeting drug abuse typically consist of various elements. These commonly include components focused on schools, families, or parenting skills, alongside mass media campaigns, public policy endeavors, and other community-based initiatives and activities. Implementing these interventions requires substantial resources and coordination due to the extensive range of activities involved. Typically, a coalition of stakeholders, including parents, educators, and community leaders, oversees the management of program components. Studies indicate that community-based programs delivering a unified, comprehensive prevention message can effectively deter adolescent substance use.

Looking ahead, it is crucial to tackle various factors that hinder the public health effectiveness of successful prevention programs. In many cases, the majority of schools implement prevention programs that lack evidence-based foundations, while family-based prevention efforts often fail to reach the families most in need. Additionally, launching community prevention programs demands significant resources (Griffin & Botwin, 2010). Psychosocial interventions such as family-based therapies, CBT, and multicomponent interventions (including MI/MET and contingency management) remain the primary strategies; the extent and effectiveness of pharmacotherapy in the treatment of adolescent SUDs remain uncertain (Fadus et al., 2019).

The omnipresent technology, social media, and digital platforms may indicate a promising avenue for treatment, one that would perhaps be more appealing for young people and could better fit their ways of interacting. This could facilitate the retention of information for the specific target group. Lately we often witness high mobility among populations, and therefore another area of focus could be the enhancement of culturally sensitive programs able to meet the needs of adolescents belonging to different racial and ethnic minorities. Emphasis should be placed on gaining a deeper understanding of accessibility disparities and challenges at individual, organizational (within treatment agencies), and systemic (in state policies and regulations) levels.

The Role of Training and Development in Addiction and Allied Mental Health Professionals for Improved SUD Treatment Outcomes

Professionals or counselors who treat people with SUDs usually have diverse training experiences and skill levels (Center for Substance Abuse Treatment, 2017). Either they are credentialed counselors or peer support workers with lived experience, they both share training and practice guidelines that are crucial to providing clients with SUDs consistent and high-quality care for optimal recovery. These common guidelines also support workforce retention among SUD counselors ensuring that highly skilled and qualified staff work in the field (Hoge et al., 2013). To ensure effective training and retention of counselors, it is pivotal that appropriate resources and sufficient training in evidence-based practices and cultural responsiveness along with appropriate supervision are available (Alagoz et al., 2017; Altschul et al., 2018).

SUD treatment has evolved significantly, recognizing that effective care is not only about therapy and medications but also the knowledge and skills of the professionals delivering these services. The importance of continuous in-service or other training and professional development cannot be overstated. The effectiveness of SUD treatment outcomes is intricately tied to the knowledge, skills, and attitudes of addiction and allied mental health professionals. The landscape of addiction treatment is ever-evolving, and staying up to date with current methodologies is essential for providing optimal care. The famous motto in the addiction recovery field that "treatment works" presupposes a strong level of professional capacity and growth mindset, in depth knowledge and analytical competences acquired for counselors, therapists, prevention specialists, or other professionals working in the field.

Extensive literature exists on what comprises an effective training and upskilling in the SUD field. Staying abreast of the latest research, evidence-based practices, and treatment modalities is vital in ensuring that professionals are adequately equipped to address the constantly changing needs of individuals with SUD.

SUD Counseling Core and Extended Competences

To support training needs and enhance program efficacy, the Center for Substance Use Treatment (2017) publication on addiction counseling competencies addresses what is essential to the effective practice of SUD counseling, as well as the knowledge, skills, and attitudes (KSAs) that counselors need to become proficient in each competency. A differentiated competency model has been generated providing a framework for understanding the relationship between core and extended competencies. The four transdisciplinary foundational competencies are (a) understanding SUDs, (b) treatment knowledge, (c) application to practice, and (d) professional readiness. They are placed at the center of a counselor's skillset and they are complemented by eight SUD-specific practice dimensions representing expanded competencies. The eight practice dimensions are clinical evaluation; treatment planning; referral; service coordination; counseling; client, family, and community education; documentation; and professional and ethical responsibilities (Center for Substance Use Treatment, 2017).

Further to identifying core or extended skills in the SUD workforce development, emphasis also needs to be put on the core pedagogy of the training's nature itself. There are diverse aspects that reflect, basically, the methodologies and rationale upon which effective immersive training designs are substantiated. The following transversal areas are significant to consider:

Adult Learning Principles

Professional education is about the education of adults. Adults think, act, and process information differently than children, in this sense they also learn differently. In fact, there's a lot of research about what makes a learning style effective for an adult. Considering that theory provides a strong grounding for trainers and learners in their professional practice (Gouthro, 2019), the most important adult learning theories, or adult learning styles, in play today are addressed next. Learning how these core adult learning principles work can improve our own education, boost our organization's performance and training, and bolster our ability to educate others. These are a) *self-directed learning*, i.e., learning at one's own pace in one's own way, b) *transformational*, i.e., learning can change our perspective on the world and vice versa, c) *experiential*, i.e., focusing on developing life experience or "hands-on" learning, d) *mentorship*, i.e., learning from a mentor (established figure) in the field, e) *motivation*, i.e., while children are motivated by parents and laws requiring their education; adults often have internal motivation, and f) *readiness for change*, i.e., adults have already been through this process and can rely on past experience or life changes to develop a renewed readiness.

Professional education programs should be designed with adult education principles in mind, recognizing that professionals are adult learners having unique needs, well-established habitual minds, and certain learning preferences. These programs should be self-directed, relevant, and draw upon the experiences and knowledge that professionals already possess, utilize, and reflect upon their own identity and understanding.

Learning Through Transformation

Everyone learns differently, and understanding the different ways that humans learn is crucial to educational success. Educators and trainees alike can greatly benefit from understanding how learning works for different individuals. By understanding how learning happens, educators can maximize their efforts and create environments where learners can thrive. Current, and aspiring, trainers can greatly benefit from understanding learning theories and learning models and implementing strategies in their training that help them appeal to more learners.

Transformative learning is one theory of learning, that focuses particularly on adult education and learning. Transformative learning is sometimes called transformation learning and focuses on the idea that learners can adjust their thinking based on new information (Mezirow & Associates, 1990; Mezirow, 1991). Mezirow's theory has developed into a larger idea that the more we learn our worldview is changing, and that helps us grasp new concepts and ideas. Mezirow's transformative learning is defined as "an orientation which holds the way learners interpret and reinterpret their sense experience is central to making meaning and hence learning." To put it in simple terms, transformative learning is the idea that learners who are getting new information are also evaluating their past ideas and understanding, and as they obtain new information and through critical reflection, they are shifting their whole worldview. It goes beyond simply acquiring knowledge and dives into the way that learners find meaning in their lives and understanding. This kind of learning experience involves a fundamental change in our perceptions—learners start to question all the things they knew or believed before and examine things from new perspectives in order to make room for new insights and information. Many learners and experts agree that this kind of learning leads to true freedom of thought and understanding. As Mezirow says, transformative learning has two basic focuses—instrumental learning and communicative learning. Instrumental learning focuses on task-oriented problem-solving, while communicative learning focuses on how people communicate their feelings, needs, and desires. Both of these elements are important in transformative learning as learners become capable of focusing on different types of their understanding and viewing new perspectives that are both logical and emotional in order to challenge their previous understanding.

SUD treatment involves helping individuals make significant life changes. To motivate people to change, often counselors need to be capable of changing themselves too. Training counselors within transformative learning principles equip them to guide patients through personal growth and transformation, promoting lasting recovery and positive change. Training should promote transformation, encouraging professionals to critically challenge their beliefs and assumptions, mostly the problematic ones, more particularly those concerning addiction and mental health. This self-reflective process can lead to a more empathetic and effective approach to treatment.

The ways that trainers can introduce this kind of learning into their classes with SUD professionals include:

- *Give learners chances to learn about new perspectives.* Trainers can provide an action event or a disorienting dilemma during their training. A picture of an active drug addict, being "stoned", a story, or a guest who does not quite meet what they have understood can help do this. For example, trainees can learn about different treatment modalities by actually meeting social workers working in diverse SUD treatment units. They may have certain beliefs about treatment, but meeting a professional who is kind and passionate about their work and health care prospects could be an action event for them. Similarly, trainers can introduce diverse experts and literature into the courses, helping participants see things from a perspective that does not match their own.

- *Help trainees identify and question their assumptions.* In the course it is important for trainees to critically challenge their assumptions. Trainers can model this behavior by sharing stories of how their own understanding and perspectives have shifted, how they have been able to challenge their own beliefs, and more. Trainers can give journal questions through which they ask students about their own assumptions, give opportunities for trainees to share, and provide writing and reflective time for trainees to analyze.
- *Create opportunities for critical discourse.* In a course that is focused on transformative learning, there must be opportunities for critical discussions. Educators can help trainees have these critical discussions by providing space for conversation, helping them create pros and cons lists of literature or articles, giving trainees the opportunity for debate, or requiring trainees to prepare both sides of a debate to understand how there are always multiple perspectives at play. Additionally, small group discussions and activities can help trainees with this critical understanding. This kind of coursework can help trainees really embrace thoughts and experiences different from their own.
- *Transformative learning examples.* There are many ways that educators can use transformative learning in their courses. For example, when trainees are learning about SUD treatment modalities, trainers can introduce the literature and articles that discuss different types of treatment, helping them gain new perspectives and understanding. Another way trainers can use transformative learning in the course is to give students contrasting articles on a topic and ask them their thoughts and insights on the different elements of each article. This helps them to question their own understanding and move through the process of learning.

Upskilling/Reskilling and Cross-Skilling/Cross-Training

In the dynamic and ever-evolving treatment workplace, an individual's commitment to adaptation and growth is indispensable. The way we discuss skills' development has expanded; words like upskilling or reskilling and cross-skilling or cross-training define how professionals adapt, grow, and reinvent themselves. Professionals in the field of SUD treatment are literally exposed to various therapeutic modalities and approaches. Cross-training ensures a broader skill set, enhances understanding, and offers therapists—or other addiction professionals—the opportunity for a growing mindset on various aspects of addictions thus leading to an increased ability to come up with treatment plans tailored to individual patient needs (Tsiboukli & Armaos, 2015). As SUD treatment becomes increasingly integrated with mental health care, cross-training modalities are essential. Mental health professionals, in general, need to have a clear understanding of addictions, and addiction specialists should be well-versed in mental health care issues. This multidisciplinary approach allows for a more comprehensive treatment plan and sustaining outcomes.

Cross-skilling of addiction and mental health allied professionals among diverse organizations highlights the notions of defensiveness (Argyris & Schön, 1978) and organizational barriers (Senge, 2006) throughout the learning process. Challenges derived from those barriers and defensive mechanisms become larger when professionals come from systems with different organizational cultures and practices. Cross-training aspires to urge professionals to question their assumptions and their practice and at the same time to develop an open communication network with peers in the field for capacity building and performance. This is highly embedded by Senge (2006) in the concept of systems thinking, founder of the concept of the "Learning Organization" and advocate of the decentralization of the role of leadership in organizations, highlighting the existence of five basic principles that should concern those who design or implement training programs in organizations. These principles include, among others, group learning mental models, i.e., the "deeply embedded assumptions, generalizations, meanings, and images that influence the way we understand the world and the way we respond, and perform in our life, and systems thinking which allows for the understanding, analysis and synthesis of interdependencies and interactions with others. (Senge, 2006).

In line with this theoretical framework, Watkins and Marsick (2003) argue that learning organizations are defined by seven characteristics. Among these characteristics, in addition to the continuous creation of learning opportunities, are team learning, shared learning systems, and the interconnectedness of each organization with its environment. However, building a learning organization requires that the latter discovers ways "to harness the commitment and capacity of employees to learn at all levels" (Senge, 1990: 4) and, at the same time, that the leaders involved in the training are given the opportunity for immediate application of the new knowledge and skills (Cross, 1981) that are expected to emerge from the learning process.

Cross-skilling is one of the most difficult tasks in adult professional education. In planning, design, and implementation of such training programs, it is important to take into account the defense mechanisms, as described by Argyris and Senge's systems thinking. Identifying systemic barriers is a first step to learn by reflection and, in turn, likely to promote development and change in these systems (Senge, 1990). Organizational or systemic barriers relate to the purpose, structure, means, culture, and hierarchy of any organization and may lead to fragmentation (Senge, 1990) as a result of lacking synergies replaced mostly by one-dimensional approaches. Fragmentation impedes any effort for collaboration, or the creation of synergies among different systems. It coexists with antagonism and reactive attitudes and leads to inertia that is caused by complacency and results in change only when problems have already arisen. Hence, the key question is how organizational barriers are to be elicited and analytically explored in cross-skilling programs for professionals coming from diverse systems and treatment modalities targeting however, addressing a similar profile of addicted individuals (Tsiboukli & Armaos, 2015).

The question that arises, however, is the possibility of capturing and addressing organizational barriers in the implementation of cross-skilling programs for staff of different systems that address the same population with seemingly the same purpose, i.e., the prevention of recidivism, but essentially approaching it with different means and tools, i.e., correction and punishment versus treatment and behavior change. As it is for most addiction specialists and practitioners, they can also be susceptible to a host of cognitive biases and be influenced by their social tendencies and social environments. These biases will activate or rely on deeply ingrained mindsets when making choices in their treatment work. Sometimes these choices seem to fit more their personal biases and less the local conditions necessary to highlight, having a reduced probability of succeeding in their therapy or counseling work.

Collaborative Learning Environments

The nature of addiction and mental health treatment often necessitates a collaborative approach. The benefits of collaboration are widely addressed in many facets of human action, including education. Collaborative learning, through peer support and interdisciplinary cooperation, can enrich the training experience. When professionals from various backgrounds share their insights and experiences, it fosters a more comprehensive understanding of SUD treatment. A training that enhances collaboration among professionals from various disciplines ensures a comprehensive and holistic treatment approach, reinforces interdisciplinary and inter-agency cooperation, and enhances a shared understanding of treatment goals and objectives among the professionals of not only one unit or organization but also among many different systems. Interdisciplinary training encourages sharing of insights, better communication, and, ultimately, a more effective patient care.

Experiential Learning

Learning is a continuous process grounded on experience; and it is conceived as a process not in terms of outcomes (Kolb, 1984). This perspective has important educational implications suggesting that all learning is *relearning*. David Kolb together with Roger Fry created the model of the experiential learning cycle, a model that has been used with great success, both in the field of corporate and in-service education and training as well as in the field of adult education and nonformal education in general. Kolb's contribution is not only limited to the field of experiential learning, but also extends to the field of organizational psychology and human resources' development.

The process of experiential learning can be described as a four-stage cycle involving four learning functions: a) concrete experience, b) reflective observation, c) generalization, and d) experimentation. The central idea of this model is that learning, and knowledge, require both the understanding and acceptance of the experience as well as its transformation into some form of knowledge (Kolb, 1984).

In Kolb's model, the four types of skills are exploited in a synthetic way, which are required for training through experience. They are as follows: a) the individual's desire to actively participate in the acquisition of new experiences, b) the individual's ability to observe, so that new experiences can be explored, c) the observational nature of the new experience, that is the new experiences can be explored from many different angles, d) the interdisciplinary approach to knowledge, e) the ability to formulate theoretical principles and generalizations, and f) the individual's ability to solve problems and make decisions so that new ideas can be put into practice.

Encouraging professionals in their learning, using their experiences and real-world scenarios, can enhance their ability to reconsider their thoughts, ideas, or beliefs; implant new ideas or ways of practice; and dispose of, or modify, old ones (Kolb, 1984). Experiential learning helps bridge the gap between academic or professional knowledge and expertise. The intricacies of SUD treatment require a deep understanding that goes beyond theory. Experiential learning provides opportunities for professionals to learn from their experience, enabling them to develop practical skills and a deeper understanding of the challenges individuals with SUD face.

Participatory Learning Approaches

Involving professionals in the development of training materials and curricula can enhance their sense of ownership and engagement. Participatory learning encourages a more proactive approach to skills development by designing and delivering professional education programs that are also based on the notion of "coproduction" of knowledge, skills, and know-how. These are programs that use participatory action research (Kemmis, 2015; Kemmis et al., 2014; Reason & Bradbury, 2001) to support trainees as "coproducers of knowledge," actively working in collaboration with the research groups in creating "developmental" interventions, rather than applying traditional "transmissive" behavior change models (Armaos et al., 2021). The approach draws on practices from radical pedagogy to support marginalized people to become architects of their own future (Freire, 1972; Vygotsky, 1978).

Participatory learning also determines the way the learning process is construed by encouraging active participation, as opposed to passive learning. Engaging professionals in problem-solving, case discussions, and ethical dilemmas enhance their critical thinking skills, ultimately leading to better decision-making and a more effective clinical practice.

Enhancing Social Action

Training programs should emphasize the role of professionals in broader societal issues related to addiction. By understanding their potential for social action and advocacy, addiction professionals can contribute to a more holistic approach to SUD treatment. This group often works with marginalized and underserved populations. Training should address the social determinants of addiction and encourage professionals to engage in advocacy, outreach, and policies that address these issues to create lasting change in communities (Cullen & Tsiboukli, 2019). Being an active member of the community around you enhances one's understanding of the challenges service recipients may be facing. In addition, liaising with the wider community also facilitates a mutual understanding between all interested parties and creates the potential for a more effective cooperation.

Immersive Curricula and Codesign

Immersive training curricula, combined with codesign principles, can create a more engaging and relevant learning experience. Professionals should be involved in the design of training programs to ensure they align with their specific needs and challenges. The complexities of SUD treatment necessitate immersive, real-world training experiences. Co-designing training programs with input from experienced professionals ensures that they are consistent with the realities of clinical practice, making them more practical and relevant.

Digital Competences

Training that is grounded in current methodologies can prevent professional burnout and subsequently lead to improved patient outcomes. Technologies also are growing fast in the health domain, but research has shown that many people still lack the knowledge and skills to use them (Hargittai & Walejko, 2008). Although almost everyone is to some extent a technology user, only a few will leverage any self-exploratory digital skills developed to transform their practice. There is, however, growing evidence of digital skills shortages in general, among front-line health professionals (OECD, 2019; OECD, 2018; The Lancet Global Health Commission, 2018).

The increased use of information and communication technologies is recognized as one of the key strategic platforms on which to build strong education and training systems. For more than a decade, the Digital Competence Framework for Citizens (DigComp) and its latest revisions in response to AI applications influx have provided a common understanding, across the European Union and beyond, of what digital competence is, and therefore provided a basis for framing a digital skills policy (Vuorikari et al., 2022). Such strategies and guidelines play a central role in achieving ambitious system's objectives with regard to the digital upskilling of the whole population as well as in developing certification in digital skills.

Innovative ways of teaching and learning are required to respond to the need for SUD professional education, tackle the shortage of health care and addiction treatment workers, and ultimately improve care provision. eLearning appears to be one such innovation (Gentry et al., 2016). eLearning, not limited only to synchronous or asynchronous computer-based eLearning, encompasses a variety of interventions characterized by their tools, contents, learning objectives, pedagogical implications, and setting of delivery. Massive Open Online Courses (MOOCs), Artificial Intelligence (AI) or other virtual environments, mLearning, simulations, and serious games are a few to report.

Today, especially in post-pandemic times, addiction and allied mental health professionals face rapidly evolving challenges, including the impact of technology, the rise of new substances, and changes in societal attitudes toward addiction. Training should address these contemporary issues, preparing professionals to adapt to the ever-changing landscape. The modern landscape of SUD treatment demands that professionals continually upskill. This includes proficiency in telehealth services, culturally sensitive care, trauma-informed treatment, and staying updated on emerging substances and trends. The ability to integrate technology and data-driven approaches into treatment becomes increasingly important. Digital upskilling enables addiction treatment professionals to leverage data analytics, predictive modeling, and machine learning (ML) algorithms to develop personalized treatment plans. Advanced technologies can aid in early intervention, relapse prevention, and identification of risk factors leading to improved treatment outcomes and long-term recovery.

In conclusion, the effectiveness of SUD treatment outcomes is intrinsically linked to the competence of addiction and mental health professionals. Ensuring that these professionals receive comprehensive, up-to-date, and innovative training is not just a matter of professional development; it is crucial for the well-being of the individuals they serve. By adhering to current methodologies and adult education principles, promoting collaborative learning, embracing experiential and transformative learning, and accommodating twenty-first-century upskilling demands within an AI new domain, we can improve the quality of care provided to those grappling with SUD and foster healthier, more resilient communities.

Clinical Supervision

Clinical supervision is widely discussed in the professional literature of several disciplines, including social work, psychology, and both mental health and SUD counseling.

Clinical supervision has taken several definitions ranging from "a disciplined, tutorial process wherein principles are transformed into practical skills, with four overlapping foci: administrative, evaluative, clinical, and supportive" (Powell, 2004: 11); "an intervention that is provided by a senior member of a profession to a more junior member or members of that same profession. This relationship is evaluative, extends over time, and has the simultaneous purposes of enhancing the professional functioning of the more junior person(s), monitoring the quality of professional services offered to the clients that she, he, or they see, and serving as a gatekeeper of those who are to enter the particular profession" (Bernard & Goodyear, 2004: 8); "a social influence process that occurs over time, in which the supervisor participates with supervisees to ensure quality clinical care. Effective supervisors observe, mentor, coach, evaluate, inspire, and create an atmosphere that promotes self-motivation, learning, and professional development. They build teams, create cohesion, resolve conflict, and shape agency culture, while attending to ethical and diversity issues in all aspects of the process" (Center for Substance Abuse Treatment, 2007: 3).

Effective supervisors are skilled, experienced clinicians. Their competencies develop not only due to their academic training or field practice; they reflect a mastery level that clinical supervisors target, which, in turn, reflects not only a new set of knowledge, skills, and abilities but also develops their assumption of a different professional role. "It is important, too, for supervisors to understand the developmental process of supervisees (e.g., how they learn and how skills are developed over time) and how to establish a spirit of learning and personal development in an organization. Service improvement occurs most effectively when service providers have the freedom to practice new skills in an environment marked more by support and mentoring than by critical judgment" (Center of Substance Abuse Treatment, 2007: 4). In any case, knowledge and skill as a counselor are not enough to ensure success as a clinical supervisor. It is well stated that specific tasks, responsibilities, and roles of supervisors vary depending on agency mission, target population, theoretical model, treatment modality, and general structure.

Clinical supervisors in SUD treatment settings are expected to be familiar with knowledge across foundational areas in clinical supervision, such as the *diverse theories, roles, and modalities of clinical supervision*; *aspects of leadership*; *supervisory alliance; critical thinking;* and *organizational management aspects*. Also, a series of competencies listed within certain performance domain areas identify the specific abilities and responsibilities that clinical supervisors must master to be effective in the essential roles they play in the service delivery system.

These five performance domains address:

- *Counselor development* which is key to the delivery of high-quality client care
- *Professional and ethical standards*
- *Program development and quality assurance*, two aspects that refer to the process of guiding a service delivery organization to maximize the potential of its staff and resources to meet the needs of the population it serves, and the process

of designing and implementing a program's activities to ensure maximum effectiveness and efficiency of services, respectively

- *Performance evaluation* which is central to the assurance of high-quality client care
- *Administration,* which reflects the clinical supervisor's responsibilities to follow the organization's policies and procedures ensuring the maintenance of case records, case documentation, assisting in financial resource development (e.g., grant proposal writing), and developing relationships with referral sources in the community (Center for Substance Abuse Treatment, 2007)

Training is designed to enhance participants' competence in terms of knowledge, skills, and attitudes related to supervision practice. It is important that training includes the best practices and evidence-supported techniques of clinical supervision. Content includes a review of targeted areas to focus on enhanced efficacy, technique, and practice. A dynamic and flexible content is necessary to respond to ongoing challenges and symptoms of psychopathology and substance use, as well as any trends in treatment modalities in psychotherapy based on subjects' profiling.

References

Abraham, O., & Chmielinski, J. (2018). Adolescents' misuse of over-the-counter medications: The need for pharmacist-led intervention. *Innovations in pharmacy, 9*(3), 1–7. https://doi.org/10.24926/iip.v9i3.979

Alagoz, E., Hartje, J., & Fitzgerald, M. (2017). *National workforce report 2017: Strategies for recruitment, retention, and development of the substance use disorder treatment and recovery services workforce: A national qualitative report.* Addiction Technology Transfer Center Network Coordinating Office.

Alarcó-Rosales, R., Sánchez-San Segundo, M., Ferrer-Cascales, R., Albaladejo-Blazquez, N., Lordan, O., & Zaragoza-Martí, A. (2021). Effects of a school-based intervention for preventing substance use among adolescents at risk of academic failure: A pilot study of the reasoning and rehabilitation V2 program. *Healthcare (Basel, Switzerland), 9*(11), 1488. https://doi.org/10.3390/healthcare9111488

Altschul, D. B., Bonham, C. A., Faulkner, M. J., Farnbach Pearson, A. W., Reno, J., Lindstrom, W., et al. (2018). State legislative approach to enumerating behavioral health workforce shortages: Lessons learned in New Mexico. *American Journal of Preventive Medicine, 54*(6, Suppl. 3), S220–S229.

Argyris, C., & Schön, D. A. (1978). *Organizational learning: A theory of action perspective.* Addison-Wesley.

Armaos, R., Cullen, J., & Tsiboukli, A. (2021). Innovative digital and experiential blended learning non-formal Programme for community animators: Lessons learned. *Adult Education Critical Issues, 1*(1), 32–40. https://doi.org/10.12681/aeci.28797v

Bachman, J., O'malley, P., Schulenberg, J., Johnston, L., Bryant, A., & Merline, A. (2014). The decline of substance use in young adulthood: Changes in social activities. *Roles, and Beliefs.* https://doi.org/10.4324/9781410606013

Bandura, A. (1977). *Social learning theory.* Prentice-Hall.

Bernard, J. M., & Goodyear, R. K. (2004). Ethical and legal considerations. In *Fundamentals of clinical supervision* (3d ed.). Allyn & Bacon.

Botvin, G. J. (2000). Preventing drug abuse in schools: Social and competence enhancement approaches targeting individual-level etiologic factors. *Addictive Behaviors, 25*(6), 887–897. https://doi.org/10.1016/s0306-4603(00)00119-2

Center for Substance Abuse Treatment. (2007). *Competencies for substance abuse treatment clinical supervisors*. Technical Assistance Publication (TAP) Series 21-A. DHHS Publication No. (SMA) 07-4243. Substance Abuse and Mental Health Services Administration.

Center for Substance Abuse Treatment. (2017). *Addiction counselling competencies: The knowledge, skills, and attitudes of professional practice* (pp. 15–4171). Technical Assistance Publication (TAP) Series 21. HHS Publication No. (SMA). Substance Abuse and Mental Health Services Administration.

Cross, K. P. (1981). *Adults as learners*. Jossey-Bass.

Cullen, J., & Tsiboukli, A. (2019). Using lifeworld analysis to co-design a new role in youth work. *Exartiseis, 32*, 27–45.

ESPAD Group. (2021). *ESPAD 2019 methodology: Methodology of the 2019 European School Survey Project on Alcohol and other Drugs*. EMCDDA Joint Publications, Publications Office of the European Union.

European Monitoring Center for Drugs and Drug Addiction. (2023). *European Drug Report 2023: Trends and Developments*. Available at https://www.emcdda.europa.eu/publications/european-drug-report/2023_en

Fadus, M. C., Squeglia, L. M., Valadez, E. A., Tomko, R. L., Bryant, B. E., & Gray, K. M. (2019). Adolescent substance use disorder treatment: An update on evidence-based strategies. *Current Psychiatry Reports, 21*(10), 96. https://doi.org/10.1007/s11920-019-1086-0

Freire, P. (1972). *Pedagogy of the oppressed*. Penguin.

Gentry, S., L'Estrade Ehrstrom, B., Gauthier, A., Alvarez, J., Wortley, D., van Rijswijk, J., Car, J., Lilienthal, A., Tudor Car, L., Nikolaou, C. K., & Zary, N. (2016). *Serious gaming and gamification interventions for health professional education*. Cochrane Database of Systematic Reviews, Issue 6. Art. No.: CD012209. https://doi.org/10.1002/14651858.CD012209

Gouthro, P. A. (2019). Taking time to learn: The importance of theory for adult education. *Adult Education Quarterly, 69*(1), 60–76. https://doi.org/10.1177/0741713618815656

Griffin, K. W., & Botvin, G. J. (2010). Evidence-based interventions for preventing substance use disorders in adolescents. *Child and Adolescent Psychiatric Clinics of North America, 19*(3), 505–526. https://doi.org/10.1016/j.chc.2010.03.005

Hargittai, Eszter & Walejko, Gina. (2008). The participation divide: Content creation and sharing in the digital age. Information. *Communication & Society*. 239–256. https://doi.org/10.1080/13691180801946150

Hoge, M. A., Stuart, G. W., Morris, J., Flaherty, M. T., Paris, M., Jr., & Goplerud, E. (2013). Mental health and addiction workforce development: Federal leadership is needed to address the growing crisis. *Health Affairs, 32*(11), 2005–2012.

International Narcotics Control Board. (2023). *Report 2022*. United Nations.

Johnston, L. D., O'Malley, P. M., Bachman, J. G., et al. (2009). *Monitoring the Future national survey results on drug use, 1975–2008. Volume I: Secondary school students*. National Institute on Drug Abuse.

Kandel, D. (2002). *Stages and pathways of drug involvement: Examining the gateway hypothesis (eds)*. University Press.

Kemmis, S. (2015). Critical theory and critical participatory action research. *The SAGE handbook of Action Research, 1*, 453–464.

Kemmis, S., McTaggart, R., & Nixon, R. (2014). *The action research planner: Doing critical participatory action research*. Springer.

Khurana, A., Romer, D., Betancourt, L. M., Brodsky, N. L., Giannetta, J. M., & Hurt, H. (2015). Experimentation versus progression in adolescent drug use: A test of an emerging neurobehavioral imbalance model. *Development and Psychopathology, 27*(3), 901–913. https://doi.org/10.1017/S0954579414000765

Kolb, D. (1984). *Experiential learning: Experience as the source of learning and development.* Prentice Hall.

Kumpfer, K. L. (2014). Family-based interventions for the prevention of substance abuse and other impulse control disorders in girls. *ISRN Addiction, 2014*, 308789. https://doi.org/10.1155/2014/308789

Lauritzen, Grethe & Nordfjærn, Trond. (2018). Changes in opiate and stimulant use through 10 years: The role of contextual factors, mental health disorders and psychosocial factors in a prospective SUD treatment cohort study. PLOS ONE. 13. https://doi.org/10.1371/journal.pone.0190381

Lochman, J., & Steenhoven, A. (2002). Family-based approaches to substance abuse prevention. *The Journal of Primary Prevention, 23*, 49–114. https://doi.org/10.1023/A:1016591216363

Marlatt, G. A., & Donovan, D. M. (Eds.). (2005) *Relapse prevention: Maintenance strategies in the treatment of addictive behaviors* (2nd ed.). The Guilford Press.

Marlatt, G. A., & George, W. H. (1984). Relapse prevention: Introduction and overview of the model. *British Journal of Addiction, 79*, 261–273.

Marsick, V. J., & Watkins, K. E. (2003). Demonstrating the value of an organization's learning culture: The dimensions of the learning organization questionnaire. *Advances in Developing Human Resources, 5*(2), 132–151.

Mezirow, J. (1991). *Transformative dimensions of adult learning*. Jossey-Bass.

Mezirow, J and Associates. (1990). *Fostering critical reflection on adulthood.* Jossey-Bass.

Miller, W. R., & Rollnick, S. (2013). *Motivational interviewing: Helping people change* (3rd ed.). Guilford Press.

Miller, W. R., & Rollnick, S. (2023). *Motivational interviewing* (4th ed.). Helping People Change and Grow. Guildford Press.

National Center for Drug Abuse Statistics. (2024, March 31). *Drug abuse statistics: Key findings*, NCDAS. Available at: https://drugabusestatistics.org/

Nawi, A. M., Ismail, R., Ibrahim, F., Hassan, M. R., Manaf, M. R. A., Amit, N., Ibrahim, N., & Shafurdin, N. S. (2021). Risk and protective factors of drug abuse among adolescents: A systematic review. *BMC Public Health, 21*(1), 2088. https://doi.org/10.1186/s12889-021-11906-2

Newcomb, M. D., & Locke, T. (2005). Health, social, and psychological consequences of drug use and abuse. In Z. Sloboda (Ed.), *Epidemiology of drug abuse* (pp. 45–59). Springer.

Parks, G. A., Anderson, B. K., & Marlatt, G. A. (2001). Relapse prevention therapy. In N. Heather, T. Peter, & T. Stockwell (Eds.), *The handbook of alcohol dependence and problems sussex.* John Wiley & Sons, Ltd..

Powell, D. J. (2004). *Clinical supervision in alcohol and drug Abuse counseling: Principles, models, methods*. Jossey-Bass.

Reason, P., & Bradbury, H. (2001). Introduction: Inquiry and participation in search of a world worthy of human aspiration. In P. Reason & H. Bradbury (Eds.), *Handbook of action research* (pp. 94–105). Sage.

Saunders, B., & Wilkinson, C. (1990). Motivation and addiction behaviour: A psychological perspective. *Drug and Alcohol Review, 13*, 133–142. https://doi.org/10.1080/09595239000185191

Sawyers, S. M., Azzopardi, P. S., Wickremarathne, D., & Patton, G. C. (2018). The age of adolescence. *The Lancet Child & Adolescent Health, 2*(3), 223–228. https://doi.org/10.1016/S2352-4642(18)30022-1

Senge, P. (1990). *The fifth discipline: The art and practice of the learning organization.* New York: Doubleday/Currency

Senge, P. (2006). *The fifth discipline: The art & practice of the learning organization.* Random House.

Substance Abuse and Mental Health Services Administration. (2021). *The substance use disorder counseling competency framework: An overview*. Advisory.

Trucco, E. M. (2020). A review of psychosocial factors linked to adolescent substance use. *Pharmacology, Biochemistry, and Behavior, 196*, 172969. https://doi.org/10.1016/j.pbb.2020.172969

Tsiboukli, A. & Armaos, R. (2015). Crostraining of professionals in diverse systems: organisational barriers, defense mechanisms, challenges and prospects (in Greek: Συν-εκπαίδευση στελεχών διαφορετικών συστημάτων: οργανωσιακοί φραγμοί, μηχανισμοί άμυνας, προκλήσεις και προοπτικές. Εξαρτήσεις, 24, 43–51. https://doi.org/10.57160/NUCR3048

Vuorikari, R., Kluzer, S., & Punie, Y. (2022). *DigComp 2.2: The digital competence framework for citizens—With new examples of knowledge, skills and attitudes*. EUR 31006 EN, Publications Office of the European Union. https://doi.org/10.2760/490274

Vygotsky, L. (1978). *Mind in society: The development of higher psychological processes*. Harvard University Press.

Watkins, K. E., & Marsick, V. J. (2003). Demonstrating the Value of an Organization's Learning Culture. *Advances in Developing Human Resources, 5*(2), 129–131. https://doi.org/10.1177/1523422303005002001

West, A. M. (1997). *Developing creativity in organizations*. The British Psychological Society.

Whitesell, M., Bachand, A., Peel, J., & Brown, M. (2013). Familial, social, and individual factors contributing to risk for adolescent substance use. *Journal of Addiction, 2013*, 579310. https://doi.org/10.1155/2013/579310

Chapter 11
Evidence-Based Practices for Treating Child and Adolescent Refugees

John-Paul Legerski, Randy Nedegaard, and Brian L. Isakson

The refugee crisis is pervasive and complex, with large number of people forcibly displaced from their homes across the globe due to persecution, war, violence, and environmental disasters. With the rise in refugees, asylum seekers, and internally displaced persons in the last decade, the number of forcibly displaced people globally exceeded 110 million (UNHCR Global Trends Report, 2023; USA for UNHCR, 2023). Children and teenagers constitute a disproportional amount of those affected by the refugee crisis, with approximately 41% of refugees being children (UNICEF, 2023). Many of these children experience trauma and other adverse experiences that can increase the risk of developing mental health disorders (Frounfelker et al., 2020). They often face disruptions in education, economic difficulties, language barriers, and resettlement challenges. Refugees and asylum seekers may also encounter hostility, anger, and mistrust after resettlement by members of their host countries, which appear to be on the rise globally, and can negatively impact their well-being (da Silva Rebelo et al., 2018). Furthermore, a disproportionate burden for hosting refugees is placed on countries with limited economic resources that often do not have the infrastructure for the long-term care of large refugee communities (International Rescue Committee, 2023).

J.-P. Legerski (✉)
Department of Psychology, University of North Dakota, Grand Forks, ND, USA
e-mail: johnpaul.legerski@und.edu

R. Nedegaard
Department of Social Work Education, California State University, Fresno, CA, USA
e-mail: nedegaard@csufresno.edu

B. L. Isakson
Department of Psychiatry and Behavioral Sciences, University of New Mexico, Albuquerque, NM, USA
e-mail: bisakson@salud.unm.edu

A. Maragakis, M. Janikian (eds.), *Evidence-Based Behavioral Health Practices in Pediatric Specialty Settings*, https://doi.org/10.1007/978-3-031-71004-9_11

The influx of refugees across the globe has coincided with increased recognition of the need for evidence-based treatment practices in psychology (Lilienfeld & Basterfield, 2020), social work (Fennig, 2021), medicine (de Bocanegra et al., 2018), and many other helping professions. Mental health professionals encounter numerous obstacles when integrating evidence-based practice into their work with traumatized refugee youth. The objectives of this chapter encompass three main aspects: (1) offering a comprehensive understanding of the refugee journey for young individuals and their families, while pinpointing factors that influence psychosocial adaptation, (2) identifying and explaining treatment methods that have been assessed within this specific population, and (3) providing practical guidelines on how to successfully implement evidence-based practices (EBPs) in trauma therapy for refugee adolescents in real-world settings.

Youth Refugees and Mental Health

The refugee experience can be divided into three distinct stages: *preflight, flight,* and *resettlement* (Betancourt et al., 2015; Sangalang et al., 2019). Each of these stages presents its own set of challenges and hardships. In the *preflight phase*, before leaving their homes, refugees can be exposed to a multitude of intense stressors, such as war, murder, torture, and rape (Goodkind et al., 2014). During the *flight phase*, when refugees choose or are forced to flee, they may encounter further violence, including the loss of loved ones, separation from family members, displacement from their homes, and the necessity to live in refugee camps. In the *resettlement phase*, refugees often experience stressors such as poverty, unfamiliar systems, separation from loved ones, language barriers, racism/stereotyping from their host community, social alienation, and/or a loss of trust in their parents' ability to provide a safe environment (Chu et al., 2013; Nickerson et al., 2015; Schick et al., 2016).

The variability in circumstances refugee youth encounter is likely an important factor influencing the variability in mental distress across different refugee youth populations. Dangmann et al. (2022) conducted a recent study examining the mental distress experienced by refugee children and adolescents. Their findings revealed that clinical assessment data indicated global rates of post-traumatic stress disorder (PTSD), depression, and anxiety among youth refugees to be 23%, 14%, and 16%, respectively. However, studies relying on self-report data from European countries reported varying rates, with PTSD ranging from 19% to 53%, depression ranging from 10% to 33%, and anxiety ranging from 9% to 32%. Collectively, these rates of poor adjustment are much higher than other child populations.

Discrepancies within and across refugee populations highlight the various trajectories in psychosocial and emotional adjustment. While several models of post-traumatic recovery trajectories have been proposed (Layne et al., 2010, La Greca et al., 2013), La Greca and her colleagues (2013) have provided evidence for the existence of three distinct trajectories following exposure to trauma: resilient, acutely distressed, and chronically distressed. Individuals in the resilient category demonstrate either stable adaptive functioning or a temporary disruption in functioning that eventually leads to full recovery over time. The acutely distressed category includes individuals who experience a more significant impairment in functioning, possibly meeting the criteria for acute stress disorder (ASD). However, these individuals gradually experience a reduction in symptoms and recover in the months that follow. On the other hand, individuals in the chronic category are at a higher risk of meeting the criteria for PTSD. They often exhibit initial signs of ASD during the early stages of recovery.

The challenges refugee youth face during the resettlement process appear to shape their recovery trajectory. Research with adult refugees has shown that resettlement difficulties, such as feelings of isolation, worries about families back home, separation from family members, and language difficulties, significantly contribute to depressive and post-traumatic stress symptoms (PTSS) (Nickerson et al., 2015). Financial difficulties, legal insecurities, immigration status, and other post-migration factors also significantly predict PTSS in immigrant survivors of political violence (Chu et al., 2013). Thus, it is perhaps not surprising that, when interviewing 30 resettled refugees from five countries about their immigration challenges, Mitschke et al., (2017) found that the most consistent themes discussed were practical needs, including social support, acquiring and maintaining employment, language and literacy training, and access to care. These factors may explain why unaccompanied minors, who are likely at the greatest risk of financial insecurity and isolation, have much higher rates of mental distress than youth refugees who migrate with family members (Dangmann et al., 2022).

Social integration into their host society also appears to be an especially important predictor of adjustment for refugee youth. A study conducted by Schick et al. (2016) in Switzerland with refugees seeking treatment for PTSD found that social integration difficulties were more strongly associated with psychological disturbances than educational, visa status, and other immigration difficulties. Furthermore, factors that have been shown to prompt well-being among refugees and asylum seekers include social support; faith, religion, and spirituality; education and training opportunities; employment and economic activities; political advocacy; and environmental conditions (Posselt et al., 2019).

Various distress responses also seem to have different associations with preflight trauma exposure and resettlement factors. While trauma experienced before migration is often linked to both PTSS and depressive symptoms, it appears that a lower socioeconomic status after migration is more consistently associated with depressive symptoms (Bogic et al., 2015; Chu et al., 2013). Consequently, depressive symptoms appear to be more closely connected to current stressors during resettlement, while PTSS is more strongly linked to past traumas encountered before flight (Layne et al., 2010). In fact, a longitudinal study of adult refugees and asylum seekers residing in the United States revealed that improvements in post-migration living conditions over time were accompanied by reductions in depressive symptoms, whereas improvements in these conditions did not predict changes in PTSS (Schick et al., 2018).

The connection between mental health and post-migration challenges is frequently bidirectional, indicating that the factors can impact and intensify one another. While the absence of social support or economic stressors might affect depressive symptoms (Schick et al., 2018), trauma and psychological distress can, in turn, impede refugee adolescents' ability to obtain support by hindering language acquisition, for instance (Iversen et al., 2014). Recognizing and tackling the mutual relationship between mental health and these post-migration social factors is imperative in fostering the general welfare of refugee adolescents and children.

Considering the numerous cultural, economic, and interpersonal stressors that refugees face during the phases of their refugee experience, it is crucial that researchers and clinicians recognize that PTSS, depressive symptoms, and other mental health difficulties experienced by refugees reflect a diverse culmination of circumstances and socioeconomic determinants rather than single events of trauma exposure (Dangmann et al., 2022; Layne et al., 2014). Therefore, initial intervention efforts with refugee youth must begin with a comprehensive and evidence-based evaluation that can account for their cultural differences, exposure to trauma, challenges in resettlement, and behavioral and emotional functioning. This evaluation can aid in monitoring progress and is necessary for clear targets for treatment. The National Child Traumatic Stress Network offers an online database of tools, accompanied by reviews, that aid in the assessment of children's experiences of trauma, their responses to it, and other mental health and trauma-related concerns. Access to this database can be obtained here: https://www.nctsn.org/treatments-and-practices/screening-and-assessments/measure-reviews/all-measure-reviews.

Interventions and Empirical Support for Treating Refugee Youth

Treatment efficacy research with refugee children and adolescents is in its early stages. Kip et al. (2020) conducted a recent meta-analysis of randomized controlled trials (RCTs) with adult refugees, which showed that psychological interventions were effective in reducing symptoms of PTSD and depression in adult refugees. Nocon et al. (2017) conducted a similar meta-analysis focusing on treatment outcome studies with children and adolescents who were refugees exposed to war. Their analysis revealed a scarcity of available high-quality research, with the heterogeneity of these studies posing challenges in interpreting the results. Fazel and Betancourt (2018) have also noted the insufficient attention in existing refugee efficacy research has been given to the dynamic interrelationship of individual mental health and broader family, school, and community systems. It is evident that more research is required to address the multifaceted nature of the refugee experience. Acknowledging these limitations, we will provide an overview of the current state of research on interventions for refugee youth.

Our objective is to support mental health service providers who have good intentions but may lack awareness of the most up-to-date treatment research when working with this distinct and extremely varied group of individuals. We will initiate our examination by exploring universal approaches like Psychological First Aid and Trauma Systems Therapy for Refugees (TST-R). Subsequently, we will concentrate on targeted interventions, including Trauma-Focused Cognitive behavioral Therapy (TF-CBT), Narrative Exposure Therapy (NET), and other interventions that may be required, when universal preventive measures alone seem inadequate in addressing severe maladjustment. After presenting an overview of these prevention and intervention approaches, we will propose additional considerations for implementing them into practice with refugee youth, as well as offer recommendations for clinical supervision and training resources.

Universal Interventions

Universal approaches can be effective when implemented early in the refugee's journey, during preflight, flight, or resettlement. It can be challenging to detect long-term recovery trajectory early on due to high prevalence of acute stress symptoms. Furthermore, early in their refugee journey, child and adolescent survivors, as well as their families, may be too overwhelmed with practical concerns (e.g., shelter and relocation stress) to engage in therapy (Legerski et al., 2012; Mitschke et al., 2017). Universal approaches can provide initial support, initiate engagement, and address early practical needs. If acute mental health problems require immediate attention, targeted interventions may be necessary.

Psychological First Aid

Psychological first aid (PFA) is an approach that aims to provide immediate care and support to individuals who have experienced a traumatic event or crisis (Vernberg et al., 2008). PFA is not a form of therapy, but rather a prevention application designed to stabilize and reduce distress in the initial stages following a crisis by emphasizing practical recommendations to offer comfort, safety, and assistance to people who are in immediate distress. It involves basic principles of human connection, safety, calming, self-efficacy, and connection to appropriate resources. PFA recognizes the importance of addressing the immediate needs of survivors, promoting normalizing activities, providing a safe and comforting environment, and establishing connections with social support. It also recognizes that some individuals and families may require more psychological services than PFA can provide. The principles, objectives, and techniques of PFA can be adapted to facilitate culturally and developmentally appropriate modifications (Grubin et al., 2022; Wang et al., 2022). PFA is often provided by trained mental health professionals, first responders, or volunteers, and can be administered in various settings, including disaster areas, hospitals, schools, or communities affected by crises, and in culturally and developmentally diverse settings (Wang et al., 2021).

A recent analysis of PFA revealed that only 12 studies have been conducted with children and adults that examine the correlation between PFA and indicators of well-being (Hermosilla et al., 2023). These studies included a wide range of populations, such as individuals affected by natural disasters, medical professionals, and law enforcement personnel. Only one study included refugees and young asylum seekers residing in Canada (de Freitas Girardi et al., 2020). The studies used various research methodologies, including quantitative approaches, with only one study using an RCT design. Despite the limited scope of these investigations, the existing evidence suggests that PFA is effective in alleviating symptoms associated with anxiety, depression, post-traumatic stress, and distress. Recent efforts have been made to increase the adaptation and use of PFA with refugees. Readers can find more information on this topic at https://www.hebpsy.net/articles.asp?id=4392.

Trauma Systems Therapy for Refugees (TST-R)

TST-R is an approach that addresses the needs of refugee youth, their families, and their communities through a multitier system approach (Benson et al., 2018). To achieve this objective, TST-R creates interdisciplinary treatment teams consisting of home-based clinicians, an outpatient clinician, school-based clinicians, a clinical supervisor, organizational support persons, and a cultural broker (National Child Traumatic Stress Network, 2016). When necessary, psychopharmacologists and social service agencies are also consulted. The treatment team meets weekly to discuss treatment goals, present cases, and provide supervision and consultation. Cultural brokers are essential members of the treatment team, nominated with the help of community leaders to ensure that they possess the same language, cultural background, and experiences as the refugees. They receive extensive training in childhood trauma and ongoing supervision as part of the TST-R treatment team. Cultural brokers are integrated throughout the TST-R treatment process.

TST-R consists of four tiers of intervention (Benson et al., 2018). In Tier 1, the focus is on community engagement, stigma reduction, and identifying families that could benefit from more intensive support. Cultural brokers and team members actively participate in cultural fairs, community gatherings, and collaborate with schools to promote community engagement. They also offer training and consultation to assist schools and health care providers in effectively addressing the needs of refugees in their communities.

Tier 2 is more structured, involving a 12-week school-based skills group led by a member of the treatment team and a cultural broker. The group sessions focus on improving verbal and nonverbal communication skills, recognizing similarities and differences between their original culture and the host culture, fostering positive peer interactions, enhancing emotional awareness and self-regulation strategies, and acquiring conflict resolution techniques. The group participants can include both refugee and non-refugee peers, which facilitates the development of relationships with individuals from the host country (Fazel, 2017).

Tiers 3 and 4 can be implemented simultaneously to help refugee youth who have experienced trauma. In Tier 3, individual therapy sessions are held in the school setting using cognitive-behavioral approaches to treat PTSS. The team members teach coping strategies to regulate emotions, use cognitive restricting techniques to address maladaptive thoughts, and guide the youth in creating a trauma narrative. In Tier 4, family-based therapy sessions are conducted within the home to focus on ensuring the child's safety. This tier is implemented when the youth refugees face heightened safety risks due to factors within their home environment or due to the severity of their emotional or behavioral dysregulation. Common risk indicators are antisocial behaviors, delinquency, or gang involvement. Both Tiers 3 and 4 involve formal assessments, the development of treatment plans, establishment of safety measures, and the promotion of emotional regulation. The duration of Tiers 3 and 4 also depends on the severity of the trauma and the specific needs of the youth. Kaplin et al. (2019) have provided a comprehensive examination of the various components of TST-R and the level of research support it has received.

Initial research support for TST-R was limited to a study conducted by the TST-R authors themselves, including a study conducted with 30 Somali, Somali-Bantu, and Bhutanese refugees residing in Boston (Benson et al., 2018). Participants were distributed across different tiers of the intervention. The results indicated that participants experienced a decrease in acculturative stress, financial hardships, and perceived discrimination, while also reporting an increase in their sense of belongingness. Additionally, it was found that resource hardships were significantly associated with PTSS. However, the impact on mental health varied. Although there was a reduction in PTSS from baseline to the 6- and 12-month follow-ups across all tiers, improvements in mental health and resources did not differ significantly across Tiers 2–4. Furthermore, there were no differences in overall depression scores over time.

More recently, Cardeli et al. (2020) conducted a study involving 103 clients and parent refugees from 17 different countries living in Maine. This study specifically focused on the crucial role of cultural brokers. The findings demonstrated that when cultural brokers were included in the treatment, there was a significant decrease in clients' self-reported emotional and behavioral problems, which was not observed in participants without cultural brokers. Despite these results, further research is needed to explore the effectiveness of TST-R with other refugee groups.

Targeted Interventions

TF-CBT

TF-CBT was created as a structured intervention for children and adolescents who experienced sexual abuse, but it has been modified and proven effective in treating PTSS resulting from various types of trauma (Cohen et al., 2004, 2012). The treatment protocol is designed based on the individual needs of the client, and it includes 12–20 90-min sessions. The treatment involves gradually and systematically addressing traumatic memories and associated triggers, while also maximizing support from caregivers. The treatment is divided into three distinct phases, which are stabilization, trauma narration and processing, and integration and consolidation. These phases encompass a total of eight components that are identified by the acronym "PRACTICE".

1. *Psychoeducation and Parenting skills:* The therapist educates the client and caregivers about trauma and its effects, and provides a treatment overview.
2. *Relaxation:* The therapist teaches the child to manage emotional distress and anxiety with relaxation techniques, deep breathing, and mindfulness.
3. *Affective Expression and Regulation*: The therapist helps children express their difficult emotions in a healthy way, rather than acting out.
4. *Cognitive Coping:* The therapist helps the child identify and challenge negative thoughts associated with trauma to manage their emotions and behaviors.
5. *Trauma Narrative Development and Processing*: The therapist guides the child toward processing the traumatic event through storytelling or written accounts. The ultimate goal is to help the child confront and integrate their memories in a safe and controlled environment.
6. In Vivo *Gradual Exposure*: Once the child's coping skills have been developed, the therapist collaborates with the client and their caregivers to create an exposure hierarchy. This hierarchy is used as a guide to gradually and systematically expose the child to trauma-related situations or reminders. The process enables the child to confront and manage their fears in a regulated manner within a secure and nurturing environment.
7. *Conjoint Parent–Child sessions:* The therapist encourages the child to share traumatic experiences and treatment gains with caregivers. Caregiver sessions are also used to establish treatment suitability, safety plans, and parenting techniques to aid progress.
8. *Enhancing Safety and Future Development:* In the final component of treatment, the therapist focuses on identifying potential hazards that children might face after the treatment is complete. Personal safety plans are frequently created in collaboration with caregivers to ensure the children's safety.

TF-CBT is a flexible therapy that works for various cultural and ethnic groups. It has been effective with African American, Hispanic, American Indian, and Alaska Native children. It has also been successful in other countries like Congo, Zambia, and Jordan. TF-CBT is useful for group and school-based therapy too (see Ennis et al., 2020 for a systematic review). Currently, there is limited research on the effectiveness of reducing trauma symptoms in refugee children.

Chipalo (2021) recently conducted a comprehensive analysis of the application of TF-CBT in treating traumatized refugee children. The study identified four published studies on the subject, three of which were conducted in Germany. These studies include a case study involving a 17-year-old girl from East Africa (Unterhitzenberger & Rosner, 2016), an uncontrolled pilot study with primarily Afghan youth (Unterhitzenberger et al., 2019), and an RCT involving Somali, Iranian, and Afghan youth (Unterhitzenberger et al., 2015). Additionally, an RCT was conducted in the United States with youth participants from various countries including Nepal, Russia, Bosnia, Iraq, Somalia, and Turkey (Schottelkorb et al., 2012). All studies demonstrated that TF-CBT was effective in reducing PTSS. However, the limited sample size, absence of active control groups, and heterogeneity of the samples suggest that these findings should be interpreted with caution. Further research is necessary to establish stronger evidence regarding the effectiveness of adapting TF-CBT for use in refugee populations.

NET

NET is a short-term variant of Cognitive Behavioral Therapy (CBT) developed by Maggie Schauer et al. (2011). It is mainly used to treat adults and children, including refugees and trauma survivors, who suffer from PTSD or related symptoms. The primary goal of NET is to help individuals create a detailed chronological narrative of their life focusing on both positive and negative experiences.

During NET, the therapist guides the client through the process of constructing a timeline of their life events, with a special emphasis on traumatic incidents. Clients are encouraged to recount their life stories in chronological order, paying particular attention to traumatic events which allows them to gradually confront and process these experiences in a safe environment. The therapist helps the client revisit past events and emotions related to those events, to reprocess traumatic memories through desensitization and reinterpretation. The therapy usually lasts for a limited number of sessions, ranging between 8 and 16.

In a meta-analysis of the efficacy of NET for treating PTSD in adults (Lely et al., 2019), NET showed greater effectiveness than both non-active and non-trauma-focused active comparators; however, the number of studies using trauma-focused comparisons was limited and more were needed. Age appeared to be an important moderator, with older adults displaying better treatment results. Individuals with migration histories also showed relatively worse treatment outcomes.

KIDNET, an 11-session child and adolescent version of NET, has been developed (Schauer et al., 2017). In this version, a therapist helps young refugees to create an illustrated and chronological narrative of their life history. This narrative helps them express their emotions, thoughts, and sensory experiences related to different life events. As an interactive component of this therapy intervention, the youth use flowers to symbolize positive events and rocks to symbolize negative events. These symbols are placed along a rope that serves as a timeline of their life journey. Additionally, parents or caregivers are involved in three sessions of the intervention where the therapist provides psychoeducation and facilitates discussions on current challenges and treatment objectives.

There has been less attention paid to the effectiveness of KIDNET in comparison to its adult counterpart, but research is starting to emerge. One of the initial studies conducted on KIDNET was a pilot study that involved six Somali children, aging from 12 to 17 years, residing in a refugee settlement in Uganda (Onyut et al., 2005). These children received four to six individual sessions of KIDNET, and their PTSD levels were assessed before, after, and 9 months after treatment. All six participants had moderate to severe PTSD at the start of the study. Although no comparison group was used, each of the participants completed the intervention and four out of the six patients no longer met the criteria for PTSD at 9 months. In another study, 26 refugee children living in Germany were randomly assigned to either the KIDNET intervention or a waiting list (Ruf et al., 2010). The results revealed significant treatment effects over time on all variables related to PTSD, indicating that the KIDNET group experienced a clinically significant improvement in symptoms and functioning, unlike the control group. Furthermore, the positive outcomes observed in the KIDNET group were sustained at the 12-month follow-up.

In a more recent study, an RCT was conducted with a group of 20 adolescent North Korean refugees residing in Seoul, South Korea (Park et al., 2020). The study aimed to compare the outcomes of using KIDNET versus treatment as usual (TAU) for addressing PTSD, depression, and sleep issues. The participants in the NET group experienced significantly greater improvements in all three areas compared to those in the TAU group. Furthermore, the reduction in PTSD symptoms was sustained after a 6-month follow-up period. The remaining two individuals had borderline PTSD scores. While results provide preliminary indicators that KIDNET may be an effective intervention for the treatment of PTSD among youth refugees in a variety of contexts, additional high-quality RCT studies are needed. Efficacy trials for KIDNET are currently underway with refugees living in Germany (Wilker et al., 2020) and in the Netherlands (Velu et al., 2022).

Mein Weg (My Way)

Mein Weg (My Way) is a brief, standardized trauma-focused group intervention specifically tailored for unaccompanied refugee minors (URMs) aged 13–21-years old (Pfeiffer & Goldbeck, 2017). Mein Weg is designed to be implemented by trained and supervised social workers for six to eight 90-min sessions and shares many key components of TF-CBT and similar interventions, such as psychoeducation, relaxation techniques, cognitive restructuring, and the use of trauma narratives. However, what sets Mein Weg apart is the intervention group format structure, which provides URMs the opportunity to openly discuss their experiences, emotions, and challenges with their peers.

Preliminary research on the effectiveness of Mein Weg, a relatively new trauma intervention, shows promising results. In a study conducted by Pfeiffer and Goldbeck (2017) in Germany, the intervention was tested on 29 URMs aged 14–18, most of whom were originally from Afghanistan. Using a pre/posttest design, the researchers observed a decrease in PTSS, particularly reexperiencing and avoidance symptoms. In a larger follow-up study, Pfeiffer et al. (2019) conducted an RCT with 50 male participants aged 14–19, who were URM originally from Middle Eastern and African countries residing in Southern Germany. The participants were randomly assigned to either the Mein Weg group or a TAU group. The results from the RCT's initial posttreatment and 3-month follow-up demonstrated significantly greater reductions in PTSS and depression symptoms among the Mein Weg group.

EMDR

Eye movement desensitization and reprocessing (EMDR) is a therapeutic technique that involves clients recalling distressing images, memories, or thoughts while simultaneously undergoing bilateral stimulation. This stimulation can be achieved through side-to-side eye movements, tapping, or auditory cues, all guided by the therapist. The underlying theory of EMDR is that these two components help the brain reprocess these distressing memories, reducing their emotional intensity. Several meta-analyses evaluating EMDR with children and adolescents have demonstrated its effectiveness (Bastien et al., 2020; Moreno-Alcazar et al., 2017). It has been found to be superior to waitlist or placebo conditions and has shown comparable efficacy to CBT in reducing post-traumatic and anxiety symptoms.

The research on EMDR therapy with refugee youth is still in its early stages and much more needs to be explored. Oras et al. (2004) conducted a study on 13 refugee youth who underwent EMDR therapy and psychodynamic art therapy in Sweden. The study found significant improvements in global functional levels and a reduction in PTSD-related symptoms after treatment. Perilli et al. (2019) conducted a more recent study on 14 refugees at a Turkey orphanage. The study used EMDR therapy in three sessions and assessed eight children posttreatment. Despite challenges like limited availability of translators and high attrition, the study showed encouraging results. There was a decrease in PTSD symptoms and an improvement in psychological well-being 45 days after the therapy sessions. Lempertz et al. (2020) conducted a similar study to evaluate the effectiveness of EMDR therapy on war refugee children living in Germany. Using a group format, the study involved 10 preschool children aged 4–6 years who were enrolled in daycare centers. At an immediate and 3-month follow-up, ratings from parents and teachers demonstrated a significant reduction in PTSD symptoms. In one EMDR RCT study with refugee youth, Banoğlu and Korkmazlar (2022) assessed the effectiveness of group EMDR therapy on Syrian refugee youth aged 6–15 years residing in Istanbul, Turkey. The study consisted of 61 participants, with 42 assigned to either the EMDR treatment group or to a waiting list control. Banoğlu and Korkmazlar discovered that the EMDR treatment group exhibited significant reductions in both depression and

PTSD symptoms compared to the waiting list group. However, no long-term follow-up results were collected.

Considerations for Implantation of Evidence-Based Refugee Treatments into Practice

Treating mental health issues in youth refugees demands a comprehensive and culturally sensitive approach. Due to the complexity of these cases, collaboration with interdisciplinary teams is essential, considering the specific cultural context of each refugee. Cultural awareness is paramount, considering that refugees and minority groups may have a limited concept of mental health, leading to underutilization of mental health services due to stigma or practical issues like transportation limitations. Therefore, understanding and respecting the cultural background, traditions, and values of the refugee population is crucial, especially considering the impact of acculturation stress and discrimination on mental health. These factors compound the trauma experienced by refugee children as they adapt to a new culture, environment, and expectations.

Examining barriers to care is vital, and once identified, advocacy becomes essential for the rights and needs of youth refugees, including access to mental health services, education, and social support. Addressing systemic barriers that may impede their well-being is crucial. Understanding obstacles associated with each agency or practice can significantly enhance engagement. Collaborating or consulting with other service providers working with this community not only shares ideas but also enhances professionals' understanding of available community services.

Understanding the needs of children and youth is key to effectiveness. Assessments must consider trauma history, the child's background, past school experience, and current social and environmental stressors. Refugee children may present with a range of symptoms and behaviors, with a propensity to report physical manifestations of trauma. Recognizing intergenerational trauma and its impact on refugee children is essential. Cultural considerations may reveal disciplinary strategies, such as corporal punishment, making education on the US child welfare system crucial to prevent potentially traumatizing involvement.

When post-traumatic stress, depression, and other related symptoms are present and intervention is necessary, we recommend adopting an evidence-based practice framework that considers client variables and preferences, clinical expertise, and the best available research. Currently, selecting formal interventions with the most available research support is challenging. Based on the efficacy studies described in this chapter, it is clear that more research is needed with larger sample sizes of diverse groups of refugee youth, active control groups, long-term follow-up analyses, and a greater emphasis on understanding the essential components of trauma interventions.

Among the limited studies that do exist, Emel Genç (2022) found that the KIDNET and EMDR appear to have the largest effect sizes and support from RCTs. KIDNET was singled out for its cultural adaptability and methodological flexibility, which allowed for its implementation across various contexts. While EMDR has large effect sizes, the use of EMDR is not without controversy. Research with adult individuals diagnosed with PTSD has shown that bilateral eye movements offer no additional benefit compared to fixating on a nonmoving stimulus (Sack et al., 2016). These and other similar findings have caused Rosenfeld and McLean (2024) to argue that EMDR is among a unique subset of interventions that take an evidence-based approach (i.e., exposure therapy) and distinguish it by adding questionable pseudoscientific components (i.e., eye movements). There are also concerns that EMDR intervention may cause important disruptions in memories. Evidence suggests that EMDR decreases the vividness of autobiographical memories and increases the production of spontaneous false memories (Otgaar et al., 2022; Kenchel et al., 2022). These negative influences may make EMDR intervention ill-advised for refugee youth, especially for those who may be involved in legal proceedings when the accuracy of memory is needed.

In a recent systematic review by Chipalo (2021), it was found that only four high-quality studies have evaluated the efficacy of TF-CBT with refugees, out of which only one was an RCT conducted by Schottelkorb et al. (2012). Despite the small sample size of 64 youth from 21 different countries, these studies provide preliminary evidence that TF-CBT can be effectively adapted and used with refugee youth. Surprisingly, there is a lack of research on TF-CBT among refugees, considering that it is one of the most commonly used and evidence-based interventions for traumatized youth in other populations and has been shown to be marginally more effective than EMDR (Lewey et al., 2018). The lack of TF-CBT research with refugee populations may be the result of logistical challenges of conducting TF-CBT RCT with refugees, namely the fact that TF-CBT requires both child and caregiver sessions, which may be a barrier for unaccompanied minors, or parents who are hesitant to participate in treatment due to language or other barriers.

There may be some instances where trauma-focused school-based inventions alone, with limited parental/caregiver involvement, may be necessary, but interventions will be most effective with family and caregiver involvement. For these and other reasons, cultural brokers and/or interpreters should be used to facilitate the integration of refugee children and families into agencies or practices. These professionals help providers understand existing roles within refugee families, fostering respect and appreciation for their culture and experiences. Refugee families exhibit resilience and strength amid numerous obstacles, and engaging them in evidence-based trauma treatment processes strengthens their support systems.

Clinical Supervision with Refugee Youth

Providing high-quality, evidence-based care to refugee youth and families is challenging and complex because of the stressors they face rebuilding their lives in a new culture (Fondacaro & Harder, 2014; Isakson et al., 2015). It can take many years to become an expert in refugee mental health. Therefore, receiving high-quality supervision and consultation is critical to providing care that meets the needs of refugee families. Supervision should include support on providing culturally sensitive services, addressing specific clinical issues that are unique to refugee families, and support the trainee to manage the stress through self-care and systems support.

It is important to have trainees do work even before meeting with a refugee client to ensure that they are as prepared as possible. This includes learning about the refugee experience and immigration process in general, local context for refugee families, and what resources there are for refugees in the community, including an understanding of the various systems with which they interact, e.g., health care systems, refugee resettlement and other refugee serving agencies, and education systems (Fondacaro & Harder, 2014; Mattar et al., 2023). Connecting trainees with cultural brokers before treatment begins can help the trainee understand the local context as well as allow them to ask questions about the culture that they would not ask a client (Fondacaro & Harder, 2014).

Supervisors should help trainees develop or enhance cultural humility and sensitivity. They should be prepared to seek to understand their client's worldview of health and healing and look for culturally specific symptoms (Isakson et al., 2015). They should develop an awareness of the importance of somatic symptoms and chronic pain and how they impact physical and mental health (Apostolidou & Schweitzer, 2017).

In addition to high-quality training in EBPs, trainees working with refugees need to have basic competencies in nonspecific common therapeutic factors, including establishing a therapeutic alliance, empathic listening, case conceptualization and role of a diagnosis in health care, and use of screening tools and how to interpret them for groups that are not normed (Fondacaro & Harder, 2014; Isakson et al., 2015). We recommend that any EBP chosen should be embedded within the ecological model to get a broad understanding of the barriers and facilitators to health and resilience and how therapy can support other interventions (Fondacaro & Harder, 2014; Mattar et al., 2023). It is critical that the therapy process start by creating a safe, non-traumatizing therapeutic environment (Isakson et al., 2015; Mattar et al., 2023; Miller et al., 2019). Finally, trainees need training on working effectively with interpreters, whether that is in person or by telephone or video (Isakson et al., 2015).

Supervision sessions should be designed to promote a safe and supportive environment that promotes self-reflection and partnership, with the expectation and allowance for mistakes to be made, to facilitate trainee competence. Provider resilience, burnout, and self-care should also be discussed before meetings with clients so that trainees are able to develop their own plans for self-care and know how to access support when needed (Mattar et al., 2023).

All of these topics should be continually discussed and monitored in supervision. Once psychotherapy begins, supervisors should help trainees have sensitive conversations with their clients about confidentiality, the purpose of therapy, and their client's view on healing from their cultural perspective (Isakson et al., 2015). Trainees should learn about the culture and political context of the countries their clients are from and be coached on how to ask questions about their client's understanding of these factors, as well as be willing to process their own cultural assumptions and biases when faced with cultural practices of the clients that do not align with their values, such as female genital mutilation, the role of women, and use of corporal punishment (Apostolidou & Schweitzer, 2017). Trainees will need to consider their clients' goals and hopes for therapy when choosing which EBPs to implement, as many refugees may be more focused on instrumental needs of safety and stability rather than symptom reduction or addressing past trauma (Apostolidou & Schweitzer, 2017; Isakson et al., 2015; Mattar et al., 2023). Once those needs are met and trust is established, clients may be ready for EBPs for symptom reduction. Trainees may be frustrated that they are not able to start right away with their EBP of choice and supervisors can help them process this frustration.

Training Opportunities and Resources

When working with refugee youth and their families, we encourage practitioners to follow guidelines provided by the National Child Traumatic Stress Network (NCTSN, 2016), which emphasize the importance of considering the trauma history of children and their families, and using culturally informed and linguistically sensitive services, strategies for access and engagement of children, and approaches for managing resettlement stressors. These guidelines are discussed in greater detail in a free online resource for mental health professionals created by the NCTSN titled, *Understanding Refugee Trauma* which can be downloaded on their website (2018). The NCTSN website serves as a clearinghouse of information and training resources for the assessment and treatment of child and adolescent traumatic stress reactions.

The British Psychological Society has also created an exhaustive report on recommendations for psychologists working with refugees and asylum seekers in the United Kingdom (Patel et al., 2018). This report provides a summary of the challenges of refugee youth, a helpful list of terminology, and detailed instructions for providing support within their communities, with specific recommendations for working with schools, nurseries, employers, and universities. Links to these resources and others are provided below.

Conclusion

The effective application of psychological interventions for refugee children and adolescents requires mental health practitioners to understand the unique and complex challenges they face. This is especially important given the influx of refugee children and their families across the globe. It is crucial to prioritize the mental health needs of refugee youth and advocate for inclusive and accessible support systems to ensure their holistic development and long-term success. By creating a safe environment for healing, incorporating culturally sensitive approaches, facilitating community social integration, and leveraging evidence-based techniques, we can help refugees alleviate symptoms of distress and empower them to thrive in their new communities.

Select Bibliography

Apostolidou, Z., & Schweitzer, R. (2017). Practitioners' perspectives on the use of clinical supervision in their therapeutic engagement with asylum seekers and refugee clients. *British Journal of Guidance & Counselling, 45*(1), 72–82. https://doi.org/10.1080/03069885.2015.1125852. An Australian study evaluating practitioners' perspectives on their use of clinical supervision when working with asylum seekers and refugees. Provides useful recommendations for practitioners who may be supervising trainees working with asylum seekers and refugees.

Frounfelker, R. L., Miconi, D., Farrar, J., Brooks, M. A., Rousseau, C., & Betancourt, T. S. (2020). Mental health of refugee children and youth: Epidemiology, interventions, and future directions. *Annual Review of Public Health, 41*, 159–176. https://doi.org/10.1146/annurev-publhealth-040119-094230. Public health review article with prevention and intervention recommendations for refugee youth. These helpful recommendations are separated by age, phase of resettlement, and setting.

National Child Traumatic Stress Network. (2018). *Understanding Refugee Trauma: For mental health professionals* [Pamphlet]. https://www.nctsn.org/resources/understanding-refugee-trauma-mental-health-professionals. A brief pamphlet that can be printed out and distributed to mental health providers and practitioners in related fields that may work with refugees and asylum seekers. This resource, available online, provides tips and brief summarizations of cultural, youth, family, and provider resources.

Patel, N., Tribe, R., & Yule, B. (2018). *Guidelines for psychologists working with Refugees and Asylum-seekers in the UK: Extended version.* The British Psychological Society. https://www.bps.org.uk/guideline/guidelines-psychologists-working-refugees-and-asylum-seekers-uk. A detailed report with important resources for directors, managers, and practitioners of organizations working with refugees and migrants. Provides instruction for accessing evidence-based psychology to aid communities in ongoing crisis.

Posselt, M., Eaton, H., Ferguson, M., Keegan, D., & Procter, N. (2019). Enablers of psychological well-being for refugees and asylum seekers living in transitional countries: A systematic review. *Health & Social Care in the Community, 27*(4), 808–823. https://doi.org/10.1111/hsc.12680. A systematic review on psychological well-being of refugees and asylum seekers, including discussion of social support; faith, religion and spirituality; cognitive strategies; education and training opportunities; employment and economic activities; behavior strategies; political advocacy; and environmental conditions. Providers and policymakers must consider these factors aid refugees in not only surviving but thriving in resettlement.

References

Banoğlu, K., & Korkmazlar, Ü. (2022). Efficacy of the eye movement desensitization and reprocessing group protocol with children in reducing posttraumatic stress disorder in refugee children. *European Journal of Trauma & Dissociation, 6*(1), 1–9. https://doi.org/10.1016/j.ejtd.2021.100241

Bastien, R. J. B., Jongsma, H. E., Kabadayi, M., & Billings, J. (2020). The effectiveness of psychological interventions for post-traumatic stress disorder in children, adolescents and young adults: A systematic review and meta-analysis. *Psychological Medicine, 50*(10), 1598–1612. https://doi.org/10.1017/s0033291720002007

Benson, M. A., Abdi, S. M., Miller, A. B., & Ellis, B. H. (2018). Trauma systems therapy for refugee children and families. In N. Morina & A. Nickerson (Eds.), *Mental health of refugee and conflict-affected populations* (pp. 243–259). Springer. https://doi.org/10.1007/978-3-319-97046-2_12

Betancourt, T. S., Abdi, S., Ito, B. S., Lilienthal, G. M., Agalab, N., & Ellis, H. (2015). We left one war and came to another: Resource loss, acculturative stress, and caregiver–child relationships in Somali refugee families. *Cultural Diversity and Ethnic Minority Psychology, 21*(1), 114–125. https://doi.org/10.1037/a0037538

Bogic, M., Njoku, A., & Priebe, S. (2015). Long-term mental health of war-refugees: A systematic literature review. *BMC International Health and Human Rights, 15*(1), 1–41. https://doi.org/10.1186/s12914-015-0064-9

Cardeli, E., Davis, S., Issa, O., & Ellis, B. H. (2020). Trauma systems therapy for refugees (TST-R) in a multi-ethnic refugee population in Maine, USA: Early findings from a programme evaluation project with children and adolescents. *The Lancet Global Health, 8*, S9. https://doi.org/10.1016/S2214-109X(20)30150-9

Chipalo, E. (2021). Is trauma focused-cognitive behavioral therapy (TF-CBT) effective in reducing trauma symptoms among traumatized refugee children? A systematic review. *Journal of Child & Adolescent Trauma, 14*(4), 545–558. https://doi.org/10.1007/s40653-021-00370-0

Chu, T., Keller, A. S., & Rasmussen, A. (2013). Effects of post-migration factors on PTSD outcomes among immigrant survivors of political violence. *Journal of Immigrant and Minority Health, 15*, 890–897. https://doi.org/10.1007/s10903-012-9696-1

Cohen, J. A., Deblinger, E., Mannarino, A. P., & Steer, R. A. (2004). A multisite, randomized controlled trial for children with sexual abuse-related PTSD symptoms. *Journal of the American Academy of Child and Adolescent Psychiatry, 43*, 393–402. https://doi.org/10.1097/00004583-200404000-00005

Cohen, J. A., Mannarino, A. P., & Deblinger, E. (2012). *Trauma-focused CBT for children and adolescents*. Guilford Press.

da Silva Rebelo, M. J., Fernández, M., & Achotegui, J. (2018). Mistrust, anger, and hostility in refugees, asylum seekers, and immigrants: A systematic review. *Canadian Psychology/Psychologie Canadienne, 59*(3), 239. https://doi.org/10.1037/cap0000131

Dangmann, C., Dybdahl, R., & Solberg, Ø. (2022). Mental health in refugee children. *Current Opinion in Psychology, 48*, 101460. https://doi.org/10.1016/j.copsyc.2022.101460

de Bocanegra, H. T., Carter-Pokras, O., Ingleby, J. D., Pottie, K., Tchangalova, N., Allen, S. I., et al. (2018). Addressing refugee health through evidence-based policies: A case study. *Annals of Epidemiology, 28*(6), 411–419. https://doi.org/10.1016/j.annepidem.2017.05.010

de Freitas Girardi, J., Miconi, D., Lyke, C., & Rousseau, C. (2020). Creative expression workshops as Psychological First Aid (PFA) for asylum-seeking children: An exploratory study in temporary shelters in Montreal. *Clinical Child Psychology and Psychiatry, 25*(2), 483–493. https://doi.org/10.1177/1359104519891760

Ennis, N., Shorer, S., Shoval-Zuckerman, Y., Freedman, S., Monson, C. M., & Dekel, R. (2020). Treating posttraumatic stress disorder across cultures: A systematic review of cultural adaptations of trauma-focused cognitive behavioral therapies. *Journal of Clinical Psychology, 76*(4), 587–611. https://doi.org/10.1002/jclp.22909

Fazel, M. (2017). Psychological and psychosocial interventions for refugee children resettled in high-income countries. *Epidemiology and Psychiatric Sciences, 27*(2), 117–123. https://doi.org/10.1017/s2045796017000695

Fazel, M., & Betancourt, T. S. (2018). Preventive mental health interventions for refugee children and adolescents in high-income settings. *The Lancet Child & Adolescent Health, 2*(2), 121–132. https://doi.org/10.1016/S2352-4642(17)30147-5

Fennig, M. (2021). Cultural adaptations of evidence-based mental health interventions for refugees: Implications for clinical social work. *The British Journal of Social Work, 51*(3), 964–981. https://doi.org/10.1093/bjsw/bcaa024

Fondacaro, K. M., & Harder, V. S. (2014). Connecting cultures: A training model promoting evidence-based psychological services for refugees. *Training and Education in Professional Psychology, 8*(4), 320. https://doi.org/10.1037/tep0000071

Genç, E. (2022). The effectiveness of trauma treatment approaches in refugee children and adolescents: A systematic review. *Children and Youth Services Review, 141*, 1–8. https://doi.org/10.1016/j.childyouth.2022.106602

Goodkind, J. R., Hess, J. M., Isakson, B., LaNoue, M., Githinji, A., Roche, N., et al. (2014). Reducing refugee mental health disparities: A community-based intervention to address postmigration stressors with African adults. *Psychological Services, 11*(3), 333. https://doi.org/10.1037/a0035081

Grubin, F., Maudrie, T. L., Neuner, S., Conrad, M., Waugh, E., Barlow, A., et al. (2022). Development and cultural adaptation of psychological first aid for COVID-19 frontline workers in American Indian/Alaska Native communities. *Journal of Prevention, 43*(5), 697–717. https://doi.org/10.1007/s10935-022-00695-y

Hermosilla, S., Forthal, S., Sadowska, K., Magill, E. B., Watson, P., & Pike, K. M. (2023). We need to build the evidence: A systematic review of psychological first aid on mental health and well-being. *Journal of Traumatic Stress, 36*(1), 5–16. https://doi.org/10.1002/jts.22888

International Rescue Committee. (2023). *Refugee facts, Statistics and FAQs*. https://www.rescue.org/article/facts-about-refugees-key-facts-faqs-and-statistics

Isakson, B. L., Legerski, J. P., & Layne, C. M. (2015). Adapting and implementing evidence-based interventions for trauma-exposed refugee youth and families. *Journal of Contemporary Psychotherapy, 45*, 245–253. https://doi.org/10.1007/s10879-015-9304-5

Iversen, V. C., Sveaass, N., & Morken, G. (2014). The role of trauma and psychological distress on motivation for foreign language acquisition among refugees. *International Journal of Culture and Mental Health, 7*(1), 59–67. https://doi.org/10.1080/17542863.2012.695384

Kaplin, D., Parente, K., & Santacroce, F. A. (2019). A review of the use of trauma systems therapy to treat refugee children, adolescents, and families. *Journal of Infant, Child, and Adolescent Psychotherapy, 18*(4), 417–431. https://doi.org/10.1080/15289168.2019.1687220

Kenchel, J. M., Domagalski, K., Butler, B. J., & Loftus, E. F. (2022). The messy landscape of eye movements and false memories. *Memory, 30*(6), 678–685. https://doi.org/10.1080/09658211.2020.1862234

Kip, A., Priebe, S., Holling, H., & Morina, N. (2020). Psychological interventions for posttraumatic stress disorder and depression in refugees: A meta-analysis of randomized controlled trials. *Clinical Psychology & Psychotherapy, 27*(4), 489–503.

La Greca, A. M., Lai, B. S., Llabre, M. M., Silverman, W. K., Vernberg, E. M., & Prinstein, M. J. (2013). Children's postdisaster trajectories of PTS symptoms: Predicting chronic distress. *Child & Youth Care Forum, 42*, 351–369. https://doi.org/10.1007/s10566-013-9206-1

Layne, C., Olsen, J. A., Baker, A., Legerski, J., Isakson, B., Pasalic, A., & Pynoos, R. S. (2010). Unpacking trauma exposure risk factors and differential pathways of influence: Predicting postwar mental distress in Bosnian adolescents. *Child Development, 81*, 1053–1076.

Layne, C. M., Steinberg, J. R., & Steinberg, A. M. (2014). Causal reasoning skills training for mental health practitioners: Promoting sound clinical judgment in evidence-based practice. *Training and Education in Professional Psychology, 8*, 292–302. https://doi.org/10.1037/tep0000037

Legerski, J. P., Vernberg, E. M., & Noland, B. J. (2012). A qualitative analysis of barriers, challenges, and successes in meeting the needs of Hurricane Katrina evacuee families. *Community Mental Health Journal, 48*, 729–740.

Lely, J. C. G., Smid, G. E., Jongedijk, R. A., Knipscheer, J. W., & Kleber, R. J. (2019). The effectiveness of narrative exposure therapy: A review, meta-analysis and meta-regression analysis. *European Journal of Psychotraumatology, 10*(1), 1550344. https://doi.org/10.1080/20008198.2018.1550344

Lempertz, D., Wichmann, M., Enderle, E., Stellermann-Strehlow, K., Pawils, S., & Metzner, F. (2020). Pre-post study to assess EMDR-based group therapy for traumatized refugee preschoolers. *Journal of EMDR Practice & Research, 14*(1), 31. https://doi.org/10.1891/1933-3196.14.1.31

Lewey, J. H., Smith, C. L., Burcham, B., Saunders, N. L., Elfallal, D., & O'Toole, S. K. (2018). Comparing the effectiveness of EMDR and TF-CBT for children and adolescents: A meta-analysis. *Journal of Child & Adolescent Trauma, 11*, 457–472. https://doi.org/10.1007/s40653-018-0212-1

Lilienfeld, S., & Basterfield, C. (2020). *History of evidence-based practice*. Oxford Research Encyclopedia of Psychology. https://doi.org/10.1093/acrefore/9780190236557.013.633

Mattar, S., Ipekci, B., Louis, E. F., & Taknint, J. T. (2023). Training considerations for providers engaged in refugee mental health care. *International Perspectives in Psychology: Research, Practice, Consultation, 12*(1), 16. https://doi.org/10.1027/2157-3891/a000057

Miller, K. K., Brown, C. R., Shramko, M., & Svetaz, M. V. (2019). Applying trauma-informed practices to the care of refugee and immigrant youth: 10 clinical pearls. *Children, 6*(8), 94. https://doi.org/10.3390/children6080094

Mitschke, D. B., Praetorius, R. T., Kelly, D. R., Small, E., & Kim, Y. K. (2017). Listening to refugees: How traditional mental health interventions may miss the mark. *International Social Work, 60*(3), 588–600. https://doi.org/10.1177/0020872816648256

Moreno-Alcazar, A., Treen, D., Valiente-Gomez, A., Sio-Eroles, A., Perez, V., Amann, B. L., & Radua, J. (2017). Efficacy of eye movement desensitization and reprocessing in children and adolescent with post-traumatic stress disorder: A meta-analysis of randomized controlled trials. *Frontiers in Psychology, 8*, 1–10. https://doi.org/10.3389/fpsyg.2017.01750

National Child Traumatic Stress Network. (2016). *TST-R culture specific fact sheet.* Retrieved from https://www.nctsn.org/sites/default/files/interventions/tstr_culture_specific_fact_sheet.pdf

Nickerson, A., Bryant, R. A., Schnyder, U., Schick, M., Mueller, J., & Morina, N. (2015). Emotion dysregulation mediates the relationship between trauma exposure, post-migration living difficulties and psychological outcomes in traumatized refugees. *Journal of Affective Disorders, 173*, 185–192. https://doi.org/10.1016/j.jad.2014.10.043

Nocon, A., Eberle-Sejari, R., Unterhitzenberger, J., & Rosner, R. (2017). The effectiveness of psychosocial interventions in war-traumatized refugee and internally displaced minors: Systematic review and meta-analysis. *European Journal of Psychotraumatology, 8*(sup2). https://doi.org/10.1080/20008198.2017.1388709

Onyut, L. P., Neuner, F., Schauer, E., Ertl, V., Odenwald, M., Schauer, M., & Elbert, T. (2005). Narrative exposure therapy as a treatment for child war survivors with posttraumatic stress disorder: Two case reports and a pilot study in an African refugee settlement. *BMC Psychiatry, 5*, 1–9. https://doi.org/10.1186/1471-244x-5-7

Oras, R., Ezpeleta, S. C. D., & Ahmad, A. (2004). Treatment of traumatized refugee children with eye movement desensitization and reprocessing in a psychodynamic context. *Nordic Journal of Psychiatry, 58*(3), 199–203. https://doi.org/10.1080/08039480410006232

Otgaar, H., Curci, A., Mangiulli, I., Battista, F., Rizzotti, E., & Sartori, G. (2022). A court ruled case on therapy-induced false memories. *Journal of Forensic Sciences, 67*(5), 2122–2129. https://doi.org/10.1111/1556-4029.15073

Park, J. K., Park, J., Elbert, T., & Kim, S. J. (2020). Effects of narrative exposure therapy on posttraumatic stress disorder, depression, and insomnia in traumatized North Korean refugee youth. *Journal of Traumatic Stress, 33*(3), 353–359. https://doi.org/10.1002/jts.22492

Perilli, S., Giuliani, A., Pagani, M., Mazzoni, G. P., Maslovaric, G., Maccarrone, B., et al. (2019). EMDR group treatment of children refugees—A field study. *Journal of EMDR Practice and Research, 13*(2), 143–155. https://doi.org/10.1891/1933-3196.13.2.143

Pfeiffer, E., & Goldbeck, L. (2017). Evaluation of a trauma-focused group intervention for unaccompanied young refugees: A pilot study. *Journal of Traumatic Stress, 30*(5), 531–536. https://doi.org/10.1002/jts.22218

Pfeiffer, E., Sachser, C., Tutus, D., Fegert, J. M., & Plener, P. L. (2019). Trauma-focused group intervention for unaccompanied young refugees: "Mein Weg"—Predictors of treatment outcomes and sustainability of treatment effects. *Child and Adolescent Psychiatry and Mental Health, 13*, 1–10. https://doi.org/10.1186/s13034-019-0277-0

Rosenfeld, E. A., & McLean, C. P. (2024). Purple hat therapies. In J. N. Stea & S. Hupp (Eds.), *Investigating clinical psychology: Pseudoscience, fringe science, and controversies* (pp. 116–126). Routledge.

Ruf, M., Schauer, M., Neuner, F., Catani, C., Schauer, E., & Elbert, T. (2010). Narrative exposure therapy for 7-to 16-year-olds: A randomized controlled trial with traumatized refugee children. *Journal of Traumatic Stress, 23*(4), 437–445. https://doi.org/10.1002/jts.20548

Sack, M., Zehl, S., Otti, A., Lahmann, C., Henningsen, P., Kruse, J., & Stingl, M. (2016). A comparison of dual attention, eye movements, and exposure only during eye movement desensitization and reprocessing for posttraumatic stress disorder: Results from a randomized clinical trial. *Psychotherapy and Psychosomatics, 85*(6), 357–365. https://doi.org/10.1159/000447671

Sangalang, C. C., Becerra, D., Mitchell, F. M., Lechuga-Peña, S., Lopez, K., & Kim, I. (2019). Trauma, post-migration stress, and mental health: A comparative analysis of refugees and immigrants in the United States. *Journal of Immigrant & Minority Health, 21*(5), 909–919. https://doi.org/10.1007/s10903-018-0826-2

Schauer, M., Neuner, F., & Elbert, T. (2011). *Narrative exposure therapy: A short-term treatment for traumatic stress disorders* (2nd ed.). Hogrefe Publishing.

Schauer, M., Neuner, F., & Elbert, T. (2017). Narrative exposure therapy for children and adolescents (KIDNET). In M. A. Landolt, M. Cloitre, & U. Schnyder (Eds.), *Evidence-based treatments for trauma related disorders in children and adolescents* (pp. 227–250). https://doi.org/10.1007/978-3-319-46138-0_11

Schick, M., Zumwald, A., Knöpfli, B., Nickerson, A., Bryant, R. A., Schnyder, U., et al. (2016). Challenging future, challenging past: The relationship of social integration and psychological impairment in traumatized refugees. *European Journal of Psychotraumatology, 7*(1). https://doi.org/10.3402/ejpt.v7.28057

Schick, M., Morina, N., Mistridis, P., Schnyder, U., Bryant, R. A., & Nickerson, A. (2018). Changes in post-migration living difficulties predict treatment outcome in traumatized refugees. *Frontiers in Psychiatry, 9*, 476. https://doi.org/10.3389/fpsyt.2018.00476

Schottelkorb, A. A., Doumas, D. M., & Garcia, R. (2012). Treatment for childhood refugee trauma: A randomized, controlled trial. *International Journal of Play Therapy, 21*(2), 57. https://doi.org/10.1037/a0027430

United Nations Children's Fund. (2023). *Child displacement*. https://data.unicef.org/topic/child-migration-and-displacement/displacement/

United Nations High Commissioner for Refugees. (2023). Global trends report 2022. https://www.unhcr.org/global-trends-report-2022

Unterhitzenberger, J., & Rosner, R. (2016). Case report: Manualized trauma-focused cognitive behavioral therapy with an unaccompanied refugee minor girl. *European Journal of Psychotraumatology, 7*(1). https://doi.org/10.3402/ejpt.v7.29246

Unterhitzenberger, J., Eberle-Sejari, R., Rassenhofer, M., Sukale, T., Rosner, R., & Goldbeck, L. (2015). Trauma-focused cognitive behavioral therapy with unaccompanied refugee minors: A case series. *BMC Psychiatry, 15*, 1–9. https://doi.org/10.1186/s12888-015-0645-0

Unterhitzenberger, J., Wintersohl, S., Lang, M., König, J., & Rosner, R. (2019). Providing manualized individual trauma-focused CBT to unaccompanied refugee minors with uncertain residence status: A pilot study. *Child and Adolescent Psychiatry and Mental Health, 13*(1), 1–10. https://doi.org/10.1186/s13034-019-0282-3

USA for UNHCR. (2023). *Refugee statistics*. https://www.unrefugees.org/refugee-facts/statistics

Velu, M. E., Martens, I., Shahab, M., de Roos, C., Jongedijk, R. A., Schok, M., & Mooren, T. (2022). Trauma-focused treatments for refugee children: Study protocol for a randomized controlled trial of the effectiveness of KIDNET versus EMDR therapy versus a waitlist control group (KIEM). *Trials, 23*(1), 1–11. https://doi.org/10.1186/s13063-022-06178-z

Vernberg, E. M., Steinberg, A. M., Jacobs, A. K., Brymer, M. J., Watson, P. J., Osofsky, J. D., et al. (2008). Innovations in disaster mental health: Psychological first aid. *Professional Psychology: Research and Practice, 39*(4), 381. https://doi.org/10.1037/a0012663

Wang, L., Norman, I., Xiao, T., Li, Y., & Leamy, M. (2021). Psychological first aid training: A scoping review of its application, outcomes and implementation. *International Journal of Environmental Research and Public Health, 18*(9), 4594. https://doi.org/10.3390/ijerph18094594

Wang, C., Liu, J. L., Marsico, K. F., & Zhu, Q. (2022). Culturally adapting youth mental health first aid training for Asian Americans. *Psychological Services, 19*(3), 551. https://doi.org/10.1037/ser0000574

Wilker, S., Catani, C., Wittmann, J., Preusse, M., Schmidt, T., May, T., et al. (2020). The efficacy of narrative exposure therapy for children (KIDNET) as a treatment for traumatized young refugees versus treatment as usual: Study protocol for a multi-center randomized controlled trial (YOURTREAT). *Trials, 21*, 1–16. https://doi.org/10.1186/s13063-020-4127-4

Index

A. Maragakis, M. Janikian (eds.), *Evidence-Based Behavioral Health Practices in Pediatric Specialty Settings*, https://doi.org/10.1007/978-3-031-71004-9

MIX
Papier aus verantwortungsvollen Quellen
Paper from responsible sources
FSC® C105338
FSC
www.fsc.org

If you have any concerns about our products,
you can contact us on
ProductSafety@springernature.com

In case Publisher is established outside the EU,
the EU authorized representative is:
Springer Nature Customer Service Center GmbH
Europaplatz 3, 69115 Heidelberg, Germany

Printed by Libri Plureos GmbH
in Hamburg, Germany